Lecture Notes in Computer Science 4136

Commenced Publication in 1973
Founding and Former Series Editors:
Gerhard Goos, Juris Hartmanis, and Jan van Leeuwen

Renate A. Schmidt (Ed.)

Relations and Kleene Algebra in Computer Science

9th International Conference on Relational
Methods in Computer Science
and 4th International Workshop on Applications
of Kleene Algebra, RelMiCS/AKA 2006
Manchester, UK, August/September, 2006
Proceedings

 Springer

Volume Editor

Renate A. Schmidt
University of Manchester
School of Computer Science
Oxford Rd, Manchester M13 9PL, UK
E-mail: Renate.Schmidt@manchester.ac.uk

Library of Congress Control Number: 2006931478

CR Subject Classification (1998): F.4, I.1, I.2.3, D.2.4

LNCS Sublibrary: SL 1 – Theoretical Computer Science and General Issues

ISSN   0302-9743
ISBN-10   3-540-37873-1 Springer Berlin Heidelberg New York
ISBN-13   978-3-540-37873-0 Springer Berlin Heidelberg New York

Springer is a part of Springer Science+Business Media

springer.com

© Springer-Verlag Berlin Heidelberg 2006
Printed in Germany

Typesetting: Camera-ready by author, data conversion by Scientific Publishing Services, Chennai, India
Printed on acid-free paper   SPIN: 11828563   06/3142   5 4 3 2 1 0

Preface

This volume contains the joint proceedings of the 9th International Conference on Relational Methods in Computer Science (RelMiCS-9) and the 4th International Workshop on Applications of Kleene Algebra (AKA 2006). The joint event was hosted by the School of Computer Science at the University of Manchester, UK, from August 29 to September 2, 2006. RelMiCS/AKA is the main forum for the relational calculus as a conceptual and methodological tool and for topics related to Kleene algebras. Within this general theme, the conference series is devoted to the theory of relation algebras, Kleene algebras and related formalisms as well as to their diverse applications in software engineering, databases and artificial intelligence. This year, special focus was on formal methods, logics of programs and links with neighboring disciplines. This diversity is reflected by the contributions to this volume.

The Programme Committee selected 25 technical contributions out of 44 initial submissions from 14 countries. Each paper was refereed by at least three reviewers on its originality, technical soundness, quality of presentation and relevance to the conference. The programme included three invited lectures by distinguished experts in the area: "Weak Kleene Algebra and Computation Trees" by Ernie Cohen (Microsoft, USA), "Finite Symmetric Integral Relation Algebras with no 3-Cycles" by Roger Maddux (Iowa State University, USA), and "Computations and Relational Bundles" by Jeff Sanders (Oxford, UK). In addition, for the first time, a PhD programme was co-organized by Georg Struth. It included the invited tutorials "Foundations of Relation Algebra and Kleene Algebra" by Peter Jipsen (Chapman University, USA), and "Relational Methods for Program Refinement" by John Derrick (Sheffield University, UK).

As in previous years, the RelMiCS Conference and the AKA Workshop were co-organized because of their considerable overlap. Previous RelMiCS meetings were held in 1994 at Dagstuhl, Germany, in 1995 at Parati, Brazil, in 1997 at Hammamet, Tunisia, in 1998 at Warsaw, Poland, in 1999 at Québec, Canada, in 2001 at Oisterwijk, The Netherlands, in 2003 at Malente, Germany and in 2005 at St.Catharines, Canada. The AKA Workshop has been held jointly with RelMiCS since 2003, after an initial Dagstuhl Seminar in 2001.

I would like to thank the many people without whom the meeting would not have been possible. First, I would like to thank all authors who submitted papers, all participants of the conference as well as the invited keynote speakers and the invited tutorial speakers for their contributions. I am very grateful to the members of the Programme Committee and the external referees for carefully reviewing and selecting the papers. I thank my colleagues on the Steering Committee for their advice and the support for the changes introduced for this year's event. Special thanks go to the members of the local organization team in the School of Computer Science at the University of Manchester for all their help: the staff in the ACSO office, especially Bryony Quick and Iain Hart, the staff of the finance office, and the technical staff, as well as Zhen Li and David Robinson. Moreover, I am extremely grateful to Georg Struth for his tremendous amount of effort—as Programme Chair he helped with every aspect of the planning and organization of RelMiCS/AKA 2006 and the PhD Programme. Finally, it is my pleasure to acknowledge the generous support by: the UK Engineering and Physical Sciences Research Council (grant EP/D079926/1), the London Mathematical Society, the British Logic Colloquium, the University of Manchester (President's Fund), and the School of Computer Science, University of Manchester.

Manchester, June 2006   Renate Schmidt
                                                         General Chair
                                                      RelMiCS/AKA 2006

Organization

Conference Chairs

Renate Schmidt (UK, General Chair)
Georg Struth (UK, Program Chair)

Steering Committee

Rudolf Berghammer (Germany, Chair)
Jules Desharnais (Canada)
Ali Jaoua (Qatar)
Bernhard Möller (Germany)
Ewa Orlowska (Poland)

Gunther Schmidt (Germany)
Renate Schmidt (UK)
Harrie de Swart (Netherlands)
Michael Winter (Canada)

Program Committee

Roland Backhouse (UK)
Rudolf Berghammer (Germany)
Stéphane Demri (France)
Jules Desharnais (Canada)
Zoltán Ésik (Hungary, Spain)
Marcelo Frías (Argentina)
Hitoshi Furusawa (Japan)
Stéphane Gaubert (France)
Steven Givant (USA)
Valentin Goranko (South Africa)
Martin Henson (UK)
Ali Jaoua (Qatar)
Peter Jipsen (USA)
Wolfram Kahl (Canada)
Yasuo Kawahara (Japan)

Zhiming Liu (China)
Bernhard Möller (Germany)
Damian Niwinski (Poland)
Ewa Orlowska (Poland)
Alban Ponse (Netherlands)
Ingrid Rewitzky (South Africa)
Ildikó Sain (Hungary)
Holger Schlingloff (Germany)
Gunther Schmidt (Germany)
Renate Schmidt (UK)
Giuseppe Scollo (Italy)
Georg Struth (UK)
Michael Winter (Canada)
Harrie de Swart (Netherlands)

External Referees

Balder ten Cate
Alexander Fronk
Marian Gheorghe

Wim Hesselink
Peter Höfner
Britta Kehden

Zhao Liang
Kamal Lodaya
Maarten Marx

Sun Meng   Andrea Schalk   Thomas Triebsees
Szabolcs Mikulas   Nikolay V. Shilov   Jeff Sanders
Venkat Murali   Kim Solin   Liang Zhao
Ian Pratt-Hartmann   Dmitry Tishkovsky

Local Organization

Renate Schmidt (Local Organization Chair)
Bryony Cook, Iain Hart (Registration, Secretarial Support)
Zhen Li (Webpages)
David Robinson (Local Organization)

Sponsoring Institutions

British Logic Colloquium
Engineering and Physical Sciences Research Council
London Mathematical Society
University of Manchester

Table of Contents

Weak Kleene Algebra and Computation Trees

Ernie Cohen

Microsoft, US
Ernie.Cohen@microsoft.com

Abstract. The Kleene algebra axioms are too strong for some program models of interest (e.g. models that mix demonic choice with angelic or probabilistic choice). This has led to proposals that weaken the right distributivity axiom to monotonicity, and possibly weaken or eliminate the right induction and left annihilation axioms (e.g. lazy Kleene algebra, probabilistic Kleene algebra, monodic tree Kleene algebra, etc.). We'll address some of the basic metatheoretic properties of these theories using rational trees modulo simulation equivalence.

R.A. Schmidt (Ed.): RelMiCS /AKA 2006, LNCS 4136, p. 1, 2006.
© Springer-Verlag Berlin Heidelberg 2006

Finite Symmetric Integral Relation Algebras with No 3-Cycles

Roger D. Maddux

Department of Mathematics
396 Carver Hall
Iowa State University
Ames, Iowa 50011
U.S.A.
maddux@iastate.edu

Abstract. The class of finite symmetric integral relation algebras with no 3-cycles is a particularly interesting and easily analyzable class of finite relation algebras. For example, it contains algebras that are not representable, algebras that are representable only on finite sets, algebras that are representable only on infinite sets, algebras that are representable on both finite and infinite sets, and there is an algorithm for determining which case holds.

Some questions raised in a preprint by Jipsen [1] are addressed in this paper. Most of the results in this paper date from 1983, but some were found and published independently by Comer [2], Jipsen [1], and Tuza [3]. The first four sections contain background material.

1 Relation Algebras and Their Relatives

A **relation algebra** is an algebraic structure

$$\mathfrak{A} = \langle A, +, ^-, ;, ^\smile, 1' \rangle, \tag{1}$$

where A is a nonempty set, $+$ and $;$ are binary operations on A, $^-$ and $^\smile$ are unary operations on A, and $1' \in A$ is a distinguished element, which satisfies these equational axioms:

R_1	$x + y = y + x,$	$+$-commutativity
R_2	$x + (y + z) = (x + y) + z,$	$+$-associativity
R_3	$\overline{\overline{x + y} + \overline{\overline{x} + y}} = x,$	Huntington's axiom
R_4	$x;(y;z) = (x;y);z,$	$;$-associativity
R_5	$(x + y);z = x;z + y;z,$	right $;$-distributivity
R_6	$x;1' = x,$	right identity law

R.A. Schmidt (Ed.): RelMiCS /AKA 2006, LNCS 4136, pp. 2–29, 2006.

R$_7$ $\breve{\breve{x}} = x,$ $\breve{}$-involution

R$_8$ $(x + y)\breve{} = \breve{x} + \breve{y},$ $\breve{}$-distributivity

R$_9$ $(x;y)\breve{} = \breve{y};\breve{x},$ $\breve{}$-involutive distributivity

R$_{10}$ $\breve{x};\overline{x;y} + \overline{y} = \overline{y}.$ Tarski/De Morgan axiom

RA is the class of relation algebras, and NA is the class of **nonassociative relation algebras**, algebras of the form (1) which satisfy all the RA axioms except ;-associativity. For every $\mathfrak{A} \in$ NA, $\langle A, +, ^- \rangle$ is a Boolean algebra by axioms R$_1$–R$_3$ (this fact is due to E. V. Huntington [4,5,6]). Because of this, $\langle A, +, ^- \rangle$ is called the **Boolean part** of $\mathfrak{A}$, and standard concepts from the theory of Boolean algebras may be applied to nonassociative relation algebras by referring to the Boolean part. BA is the class of Boolean algebras. For any algebra $\mathfrak{A} \in$ NA, the **identity element** of $\mathfrak{A}$ is 1', and other operations and elements, which correspond to intersection, difference, empty relation, diversity relation, and universal relation, are defined by

$$x \cdot y := \overline{\overline{x} + \overline{y}}, \tag{2}$$
$$x - y := \overline{\overline{x} + y}, \tag{3}$$
$$0 := \overline{1' + \overline{1'}}, \tag{4}$$
$$0' := \overline{1'}, \tag{5}$$
$$1 := 1' + \overline{1'}. \tag{6}$$

The **zero element** is 0, the **diversity element** is 0', and the **unit element** is 1. Every algebra $\mathfrak{A} \in$ NA satisfies the **cycle law**:

$$\breve{x};z \cdot y = 0 \quad \text{iff} \quad x;y \cdot z = 0 \quad \text{iff} \quad z;\breve{y} \cdot x = 0 \tag{7}$$
$$\text{iff} \quad y;\breve{z} \cdot \breve{x} = 0 \quad \text{iff} \quad \breve{y};\breve{x} \cdot \breve{z} = 0 \quad \text{iff} \quad \breve{z};x \cdot \breve{y} = 0$$

and many other laws, a few of which we gather here:

$$\breve{1'} = 1', \quad \breve{0'} = 0', \tag{8}$$
$$1';x = x, \tag{9}$$
$$0;x = x;0 = 0, \tag{10}$$
$$1;1 = 1. \tag{11}$$

An algebra $\mathfrak{A} \in$ NA is **symmetric** if it satisfies the equation

$$\breve{x} = x. \tag{12}$$

If $\mathfrak{A} \in$ NA is symmetric then $\mathfrak{A}$ is also **commutative**, *i.e.*, it satisfies the equation

$$x;y = y;x. \tag{13}$$

WA is the class of **weakly associative relation algebras**, those algebras in NA which satisfy the **weak associative law**

$$((x \cdot 1');1);1 = (x \cdot 1');(1;1). \tag{14}$$

Finally, SA is the class of **semiassociative relation algebras**, those algebras in NA which satisfy the **semiassociative law**

$$(x\,;1)\,;1 = x\,;(1\,;1). \tag{15}$$

Clearly $\mathsf{NA} \subseteq \mathsf{WA} \subseteq \mathsf{SA} \subseteq \mathsf{RA}$. Since every NA satisfies $1\,;1 = 1$, the weak associative law and semiassociative law can be simplified by replacing $1\,;1$ with 1.

The set of atoms of an algebra $\mathfrak{A} \in \mathsf{NA}$ is $At\mathfrak{A}$. An atom $x \in At\mathfrak{A}$ is an **identity atom** if $x \leq 1$' and a **diversity atom** if $x \leq 0$'. An algebra $\mathfrak{A} \in \mathsf{NA}$ is **integral** if $0 \neq 1$ ($\mathfrak{A}$ is nontrivial) and $x\,;y = 0$ implies $x = 0$ or $y = 0$. If $\mathfrak{A} \in \mathsf{NA}$ and 1' $\in At\mathfrak{A}$ then $\mathfrak{A}$ is integral. The converse holds if $\mathfrak{A} \in \mathsf{SA}$. However, it fails for some $\mathfrak{A} \in \mathsf{WA}$. Around 1940 J. C. C. McKinsey invented a nontrivial WA with zero-divisors in which 1' is an atom. This algebra shows that $\mathsf{WA} \supset \mathsf{SA}$. In fact, all the inclusions are proper (see [7, Cor. 2.6, Th. 3.7] or [8, Th. 450]), so we have

$$\mathsf{NA} \supset \mathsf{WA} \supset \mathsf{SA} \supset \mathsf{RA}.$$

2   Representable Relation Algebras

For every set U, let $Sb\,(U)$ be the set of subsets of U, and let

$$\mathfrak{Bl}\,(U) := \langle Sb\,(U), \cup, {}^{-} \rangle,$$

where $\cup$ is the binary operation on $Sb\,(U)$ of forming the union of any two subsets of U, and $^{-}$ is the unary operation of complementation with respect to U (so $\overline{X} = U \sim X$, where $\sim$ is the operation of forming the set-theoretic difference of two sets). $\mathfrak{Bl}\,(U)$ is the **Boolean algebra of subsets of** U. For every equivalence relation E, let

$$\mathfrak{Sb}\,(E) := \langle Sb\,(E), \cup, {}^{-}, |, {}^{-1}, \mathsf{Id} \cap E \rangle, \tag{16}$$

where $|$ is relative multiplication of binary relations, defined for binary relations R and S by

$$R\,|\,S := \{ \langle a, c \rangle : \exists_b (\langle a, b \rangle \in R, \ \langle b, c \rangle \in S) \}, \tag{17}$$

$^{-1}$ is conversion of binary relations, defined for any binary relation R by

$$R^{-1} := \{ \langle b, a \rangle : \langle a, b \rangle \in R \}, \tag{18}$$

and $\mathsf{Id} \cap E$ is the identity relation on the field $Fd\,(E) = \{ x : \exists_y (xEy) \}$ of E (Id is the class of pairs of sets of the form $\langle x, x \rangle$). $\mathfrak{Sb}\,(E)$ is the **relation algebra of subrelations of** E, and we refer to $\mathfrak{Sb}\,(E)$ as an **equivalence relation algebra**. It is necessary to assume that E is an equivalence relation in order to ensure that $Sb\,(E)$ is closed under relative multiplication and conversion. $Sb\,(U^2)$, the powerset of U^2, is the **set of binary relations on** U. For every set U, $\mathfrak{Re}\,(U)$ is the **square relation algebra on** U, defined by

$$\mathfrak{Re}\,(U) := \mathfrak{Sb}\,(U^2). \tag{19}$$

Every square relation algebra is an equivalence relation algebra, but not every equivalence relation algebra is a square relation algebra (or even isomorphic to one). In fact, $\mathfrak{Sb}\,(E)$ is not isomorphic to a square relation algebra whenever E is an equivalence relation with two or more equivalence classes. For example, if U and V are nonempty disjoint sets and $E = U^2 \cup V^2$, then $\mathfrak{Sb}\,(E)$ is an equivalence relation algebra which is isomorphic to the direct product of the two square relation algebras $\mathfrak{Re}\,(U)$ and $\mathfrak{Re}\,(V)$. The projection functions from $\mathfrak{Sb}\,(E)$ onto the two factor algebras are nontrivial homomorphisms since U and V are not empty. However, nontrivial square relation algebras are simple and have no nontrivial homomorphisms. Consequently $\mathfrak{Sb}\,(E)$ is not isomorphic to any square relation algebra.

We say $\mathfrak{A}$ is a **proper relation algebra** if there is an equivalence relation E such that $\mathfrak{A}$ is a subalgebra of $\mathfrak{Sb}\,(E)$. An algebra $\mathfrak{A}$ is a **representable relation algebra** if it is isomorphic to a proper relation algebra. RRA is the class of **representable relation algebras**. We say that ρ is a **representation of** $\mathfrak{A}$ **over** E and that the field of E is the **base set** of ρ if E is an equivalence relation and ρ is an embedding of $\mathfrak{A}$ into $\mathfrak{Sb}\,(E)$. Thus $\mathfrak{A} \in$ RRA iff there is a representation of $\mathfrak{A}$ over some equivalence relation. We say that ρ is a **square representation of** $\mathfrak{A}$ **on** U (and that U is the **base set** of ρ) if ρ is a representation of $\mathfrak{A}$ over U^2. Let fRRA be the class of **finitely representable relation algebras**, those algebras in RRA which have a representation with a finite base set.

It is easy to see that RRA is closed under the formation of subalgebras and direct products. For subalgebras this is immediate from the relevant definitions. As part of the proof for direct products, note that if E is an I-indexed system of nonempty pairwise disjoint equivalence relations then they also have disjoint fields, $\bigcup_{i \in I} E_i$ is an equivalence relation, and

$$\mathfrak{Sb}\,\left(\textstyle\bigcup_{i \in I} E_i\right) \cong \prod_{i \in I} \mathfrak{Sb}\,(E_i)$$

via the isomorphism which sends each $R \in Sb\,\left(\bigcup_{i \in I} E_i\right)$ to $\langle R \cap E_i : i \in I \rangle$. A special case of this observation is that if E is a nonempty equivalence relation then, letting $Fd\,(E)/E$ be the set of E-equivalence classes of elements in the field $Fd\,(E)$ of E, we have

$$\mathfrak{Sb}\,(E) \cong \prod_{U \in Fd(E)/E} \mathfrak{Re}\,(U)$$

via the isomorphism which sends each $R \in Sb\,(E)$ to $\langle R \cap U^2 : U \in Fd\,(E)/E \rangle$.

Suppose ρ is a square representation of $\mathfrak{A}$ on a finite set U. Create a system of pairwise disjoint sets V_i and bijections $\sigma_i : U \to V_i$ for every $i \in \omega$. For every $a \in A$, let

$$\tau(a) := \bigcup_{i \in \omega} \{ \langle \sigma_i(x), \sigma_i(y) \rangle : \langle x, y \rangle \in \rho(a) \},$$

and set $E := \bigcup_{i \in \omega} V_i \times V_i$. Then E is an equivalence relation and τ is a representation of $\mathfrak{A}$ over E. Since τ has an infinite base set, this shows that if a representable relation algebra has a square representation (or, in fact, any representation) with a finite base set, then it also has a representation with an infinite base set (but not necessarily a *square* representation with an infinite base set).

It is much harder to show that RRA is closed under the formation of homomorphic images. This was first proved by Tarski [9], and it has been reproved in several different ways; see [8,10], Jónsson [11,12], and Hirsch-Hodkinson [13]. It follows by Birkhoff's HSP-Theorem [14] that RRA has an equational axiomatization. However, Monk [15] proved that RRA does not have a finite equational (nor even first-order) axiomatization, and Jónsson [12] proved that RRA does not have an equational basis containing only finitely many variables (see [8, Th. 466–7]).

If ρ is a representation of $\mathfrak{A}$ over an equivalence relation E, then, for all $a, b \in A$, we have

$$\rho(a + b) = \rho(a) \cup \rho(b), \tag{20}$$

$$\rho(\overline{a}) = E \sim \rho(a), \tag{21}$$

$$\rho(a \cdot b) = \rho(a) \cap \rho(b), \tag{22}$$

$$\rho(0) = \emptyset, \tag{23}$$

$$\rho(a\,;b) = \rho(a)|\rho(b), \tag{24}$$

$$\rho(\breve{a}) = (\rho(a))^{-1}, \tag{25}$$

$$\rho(1') = \mathsf{Id} \cap E. \tag{26}$$

The concept of weak representation introduced by Jónsson [16] is obtained by dropping conditions (20) and (21). We say that ρ is a **weak representation** of $\mathfrak{A}$ **over** E if (22)–(26) hold for all $a, b \in A$. An algebra $\mathfrak{A} \in \mathsf{NA}$ is **weakly representable** if it has a weak representation over some equivalence relation, and wRRA is the class of algebras in NA that have a weak representation.

3   5-Dimensional Relation Algebras

Let $\mathfrak{A} \in \mathsf{NA}$ and assume $3 \leq n \leq \omega$. A function that maps n^2 into the universe A of $\mathfrak{A}$ is called an n-**by-**n **matrix** of $\mathfrak{A}$. Let $B_n\mathfrak{A}$ be the set of those n-by-n matrices of atoms of $\mathfrak{A}$ such that, for all $i, j, k < n$, $a_{ii} \leq 1'$, $\breve{a}_{ij} = a_{ji}$, and $a_{ik} \leq a_{ij}\,;a_{jk}$. Let $k, l < n$. We say that two matrices $a, b \in B_n\mathfrak{A}$ **agree up to** k if $a_{ij} = b_{ij}$ whenever $k \neq i, j < n$, and we say that they **agree up to** k, l if $a_{ij} = b_{ij}$ whenever $k, l \neq i, j < n$. We say that N is an n-**dimensional relational basis for** $\mathfrak{A} \in \mathsf{NA}$ if

1. $\emptyset \neq N \subseteq B_n\mathfrak{A}$,
2. for every atom $x \in At\mathfrak{A}$ there is some $a \in N$ such that $a_{01} = x$,
3. if $a \in N$, $i, j, k < n$, $i, j \neq k$, $x, y \in At\mathfrak{A}$, and $a_{ij} \leq x\,;y$, then for some $b \in N$, a and b agree up to i, $b_{ik} = x$, and $b_{kj} = y$.

For example, if U is any set then $B_n\mathfrak{Re}(U)$ is a relational basis for $\mathfrak{Re}(U)$. An algebra $\mathfrak{A} \in \mathsf{NA}$ is a **relation algebra of dimension** n if $\mathfrak{A}$ is a subalgebra of a complete atomic NA that has an n-dimensional relational basis. RA_n is the **class of relation algebras of dimension** n. It happens (see [8]) that

$$\mathsf{SA} = \mathsf{RA}_3 \supset \mathsf{RA} = \mathsf{RA}_4 \supset \mathsf{RA}_5 \supset \mathsf{RA}_\omega = \mathsf{RRA}. \tag{27}$$

The following equation (called (M) in [8]) is true in every RA_5. The notational convention in this equation is that $x_{ij} = (x_{ji})^\smile$.

$$x_{01} \cdot (x_{02} \cdot x_{03}\,;x_{32})\,;(x_{21} \cdot x_{24}\,;x_{41}) \leq \tag{28}$$
$$x_{03}\,;((x_{30}\,;x_{01} \cdot x_{32}\,;x_{21})\,;x_{14} \cdot x_{32}\,;x_{24} \cdot x_{30}\,;(x_{01}\,;x_{14} \cdot x_{02}\,;x_{24}))\,;x_{41}.$$

Equation (28) is part of the axiom set in Jónsson [16] and it is an equational form of a condition on atoms given by Lyndon [17]. For a relation algebra $\mathfrak{A} \in \mathsf{RA}$, failure of (28) is a simple test for nonrepresentability that implies something stronger, namely, nonmembership in RA_5.

4   Cycle Structures and Complex Algebras

The **cycle structure** of $\mathfrak{A} \in \mathsf{NA}$ is the ternary relation

$$Cy(\mathfrak{A}) := \{\langle x, y, z \rangle : x, y, z \in At\mathfrak{A},\ x\,;y \geq z\}.$$

For any atoms $x, y, z \in At\mathfrak{A}$, let

$$[x, y, z] := \{\langle x, y, z \rangle, \langle \breve{x}, z, y \rangle, \langle y, \breve{z}, \breve{x} \rangle, \langle \breve{y}, \breve{x}, \breve{z} \rangle, \langle \breve{z}, x, \breve{y} \rangle, \langle z, \breve{y}, x \rangle\}. \tag{29}$$

The set $[x, y, z]$ of triples of atoms is called a **cycle**. By the cycle law (7), the cycle structure of $\mathfrak{A}$ is a disjoint union of cycles. We say that $[x, y, z]$ is a **forbidden cycle** of $\mathfrak{A}$ if $[x, y, z] \cap Cy(\mathfrak{A}) = \emptyset$, that $[x, y, z]$ is a **cycle** of $\mathfrak{A}$ if $[x, y, z] \subseteq Cy(\mathfrak{A})$, that $[x, y, z]$ is an **identity cycle** if one (or, equivalently, all) of its triples contains an identity atom, and that $[x, y, z]$ is a **diversity cycle** if all of the elements in its triples are diversity atoms. In case $\mathfrak{A}$ is symmetric, we say that $[x, y, z]$ is a **3-cycle** (or **2-cycle** or **1-cycle**) of $\mathfrak{A}$ if $[x, y, z] \subseteq Cy(\mathfrak{A})$ and $|\{x, y, z\}| = 3$ (or 2 or 1, respectively). In case $\mathfrak{A} \in \mathsf{NA}$ is symmetric and 1' is an atom of $\mathfrak{A}$, we say that $\mathfrak{A}$ **has no 3-cycles** if every 3-cycle is forbidden.

Suppose that T is a ternary relation and U is the field of T, i.e.,

$$U := \{x : \exists_y \exists_z (Txyz \text{ or } Tyxz \text{ or } Tyzx)\}. \tag{30}$$

We will use T to construct an algebra whose universe is $Sb\,(U)$. First, define a binary operation $;$ on the powerset of U, by letting, for any $X, Y \subseteq U$,

$$X\,;Y := \{c : \exists_x \exists_y (x \in X, y \in Y, Txyc)\}.$$

Define the binary relation $S \subseteq U^2$ by

$$S := \{\langle a, b \rangle : a, b \in U, \forall_x \forall_y ((Taxy \iff Tbyx), (Txay \iff Tybx))\}. \tag{31}$$

Note that S must be a symmetric relation because of the form of its definition. Use S to define $\check{X} \subseteq U$ for every subset $X \subseteq U$ by

$$\check{X} := \{b : \exists_x (Sxb, x \in X)\}.$$

Finally, define the subset $I \subseteq U$ by

$$I := \{a : a \in U, \forall_x \forall_y ((Taxy \text{ or } Txay) \implies x = y)\}. \tag{32}$$

The operations $;$ and $\check{\ }$ along with the distinguished subset I are enough to define, starting from the Boolean algebra of all subsets of U, an algebra called the **complex algebra** of T, namely,

$$\mathfrak{Cm}(T) := \langle Sb(U), \cup, {}^-, ;, {}^\vee, I \rangle.$$

The Boolean part of $\mathfrak{Cm}(T)$ is $\mathfrak{Bl}(U)$, the complete atomic Boolean algebra of all subsets of the field of T. The complex algebra $\mathfrak{Cm}(T)$ is a relation algebra when certain elementary conditions are satisfied by T, as stated in the next theorem.

Theorem 1 ([7, Th. 2.2, 2.6]). *Suppose T is a ternary relation. Define U, S, and I by (30), (31), and (32). Consider the following six statements.*

$$\forall_a (a \in U \implies \exists_b Sab), \tag{33}$$

$$\forall_a (a \in U \implies \exists_i (i \in I, \ Tiaa)), \tag{34}$$

$$\forall_x \forall_y \forall_z \forall_a \forall_b (Txyz, \ Tzab \implies \exists_c (Txcb, \ Tyac)), \tag{35}$$

$$\forall_x \forall_y \forall_z \forall_a \forall_b (Txyz, \ Tzab \implies \exists_c Txcb), \tag{36}$$

$$\forall_x \forall_y \forall_z \forall_a \forall_b (Txyz, \ Tzab, \ Ix \implies \exists_c Txcb), \tag{37}$$

$$\forall_x \forall_z \forall_a \forall_b (Txzz, \ Tzab, \ Ix \implies Txbb). \tag{38}$$

1. *If (33) and (34) then S is an involution, i.e., $S : U \to U$ and $S(S(x)) = x$ for all $x \in U$.*
2. *$\mathfrak{Cm}(T) \in \mathsf{NA}$ iff (33) and (34).*
3. *$\mathfrak{Cm}(T) \in \mathsf{RA}$ iff (33), (34), and (35).*
4. *$\mathfrak{Cm}(T) \in \mathsf{SA}$ iff (33), (34), and (36).*
5. *$\mathfrak{Cm}(T) \in \mathsf{WA}$ iff (33), (34), and either (37) or (38).*

Statement (33) says that every atom has a converse and statement (34) says that every atom has a left identity element. Statement (35) expresses $;$-associativity for atoms (and has the same form as Pasch's Axiom). Statement (36), which expresses the semiassociative law applied to atoms, is a strict weakening of (35), obtained by deleting one of the conclusions. Statements (37) and (38) are obtained from (36) by weakening the hypotheses, and each of them expresses the weak associative law applied to atoms.

The identity element of the complex algebra of T is an atom just in case $I = \{e\}$ for some $e \in U$. Whenever this is the case, (34) takes on the following simpler form,

$$\forall_a (a \in U \implies Teaa).$$

Every square relation algebra on a set is a complex algebra, for if U is an arbitrary set and $T = \{\langle\langle a, b\rangle, \langle b, c\rangle, \langle a, c\rangle\rangle : a, b, c \in U\}$, then the complex algebra $\mathfrak{Cm}(T)$ is the square relation algebra on U:

$$\mathfrak{Re}(U) = \mathfrak{Cm}(\{\langle\langle a, b\rangle, \langle b, c\rangle, \langle a, c\rangle\rangle : a, b, c \in U\}).$$

Let $\mathfrak{G} = \langle G, \circ\rangle$ be a group. Treat the group multiplication $\circ$ as a ternary relation, *i.e.*,

$$\circ = \{\langle x, y, z\rangle : x, y, z \in G, \ x \circ y = z\},$$

and define $\mathfrak{Cm}(\mathfrak{G})$ to be $\mathfrak{Cm}(\circ)$. GRA is the **class of group relation algebras**, the class of algebras that are isomorphic to a subalgebra of $\mathfrak{Cm}(\mathfrak{G})$ for some group $\mathfrak{G}$. If, for every $X \subseteq G$, we let

$$\sigma(X) := \{\langle g, g \circ x\rangle : g \in G, \ x \in X\}$$

then σ is a square representation of $\mathfrak{Cm}(\mathfrak{G})$ on G. Thus every group relation algebra is representable:

$$\text{GRA} \subseteq \text{RRA}. \tag{39}$$

5   Cycle Structures of Algebras Without 3-Cycles

Let $\mathfrak{A} \in \text{RA}$. Define binary relations $\rightarrow$, $\Rightarrow$, and $\Leftrightarrow$, on diversity atoms $x, y \in At\mathfrak{A}$ as follows:

1. $x \rightarrow y$ iff $x \neq y$ and $x \leq y;y$,
2. $x \Rightarrow y$ iff $x = y$ or $x \rightarrow y$,
3. $x \Leftrightarrow y$ iff $x \Rightarrow y$ and $y \Rightarrow x$.

For every diversity atom x, let

$$[x] = \{y : 0' \geq y \in At\mathfrak{A}, \ x \Leftrightarrow y\},$$
$$D = \{[x] : 0' \geq x \in At\mathfrak{A}\}.$$
$$[\Rightarrow] = \{\langle[x], [y]\rangle : 0' \geq x, y \in At\mathfrak{A}, \ x \Rightarrow y\}$$

The following theorem includes some elementary facts about cycle structures of finite symmetric integral relation algebras, noticed by those attempting to enumerate small finite relation algebras, such as Lyndon [17, fn. 13], Backer [18], McKenzie [19], Wostner [20], Maddux [7,21], Comer [22,23], Jipsen [24], Jipsen-Lukács [25,26], and Andréka-Maddux [27], and explicitly mentioned in at least Jipsen [1, Th. 1] and Tuza [3, Th. 2.1].

Theorem 2. *Assume $\mathfrak{A} \in \text{RA}$, $\mathfrak{A}$ is symmetric, atomic, integral, and has no 3-cycles. Then, for all diversity atoms $x, y, z \in At\mathfrak{A} \sim \{1'\}$, we have*

1. *if $x \rightarrow y$, $y \rightarrow z$, and $x \neq z$, then $x \rightarrow z$,*
2. *$\Rightarrow$ is reflexive and transitive,*
3. *either $x \Rightarrow y$ or $y \Rightarrow x$,*

4. ⇔ *is an equivalence relation,*
5. [⇒] *is a linear ordering of D.*

Proof. For part 1, assume $x \to y$, $y \to z$, and $x \neq z$. Then $x \neq y$, $x \leq y;y$, $y \neq z$, and $y \leq z;z$. Note that $y \cdot z = 0$ because y and z are distinct atoms and $y;z \leq y + z$ because $\mathfrak{A}$ has no 3-cycles. Hence

$$
\begin{aligned}
x &\leq y;y & & x \to y \\
&\leq y;(z;z) & & y \to z \\
&= (y;z);z & & \text{R}_4 \\
&\leq (y+z);z & & \\
&= y;z + z;z & & \text{R}_5 \\
&= y + z + z;z & &
\end{aligned}
$$

but $x \cdot y = 0 = x \cdot z$, so $x \leq z;z$. From this and $x \neq z$ we get $x \to z$.

Part 2 is trivial.

For part 3, first note that if $x = y$ then both $y \Rightarrow x$ and $x \Rightarrow y$, so we may assume $x \neq y$. Then $x \cdot y = 0$ since distinct atoms are disjoint and $x;y \leq= x + y$ since $\mathfrak{A}$ has no 3-cycles. We must therefore have either $x;y \cdot x \neq 0$ or $x;y \cdot y \neq 0$, since otherwise we would have $x;y = 0$, which implies since $\mathfrak{A}$ is integral that either $x = 0$ or $y = 0$, contrary to the assumption that x and y are (nonzero) atoms. If $x;y \cdot x \neq 0$, then $\breve{x};x \cdot y \neq 0$ by the cycle law, hence $y \leq \breve{x};x = x;x$ by symmetry and $y \in At\mathfrak{A}$. In this case, $y \to x$. On the other hand, if $x;y \cdot y \neq 0$ then $x \leq y;y$ and $x \to y$.

For part 4, notice that the relation ⇔ is transitive and reflexive by its definition and part 1, and that ⇔ is symmetric just by its definition.

Part 5 follows from parts 3 and 4.   □

Assume $\mathfrak{A} \in \mathsf{RA}$ and $\mathfrak{A}$ is symmetric, integral, and finite. By Th. 2, D is linearly ordered by ⇒. D is finite since $\mathfrak{A}$ is finite. Let $n = |D|$. We may choose representatives $a_1, \ldots, a_n \in At\mathfrak{A}$ from the equivalence classes of diversity atoms so that

$$a_1 \to a_2 \to a_3 \to \cdots \to a_{n-2} \to a_{n-1} \to a_n,$$
$$[a_1] \cup \cdots \cup [a_n] = At\mathfrak{A} \sim \{1'\}.$$

For each $i \in \{1, \ldots, n\}$, let s_i be the number of atoms in $[a_i]$ that appear in a 1-cycle of $\mathfrak{A}$, and let t_i be the number of atoms in $[a_i]$ that do not appear in a 1-cycle of $\mathfrak{A}$, *i.e.*,

$$s_i = |[a_i] \cap \{a : a \leq a;a\}|, \tag{40}$$
$$t_i = |[a_i] \cap \{a : 0 = a \cdot a;a\}|. \tag{41}$$

We refer to these numbers as the **cycle parameters** of $\mathfrak{A}$, and define

$$\mathrm{Cp}(\mathfrak{A}) := \begin{pmatrix} s_1 & \cdots & s_n \\ t_1 & \cdots & t_n \end{pmatrix}.$$

Notice that $\sum_{i=1}^{n} t_i$ is the number of 1-cycles other than $[1',1',1']$, and the number of diversity atoms of $\mathfrak{A}$ is $\sum_{i=1}^{n}(s_i + t_i)$. In case $\mathfrak{A}$ has no diversity atoms, we set $\mathrm{Cp}(\mathfrak{A}) := \begin{pmatrix} 0 \\ 0 \end{pmatrix}$.

Two basic observations, included in the following theorem, are that the isomorphism type of $\mathfrak{A}$ is determined by $\mathrm{Cp}(\mathfrak{A})$, and that (almost) any two sequences of nonnegative integers with the same length determine a finite symmetric integral relation algebra with no 3-cycles.

Theorem 3.

1. *If $\mathfrak{A}$ and $\mathfrak{B}$ are finite symmetric integral relation algebras with no 3-cycles and $\mathrm{Cp}(\mathfrak{A}) = \mathrm{Cp}(\mathfrak{B})$ then $\mathfrak{A} \cong \mathfrak{B}$.*
2. *If $n \in \omega$, $s_1, \ldots, s_n \in \omega$, $t_1, \ldots, t_n \in \omega$, and $0 < s_1 + t_1, \ldots, s_n + t_n$, then there is some finite symmetric integral relation algebra $\mathfrak{A}$ with no 3-cycles such that*

$$\mathrm{Cp}(\mathfrak{A}) = \begin{pmatrix} s_1 & \cdots & s_n \\ t_1 & \cdots & t_n \end{pmatrix}.$$

6   2-Cycle Products of Algebras

Next we describe a special kind of product $\mathfrak{A}[\mathfrak{B}]$ of two finite algebras $\mathfrak{A}, \mathfrak{B} \in \mathsf{NA}$ in which the identity element is an atom and both algebras have at least one diversity atom. Since we only need to describe this product up to isomorphism, we make the convenient assumption that 1' is the same atom in both $\mathfrak{A}$ and $\mathfrak{B}$, and that otherwise the sets of atoms of these algebras are disjoint, that is, $\{1'\} = At\mathfrak{A} \cap At\mathfrak{B}$. We may then define the **2-cycle product** $\mathfrak{A}[\mathfrak{B}]$ of $\mathfrak{A}$ and $\mathfrak{B}$ as the complex algebra of the ternary relation T, where

$$T := Cy(\mathfrak{A}) \cup Cy(\mathfrak{B}) \cup \{[a,b,b] : a \in At\mathfrak{A} \sim \{1'\}, b \in At\mathfrak{B} \sim \{1'\}\}. \qquad (42)$$

(The name comes from the symmetric case, in which the cycles added to those of $\mathfrak{A}$ and $\mathfrak{B}$ are all 2-cycles.) Comer [2] proved that $\mathfrak{A}[\mathfrak{B}] \in \mathsf{RRA}$ iff $\mathfrak{A}, \mathfrak{B} \in \mathsf{RRA}$, and $\mathfrak{A}[\mathfrak{B}] \in \mathsf{GRA}$ iff $\mathfrak{A}, \mathfrak{B} \in \mathsf{GRA}$. This is proved below, but first we note the connection between this operation and the cycle parameters introduced above.

Theorem 4. *Assume $\mathfrak{A}, \mathfrak{B} \in \mathsf{NA}$, $\mathfrak{A}$ and $\mathfrak{B}$ are finite, $1' \in At\mathfrak{A}$, $1' \in At\mathfrak{B}$, $\{1'\} = At\mathfrak{A} \cap At\mathfrak{B}$, $At\mathfrak{A} \sim \{1'\} \neq \emptyset \neq At\mathfrak{B} \sim \{1'\}$, and $\mathfrak{A}[\mathfrak{B}] := \mathfrak{Cm}(T)$ where T is defined in (42). If $\mathfrak{A}$ and $\mathfrak{B}$ are symmetric, have no 3-cycles, and*

$$\mathrm{Cp}(\mathfrak{A}) = \begin{pmatrix} s_1 & \cdots & s_n \\ t_1 & \cdots & t_n \end{pmatrix}, \qquad \mathrm{Cp}(\mathfrak{B}) = \begin{pmatrix} s_1' & \cdots & s_n' \\ t_1' & \cdots & t_n' \end{pmatrix},$$

where $0 < s_1 + t_1, \ldots, s_n + t_n, s_1' + t_1', \ldots, s_n' + t_n'$, then

$$\mathrm{Cp}(\mathfrak{A}[\mathfrak{B}]) = \begin{pmatrix} s_1 & \cdots & s_n & s_1' & \cdots & s_n' \\ t_1 & \cdots & t_n & t_1' & \cdots & t_n' \end{pmatrix}.$$

Next is the part of Comer's theorem that we need later.

Theorem 5 (Comer [2]). *Suppose* $\mathfrak{A}, \mathfrak{B} \in \mathsf{NA}$, $\mathfrak{A}$ *and* $\mathfrak{B}$ *are finite,* $1' \in At\mathfrak{A}$, $1' \in At\mathfrak{B}$, $\{1'\} = At\mathfrak{A} \cap At\mathfrak{B}$, *and* $At\mathfrak{A} \sim \{1'\} \neq \emptyset \neq At\mathfrak{B} \sim \{1'\}$.

1. *If* σ *is a square representation of* $\mathfrak{A}$ *on* U *and* τ *is a square representation of* $\mathfrak{B}$ *on* V, *then there is a square representation* φ *of* $\mathfrak{A}[\mathfrak{B}]$ *on* $U \times V$ *such that, for all* $a \in At\mathfrak{A} \sim \{1'\}$ *and all* $b \in At\mathfrak{B} \sim \{1'\}$,

$$\varphi(1') = \{\langle\langle u_0, v_0\rangle, \langle u_1, v_1\rangle\rangle : u_0 = u_1, v_0 = v_1\},$$
$$\varphi(a) = \{\langle\langle u_0, v_0\rangle, \langle u_1, v_1\rangle\rangle : \langle u_0, u_1\rangle \in \sigma(a), v_0 = v_1\},$$
$$\varphi(b) = \{\langle\langle u_0, v_0\rangle, \langle u_1, v_1\rangle\rangle : \langle v_0, v_1\rangle \in \sigma(b)\}.$$

2. *If* $\mathfrak{G}$ *and* $\mathfrak{H}$ *are groups with identity elements* e_G *and* e_H, *respectively,* σ *is an embedding of* $\mathfrak{A}$ *into* $\mathfrak{Cm}(\mathfrak{G})$, *and* τ *is an embedding of* $\mathfrak{B}$ *into* $\mathfrak{Cm}(\mathfrak{H})$, *then there is an embedding* φ *of* $\mathfrak{A}[\mathfrak{B}]$ *into* $\mathfrak{Cm}(\mathfrak{G} \times \mathfrak{H})$ *such that, for all* $a \in At\mathfrak{A} \sim \{1'\}$ *and all* $b \in At\mathfrak{B} \sim \{1'\}$,

$$\varphi(1') = \{\langle e_G, e_H\rangle\},$$
$$\varphi(a) = \sigma(a) \times \{e_H\},$$
$$\varphi(b) = G \times \tau(b).$$

Proof. The statement of part 1 describes the action of φ on the atoms of $\mathfrak{A}[\mathfrak{B}]$. What remains is to extend φ to all elements of $\mathfrak{A}[\mathfrak{B}]$ by setting

$$\varphi(x) = \bigcup_{x \geq c \in At\mathfrak{A}[\mathfrak{B}]} \varphi(c),$$

and check that the extended φ really is a square representation as claimed. Part 2 is handled similarly. $\square$

The 2-cycle product can also be defined for linearly ordered sets of algebras. Suppose that $\mathfrak{A}_i \in \mathsf{NA}$ and $\mathfrak{A}_i$ is atomic for every $i \in I$, that there is a single fixed element $1'$ which is the identity element and also an atom of $\mathfrak{A}_i$ for every $i \in I$, that the sets of diversity atoms of algebras in $\{\mathfrak{A}_i : i \in I\}$ are pairwise disjoint, and that $<$ is a strict linear ordering of I. Let

$$T := \bigcup_{i \in I} Cy(\mathfrak{A}_i) \cup$$
$$\left\{ [a_i, a_j, a_j] : i, j \in I, \ i < j, \ a_i \in At\mathfrak{A}_i \sim \{1'\}, \ a_j \in At\mathfrak{A}_j \sim \{1'\} \right\}.$$

Then the 2-cycle product of the $<$-ordered system $\langle \mathfrak{A}_i : i \in I\rangle$ is $\mathfrak{Cm}(T)$.

7 Algebras with Parameters $\binom{\alpha}{0}$

The next theorem shows that $\mathfrak{A}$ is group representable if $\mathrm{Cp}(\mathfrak{A})$ is $\binom{1}{0}$, $\binom{2}{0}$, $\binom{3}{0}$, $\binom{4}{0}$, ... The proof shows that $\mathfrak{A}$ can be embedded in the complex algebra of an infinite group in every such case. As we shall see, $\mathfrak{A}$ can be embedded in the complex algebra finite group in the first two cases, but not in any of the remaining cases.

Theorem 6. *Assume $\mathfrak{A} \in$ RA, $\mathfrak{A}$ is complete, atomic, and $\mathrm{Cp}(\mathfrak{A}) = \begin{pmatrix} \alpha \\ 0 \end{pmatrix}$ for some nonzero cardinal $\alpha > 0$. Then $\mathfrak{A} \in$ GRA.*

Proof. Let $\mathbb{Z} = \{\ldots, -1, 0, 1, \ldots\}$ be the integers and let $\leq^*$ be a lexicographical ordering of $\mathbb{Z} \times \alpha$ such that, for all $a, b \in \mathbb{Z}$ and all $\kappa, \lambda < \alpha$, $\langle a, \kappa \rangle \leq^* \langle b, \lambda \rangle$ iff $a < b$ or else $a = b$ and $\kappa \leq \lambda$. Also, $\langle a, \kappa \rangle <^* \langle b, \lambda \rangle$ iff $a \leq^* b$ and $\langle a, \kappa \rangle \neq \langle b, \lambda \rangle$. Let G be the set of functions $f : \mathbb{Z} \times \alpha \to \{0, 1\}$ such that $f(a, \kappa) = 0$ for all but finitely many pairs $\langle a, \kappa \rangle \in \mathbb{Z} \times \alpha$. Let $\mathbf{0}$ be the function in G which maps every pair $\langle a, \kappa \rangle$ to 0. The functions in G form a group $\mathfrak{G}$ under addition modulo 2: if $f, g \in G$ then, for all $\langle a, \kappa \rangle \in \mathbb{Z} \times \alpha$, $(f + g)(a, \kappa) = f(a, \kappa) +_2 g(a, \kappa)$, where $+_2$ is addition *modulo* 2. Thus G is isomorphic to the direct sum of $\mathbb{Z} \times \alpha$ copies of $\mathbb{Z}_2$ (the cyclic group of order 2, with elements $\{0, 1\}$ and operation $+_2$ of addition *modulo* 2.). The cardinality of G is α if $\alpha \geq \omega$ and it is ω if $\alpha \leq \omega$.

Suppose $f \in G$. Let $\mathsf{supp}(f) = \{\langle a, \kappa \rangle : f(a, \kappa) = 1\}$. Note that $\mathsf{supp}(f)$ is finite, so if $\mathsf{supp}(f)$ is nonempty, then there is some pair $\langle a, \kappa \rangle \in \mathsf{supp}(f)$ which is smallest with respect to the ordering $\leq^*$. Let $L(f)$ be this smallest pair. For every $\kappa < \alpha$, let $G_\kappa = \{f : f \in G, \ L(f) = \kappa\}$. This yields a partition of G into the sets $\{\mathbf{0}\}$ and G_κ for $\kappa < \alpha$. Note that $L(f + g) = L(f)$ if $L(f) \leq^* L(g)$, and $L(f + g) = L(g)$ if $L(g) \leq^* L(f)$.

We show next that the partition of G is the set of atoms of a subalgebra of $\mathfrak{Cm}(\mathfrak{G})$ isomorphic to $\mathfrak{A}$. First, $\{\mathbf{0}\}$ is the identity element in $\mathfrak{Cm}(\mathfrak{G})$ since

$$G_\kappa ; \{\mathbf{0}\} = \{f + \mathbf{0} : f \in G_\kappa\} = G_\kappa,$$

so we need only show, for distinct $\kappa, \lambda < \alpha$, that

$$G_\kappa ; G_\lambda = G_\kappa \cup G_\lambda, \tag{43}$$
$$G_\kappa ; G_\kappa = G. \tag{44}$$

Suppose $f \in G_\kappa$, $g \in G_\lambda$, $L(f) = \langle a, \kappa \rangle$, and $L(g) = \langle b, \lambda \rangle$. Then $L(f) \neq L(g)$ since $\kappa \neq \lambda$. If $L(f) <^* L(g)$, then $L(f) = L(f + g)$, so $f + g \in G_\kappa$, and if $L(g) <^* L(f)$, then $L(g) = L(f+g)$, so $f+g \in G_\lambda$. This shows $G_\kappa ; G_\lambda \subseteq G_\kappa \cup G_\lambda$.

To show $G_\kappa \subseteq G_\kappa ; G_\lambda$ when $\kappa \neq \lambda$, suppose $f \in G_\kappa$ and $L(f) = \langle a, \kappa \rangle$. Let g be the function in G whose output is 0 at every value with one exception, namely $g(a+1, \lambda) = 1$. Let $h = f+g$. Clearly $g \in G_\lambda$. Also, $L(h) = L(f+g) = L(f)$ since $L(f) = \langle a, \kappa \rangle <^* \langle a + 1, \lambda \rangle = L(g)$, so $h \in G_\lambda$. Finally, $g + h = g + (f + g) = f$. Similarly, $G_\lambda \subseteq G_\kappa ; G_\kappa$. This completes the proof of (43).

To show $G \subseteq G_\kappa ; G_\kappa$, consider any $g \in G$ with $L(g) = \langle a, \lambda \rangle$. Let f be the function in G whose output is 0 at every value with one exception, namely $f(a - 1, \kappa) = 1$. Let $h = g + f$. Then $f \in G_\kappa$, $L(h) = L(g + f) = L(f)$ since $\langle a - 1, \kappa \rangle = L(f) <^* L(g) = \langle a, \lambda \rangle$, hence $h \in G_\kappa$, and $g = f + h$. □

There is exactly one relation algebra with a single atom. In the numbering system of [8], this is the algebra 1_1. Its sole atom is 1'. It has a single cycle, the identity cycle [1', 1', 1']. The cycle parameters of this algebra are

$$\mathrm{Cp}(1_1) = \begin{pmatrix} 0 \\ 0 \end{pmatrix}.$$

The algebra 1_1 is group representable. In fact, it is isomorphic to the complex algebra $\mathfrak{Cm}\,(\mathbb{Z}_1)$ of the one-element group $\mathbb{Z}_1$:

$$1_1 \cong \mathfrak{Cm}\,(\mathbb{Z}_1).$$

The two relation algebras with two atoms are called 1_2 and 2_2 in the numbering system of [8]. The cycles of 1_2 are just the identity cycles [1', 1', 1'] and [1', 0', 0'], while the cycles of 2_2 are the identity cycles [1', 1', 1'], [1', 0', 0'], and also the diversity cycle [0', 0', 0']. The multiplication tables for the atoms of these two algebras are

1_2	1'	0'
1'	1'	0'
0'	0'	1'

2_2	1'	0'
1'	1'	0'
0'	0'	1'0'

The second table illustrates our notational convention of omitting + signs and avoiding abbreviations by listing all the atoms in a given product, so that, for example, we put "1'0'" instead of "1' + 0'" or simply "1". This notational convention is followed in the tables below. The cycle parameters of 1_2 and 2_2 are

$$\mathrm{Cp}(1_2) = \begin{pmatrix} 0 \\ 1 \end{pmatrix}, \quad \mathrm{Cp}(2_2) = \begin{pmatrix} 1 \\ 0 \end{pmatrix}.$$

The algebras 1_2 and 2_2 are group representable. In fact, 1_2 is already the complex algebra of a group since $1_2 \cong \mathfrak{Cm}\,(\mathbb{Z}_2)$. Thus 1_2 has a square representation on a 2-element set, but it does not have a square representation on a set of any other cardinality. Th. 6 applies to 2_2, but we may also embed 2_2 into the complex algebra of the cyclic group $\mathbb{Z}_n$ of order $n \geq 3$ (whose elements are $\{0, 1, 2, \ldots, n-1\}$ and whose operation is $+_n$, addition *modulo n*) by mapping 1' to $\{0\}$ and 0' to $\{1, 2, \ldots, n-1\}$. Thus, using "$\cong|\subseteq$" to mean, "is isomorphic to a subalgebra of", we have

$$2_2 \cong|\subseteq \mathfrak{Cm}\,(\mathbb{Z}_n) \quad \text{for all } n \geq 3.$$

Thus 2_2 has square representations on sets of every cardinality from 3 on up.

There are seven symmetric integral relation algebras with exactly three atoms. The three atoms are 1', a, and b, and the algebras are 1_1–1_7 in the notational system of [8]. The cycles of these seven algebras are given in the following table (in which we write, for example, simply "aaa" instead of "$[a, a, a]$"):

	1'1'1'	1'aa	1'bb	aaa	bbb	abb	baa
1_7	1'1'1'	1'aa	1'bb	...	...	abb	...
2_7	1'1'1'	1'aa	1'bb	aaa	...	abb	...
3_7	1'1'1'	1'aa	1'bb	...	bbb	abb	...
4_7	1'1'1'	1'aa	1'bb	aaa	bbb	abb	...
5_7	1'1'1'	1'aa	1'bb	...	...	abb	baa
6_7	1'1'1'	1'aa	1'bb	aaa	...	abb	baa
7_7	1'1'1'	1'aa	1'bb	aaa	bbb	abb	baa

Here are the multiplication tables for the atoms of the seven algebras 1_1–1_7.

1_7	1'	a	b
1'	1'	a	b
a	a	1'	b
b	b	b	1'a

2_7	1'	a	b
1'	1'	a	b
a	a	1'a	b
b	b	b	1'a

3_7	1'	a	b
1'	1'	a	b
a	a	1'	b
b	b	b	1'ab

4_7	1'	a	b
1'	1'	a	b
a	a	1'a	b
b	b	b	1'ab

5_7	1'	a	b
1'	1'	a	b
a	a	1'b	ab
b	b	ab	1'a

6_7	1'	a	b
1'	1'	a	b
a	a	1'ab	ab
b	b	ab	1'a

7_7	1'	a	b
1'	1'	a	b
a	a	1'ab	ab
b	b	ab	1'ab

The cycle parameters of algebras 1_1–1_7 are

$$\mathrm{Cp}(1_7) = \begin{pmatrix} 0 & 0 \\ 1 & 1 \end{pmatrix}, \quad \mathrm{Cp}(2_7) = \begin{pmatrix} 1 & 0 \\ 0 & 1 \end{pmatrix}, \quad \mathrm{Cp}(3_7) = \begin{pmatrix} 0 & 1 \\ 1 & 0 \end{pmatrix}, \quad \mathrm{Cp}(4_7) = \begin{pmatrix} 1 & 1 \\ 0 & 0 \end{pmatrix},$$

$$\mathrm{Cp}(5_7) = \begin{pmatrix} 0 \\ 2 \end{pmatrix}, \quad \mathrm{Cp}(6_7) = \begin{pmatrix} 1 \\ 1 \end{pmatrix}, \quad \mathrm{Cp}(7_7) = \begin{pmatrix} 2 \\ 0 \end{pmatrix}.$$

From Th. 4 and Th. 5 we get

$$1_7 \cong|\subseteq \mathfrak{Cm}\left(\mathbb{Z}_2^2\right),$$
$$2_7 \cong|\subseteq \mathfrak{Cm}\left(\mathbb{Z}_3 \times \mathbb{Z}_2\right) \cong \mathfrak{Cm}\left(\mathbb{Z}_6\right),$$
$$3_7 \cong|\subseteq \mathfrak{Cm}\left(\mathbb{Z}_2 \times \mathbb{Z}_3\right) \cong \mathfrak{Cm}\left(\mathbb{Z}_6\right),$$
$$4_7 \cong|\subseteq \mathfrak{Cm}\left(\mathbb{Z}_3^2\right).$$

Next we show that 5_7, 6_7, and 7_7 can be embedded in the complex algebras of the finite groups $\mathbb{Z}_5$, $\mathbb{Z}_8$, and $\mathbb{Z}_3^2$, respectively. We have $5_7 \cong|\subseteq \mathfrak{Cm}\left(\mathbb{Z}_5\right)$ via ρ if

$$\rho(1') = \{\langle i, i\rangle : i \in 5\},$$
$$\rho(a) = \{\langle i, i +_5 j\rangle : i \in 5, \, j \in \{1, 4\}\},$$
$$\rho(b) = \{\langle i, i +_5 j\rangle : i \in 5, \, j \in \{2, 3\}\},$$

$6_7 \cong|\subseteq \mathfrak{Cm}\left(\mathbb{Z}_8\right)$ via ρ if

$$\rho(1') = \{\langle i, i\rangle : i \in 8\},$$
$$\rho(a) = \{\langle i, i +_8 j\rangle : i \in 8, \, j \in \{2, 3, 5, 6\}\},$$
$$\rho(b) = \{\langle i, i +_8 j\rangle : i \in 8, \, j \in \{1, 4, 7\}\},$$

and $7_7 \cong | \subseteq \mathfrak{Cm}\left(\mathbb{Z}_3^2\right)$ via ρ if

$$\rho(1') = \{\langle\langle i,j\rangle,\langle i,j\rangle\rangle : i \in 3, \ j \in 3\},$$
$$\rho(a) = \{\langle\langle i,j\rangle,\langle i +_3 k,j\rangle\rangle : i,j \in 3, \ k \in \{1,2\}\}$$
$$\cup \ \{\langle\langle i,j\rangle,\langle i,j +_3 k\rangle\rangle : i,j \in 3, \ k \in \{1,2\}\},$$
$$\rho(b) = 9^2 \sim \rho(a) \sim \rho(1').$$

All these representations of algebras 1_1–7_7 have been known to many mathematicians, beginning with Lyndon [17]. (However, the representations of 6_7 and 7_7 given in Tuza [3, p. 680] are incorrect.) We have supplemented Th. 6 by showing that $\mathfrak{A}$ is embeddable in the complex algebra of a finite group whenever the cycle parameters of $\mathfrak{A}$ are $\begin{pmatrix}0\\0\end{pmatrix}$, $\begin{pmatrix}1\\0\end{pmatrix}$, or $\begin{pmatrix}2\\0\end{pmatrix}$. On the other hand, Th. 7 below shows that $\mathfrak{A}$ has no square representation (and no representation) on a finite set if $n \geq 3$ and $\mathrm{Cp}(\mathfrak{A}) = \begin{pmatrix}n\\0\end{pmatrix}$. This was first proved in [21, pp. 65–66] (see [8, Th. 453] or Tuza [3, Th. 2.3]), and is generalized here to cover *weak* representations of a larger class of (not necessarily atomic) algebras.

Theorem 7. *Let* $\mathfrak{A} \in \mathsf{NA}$. *Suppose there are distinct nonzero elements* $a,b,c \in A$ *such that*

$$0 = a \cdot 1' = b \cdot 1' = c \cdot 1', \tag{45}$$
$$c;b \cdot a;b \leq b \leq c;\breve{c}, \tag{46}$$
$$a;c \cdot b;c \leq c \leq a;\breve{a}, \tag{47}$$
$$b;a \cdot c;a \leq a \leq b;\breve{b}. \tag{48}$$

Then every weak representation of $\mathfrak{A}$ *must have an infinite base set.*

Proof. Suppose ρ is a weak representation with base set $U = Fd(E)$. Then $\rho(a) \neq \emptyset$ since $\emptyset = \rho(0)$, $a \neq 0$, and ρ is one-to-one. Consequently we may choose some $v,w \in U$ such that $\langle v,w\rangle \in \rho(a)$. If $v = w$ then $\langle v,w\rangle = \mathsf{Id} \cap E = \rho(1')$, so $\langle v,w\rangle = \rho(1') \cap \rho(a) = \rho(1' \cdot a) = \rho(0) = \emptyset$, a contradiction. Hence $v \neq w$. Let $V_1 = \{v\}$. Then $w \notin V_1$ and

$$V_0 \times \{w\} \subseteq \rho(a). \tag{49}$$

Next, from our assumption that $a \leq b;\breve{b}$ we conclude that $\langle v,w\rangle \in \rho(a) \subseteq \rho(b)|(\rho(b))^{-1}$, so there is some $x \in U$ such that $\langle v,x\rangle \in \rho(b)$ and $\langle w,x\rangle \in \rho(b)$. We therefore have $v \neq x$ and $w \neq x$ since $b \cdot 1' = 0$, and

$$(V_1 \cup \{w\}) \times \{x\} \subseteq \rho(b). \tag{50}$$

Note that $|V_1| = 1$. Assume that we have constructed a set $V_i \subseteq U$ such that $|V_i| = i \geq 1$ and that we have found distinct elements $w,x \in U \sim V_i$ such that

$$V_i \times \{w\} \subseteq \rho(a), \tag{51}$$

$$(V_i \cup \{w\}) \times \{x\} \subseteq \rho(b). \tag{52}$$

From (52) and the assumption that $b \le c;\check{c}$, we get $\langle w, x \rangle \in \rho(b) \subseteq \rho(c)|(\rho(c))^{-1}$, so there is some $y \in U$ such that $\langle w, y \rangle \in \rho(c)$ and $\langle x, y \rangle \in \rho(c)$. Note that $w \ne y$ and $x \ne y$ since $c \cdot 1' = 0$. For every $v \in V_i$, we have $\langle v, w \rangle \in \rho(a)$ by (51) and $\langle v, x \rangle \in \rho(b)$ by (52), so

$$\langle v, y \rangle \in \rho(a)|\rho(c) \cap \rho(b)|\rho(c) = \rho(a;c \cdot b;c).$$

By the assumption that $a;c \cdot b;c \le c$, this gives us $\langle v, y \rangle \in \rho(c)$, hence $v \ne y$ since $c \cdot 1' = 0$. We have therefore proved that $y \notin V_i$ and

$$(V_i \cup \{w, x\}) \times \{y\} \subseteq \rho(c). \tag{53}$$

Letting $V_{i+1} = V_i \cup \{w\}$, we have $|V_{i+1}| = |V_i| + 1 = i + 1$, $x, y \in U \sim V_{i+1}$, $x \ne y$, and we may restate (52) and (53) as

$$V_{i+1} \times \{x\} \subseteq \rho(b), \tag{54}$$
$$(V_{i+1} \cup \{x\}) \times \{y\} \subseteq \rho(c). \tag{55}$$

By a similar argument, starting from (54) and (55) and using the assumptions $c \le a;\check{a}$, $b;a \cdot c;a \le a$, and $a \cdot 1' = 0$, we conclude that there is some $z \in U \sim (V_{i+1} \cup \{x, y\})$ such that

$$(V_{i+1} \cup \{x, y\}) \times \{z\} \subseteq \rho(a). \tag{56}$$

Let $V_{i+2} = V_{i+1} \cup \{x\}$. Then $|V_{i+2}| = i + 2$, $y, z \in U \sim V_{i+2}$, $y \ne z$, and from (55) and (56) we have

$$V_{i+2} \times \{y\} \subseteq \rho(c), \tag{57}$$
$$(V_{i+2} \cup \{y\}) \times \{z\} \subseteq \rho(a). \tag{58}$$

By repeating the argument once more, using assumptions $a \le b;\check{b}$, $c;b \cdot a;b \le b$, and $a \cdot 1' = 0$, we conclude that there is some $u \in U \sim V_{i+2}$ such that if $V_{i+3} = V_{i+2} \cup \{y\}$ then $z, u \in U \sim V_{i+3}$, $z \ne u$, and

$$V_{i+3} \times \{z\} \subseteq \rho(a), \tag{59}$$
$$(V_{i+3} \cup \{z\}) \times \{u\} \subseteq \rho(b). \tag{60}$$

We have now completed a cycle consisting of three similar steps and have used all the assumptions. Starting from (51) and (52), we found that there are three distinct $y, z, u \in U \sim V_i$ such that (59) and (60) hold with $V_{i+3} = V_i \cup \{y, z, u\}$ and $|V_{i+3}| = i + 3$. This cycle may be repeated indefinitely, so it follows that U must be infinite. $\qquad\square$

8   A Subvariety of RA

In this section we define a variety whose finite algebras are symmetric integral relation algebras with no 3-cycles.

Theorem 8. *Let* $\mathfrak{A} \in$ NA. *Then* $\mathfrak{A}$ *satisfies all or none of the following conditions.*

$$x;(\overline{x} \cdot y) \le x + y, \tag{61}$$

$$x;y \le x;(x \cdot y) + x + y, \tag{62}$$

$$x \cdot y = 0 \implies x;y \le x + y. \tag{63}$$

Proof. Assume $\mathfrak{A}$ satisfies (61). We show that $\mathfrak{A}$ also satisfies (62). By elementary laws from the theory of Boolean algebras we have

$$x;y = x;(x \cdot y + \overline{x} \cdot y).$$

By applying left ;-distributivity, which is easily derived from axioms R_5, R_7–R_9, we obtain
$$x;y = x;(x \cdot y) + x;(\overline{x} \cdot y).$$

To this we apply (61) and some elementary Boolean algebraic laws to get

$$x;y \le x;(x \cdot y) + x + y,$$

as desired. Next we assume $\mathfrak{A}$ satisfies (62) and show that $\mathfrak{A}$ also satisfies (63). Assume $x \cdot y = 0$. Then

$$
\begin{aligned}
x;y &\le x;(x \cdot y) + x + y && \text{(62)} \\
&= x;0 + x + y && \text{hypothesis} \\
&= x + y && \text{(10)}
\end{aligned}
$$

Finally, we assume $\mathfrak{A}$ satisfies (62) and show that $\mathfrak{A}$ satisfies (63). By Boolean algebra we have $x \cdot \overline{x} \cdot y = 0$, so from (63) we obtain

$$x;(\overline{x} \cdot y) \le x + \overline{x} \cdot y.$$

But the absorption law of Boolean algebra states that $x + \overline{x} \cdot y = x + y$, so we have
$$x;(\overline{x} \cdot y) \le x + y,$$

as desired. $\qquad\square$

Following Jipsen [1], we say an algebra $\mathfrak{A} \in$ NA is **subadditive** if it satisfies any (or all) of the conditions (61)–(63). Let V be the variety generated by the symmetric subadditive integral relation algebras. We proceed to obtain an equational basis for V.

Theorem 9. *Assume* $\mathfrak{A} \in$ SA *and* $\mathfrak{A}$ *satisfies*

$$(x \cdot 1');1;(y \cdot 1') = (x \cdot y \cdot 1');1;(x \cdot y \cdot 1') \tag{64}$$

Then $\mathfrak{A}$ *is trivial or is a subdirect product of integral algebras in* SA.

Proof. Suppose $\mathfrak{A}$ in nontrivial. By Birkhoff's Theorem [28], $\mathfrak{A}$ is a subdirect product of a system $\langle \mathfrak{B}_i : i \in I \rangle$, where each $\mathfrak{B}_i \in \mathsf{SA}$ is a subdirectly indecomposable homomorphic image of $\mathfrak{A}$ that satisfies (64) because $\mathfrak{A}$ does so. Every subdirectly indecomposable algebra in SA is simple (see [8, Th. 386]). Thus, for every $i \in I$, $\mathfrak{B}_i$ is a *simple* homomorphic image of $\mathfrak{A}$ that satisfies (64). A simple algebra in SA is integral iff 1' is an atom (see [8, Th. 353]). Therefore, if $\mathfrak{B}_i$ is *not* integral, there must be some $x, y \in B_i$ such that $x \cdot 1' \neq 0$, $y \cdot 1' \neq 0$, and $x \cdot y = 0$. In a simple SA, $u \neq 0$ iff $1;u;1 = 1$ (see [8, Th. 379]), so, by the simplicity of $\mathfrak{B}_i$, we have

$$1 = 1;(x \cdot 1');1, \quad 1 = 1;(y \cdot 1');1. \tag{65}$$

Many special cases of the associative law hold in every SA (see [29, Th. 25] or [8, Th. 365]). Associativity may be freely applied to any relative product in which one of the factors is 1. We therefore get

$$
\begin{aligned}
1 &= 1;1 \\
&= (1;(x \cdot 1');1);(1;(y \cdot 1');1) && (65) \\
&= 1;((x \cdot 1');1;(y \cdot 1'));1 && \text{semiassociativity} \\
&= 1;((x \cdot y \cdot 1');1;(x \cdot y \cdot 1'));1 && (64) \\
&= 1;(0;1;0);1 && x \cdot y = 0 \\
&= 0,
\end{aligned}
$$

contradicting the simplicity of $\mathfrak{B}_i$ (simple algebras have at least two elements). It follows that $\mathfrak{B}_i$ is actually integral. $\qquad\square$

Corollary 1. *The subvariety of* SA *generated by the integral algebras in* SA *has the following equational basis:* R_1–R_3, (15), R_5–R_{10}, (64).

Proof. If an algebra satisfies the equations, then by Th. 9 it is in the variety generated by the integral algebras in SA. For the converse, suppose $\mathfrak{A} \in \mathsf{SA}$ is integral. Then 1' is an atom of $\mathfrak{A}$, so for all $x, y \in A$, either $x \cdot 1' = 0$ or $x \cdot 1' = 1'$, and either $y \cdot 1' = 0$ or $y \cdot 1' = 1'$. In all these cases, (64) is satisfied. $\qquad\square$

Corollary 2. V *has the following equational basis:* R_1–R_{10}, (12), (62), (64).

9   Classifying Finite Algebras

Now we take up the problem of classifying finite algebras in V. The next theorem says that several otherwise distinct classes coincide on finite algebras in V. A finite algebra $\mathfrak{A} \in \mathsf{V}$ is the direct product of finite simple algebras in V, each of which is integral and has no 3-cycles because of subadditivity. The next theorem shows that such an algebra is either very representable (in GRA) or very non-representable (not in RA_5). The equivalence of (b) and (f) in part 1 appears in Tuza [3, Th. 2.2], while part 2 is in Tuza [3, Th. 2.3].

Theorem 10. *Let* $\mathfrak{A}$ *be a finite symmetric integral* RA *with no 3-cycles. Suppose*
$$\mathrm{Cp}(\mathfrak{A}) = \begin{pmatrix} s_1 & \cdots & s_n \\ t_1 & \cdots & t_n \end{pmatrix}.$$

1. *The following statements are equivalent:*
   (a) $\mathfrak{A} \in$ GRA,
   (b) $\mathfrak{A} \in$ RRA,
   (c) $\mathfrak{A} \in$ wRRA,
   (d) $\mathfrak{A} \in$ RA$_5$,
   (e) $\mathfrak{A}$ *satisfies* (28),
   (f) *for every* $i \in \{1, \ldots, n\}$, *if* $t_i > 0$ *then* $s_i + t_i \leq 2$.
2. *If* (a)–(f) *hold, then the following statements are equivalent:*
   (g) $\mathfrak{A}$ *is representable over a finite set,*
   (h) *for every* $i \in \{1, \ldots, n\}$, *if* $s_i = 0$ *then* $t_i \leq 2$.

Proof. Part 1: (a) implies (b) by (39).

(b) implies (c) since RRA $\subseteq$ wRRA (every representation is a weak representation).

(b) also implies (d) by (27). To prove that RRA $\subseteq$ RA$_5$, the key step is to check that if E is an equivalence relation, then $B_5\mathfrak{Sb}\,(E)$ is a 5-dimensional relational basis for $\mathfrak{Sb}\,(E)$.

(c) and (d) imply part (e), and the proofs are closely related. First, one should check that (28) holds in every equivalence relation algebra $\mathfrak{Sb}\,(E)$. If (c) holds then $\mathfrak{A}$ has a weak representation mapping $\mathfrak{A}$ into some $\mathfrak{Sb}\,(E)$. A weak representation preserves all the operations involved in (28), so (28) holds in $\mathfrak{A}$. If (d) holds then $\mathfrak{A}$ is a subalgebra of a complete atomic algebra that has a 5-dimensional basis. The proof that (28) holds in every RRA can be translated into a proof that (28) holds in every algebra with a 5-dimensional relational basis. For details, see [8, Th. 341].

To show that (e) implies (f), assume that (f) fails, that is, there is some $i \in \{1, \ldots, n\}$ such that $t_i > 0$ and $s_i + t_i \geq 3$. These hypotheses imply there are distinct diversity atoms $a, b, c \in At\mathfrak{A} \sim \{1'\}$ such that $a \cdot a; a = 0$ and $a \Leftrightarrow b \Leftrightarrow c$. We will show that (28) fails when $x_{01} = c$, $x_{02} = b$, $x_{03} = b$, $x_{32} = a$, $x_{21} = c$, $x_{24} = a$, and $x_{41} = a$. First, we have $b \cdot b; a = b$, $c \cdot a; a = c$, and $c \cdot b; c = c$ since $a \to b$, $c \to a$, and $b \to c$, so

$$x_{01} \cdot (x_{02} \cdot x_{03}; x_{32}); (x_{21} \cdot x_{24}; x_{41}) = c \cdot (b \cdot b; a); (c \cdot a; a) = c \cdot b; c = c.$$

By subadditivity we have

$$b; c \cdot a; c \leq (b + c) \cdot (a + c) = c,$$
$$c; a \cdot b; a \leq (c + a) \cdot (b + a) = a,$$

so

$$x_{03}; \big((x_{30}; x_{01} \cdot x_{32}; x_{21}); x_{14} \cdot x_{32}; x_{24} \cdot x_{30}; (x_{01}; x_{14} \cdot x_{02}; x_{24})\big); x_{41}$$
$$= b; \big((b; c \cdot a; c); a \cdot a; a \cdot b; (c; a \cdot b; a)\big); a$$

$$\leq b;(c;a \cdot a;a \cdot b;a);a$$
$$\leq b;(a \cdot a;a);a$$
$$= b;0;a$$
$$= 0.$$

Thus the left side of (28) evaluates to c, while the right side is 0. But $c \not\leq 0$, so (28) fails. This shows that (e) implies (f).

Finally, we show that (f) implies (a). By Th. 4, $\mathfrak{A}$ is the 2-cycle product of algebras $\mathfrak{B}_i$ where $\mathrm{Cp}(\mathfrak{B}_i) = \begin{pmatrix} s_i \\ t_i \end{pmatrix}$ for $1 \leq i \leq n$. By (f), $\mathrm{Cp}(\mathfrak{B}_i)$ appears in the following list.

$$\begin{pmatrix} 0 \\ 0 \end{pmatrix}, \begin{pmatrix} 0 \\ 1 \end{pmatrix}, \begin{pmatrix} 1 \\ 0 \end{pmatrix}, \begin{pmatrix} 0 \\ 2 \end{pmatrix}, \begin{pmatrix} 1 \\ 1 \end{pmatrix}, \begin{pmatrix} 2 \\ 0 \end{pmatrix}, \begin{pmatrix} 3 \\ 0 \end{pmatrix}, \begin{pmatrix} 4 \\ 0 \end{pmatrix}, \cdots \tag{66}$$

From Th. 6 and the remarks that follow Th. 6 we know that $\mathfrak{B}_i \in \mathsf{GRA}$ for $1 \leq i \leq n$. By Th. 5, we may conclude that $\mathfrak{A} \in \mathsf{GRA}$.

For part 2, note that if $\mathrm{Cp}(\mathfrak{B}_i)$ is one of the first six cases in (66), then $\mathfrak{B}_i$ can be embedded in the complex algebra of one of the finite groups $\mathbb{Z}_1$, $\mathbb{Z}_2$, $\mathbb{Z}_3$, $\mathbb{Z}_5$, $\mathbb{Z}_8$, or $\mathbb{Z}_3^2$, and hence $\mathfrak{B}_i$ has a square representation on a set of cardinality 1, 2, 3, 5, 8, or 9. Since (h) rules out all but the first six cases of (66), it follows that (h) implies (g) whenever (a)–(f) hold. Finally, if (a)–(g) hold then by Th. 7 $\mathrm{Cp}(\mathfrak{B}_i)$ must be one of the first six cases in (66), $i.e.$, (h) must also hold.   □

Theorem 11. *There is an algorithm for classifying each finite algebra in* V *as either in* fRRA, *in* RRA $\sim$ fRRA, *or in* SA $\sim$ RA$_5$.

Proof. Let $\mathfrak{A}$ be a finite algebra in V, specified by its cycle structure $T = Cy(\mathfrak{A})$ in the sense that $\mathfrak{Cm}\,(T) \cong \mathfrak{A}$. Divide the atoms of $\mathfrak{A}$ (the field of T) into equivalence classes with respect to the equivalence relation

$$E := \{\langle x, y \rangle : \exists_z (z \in Fd\,(T), \langle x, y, z \rangle \in T)\}.$$

(That E is an equivalence relation follows from $\mathfrak{A} \in \mathsf{SA}$.) Then the complex algebra $\mathfrak{Cm}\,(T)$ is isomorphic to the direct product of the complex algebras of the E-equivalence classes, each of which is a simple homomorphic image of $\mathfrak{A}$, $i.e.$,

$$\mathfrak{Cm}\,(T) \cong \prod_{C \in Fd(T)/E} \mathfrak{Cm}\,(T \cap C^2). \tag{67}$$

Each factor in (67) is integral by Th. 9 because it satisfies the equation (64), is symmetric by (12), and is subadditive by (62). Therefore each factor in (67) has no 3-cycles and it has a list of cycle parameters. Compute this list of parameters, and proceed as follows.

1. Provisionally classify $\mathfrak{A}$ as representable on a finite set:

$$\mathfrak{A} \in \mathsf{fRRA}. \tag{68}$$

   In some order, check each column that appears in the list of cycle parameters of a factor in (67). The are two things to check, and they can be done in either order.

2. If the column indicates 3 or more atoms with all their 1-cycles, *i.e.*, the column is one of these: $\begin{pmatrix}3\\0\end{pmatrix}$, $\begin{pmatrix}4\\0\end{pmatrix}$, ..., then the factor in (67) (and $\mathfrak{A}$ itself) cannot have a representation on a finite set. Change the provisional classification of $\mathfrak{A}$ to representable but not on finite sets,

$$\mathfrak{A} \in \mathsf{RRA} \sim \mathsf{fRRA}, \tag{69}$$

and CONTINUE.

3. If the column indicates at least 3 atoms and at least one missing 1-cycle, *i.e.*, it is one of

$$\begin{pmatrix}2\\1\end{pmatrix}, \begin{pmatrix}3\\1\end{pmatrix}, \begin{pmatrix}4\\1\end{pmatrix}, \cdots, \begin{pmatrix}1\\2\end{pmatrix}, \begin{pmatrix}2\\2\end{pmatrix}, \begin{pmatrix}3\\2\end{pmatrix}, \cdots, \begin{pmatrix}0\\3\end{pmatrix}, \begin{pmatrix}1\\3\end{pmatrix}, \begin{pmatrix}2\\3\end{pmatrix}, \cdots,$$

then the factor in (67) does not satisfy (28), and $\mathfrak{A}$ is not in RA_5 (nor in RRA, GRA, or wRRA). Change the classification of $\mathfrak{A}$ to

$$\mathfrak{A} \in \mathsf{SA} \sim \mathsf{RA}_5, \tag{70}$$

and STOP.

If no column summing to 3 or more is met, then (68) remains in force at the end. The algorithm stops and the classification of $\mathfrak{A}$ is changed to (70) if (28) fails at some column. If that does not happen, then the ultimate classification is one of two types of representability. (68) holds unless a column is found that shifts the classification to (69).   □

Now we apply the algorithm to the symmetric integral relation algebras that have no 3-cycles and either four or five atoms. (The data on finite algebras in this paper and in [8] have been checked or obtained with [30].) There are 65 symmetric integral relation algebras that have four atoms. They are the algebras 1_{65}–65_{65} in the numbering system of [8]. The atoms of these algebras are 1', a, b, and c. The algebras in this set of 65 which have no 3-cycles are 1_{65}–24_{65}. Here are their cycles.

	aaa	bbb	ccc	abb	baa	acc	caa	bcc	cbb
1_{65}	$\cdots$	$\cdots$	$\cdots$	abb	$\cdots$	acc	$\cdots$	bcc	$\cdots$
2_{65}	aaa	$\cdots$	$\cdots$	abb	$\cdots$	acc	$\cdots$	bcc	$\cdots$
3_{65}	$\cdots$	bbb	$\cdots$	abb	$\cdots$	acc	$\cdots$	bcc	$\cdots$
4_{65}	aaa	bbb	$\cdots$	abb	$\cdots$	acc	$\cdots$	bcc	$\cdots$
5_{65}	$\cdots$	$\cdots$	ccc	abb	$\cdots$	acc	$\cdots$	bcc	$\cdots$
6_{65}	aaa	$\cdots$	ccc	abb	$\cdots$	acc	$\cdots$	bcc	$\cdots$
7_{65}	$\cdots$	bbb	ccc	abb	$\cdots$	acc	$\cdots$	bcc	$\cdots$
8_{65}	aaa	bbb	ccc	abb	$\cdots$	acc	$\cdots$	bcc	$\cdots$
9_{65}	$\cdots$	$\cdots$	$\cdots$	abb	baa	acc	$\cdots$	bcc	$\cdots$
10_{65}	aaa	$\cdots$	$\cdots$	abb	baa	acc	$\cdots$	bcc	$\cdots$
11_{65}	aaa	bbb	$\cdots$	abb	baa	acc	$\cdots$	bcc	$\cdots$

	aaa	bbb	ccc	abb	baa	acc	caa	bcc	cbb
12_{65}	$\cdots$	$\cdots$	ccc	abb	baa	acc	$\cdots$	bcc	$\cdots$
13_{65}	aaa	$\cdots$	ccc	abb	baa	acc	$\cdots$	bcc	$\cdots$
14_{65}	aaa	bbb	ccc	abb	baa	acc	$\cdots$	bcc	$\cdots$
15_{65}	$\cdots$	$\cdots$	$\cdots$	$\cdots$	baa	acc	caa	bcc	$\cdots$
16_{65}	aaa	$\cdots$	$\cdots$	$\cdots$	baa	acc	caa	bcc	$\cdots$
17_{65}	$\cdots$	bbb	$\cdots$	$\cdots$	baa	acc	caa	bcc	$\cdots$
18_{65}	aaa	bbb	$\cdots$	$\cdots$	baa	acc	caa	bcc	$\cdots$
19_{65}	aaa	$\cdots$	ccc	$\cdots$	baa	acc	caa	bcc	$\cdots$
20_{65}	aaa	bbb	ccc	$\cdots$	baa	acc	caa	bcc	$\cdots$
21_{65}	$\cdots$	$\cdots$	$\cdots$	abb	baa	acc	caa	bcc	cbb
22_{65}	aaa	$\cdots$	$\cdots$	abb	baa	acc	caa	bcc	cbb
23_{65}	aaa	bbb	$\cdots$	abb	baa	acc	caa	bcc	cbb
24_{65}	aaa	bbb	ccc	abb	baa	acc	caa	bcc	cbb

Next are the multiplication tables for the atoms of algebras 1_{65}–24_{65}. (By the way, the table for the nonrepresentable relation algebra 21_{65} given in Tuza [3, p. 683] is incorrect; it is instead the table for an algebra in $\mathsf{SA} \sim \mathsf{RA}$.)

1_{65}	1'	a	b	c
1'	1'	a	b	c
a	a	1'	b	c
b	b	b	1'a	c
c	c	c	c	1'ab

2_{65}	1'	a	b	c
1'	1'	a	b	c
a	a	1'a	b	c
b	b	b	1'a	c
c	c	c	c	1'ab

3_{65}	1'	a	b	c
1'	1'	a	b	c
a	a	1'	b	c
b	b	b	1'ab	c
c	c	c	c	1'ab

4_{65}	1'	a	b	c
1'	1'	a	b	c
a	a	1'a	b	c
b	b	b	1'ab	c
c	c	c	c	1'ab

5_{65}	1'	a	b	c
1'	1'	a	b	c
a	a	1'	b	c
b	b	b	1'a	c
c	c	c	c	1'abc

6_{65}	1'	a	b	c
1'	1'	a	b	c
a	a	1'a	b	c
b	b	b	1'a	c
c	c	c	c	1'abc

7_{65}	1'	a	b	c
1'	1'	a	b	c
a	a	1'	b	c
b	b	b	1'ab	c
c	c	c	c	1'abc

8_{65}	1'	a	b	c
1'	1'	a	b	c
a	a	1'a	b	c
b	b	b	1'ab	c
c	c	c	c	1'abc

9_{65}	1'	a	b	c
1'	1'	a	b	c
a	a	1'b	ab	c
b	b	ab	1'a	c
c	c	c	c	1'ab

10_{65}	1'	a	b	c
1'	1'	a	b	c
a	a	1'ab	ab	c
b	b	ab	1'a	c
c	c	c	c	1'ab

11_{65}	1'	a	b	c
1'	1'	a	b	c
a	a	1'ab	ab	c
b	b	ab	1'ab	c
c	c	c	c	1'ab

12_{65}	1'	a	b	c
1'	1'	a	b	c
a	a	1'b	ab	c
b	b	ab	1'a	c
c	c	c	c	1'abc

13_{65}	1'	a	b	c
1'	1'	a	b	c
a	a	1'ab	ab	c
b	b	ab	1'a	c
c	c	c	c	1'abc

14_{65}	1'	a	b	c
1'	1'	a	b	c
a	a	1'ab	ab	c
b	b	ab	1'ab	c
c	c	c	c	1'abc

15_{65}	1'	a	b	c
1'	1'	a	b	c
a	a	1'bc	a	ac
b	b	a	1'	c
c	c	ac	c	1'ab

16_{65}	1'	a	b	c
1'	1'	a	b	c
a	a	1'abc	a	ac
b	b	a	1'	c
c	c	ac	c	1'ab

17_{65}	1'	a	b	c
1'	1'	a	b	c
a	a	1'bc	a	ac
b	b	a	1'b	c
c	c	ac	c	1'ab

18_{65}	1'	a	b	c
1'	1'	a	b	c
a	a	1'abc	a	ac
b	b	a	1'b	c
c	c	ac	c	1'ab

19_{65}	1'	a	b	c
1'	1'	a	b	c
a	a	1'abc	a	ac
b	b	a	1'	c
c	c	ac	c	1'abc

20_{65}	1'	a	b	c
1'	1'	a	b	c
a	a	1'abc	a	ac
b	b	a	1'b	c
c	c	ac	c	1'abc

21_{65}	1'	a	b	c
1'	1'	a	b	c
a	a	1'bc	ab	ac
b	b	ab	1'ac	bc
c	c	ac	bc	1'ab

22_{65}	1'	a	b	c
1'	1'	a	b	c
a	a	1'abc	ab	ac
b	b	ab	1'ac	bc
c	c	ac	bc	1'ab

23_{65}	1'	a	b	c
1'	1'	a	b	c
a	a	1'abc	ab	ac
b	b	ab	1'abc	bc
c	c	ac	bc	1'ab

24_{65}	1'	a	b	c
1'	1'	a	b	c
a	a	1'abc	ab	ac
b	b	ab	1'abc	bc
c	c	ac	bc	1'abc

$$\mathrm{Cp}(1_{65}) = \begin{pmatrix} 0 & 0 & 0 \\ 1 & 1 & 1 \end{pmatrix} \quad \mathrm{Cp}(2_{65}) = \begin{pmatrix} 1 & 0 & 0 \\ 0 & 1 & 1 \end{pmatrix} \quad \mathrm{Cp}(3_{65}) = \begin{pmatrix} 0 & 1 & 0 \\ 1 & 0 & 1 \end{pmatrix} \quad \mathrm{Cp}(4_{65}) = \begin{pmatrix} 1 & 1 & 0 \\ 0 & 0 & 1 \end{pmatrix}$$

$$\mathrm{Cp}(5_{65}) = \begin{pmatrix} 0 & 0 & 1 \\ 1 & 1 & 0 \end{pmatrix} \quad \mathrm{Cp}(6_{65}) = \begin{pmatrix} 1 & 0 & 1 \\ 0 & 1 & 0 \end{pmatrix} \quad \mathrm{Cp}(7_{65}) = \begin{pmatrix} 0 & 1 & 1 \\ 1 & 0 & 0 \end{pmatrix} \quad \mathrm{Cp}(8_{65}) = \begin{pmatrix} 1 & 1 & 1 \\ 0 & 0 & 0 \end{pmatrix}$$

$$\mathrm{Cp}(9_{65}) = \begin{pmatrix} 0 & 0 \\ 2 & 1 \end{pmatrix} \quad \mathrm{Cp}(10_{65}) = \begin{pmatrix} 1 & 0 \\ 1 & 1 \end{pmatrix} \quad \mathrm{Cp}(11_{65}) = \begin{pmatrix} 2 & 0 \\ 0 & 1 \end{pmatrix} \quad \mathrm{Cp}(12_{65}) = \begin{pmatrix} 0 & 1 \\ 2 & 0 \end{pmatrix}$$

$$\mathrm{Cp}(13_{65}) = \begin{pmatrix} 1 & 1 \\ 1 & 0 \end{pmatrix} \quad \mathrm{Cp}(14_{65}) = \begin{pmatrix} 2 & 1 \\ 0 & 0 \end{pmatrix} \quad \mathrm{Cp}(15_{65}) = \begin{pmatrix} 0 & 0 \\ 1 & 2 \end{pmatrix} \quad \mathrm{Cp}(16_{65}) = \begin{pmatrix} 0 & 1 \\ 1 & 1 \end{pmatrix}$$

$$\mathrm{Cp}(17_{65}) = \begin{pmatrix} 1 & 0 \\ 0 & 2 \end{pmatrix} \quad \mathrm{Cp}(18_{65}) = \begin{pmatrix} 1 & 1 \\ 0 & 1 \end{pmatrix} \quad \mathrm{Cp}(19_{65}) = \begin{pmatrix} 0 & 2 \\ 1 & 0 \end{pmatrix} \quad \mathrm{Cp}(20_{65}) = \begin{pmatrix} 1 & 2 \\ 0 & 0 \end{pmatrix}$$

$$\mathrm{Cp}(21_{65}) = \begin{pmatrix} 0 \\ 3 \end{pmatrix} \quad \mathrm{Cp}(22_{65}) = \begin{pmatrix} 1 \\ 2 \end{pmatrix} \quad \mathrm{Cp}(23_{65}) = \begin{pmatrix} 2 \\ 1 \end{pmatrix} \quad \mathrm{Cp}(24_{65}) = \begin{pmatrix} 3 \\ 0 \end{pmatrix}$$

Applying the algorithm of Th. 11 to algebras 1_{65}–24_{65} produces the following results.

1. Algebras 1_{65}–20_{65} are group representable relation algebras that have square representations on finite sets.
2. Algebras 21_{65}–23_{65} fail to satisfy equation (28), and hence are nonrepresentable relation algebras that are also not in RA_5 and are not weakly representable.
3. The algebra 24_{65} is a representable relation algebra that has no square representation on a finite set.

There are 3013 symmetric integral relation algebras that have five atoms. They are the algebras 1_{3013}–823_{3013} in the numbering system of [8]. The atoms of these algebras are 1', a, b, c, and d. Among these algebras, the ones that have no 3-cycles are 1_{3013}–823_{3013}. Their cycles and cycle parameters are given in the following tables.

	aaa	bbb	ccc	ddd	abb	baa	acc	caa	add	daa	bcc	cbb	bdd	dbb	cdd	dcc
1_{3013}	···	···	···	···	abb	···	acc	···	add	···	bcc	···	bdd	···	cdd	···
2_{3013}	aaa	···	···	···	abb	···	acc	···	add	···	bcc	···	bdd	···	cdd	···
3_{3013}	···	bbb	···	···	abb	···	acc	···	add	···	bcc	···	bdd	···	cdd	···
4_{3013}	aaa	bbb	···	···	abb	···	acc	···	add	···	bcc	···	bdd	···	cdd	···
5_{3013}	···	···	ccc	···	abb	···	acc	···	add	···	bcc	···	bdd	···	cdd	···
6_{3013}	aaa	···	ccc	···	abb	···	acc	···	add	···	bcc	···	bdd	···	cdd	···
7_{3013}	···	bbb	ccc	···	abb	···	acc	···	add	···	bcc	···	bdd	···	cdd	···
8_{3013}	aaa	bbb	ccc	···	abb	···	acc	···	add	···	bcc	···	bdd	···	cdd	···
9_{3013}	···	···	···	ddd	abb	···	acc	···	add	···	bcc	···	bdd	···	cdd	···
10_{3013}	aaa	···	···	ddd	abb	···	acc	···	add	···	bcc	···	bdd	···	cdd	···
11_{3013}	···	bbb	···	ddd	abb	···	acc	···	add	···	bcc	···	bdd	···	cdd	···
12_{3013}	aaa	bbb	···	ddd	abb	···	acc	···	add	···	bcc	···	bdd	···	cdd	···
13_{3013}	···	···	ccc	ddd	abb	···	acc	···	add	···	bcc	···	bdd	···	cdd	···
14_{3013}	aaa	···	ccc	ddd	abb	···	acc	···	add	···	bcc	···	bdd	···	cdd	···
15_{3013}	···	bbb	ccc	ddd	abb	···	acc	···	add	···	bcc	···	bdd	···	cdd	···
16_{3013}	aaa	bbb	ccc	ddd	abb	···	acc	···	add	···	bcc	···	bdd	···	cdd	···
17_{3013}	···	···	···	···	abb	baa	acc	···	add	···	bcc	···	bdd	···	cdd	···
18_{3013}	aaa	···	···	···	abb	baa	acc	···	add	···	bcc	···	bdd	···	cdd	···
19_{3013}	aaa	bbb	···	···	abb	baa	acc	···	add	···	bcc	···	bdd	···	cdd	···
20_{3013}	···	···	ccc	···	abb	baa	acc	···	add	···	bcc	···	bdd	···	cdd	···
21_{3013}	aaa	···	ccc	···	abb	baa	acc	···	add	···	bcc	···	bdd	···	cdd	···
22_{3013}	aaa	bbb	ccc	···	abb	baa	acc	···	add	···	bcc	···	bdd	···	cdd	···
23_{3013}	···	···	···	ddd	abb	baa	acc	···	add	···	bcc	···	bdd	···	cdd	···
24_{3013}	aaa	···	···	ddd	abb	baa	acc	···	add	···	bcc	···	bdd	···	cdd	···
25_{3013}	aaa	bbb	···	ddd	abb	baa	acc	···	add	···	bcc	···	bdd	···	cdd	···
26_{3013}	···	···	ccc	ddd	abb	baa	acc	···	add	···	bcc	···	bdd	···	cdd	···
27_{3013}	aaa	···	ccc	ddd	abb	baa	acc	···	add	···	bcc	···	bdd	···	cdd	···
28_{3013}	aaa	bbb	ccc	ddd	abb	baa	acc	···	add	···	bcc	···	bdd	···	cdd	···
29_{3013}	···	···	···	···	···	baa	acc	caa	add	···	bcc	···	bdd	···	cdd	···
30_{3013}	aaa	···	···	···	···	baa	acc	caa	add	···	bcc	···	bdd	···	cdd	···
31_{3013}	···	bbb	···	···	···	baa	acc	caa	add	···	bcc	···	bdd	···	cdd	···
32_{3013}	aaa	bbb	···	···	···	baa	acc	caa	add	···	bcc	···	bdd	···	cdd	···
33_{3013}	aaa	···	ccc	···	···	baa	acc	caa	add	···	bcc	···	bdd	···	cdd	···
34_{3013}	aaa	bbb	ccc	···	···	baa	acc	caa	add	···	bcc	···	bdd	···	cdd	···
35_{3013}	···	···	···	ddd	···	baa	acc	caa	add	···	bcc	···	bdd	···	cdd	···
36_{3013}	aaa	···	···	ddd	···	baa	acc	caa	add	···	bcc	···	bdd	···	cdd	···
37_{3013}	···	bbb	···	ddd	···	baa	acc	caa	add	···	bcc	···	bdd	···	cdd	···
38_{3013}	aaa	bbb	···	ddd	···	baa	acc	caa	add	···	bcc	···	bdd	···	cdd	···
39_{3013}	aaa	···	ccc	ddd	···	baa	acc	caa	add	···	bcc	···	bdd	···	cdd	···
40_{3013}	aaa	bbb	ccc	ddd	···	baa	acc	caa	add	···	bcc	···	bdd	···	cdd	···
41_{3013}	···	···	···	···	···	baa	···	caa	add	daa	bcc	···	bdd	···	cdd	···
42_{3013}	aaa	···	···	···	···	baa	···	caa	add	daa	bcc	···	bdd	···	cdd	···
43_{3013}	···	bbb	···	···	···	baa	···	caa	add	daa	bcc	···	bdd	···	cdd	···
44_{3013}	aaa	bbb	···	···	···	baa	···	caa	add	daa	bcc	···	bdd	···	cdd	···

	aaa	bbb	ccc	ddd	abb	baa	acc	caa	add	daa	bcc	cbb	bdd	dbb	cdd	dcc
45_{3013}	···	···	ccc	···	···	baa	···	caa	add	daa	bcc	···	bdd	···	cdd	···
46_{3013}	aaa	···	ccc	···	···	baa	···	caa	add	daa	bcc	···	bdd	···	cdd	···
47_{3013}	···	bbb	ccc	···	···	baa	···	caa	add	daa	bcc	···	bdd	···	cdd	···
48_{3013}	aaa	bbb	ccc	···	···	baa	···	caa	add	daa	bcc	···	bdd	···	cdd	···
49_{3013}	aaa	···	···	ddd	···	baa	···	caa	add	daa	bcc	···	bdd	···	cdd	···
50_{3013}	aaa	bbb	···	ddd	···	baa	···	caa	add	daa	bcc	···	bdd	···	cdd	···
51_{3013}	aaa	···	ccc	ddd	···	baa	···	caa	add	daa	bcc	···	bdd	···	cdd	···
52_{3013}	aaa	bbb	ccc	ddd	···	baa	···	caa	add	daa	bcc	···	bdd	···	cdd	···
53_{3013}	···	···	···	···	abb	baa	acc	caa	add	···	bcc	cbb	bdd	···	cdd	···
54_{3013}	aaa	···	···	···	abb	baa	acc	caa	add	···	bcc	cbb	bdd	···	cdd	···
55_{3013}	aaa	bbb	···	···	abb	baa	acc	caa	add	···	bcc	cbb	bdd	···	cdd	···
56_{3013}	aaa	bbb	ccc	···	abb	baa	acc	caa	add	···	bcc	cbb	bdd	···	cdd	···
57_{3013}	···	···	···	ddd	abb	baa	acc	caa	add	···	bcc	cbb	bdd	···	cdd	···
58_{3013}	aaa	···	···	ddd	abb	baa	acc	caa	add	···	bcc	cbb	bdd	···	cdd	···
59_{3013}	aaa	bbb	···	ddd	abb	baa	acc	caa	add	···	bcc	cbb	bdd	···	cdd	···
60_{3013}	aaa	bbb	ccc	ddd	abb	baa	acc	caa	add	···	bcc	cbb	bdd	···	cdd	···
61_{3013}	···	···	···	···	···	baa	···	caa	add	daa	bcc	cbb	bdd	···	cdd	···
62_{3013}	aaa	···	···	···	···	baa	···	caa	add	daa	bcc	cbb	bdd	···	cdd	···
63_{3013}	···	bbb	···	···	···	baa	···	caa	add	daa	bcc	cbb	bdd	···	cdd	···
64_{3013}	aaa	bbb	···	···	···	baa	···	caa	add	daa	bcc	cbb	bdd	···	cdd	···
65_{3013}	···	bbb	ccc	···	···	baa	···	caa	add	daa	bcc	cbb	bdd	···	cdd	···
66_{3013}	aaa	bbb	ccc	···	···	baa	···	caa	add	daa	bcc	cbb	bdd	···	cdd	···
67_{3013}	aaa	···	···	ddd	···	baa	···	caa	add	daa	bcc	cbb	bdd	···	cdd	···
68_{3013}	aaa	bbb	···	ddd	···	baa	···	caa	add	daa	bcc	cbb	bdd	···	cdd	···
69_{3013}	aaa	bbb	ccc	ddd	···	baa	···	caa	add	daa	bcc	cbb	bdd	···	cdd	···
70_{3013}	···	···	···	···	abb	baa	···	caa	add	daa	···	cbb	bdd	dbb	cdd	···
71_{3013}	aaa	···	···	···	abb	baa	···	caa	add	daa	···	cbb	bdd	dbb	cdd	···
72_{3013}	aaa	bbb	···	···	abb	baa	···	caa	add	daa	···	cbb	bdd	dbb	cdd	···
73_{3013}	···	···	ccc	···	abb	baa	···	caa	add	daa	···	cbb	bdd	dbb	cdd	···
74_{3013}	aaa	···	ccc	···	abb	baa	···	caa	add	daa	···	cbb	bdd	dbb	cdd	···
75_{3013}	aaa	bbb	ccc	···	abb	baa	···	caa	add	daa	···	cbb	bdd	dbb	cdd	···
76_{3013}	aaa	bbb	···	ddd	abb	baa	···	caa	add	daa	···	cbb	bdd	dbb	cdd	···
77_{3013}	aaa	bbb	ccc	ddd	abb	baa	···	caa	add	daa	···	cbb	bdd	dbb	cdd	···
78_{3013}	···	···	···	···	abb	baa	acc	caa	add	daa	bcc	cbb	bdd	dbb	cdd	dcc
79_{3013}	aaa	···	···	···	abb	baa	acc	caa	add	daa	bcc	cbb	bdd	dbb	cdd	dcc
80_{3013}	aaa	bbb	···	···	abb	baa	acc	caa	add	daa	bcc	cbb	bdd	dbb	cdd	dcc
81_{3013}	aaa	bbb	ccc	···	abb	baa	acc	caa	add	daa	bcc	cbb	bdd	dbb	cdd	dcc
82_{3013}	aaa	bbb	ccc	ddd	abb	baa	acc	caa	add	daa	bcc	cbb	bdd	dbb	cdd	dcc

$$\mathrm{Cp}(1_{3013}) = \begin{pmatrix} 0 & 0 & 0 & 0 \\ 1 & 1 & 1 & 1 \end{pmatrix} \quad \mathrm{Cp}(2_{3013}) = \begin{pmatrix} 1 & 0 & 0 & 0 \\ 0 & 1 & 1 & 1 \end{pmatrix} \quad \mathrm{Cp}(3_{3013}) = \begin{pmatrix} 0 & 1 & 0 & 0 \\ 1 & 0 & 1 & 1 \end{pmatrix}$$

$$\mathrm{Cp}(4_{3013}) = \begin{pmatrix} 1 & 1 & 0 & 0 \\ 0 & 0 & 1 & 1 \end{pmatrix} \quad \mathrm{Cp}(5_{3013}) = \begin{pmatrix} 0 & 0 & 1 & 0 \\ 1 & 1 & 0 & 1 \end{pmatrix} \quad \mathrm{Cp}(6_{3013}) = \begin{pmatrix} 1 & 0 & 1 & 0 \\ 0 & 1 & 0 & 1 \end{pmatrix}$$

$$\mathrm{Cp}(7_{3013}) = \begin{pmatrix} 0 & 1 & 1 & 0 \\ 1 & 0 & 0 & 1 \end{pmatrix} \quad \mathrm{Cp}(8_{3013}) = \begin{pmatrix} 1 & 1 & 1 & 0 \\ 0 & 0 & 0 & 1 \end{pmatrix} \quad \mathrm{Cp}(9_{3013}) = \begin{pmatrix} 0 & 0 & 0 & 1 \\ 1 & 1 & 1 & 0 \end{pmatrix}$$

$$\mathrm{Cp}(10_{3013}) = \begin{pmatrix} 1 & 0 & 0 & 1 \\ 0 & 1 & 1 & 0 \end{pmatrix} \quad \mathrm{Cp}(11_{3013}) = \begin{pmatrix} 0 & 1 & 0 & 1 \\ 1 & 0 & 1 & 0 \end{pmatrix} \quad \mathrm{Cp}(12_{3013}) = \begin{pmatrix} 1 & 1 & 0 & 1 \\ 0 & 0 & 1 & 0 \end{pmatrix}$$

$$\mathrm{Cp}(13_{3013}) = \begin{pmatrix} 0 & 0 & 1 & 1 \\ 1 & 1 & 0 & 0 \end{pmatrix} \quad \mathrm{Cp}(14_{3013}) = \begin{pmatrix} 1 & 0 & 1 & 1 \\ 0 & 1 & 0 & 0 \end{pmatrix} \quad \mathrm{Cp}(15_{3013}) = \begin{pmatrix} 0 & 1 & 1 & 1 \\ 1 & 0 & 0 & 0 \end{pmatrix}$$

$$\mathrm{Cp}(16_{3013}) = \begin{pmatrix} 1 & 1 & 1 & 1 \\ 0 & 0 & 0 & 0 \end{pmatrix} \quad \mathrm{Cp}(17_{3013}) = \begin{pmatrix} 0 & 0 & 0 \\ 2 & 1 & 1 \end{pmatrix} \quad \mathrm{Cp}(18_{3013}) = \begin{pmatrix} 1 & 0 & 0 \\ 1 & 1 & 1 \end{pmatrix}$$

$$\mathrm{Cp}(19_{3013}) = \begin{pmatrix} 2 & 0 & 0 \\ 0 & 1 & 1 \end{pmatrix} \quad \mathrm{Cp}(20_{3013}) = \begin{pmatrix} 0 & 1 & 0 \\ 2 & 0 & 1 \end{pmatrix} \quad \mathrm{Cp}(21_{3013}) = \begin{pmatrix} 1 & 1 & 0 \\ 1 & 0 & 1 \end{pmatrix}$$

$$\mathrm{Cp}(22_{3013}) = \begin{pmatrix} 2 & 1 & 0 \\ 0 & 0 & 1 \end{pmatrix} \quad \mathrm{Cp}(23_{3013}) = \begin{pmatrix} 0 & 0 & 1 \\ 2 & 1 & 0 \end{pmatrix} \quad \mathrm{Cp}(24_{3013}) = \begin{pmatrix} 1 & 0 & 1 \\ 1 & 1 & 0 \end{pmatrix}$$

$$\mathrm{Cp}(25_{3013}) = \begin{pmatrix} 2 & 0 & 1 \\ 0 & 1 & 0 \end{pmatrix} \quad \mathrm{Cp}(26_{3013}) = \begin{pmatrix} 0 & 1 & 1 \\ 2 & 0 & 0 \end{pmatrix} \quad \mathrm{Cp}(27_{3013}) = \begin{pmatrix} 1 & 1 & 1 \\ 1 & 0 & 0 \end{pmatrix}$$

$$\mathrm{Cp}(28_{3013}) = \begin{pmatrix} 2 & 1 & 1 \\ 0 & 0 & 0 \end{pmatrix} \quad \mathrm{Cp}(29_{3013}) = \begin{pmatrix} 0 & 0 & 0 \\ 1 & 2 & 1 \end{pmatrix} \quad \mathrm{Cp}(30_{3013}) = \begin{pmatrix} 0 & 1 & 0 \\ 1 & 1 & 1 \end{pmatrix}$$

$$\mathrm{Cp}(31_{3013}) = \begin{pmatrix} 1 & 0 & 0 \\ 0 & 2 & 1 \end{pmatrix} \quad \mathrm{Cp}(32_{3013}) = \begin{pmatrix} 1 & 1 & 0 \\ 0 & 1 & 1 \end{pmatrix} \quad \mathrm{Cp}(33_{3013}) = \begin{pmatrix} 0 & 2 & 0 \\ 1 & 0 & 1 \end{pmatrix}$$

$$\mathrm{Cp}(34_{3013}) = \begin{pmatrix} 1 & 2 & 0 \\ 0 & 0 & 1 \end{pmatrix} \quad \mathrm{Cp}(35_{3013}) = \begin{pmatrix} 0 & 0 & 1 \\ 1 & 2 & 0 \end{pmatrix} \quad \mathrm{Cp}(36_{3013}) = \begin{pmatrix} 0 & 1 & 1 \\ 1 & 1 & 0 \end{pmatrix}$$

$$\mathrm{Cp}(37_{3013}) = \begin{pmatrix} 1 & 0 & 1 \\ 0 & 2 & 0 \end{pmatrix} \quad \mathrm{Cp}(38_{3013}) = \begin{pmatrix} 1 & 1 & 1 \\ 0 & 1 & 0 \end{pmatrix} \quad \mathrm{Cp}(39_{3013}) = \begin{pmatrix} 0 & 2 & 1 \\ 1 & 0 & 0 \end{pmatrix}$$

$$\mathrm{Cp}(40_{3013}) = \begin{pmatrix} 1 & 2 & 1 \\ 0 & 0 & 0 \end{pmatrix} \quad \mathrm{Cp}(41_{3013}) = \begin{pmatrix} 0 & 0 & 0 \\ 1 & 1 & 2 \end{pmatrix} \quad \mathrm{Cp}(42_{3013}) = \begin{pmatrix} 0 & 0 & 1 \\ 1 & 1 & 1 \end{pmatrix}$$

$$\mathrm{Cp}(43_{3013}) = \begin{pmatrix} 1 & 0 & 0 \\ 0 & 1 & 2 \end{pmatrix} \quad \mathrm{Cp}(44_{3013}) = \begin{pmatrix} 1 & 0 & 1 \\ 0 & 1 & 1 \end{pmatrix} \quad \mathrm{Cp}(45_{3013}) = \begin{pmatrix} 0 & 1 & 0 \\ 1 & 0 & 2 \end{pmatrix}$$

$$\mathrm{Cp}(46_{3013}) = \begin{pmatrix} 0 & 1 & 1 \\ 1 & 0 & 1 \end{pmatrix} \quad \mathrm{Cp}(47_{3013}) = \begin{pmatrix} 1 & 1 & 0 \\ 0 & 0 & 2 \end{pmatrix} \quad \mathrm{Cp}(48_{3013}) = \begin{pmatrix} 1 & 1 & 1 \\ 0 & 0 & 1 \end{pmatrix}$$

$$\mathrm{Cp}(49_{3013}) = \begin{pmatrix} 0 & 0 & 2 \\ 1 & 1 & 0 \end{pmatrix} \quad \mathrm{Cp}(50_{3013}) = \begin{pmatrix} 1 & 0 & 2 \\ 0 & 1 & 0 \end{pmatrix} \quad \mathrm{Cp}(51_{3013}) = \begin{pmatrix} 0 & 1 & 2 \\ 1 & 0 & 0 \end{pmatrix}$$

$$\mathrm{Cp}(52_{3013}) = \begin{pmatrix} 1 & 1 & 2 \\ 0 & 0 & 0 \end{pmatrix} \quad \mathrm{Cp}(53_{3013}) = \begin{pmatrix} 0 & 0 \\ 3 & 1 \end{pmatrix} \quad \mathrm{Cp}(54_{3013}) = \begin{pmatrix} 1 & 0 \\ 2 & 1 \end{pmatrix}$$

$$\mathrm{Cp}(55_{3013}) = \begin{pmatrix} 2 & 0 \\ 1 & 1 \end{pmatrix} \quad \mathrm{Cp}(56_{3013}) = \begin{pmatrix} 3 & 0 \\ 0 & 1 \end{pmatrix} \quad \mathrm{Cp}(57_{3013}) = \begin{pmatrix} 0 & 1 \\ 3 & 0 \end{pmatrix}$$

$$\mathrm{Cp}(58_{3013}) = \begin{pmatrix} 1 & 1 \\ 2 & 0 \end{pmatrix} \quad \mathrm{Cp}(59_{3013}) = \begin{pmatrix} 2 & 1 \\ 1 & 0 \end{pmatrix} \quad \mathrm{Cp}(60_{3013}) = \begin{pmatrix} 3 & 1 \\ 0 & 0 \end{pmatrix}$$

$$\mathrm{Cp}(61_{3013}) = \begin{pmatrix} 0 & 0 \\ 2 & 2 \end{pmatrix} \quad \mathrm{Cp}(62_{3013}) = \begin{pmatrix} 0 & 1 \\ 2 & 1 \end{pmatrix} \quad \mathrm{Cp}(63_{3013}) = \begin{pmatrix} 1 & 0 \\ 1 & 2 \end{pmatrix}$$

$$\mathrm{Cp}(64_{3013}) = \begin{pmatrix} 1 & 1 \\ 1 & 1 \end{pmatrix} \quad \mathrm{Cp}(65_{3013}) = \begin{pmatrix} 2 & 0 \\ 0 & 2 \end{pmatrix} \quad \mathrm{Cp}(66_{3013}) = \begin{pmatrix} 2 & 1 \\ 0 & 1 \end{pmatrix}$$

$$\mathrm{Cp}(67_{3013}) = \begin{pmatrix} 0 & 2 \\ 2 & 0 \end{pmatrix} \quad \mathrm{Cp}(68_{3013}) = \begin{pmatrix} 1 & 2 \\ 1 & 0 \end{pmatrix} \quad \mathrm{Cp}(69_{3013}) = \begin{pmatrix} 2 & 2 \\ 0 & 0 \end{pmatrix}$$

$$\mathrm{Cp}(70_{3013}) = \begin{pmatrix} 0 & 0 \\ 1 & 3 \end{pmatrix} \quad \mathrm{Cp}(71_{3013}) = \begin{pmatrix} 0 & 1 \\ 1 & 2 \end{pmatrix} \quad \mathrm{Cp}(72_{3013}) = \begin{pmatrix} 0 & 2 \\ 1 & 1 \end{pmatrix}$$

$$\mathrm{Cp}(73_{3013}) = \begin{pmatrix} 1 & 0 \\ 0 & 3 \end{pmatrix} \quad \mathrm{Cp}(74_{3013}) = \begin{pmatrix} 1 & 1 \\ 0 & 2 \end{pmatrix} \quad \mathrm{Cp}(75_{3013}) = \begin{pmatrix} 1 & 2 \\ 0 & 1 \end{pmatrix}$$

$$\mathrm{Cp}(76_{3013}) = \begin{pmatrix} 0 & 3 \\ 1 & 0 \end{pmatrix} \quad \mathrm{Cp}(77_{3013}) = \begin{pmatrix} 1 & 3 \\ 0 & 0 \end{pmatrix} \quad \mathrm{Cp}(78_{3013}) = \begin{pmatrix} 0 \\ 4 \end{pmatrix}$$

$$\mathrm{Cp}(79_{3013}) = \begin{pmatrix} 1 \\ 3 \end{pmatrix} \qquad \mathrm{Cp}(80_{3013}) = \begin{pmatrix} 2 \\ 2 \end{pmatrix} \qquad \mathrm{Cp}(81_{3013}) = \begin{pmatrix} 3 \\ 1 \end{pmatrix}$$

$$\mathrm{Cp}(82_{3013}) = \begin{pmatrix} 4 \\ 0 \end{pmatrix}$$

Applying the algorithm of Th. 11 to algebras 1_{3013}–82_{3013} gives the following results.

1. Algebras 1_{3013}–52_{3013} and 61_{3013}–69_{3013} are group representable relation algebras that have square representations on finite sets.
2. Algebras 53_{3013}–55_{3013}, 57_{3013}–59_{3013}, 70_{3013}–75_{3013}, and 78_{3013}–81_{3013} fail to satisfy equation (28), and hence are nonrepresentable relation algebras that are also not in RA_5 and are not weakly representable.
3. Algebras 56_{3013}, 60_{3013}, 76_{3013}, 77_{3013}, and 82_{3013} are representable relation algebras that have no square representation on a finite set.

References

1. Jipsen, P.: Varieties of symmetric subadditive relation algebras. Preprint, pp. 3 (1990)
2. Comer, S.D.: Extension of polygroups by polygroups and their representations using color schemes. In: Universal algebra and lattice theory (Puebla, 1982). Volume 1004 of Lecture Notes in Math. Springer, Berlin (1983) 91–103
3. Tuza, Z.: Representations of relation algebras and patterns of colored triplets. In: Algebraic logic (Budapest, 1988). Volume 54 of Colloq. Math. Soc. János Bolyai. North-Holland, Amsterdam (1991) 671–693
4. Huntington, E.V.: New sets of independent postulates for the algebra of logic, with special reference to Whitehead and Russell's *Principia Mathematica*. Trans. Amer. Math. Soc. **35**(1) (1933) 274–304
5. Huntington, E.V.: Boolean algebra. A correction to: "New sets of independent postulates for the algebra of logic, with special reference to Whitehead and Russell's *Principia Mathematica*" [Trans. Amer. Math. Soc. **35** (1933), no. 1, 274–304; 1501684]. Trans. Amer. Math. Soc. **35**(2) (1933) 557–558
6. Huntington, E.V.: A second correction to: "New sets of independent postulates for the algebra of logic, with special reference to Whitehead and Russell's *Principia Mathematica*" [Trans. Amer. Math. Soc. **35** (1933), no. 1, 274–304; 1501684]. Trans. Amer. Math. Soc. **35**(4) (1933) 971
7. Maddux, R.D.: Some varieties containing relation algebras. Trans. Amer. Math. Soc. **272**(2) (1982) 501–526
8. Maddux, R.D.: Relation Algebras. Volume 150 of Studies in Logic and the Foundations of Mathematics. Elsevier, Amsterdam (2006)
9. Tarski, A.: Contributions to the theory of models. III. Nederl. Akad. Wetensch. Proc. Ser. A. **58** (1955) 56–64 = Indagationes Math. **17**, 56–64 (1955)
10. Maddux, R.D.: Some sufficient conditions for the representability of relation algebras. Algebra Universalis **8**(2) (1978) 162–172
11. Jónsson, B.: Varieties of relation algebras. Algebra Universalis **15**(3) (1982) 273–298

12. Jónsson, B.: The theory of binary relations. In Andréka, H., Monk, J.D., Németi, I., eds.: Algebraic Logic (Budapest, 1988). Volume 54 of Colloquia Mathematica Societatis János Bolyai. North-Holland, Amsterdam (1991) 245–292

13. Hirsch, R., Hodkinson, I.: Relation algebras by games. Volume 147 of Studies in Logic and the Foundations of Mathematics. North-Holland Publishing Co., Amsterdam (2002). With a foreword by Wilfrid Hodges.

14. Birkhoff, G.: On the structure of abstract algebras. Proc. Cambridge Philos. Soc. **31** (1935) 433–454

15. Monk, J.D.: On representable relation algebras. Michigan Math. J. **11** (1964) 207–210

16. Jónsson, B.: Representation of modular lattices and of relation algebras. Trans. Amer. Math. Soc. **92** (1959) 449–464

17. Lyndon, R.C.: The representation of relational algebras. Ann. of Math. (2) **51** (1950) 707–729

18. Backer, F.: Representable relation algebras. Report for a seminar on relation algebras conducted by A. Tarski, mimeographed, University of California, Berkeley (Spring, 1970)

19. McKenzie, R.N.: The representation of relation algebras. PhD thesis, University of Colorado, Boulder (1966)

20. Wostner, U.: Finite relation algebras. Notices of the AMS **23** (1976) A–482

21. Maddux, R.D.: Topics in Relation Algebras. PhD thesis, University of California, Berkeley (1978)

22. Comer, S.D.: Multivalued loops and their connection with algebraic logic (1979), monograph, 173 pp.

23. Comer, S.D.: Multi-Valued Algebras and their Graphical Representation (July 1986), monograph, 103 pp.

24. Jipsen, P.: Computer-aided investigations of relation algebras. PhD thesis, Vanderbilt University (1992)

25. Jipsen, P., Lukács, E.: Representability of finite simple relation algebras with many identity atoms. In: Algebraic logic (Budapest, 1988). Volume 54 of Colloq. Math. Soc. János Bolyai. North-Holland, Amsterdam (1991) 241–244

26. Jipsen, P., Lukács, E.: Minimal relation algebras. Algebra Universalis **32**(2) (1994) 189–203

27. Andréka, H., Maddux, R.D.: Representations for small relation algebras. Notre Dame J. Formal Logic **35**(4) (1994) 550–562

28. Birkhoff, G.: Subdirect unions in universal algebra. Bull. Amer. Math. Soc. **50** (1944) 764–768

29. Maddux, R.D.: Pair-dense relation algebras. Trans. Amer. Math. Soc. **328**(1) (1991) 83–131

30. The GAP Group (`http://www-gap.dcs.st-and.ac.uk/~gap`) Aachen, St Andrews: GAP – Groups, Algorithms, and Programming, Version 4.2. (1999)

Computations and Relational Bundles

J.W. Sanders

Programming Research Group
Oxford University Computing Laboratory
Wolfson Building, Parks Road, Oxford, OX1 3QD
`jeff@comlab.ox.ac.uk`

Abstract. We explore the view of a computation as a relational section of a (trivial) fibre bundle: initial states lie in the base of the bundle and final states lie in the fibres located at their initial states. This leads us to represent a computation in 'fibre-form' as the angelic choice of its behaviours from each initial state. That view is shown to have the advantage also of permitting final states to be of different types, as might be used for example in a semantics of probabilistic computations, and of providing a natural setting for refinement of computations.

However we apply that view in a different direction. On computations more general than code the two standard models, the relational and the predicate-transformer models, obey different laws. One way to understand that difference is to study the laws of more refined models, like the semantics of probabilistic computations. Another is to characterise each model by its laws. In spite of their differences, the relational model is embedded in the transformer model by a Galois connection which can be used to transfer much of the structure on transformers to the relational model. We investigate the extent to which the conjugate on predicate transformers translates to relations and use the result to motivate a characterisation of relational computations, achieved by using fibre-forms.

1 Introduction

The binary-relation and predicate-transformer models of (sequential) programs have different flavours and different properties but each satisfies all the laws required of programs. Indeed consistency of the two models is maintained by the wp-based Galois connection between them. But the extension from programs to more general commands, specifications, or 'computations' as we shall call them, includes arbitrary (not merely finite) demonic nondeterminism and its 'dual', angelic nondeterminism. Thus whilst programs may not terminate, computations may (dually) not be enabled. In this extension the equivalence of the two semantic models is lost: the Galois connection does not preserve angelic nondeterminism.

Differences in the way sequential composition interacts with the two forms of nondeterminism (demonic and angelic) are, for arbitrary computations, summarised in Fig. 1. Recall that in the relational model demonic nondeterminism

R.A. Schmidt (Ed.): RelMiCS /AKA 2006, LNCS 4136, pp. 30–62, 2006.

$$(\cup \mathcal{R}) \mathbin{\mathring{,}} S \;\; = \;\; \cup \{ R \mathbin{\mathring{,}} S \mid R \in \mathcal{R} \} \tag{1}$$

$$S \mathbin{\mathring{,}} (\cup \mathcal{R}) \;\; = \;\; \cup \{ S \mathbin{\mathring{,}} R \mid R \in \mathcal{R} \} \qquad \text{if } \mathcal{R} \text{ nonempty} \tag{2}$$

$$(\cap \mathcal{R}) \mathbin{\mathring{,}} S \;\; \subseteq \;\; \cap \{ R \mathbin{\mathring{,}} S \mid R \in \mathcal{R} \}, \qquad = \text{ if } S \text{ is injective} \tag{3}$$

$$S \mathbin{\mathring{,}} (\cap \mathcal{R}) \;\; \subseteq \;\; \cap \{ S \mathbin{\mathring{,}} R \mid R \in \mathcal{R} \}, \qquad = \text{ if } S \text{ is predeterministic} \tag{4}$$

$$(\wedge T) \circ u \;\; = \;\; \wedge \{ t \circ u \mid t \in T \} \tag{5}$$

$$u \circ (\wedge T) \;\; \leq \;\; \wedge \{ u \circ t \mid t \in T \}, \qquad = \text{ if } u \text{ is positively conjunctive} \tag{6}$$

$$(\vee T) \circ u \;\; = \;\; \vee \{ t \circ u \mid t \in T \} \tag{7}$$

$$u \circ (\vee T) \;\; \geq \;\; \vee \{ u \circ t \mid t \in T \}, \qquad = \text{ if } u \text{ is disjunctive.} \tag{8}$$

Fig. 1. Differences between the relational model—Laws (1) to (4)—and the transformer model—Laws (5) to (8)—for arbitrary computations. In the former, demonic choice, angelic choice and refinement are respectively $\cup$, $\cap$ and $\supseteq$, whilst in the latter they are $\wedge$, $\vee$ and $\leq$. Only the first Laws ((1) and (5)) and last ((4) and (8)) coincide.

$\sqcap$ is given by union, angelic nondeterminism $\sqcup$ by intersection and refinement $\sqcap$ by containment $\supseteq$, whilst in the transformer model demonic nondeterminism is given by pointwise conjunction $\wedge$, angelic nondeterminism by pointwise disjunction $\vee$ and refinement by pointwise implication $\leq$. In the following we translate the laws of Fig. 1 from semantic to algebraic (i.e. $\sqcap$, $\sqcup$, $\sqsubseteq$) notation.

Thus the two models agree, for general computations, on only two of the four laws. Firstly, by Laws (1) and (5), sequential composition distributes initial demonic nondeterminism

$$(\sqcap \mathcal{A}) \mathbin{\mathring{,}} B \;\; = \;\; \sqcap \{ A \mathbin{\mathring{,}} B \mid A \in \mathcal{A} \}.$$

Operationally it might be reasoned that the demon resolving the nondeterminism (having memory but not prescience) does so initially on both sides hence, confronted with the same choices, produces the same behaviours.

Distribution of final demonic nondeterminism is valid in the relational model (Law (2)) but in the transformer model is valid in only one direction (Law (6))

$$B \mathbin{\mathring{,}} (\sqcap \mathcal{A}) \;\; \sqsubseteq \;\; \sqcap \{ B \mathbin{\mathring{,}} A \mid A \in \mathcal{A} \}.$$

Operationally, the demon has more choice the later it acts. Thus on the right, where the choice is made initially, there are fewer choices and so fewer behaviours than on the left. However the choices coincide if execution of B results in no further choices for the demon: if B is free of angelic choice or, in other words, forms a program (i.e. lies in the range of the Galois embedding wp and is thus positively conjunctive as a transformer).

And secondly the two models agree, by Laws (4) and (8), on the refinement in which sequential composition distributes final angelic nondeterminism

$$B \mathbin{\mathring{,}} (\sqcup \mathcal{A}) \;\; \sqsupseteq \;\; \sqcup \{ B \mathbin{\mathring{,}} A \mid A \in \mathcal{A} \}.$$

Operationally, angelic choice is dual to demonic choice: the angel resolving the choice has prescience but not memory. Thus on the right the angel has an initial choice, and hence the entire computation of B in which to make it; but on the left it makes a choice after B, with fewer alternatives and so fewer resulting behaviours. Equality holds if execution of B offers no further demonic choice by which the angel might profit: from each initial state computation B is either un-enabled, aborts or is deterministic (i.e. is predeterministic and is thus disjunctive as a transformer).

Distribution of initial angelic choice is valid in the transformer model (Law (7)) (operationally, the angel has the same choices on each side) but in the relational model is valid in only one direction (Law (3)) unless computation B terminates from distinct initial states in distinct final states.

For theoretical purposes, a richer model is better and for that reason most semantic study has taken place inside the transformer model. For instance the operational justification of the laws just outlined comes from the angel/demon game interpretation supported by the transformer model [BvW98]. One vital feature of that model has been its involution underlying the duality mentioned above, which conflates the two simulation rules for data refinement which in the relational model are distinct, with the result that one rule alone is complete in the transformer model though two are required in the relational model.

But what of the system designer who is committed to formal methods only to be confronted by an inconsistency in the laws satisfied by what he might regard as the most intuitive model (binary relations) and the more studied model (transformers)? The distinction may become apparent as soon as a refinement is attempted from specification to code. Or what of the implementor who wishes to document her clever implementation more abstractly for comparison with others; does she allow demonic choice to 'look ahead'? It would be perverse of specialists in formal methods to confront practitioners of their subject with an inconsistent array of techniques. And the laws of computations become important as soon as derivation of code from specification is practiced. How should a software engineer be expected to express a preference between the relational and transformer laws? And what exactly are the consequences of that distinction, anyway?

One way to understand better the difference between those models of computations more general than programs is to investigate stronger paradigms of computation. For example inclusion of a binary combinator $_p\oplus$ for probabilistic choice leads to more subtle behaviours, even of the standard combinators, and so to more detailed relational and transformer models. It provides, for example, insight into memory and prescience of demon and angel.

But in this paper we take an alternative approach and characterise what it means for a model of computation to 'look relational'. By starting with the laws satisfied by the binary-relation model we adopt a fibre-wise approach that enables us to express a computation in terms of its fibres and hence to construct an isomorphism—that preserves computational structure—between a model of those laws and the relational model itself. A model that looks relational actu-

$$
\begin{array}{ll}
\textbf{skip} & \text{no op} \\
\textbf{abort} & \text{divergent computation} \\
x := e & \text{assignment} \\
A \lhd b \rhd B & \text{binary conditional} \\
A \,\mathring{,}\, B & \text{sequential composition} \\
\mu F & \text{recursion} \\
A \sqcap B & \text{demonic choice} \\
\\
A \sqsubseteq B & \text{refinement: } A \sqcap B = A
\end{array}
$$

Fig. 2. Syntax for the space $(gcl.X, \sqsubseteq)$ of programs over state space X

ally is relational; given a state space, the laws of relational computations are categorical.

The underlying ingredient in our approach is the fibre-wise view of computations. To express it, we recall the definition of a (discrete) bundle from differential geometry and topology, and consider the structure it offers for a theory of computation. But first we need to recall the relational and transformer models of computations and the Galois connection between them (Section 3); and before that we must recall the notion of computation itself and the special case of programs (Section 2).

Notation

This paper uses the following general notation.

If f is a function then $f.x$ denotes its application to argument x. Function application binds to the left.

For any set X, the set of all subsets of X is denoted $\mathbb{P}.X$, the set of all finite subsets of X is denoted $\mathbb{F}.X$ and the set of all predicates on X is denoted $\mathbb{P}\mathrm{r}.X$. The cardinality of a set E is denoted $\#E$.

The set of all relations between sets X and Y is denoted $X \leftrightarrow Y$. If its field is evident from the context we write id for the identity relation; when the field A requires emphasis we write id $.A$. The converse of relation r is written $r^{\sim}$. The image of set E by relation r is written $r.(\!| E |\!)$. To express a relation r pointwise we use infix notation: $x\, r\, y$.

2   Programs and Computations

2.1   The Space of Programs

As space of programs we adopt a commonly-used mild extension $gcl.X$ of Dijkstra's guarded-command language,[1] that is closed under nonempty finite infima

---

[1] Extended to contain recursion rather than merely iteration; and syntactically different by replacing general conditional with binary conditional and explicit binary demonic choice.

$$(gcl.X, \sqsubseteq) \text{ is a complete partial order with min } \mathbf{abort} \qquad (9)$$
$$A = \sqcup\{A \vartriangleleft x \in F \vartriangleright \mathbf{abort} \mid F \in \mathbb{F}.X\} \qquad (10)$$

$$A \,\fatsemi\, (B \,\fatsemi\, C) = (A \,\fatsemi\, B) \,\fatsemi\, C \qquad (11)$$
$$A \,\fatsemi\, \mathbf{skip} = A = \mathbf{skip} \,\fatsemi\, A \qquad (12)$$

$$\mathbf{abort} \,\fatsemi\, A = \mathbf{abort} = A \,\fatsemi\, \mathbf{abort} \qquad (13)$$
$$(\sqcap \mathcal{A}) \,\fatsemi\, B = \sqcap\{A \,\fatsemi\, B \mid A \in \mathcal{A}\} \qquad \text{if } \mathcal{A} \text{ nonempty finite} \qquad (14)$$
$$B \,\fatsemi\, (\sqcap \mathcal{A}) = \sqcap\{B \,\fatsemi\, A \mid A \in \mathcal{A}\} \qquad \text{if } \mathcal{A} \text{ nonempty finite} \qquad (15)$$

Fig. 3. Laws concerning order and sequential composition for the space $gcl.X$ of (relational) programs with state space X

(representing demonic nondeterminism) and suprema of directed sets (representing common refinement). Since it also contains a least element, **abort**, it forms a complete partial order: Law (9). The compact elements are those programs that abort off some finite set of initial states, and each program is the supremum of the directed set of its approximating compacts (cf. Law (10), which states a special case in which the approximations from any initial state reflect the full behaviour of the computation there).

Each assignment $x := e$ is assumed to be terminating and deterministic; in other words the expression e provides a total function on state space X. A demonically-nondeterministic assignment (like $x := \pm\sqrt{y}$) is expressed using demonic choice (i.e. $(x := \sqrt{y}) \sqcap (x := -\sqrt{y})$). Nondefinedness (like $x := 1/y$) is achieved using conditional (e.g. $1/y \vartriangleleft y \neq 0 \vartriangleright \mathbf{abort}$).

Sequential composition is associative, Law (11), with **skip** a left unit and a right unit (Law (12)) and **abort** a left zero and a right zero (Law (13)). However sequential composition is distributed, on either side, only by the $\sqcap$ of nonempty (finite, of course) sets: Laws (14) and (15). Indeed an empty $\sqcap$ would be a greatest element of $(gcl.X, \sqsubseteq)$, a program refining every program, and no such *program* exists. (In fact, it is the more general *computation* **magic** of Laws (31) and (32). It provides an alternative description of nondefinedness: $1/y \vartriangleleft y \neq 0 \vartriangleright \mathbf{magic}$.)

Laws for conditional and assignment appear in Fig. 4. In this paper we suppress the treatment of recursion (least fixed point) and local block (**var** $y :$ $Y \cdot A$ **rav**).

2.2 The Space of Computations

The more comprehensive space $Gcl.X$ of computations includes that of programs (i.e. $gcl.X$); see Fig. 5. It is closed under the combinators of $gcl.X$ but also under arbitrary infima and hence also (by a standard general result) under arbitrary suprema. It thus forms a *complete lattice*: Law (27).

$$A \lhd true \rhd B \;=\; A \tag{16}$$

$$A \lhd b \rhd B \;=\; B \lhd \neg b \rhd A \tag{17}$$

$$A \lhd b \rhd A \;=\; A \tag{18}$$

$$(A \lhd b \rhd B) \lhd c \rhd C \;=\; A \lhd b \wedge c \rhd (B \lhd c \rhd C) \tag{19}$$

$$A \lhd b \rhd (B \lhd c \rhd C) \;=\; (A \lhd b \rhd B) \lhd c \rhd (A \lhd b \rhd C) \tag{20}$$

$$(A \lhd b \rhd B) \,\fatsemi\, C \;=\; (A \,\fatsemi\, C) \lhd b \rhd (B \,\fatsemi\, C) \tag{21}$$

$$A \sqcap (B \lhd b \rhd C) \;=\; (A \sqcap B) \lhd b \rhd (A \sqcap C) \tag{22}$$

$$x := x \;=\; \mathbf{skip} \tag{23}$$

$$(x := e) \,\fatsemi\, (x := f) \;=\; x := (e \,;\, f) \tag{24}$$

$$(x := e) \lhd b \rhd (x := f) \;=\; x := (e \lhd b \rhd f) \tag{25}$$

$$(x := e) \,\fatsemi\, (A \lhd b \rhd B) \;=\; (x := e) \,\fatsemi\, A$$
$$\lhd \; e \,;\, b \; \rhd$$
$$(x := e) \,\fatsemi\, B \tag{26}$$

Fig. 4. Laws concerning conditional and (terminating deterministic) assignment for the space $gcl.X$ of (relational) programs over X

$gcl.X$	programs and their combinators
magic	always-unenabled computation
$\sqcap \mathcal{A}$	arbitrary demonic choice
$\sqcup \mathcal{A}$	arbitrary angelic choice

Fig. 5. Syntax for the space $Gcl.X$ of (general) computations over state space X

The empty $\sqcap$, the greatest element of $Gcl.X$, is called **magic**. The least element of $(Gcl.X, \sqsubseteq)$ remains the least element, **abort**, of $(gcl.X, \sqsubseteq)$; it is the empty $\sqcup$. A computation is compact iff there is some finite set of states off which it aborts (and on which—state by state—it is either unenabled or exhibits arbitrary behaviour). Again, each computation is the $\sqcup$ of the directed set of its compact approximations (cf. Law (28)). The difference between Laws (10) and (28) is that the latter may display unenabled behaviour at some initial states.

The following shorthand is standard and convenient. For any subset E of state space, $x :\in E$ denotes the demonic choice $\sqcap\{x := e \mid e \in E\}$. If E is nonempty and finite it is a program; otherwise it is just a computation, which if E is empty is **magic**.

2.3   Termination, Enabledness and Determinism

The important computational concepts of termination, enabledness and determinism are expressed algebraically (using conditionals) as follows. More

$$(Gcl.X, \sqsubseteq) \text{ is a complete lattice with min } \mathbf{abort} \text{ and max } \mathbf{magic} \quad (27)$$

$$A \;=\; \sqcup\{A \lhd x \in F \rhd \mathbf{abort} \mid F \in \mathbb{F}.X\} \quad (28)$$

$$\mathbf{abort} \, \mathbin{\substack{\circ\\\circ}} \, A \;=\; \mathbf{abort} \quad (29)$$

$$A \, \mathbin{\substack{\circ\\\circ}} \, \mathbf{abort} \;=\; \mathbf{abort} \qquad \text{if } A \text{ always enabled} \quad (30)$$

$$\mathbf{magic} \, \mathbin{\substack{\circ\\\circ}} \, A \;=\; \mathbf{magic} \quad (31)$$

$$A \, \mathbin{\substack{\circ\\\circ}} \, \mathbf{magic} \;=\; \mathbf{magic} \qquad \text{if } A \text{ terminating} \quad (32)$$

$$A \sqcup (B \lhd b \rhd C) \;=\; (A \sqcup B) \lhd b \rhd (A \sqcup C) \quad (33)$$

$$(\sqcap \mathcal{A}) \, \mathbin{\substack{\circ\\\circ}} \, B \;=\; \sqcap\{A \, \mathbin{\substack{\circ\\\circ}} \, B \mid A \in \mathcal{A}\} \quad (34)$$

$$B \, \mathbin{\substack{\circ\\\circ}} \, (\sqcap \mathcal{A}) \;=\; \sqcap\{B \, \mathbin{\substack{\circ\\\circ}} \, A \mid A \in \mathcal{A}\} \qquad \text{if } \mathcal{A} \text{ nonempty} \quad (35)$$

$$(\sqcup \mathcal{A}) \, \mathbin{\substack{\circ\\\circ}} \, B \;\sqsupseteq\; \sqcup\{A \, \mathbin{\substack{\circ\\\circ}} \, B \mid A \in \mathcal{A}\}, \quad (36)$$

$$\;=\; \sqcup\{A \, \mathbin{\substack{\circ\\\circ}} \, B \mid A \in \mathcal{A}\} \qquad \text{if } B \text{ is injective} \quad (37)$$

$$B \, \mathbin{\substack{\circ\\\circ}} \, (\sqcup \mathcal{A}) \;\sqsupseteq\; \sqcup\{B \, \mathbin{\substack{\circ\\\circ}} \, A \mid A \in \mathcal{A}\}, \quad (38)$$

$$\;=\; \sqcup\{B \, \mathbin{\substack{\circ\\\circ}} \, A \mid A \in \mathcal{A}\} \qquad \text{if } B \text{ is predeterministic} \quad (39)$$

Fig. 6. Laws concerning order and sequential composition for the space $Gcl.X$ of relational computations over X

succinct equivalents will be available after introduction of the fibre notation in Sec. 4.1.

Suppose that A is a computation and $x_0 : X$ is a state. Then A is said to *abort at x_0* iff it might not (equivalently 'will not' in the standard (Hoare/Dijkstra) model we adopt) terminate there:

$$(A \lhd x = x_0 \rhd \mathbf{abort}) \;=\; \mathbf{abort}.$$

Computation A is said to be *enabled at x_0* iff it may (equivalently 'does') begin there

$$(A \lhd x = x_0 \rhd \mathbf{magic}) \;\neq\; \mathbf{magic}.$$

That inequality is equivalent (in view of Laws (30) and (31)) to the identity

$$(A \lhd x = x_0 \rhd \mathbf{magic}) \, \mathbin{\substack{\circ\\\circ}} \, \mathbf{abort} \;=\; (\mathbf{abort} \lhd x = x_0 \rhd \mathbf{magic}).$$

And computation A is said to *terminate at x_0* iff it does not abort whenever it is enabled there. In defining termination to permit non-enabledness, the concern has been foremost to ensure Law (4) holds and secondarily to maintain consistency with the transformer characterisation of termination.

Computaton A is 'deterministic' at x_0 iff it is enabled there and 'terminates in only a single final state'. To define that term: computation A is *co-atomic* at

x_0 iff the computation $(A \lhd x = x_0 \rhd \textbf{magic})$ is co-atomic: namely **magic** is the only computation that strictly refines it:

$$(A \lhd x = x_0 \rhd \textbf{magic}) \neq \textbf{magic}$$
$$\wedge$$
$$\forall B : Gcl.X \cdot (A \lhd x = x_0 \rhd \textbf{magic}) \sqsubset B \;\Rightarrow\; B = \textbf{magic}.$$

Then A is defined to be *deterministic* at x_0 iff $(A \lhd x = x_0 \rhd \textbf{magic})$ is co-atomic. Finally A is *predeterministic* at x_0 iff either it does not terminate or is deterministic at x_0, whenever it is enabled at x_0.

A computation is terminating [always-enabled, deterministic, predeterministic] means that it is terminating [enabled, deterministic, predeterministic] at each initial state x_0. Code is always enabled; and the celebrated loop rule ensures termination of code in the form of an iteration. But **magic**, for example, is not enabled and (hence is) terminating.

Laws for sequential composition are more subtle for $Gcl.X$ than $gcl.X$, because of partially-enabled computations like coercions; see Laws (30), (31) and (32).

2.4   Assertions and Coercions

For any predicate b on state space the *assertion* at b is the program that skips at initial states satisfying b and otherwise aborts. Its definition and basic properties are summarised as follows; the proof is routine from the axioms above.

Lemma (assertions).   The assertion function from predicates to assertions

$$ass : (\mathbb{Pr}.X, \leq) \to (gcl.X, \sqsubseteq)$$
$$ass.b \;\widehat{=}\; \textbf{skip} \lhd b \rhd \textbf{abort}$$

is injective and satisfies

1. $ass.true = \textbf{skip}$;
2. $ass.(b \wedge c) = (ass.b) \,\mathring{,}\, (ass.c)$ and, in particular, assertions commute;
3. for any finite nonempty set B of predicates, $ass.(\wedge B) = \sqcap\{ass.b \mid b \in B\}$ and, in particular, ass is monotone;
4. for any finite set B of predicates, $ass.(\vee B) = \sqcup\{ass.b \mid b \in B\}$ and, in particular, $ass.false = \textbf{abort}$.

Conditional is regained from assertions using angelic choice:

$$(ass.b \,\mathring{,}\, A) \sqcup (ass.\neg b \,\mathring{,}\, B) \;=\; A \lhd b \rhd B. \tag{40}$$

For any predicate b on state space the *coercion* at b is the computation, $coer.b$, that skips at initial states satisfying b and otherwise is magic. In the transformer model (see Sec. 3.2) assertions and coercions are dual. In the current context, that duality manifests itself in the following analogue of the 'assertions' Lemma.

Lemma (coercions). The coercion function from predicates to coercions

$$coer : (\mathbb{P}r.X, \leq) \rightarrow (Gcl.X, \sqsubseteq)$$

$$coer.b \;\hat{=}\; \mathbf{skip} \lhd b \rhd \mathbf{magic}$$

is injective and satisfies

1. $coer.true = \mathbf{skip}$;
2. $coer.(b \wedge c) = (coer.b) \, \mathbin{\raise0.3ex\hbox{$\scriptstyle\mathfrak{g}$}} \, (coer.c)$ and, in particular, coercions commute;
3. for any finite nonempty set B of predicates, $coer.(\wedge B) = \sqcup \{coer.b \mid b \in B\}$ and, in particular, $coer$ is antitone;
4. for any finite set B of predicates, $coer.(\vee B) = \sqcap \{coer.b \mid b \in B\}$ and, in particular, $coer.false = \mathbf{magic}$.

Conditional is regained from coercions by the daul of (40):

$$(coer.b \, \mathbin{\raise0.3ex\hbox{$\scriptstyle\mathfrak{g}$}} \, A) \sqcap (coer.\neg b \, \mathbin{\raise0.3ex\hbox{$\scriptstyle\mathfrak{g}$}} \, B) = A \lhd b \rhd B. \tag{41}$$

Moreover the two are dually interdefinable using the laws above:

$$coer.b = \mathbf{skip} \sqcup (ass.\neg b \, \mathbin{\raise0.3ex\hbox{$\scriptstyle\mathfrak{g}$}} \, \mathbf{magic})$$

$$ass.b = \mathbf{skip} \sqcap (coer.\neg b \, \mathbin{\raise0.3ex\hbox{$\scriptstyle\mathfrak{g}$}} \, \mathbf{magic}).$$

We must be careful not to allow such simply duality to raise our hopes concerning the degree to which there is a transformer-like dual on the space of (relational) computations.

3   Models

In this section we recall the relational and transformer models of computations (and programs in particular) and the Galois connection between them. In the remainder of the paper, our interest will be primarily relational. That is why when, in the previous section, we have had to choose between laws satisfied by the relational and transformer models, we have opted for the former.

Definition (computation structure). A (relational) *computation structure* is a model satisfying the laws of $Gcl.X$ namely those of Fig. 3, without Law (13), and those of Figs. 4 and 6. (Note that Law (14) is superseded by Law (34) and so need not be explicitly suppressed.) A (relational) *program structure* is a model of the laws of $gcl.X$, namely those of Figs. 3 and 4.

3.1   Relational Semantics

In the presence of both demonic nondeterminism and divergence, the relational model of computation employs a virtual state $\bot$, distinct from each (actual) state in X, to encode divergence. Thus state space becomes $X_\bot \;\hat{=}\; X \cup \{\bot\}$. If A is a computation then its relational semantics $[A]_{\mathcal{R}}$ is a binary relation on

$$
\begin{aligned}
[\mathbf{abort}]_{\mathcal{R}} &\;\hat{=}\; \omega_{\perp} \\
[\mathbf{magic}]_{\mathcal{R}} &\;\hat{=}\; \{\}_{\perp} \\
[\mathbf{skip}]_{\mathcal{R}} &\;\hat{=}\; id_{\perp} \\
[x := e]_{\mathcal{R}} &\;\hat{=}\; \{(x, e.x) : X_{\perp} \times X_{\perp} \mid e.x \text{ terminates}\}_{\perp} \qquad e \text{ finitary} \\
[A \,{}^{\circ}_{\circ}\, B]_{\mathcal{R}} &\;\hat{=}\; [A]_{\mathcal{R}} \,{}^{\circ}_{\circ}\, [B]_{\mathcal{R}} \\
[A \triangleleft b \triangleright B]_{\mathcal{R}} &\;\hat{=}\; \{(x,y) \mid x[A]_{\mathcal{R}} y \;\triangleleft\; b.x \;\triangleright\; x[B]_{\mathcal{R}} y\} \\
[A \sqcap B]_{\mathcal{R}} &\;\hat{=}\; [A]_{\mathcal{R}} \cup [B]_{\mathcal{R}} \\
[A \sqcup B]_{\mathcal{R}} &\;\hat{=}\; [A]_{\mathcal{R}} \cap [B]_{\mathcal{R}} \\
[\mu F]_{\mathcal{R}} &\;\hat{=}\; \cup\{r : Rel.X \mid F.r = r\}, \qquad F \text{ monotone on } (Rel.X, \supseteq) \\
A \sqsubseteq B &\;\hat{=}\; [A]_{\mathcal{R}} \supseteq [B]_{\mathcal{R}}
\end{aligned}
$$

Fig. 7. Relational semantics $[A]_{\mathcal{R}}$ of computation $A : Gcl.X$

$X_{\perp}$ that relates: the virtual state $\perp$ to each state in $X_{\perp}$; an initial state $x : X$ to $\perp$ if computation A may diverge from x; and initial x to a final state $y : X$ if execution of A from x may terminate in y. For convenience, if s is a relation on X, we let $s_{\perp}$ denote the relation on $X_{\perp}$ that relates $\perp$ to everything but on X behaves just like s (i.e. $x\, s_{\perp} y \;\hat{=}\; x =\perp \lor x\, s\, y$).

Definition (relational model). The *relational model of computations* is the (relational) computation structure $Rel.X$ whose elements are the binary relations on $X_{\perp}$ that are strict (i.e. $\perp r \perp$) and upclosed with the flat order on X (i.e. $x\, r \perp \;\Rightarrow\; r.(\!| x |\!) = X_{\perp}$), and with the computations and combinators identified in Fig. 7. The *relational model of programs* is the (relational) program substructure $rel.X$ of $Rel.X$ consisting of the set of relations r on $X_{\perp}$ that are also finitary: $r.(\!| x |\!) \neq X_{\perp} \;\Rightarrow\; 0 < \#r.(\!| x |\!) < \infty$.

The fact that those really are computation structures, i.e. satisfy the appropriate laws, seems largely to be folklore [H87]:

Theorem (relational model). The semantic function $[-]_{\mathcal{R}} : Gcl.X \to Rel.X$ of Fig. 7 is a bijection, confirming that $Rel.X$ is a (relational) computation structure and $rel.X$ is a (relational) program structure.

3.2   Transformer Semantics

If A is a computation then its predicate-transformer semantics is a function $[A]_{\mathcal{T}}$ on predicates over X that maps a postcondition q to the weakest precondition at which termination of A is sure to hold in a state satisfying q.

Definition (transformer model). The *transformer model of computations* is the space $\mathcal{T}.X$ of monotone functions on the space of predicates over X with implication ordering, and with the computations and combinators identified in

$$[\mathbf{abort}]_{\mathcal{T}} \;\widehat{=}\; \mathit{false}$$
$$[\mathbf{magic}]_{\mathcal{T}} \;\widehat{=}\; \mathit{true}$$
$$[\mathbf{skip}]_{\mathcal{T}} \;\widehat{=}\; \{(q,q) \mid q \in \mathbb{P}\mathrm{r}.X\}$$
$$[x := e]_{\mathcal{T}}.q.x \;\widehat{=}\; q.(e.x) \quad (= q[e/x])$$
$$[A \,\overset{\circ}{,}\, B]_{\mathcal{T}} \;\widehat{=}\; [A]_{\mathcal{T}} \circ [B]_{\mathcal{T}}$$
$$[A \vartriangleleft b \vartriangleright B]_{\mathcal{T}} \;\widehat{=}\; [A]_{\mathcal{T}} \vartriangleleft b \vartriangleright [B]_{\mathcal{T}}$$
$$[A \sqcap B]_{\mathcal{T}} \;\widehat{=}\; [A]_{\mathcal{T}} \wedge [B]_{\mathcal{T}}$$
$$[A \sqcup B]_{\mathcal{T}} \;\widehat{=}\; [A]_{\mathcal{T}} \vee [B]_{\mathcal{T}}$$
$$[\mu F]_{\mathcal{T}} \;\widehat{=}\; \wedge\{t : \mathcal{T}.X \mid F.t = t\}, \; F \text{ monotone on } \mathcal{T}.X$$
$$A \sqsubseteq B \;\widehat{=}\; [A]_{\mathcal{T}} \leq [B]_{\mathcal{T}}$$

Fig. 8. Transformer semantics $[A]_{\mathcal{T}}$ of computation $A : Gcl.X$

Fig. 8. The *transformer model of programs* is the subspace consisting of strict, positively conjunctive and continuous transformers.

The fact that $\mathcal{T}.X$ models the appropriate laws is of course standard and well documented [D76, H92, N89]. The transformer semantics is given in Fig. 8. But for implicit consistency with the relational semantics we prefer to deduce it—as much as is possible—from the Galois connection between the relational and transformer models (in the next section). For now, we record:

Theorem (transformer model). The semantic function $[-]_{\mathcal{T}} : Gcl.X \rightarrow \mathcal{T}.X$ of Fig. 8 is a bijection, confirming that $\mathcal{T}.X$ satisfies the laws of Fig. 3 without the last three ((13), (14), (15)), the laws of Fig. 4 and the laws of Fig. 6 with Laws (35), (36), (37) replaced by Laws (6) and (7). Furthermore the subspace corresponding to programs satisfies the laws of Figs. 3 and 4.

Perhaps the most important distinction between the relational and transformers models is that the latter has a notion of dual, whilst the former does not.

Definition (transformer duality). Duality is defined on predicate transformers by $t^{*}.q \;\widehat{=}\; \neg t.\neg q$, and is readily shown to be well-defined on the subspace $\mathcal{T}.X$ of monotone transformers.

The fundamental properties of transformer duality [BvW98] are established by straightforward calculation:

Theorem (transformer duality). The duality function on predicate transformers is a bijection satisfying

$$u \sqsubseteq t^{*} \;=\; t \sqsubseteq u^{*} \tag{42}$$
$$(t \circ u)^{*} \;=\; t^{*} \circ u^{*} \tag{43}$$

$$t^{**} = t \tag{44}$$

$$[\mathbf{skip}]_{\mathcal{T}}^{*} = [\mathbf{skip}]_{\mathcal{T}} \tag{45}$$

$$[\mathbf{abort}]_{\mathcal{T}}^{*} = [\mathbf{magic}]_{\mathcal{T}} \tag{46}$$

$$(t \sqcap u)^{*} = t^{*} \sqcup u^{*} \tag{47}$$

$$(t \sqcup u)^{*} = t^{*} \sqcap u^{*} \tag{48}$$

$$(t \lhd b \rhd u)^{*} = t^{*} \lhd b \rhd u^{*} \tag{49}$$

$$[ass.b]_{\mathcal{T}}^{*} = [coer.b]_{\mathcal{T}} \tag{50}$$

It is easily shown that there is no dual on relations having those properties. For example the best-behaved candidate, the translation of $*$ using the Galois connection (see Sec. 3.3) between relations and transformers (i.e. $r^{\dagger} \mathrel{\hat{=}} rp.((wp.r)^{*})$), satisfies (42), (43), (45), (46), (47), (49), (50) although it fails to be bijective and satisfies merely $r^{\dagger\dagger} \supseteq r$ and $(\cap R)^{\dagger} \supseteq \cup R^{\dagger}$.

3.3 Galois Connection

Recall that the function $wp : Rel.X \rightarrow \mathcal{T}.X$ is defined, for relational computation $r : Rel.X$, postcondition $q : \mathbb{P}r.X$ and state $x : X$:

$$wp.r.q.x \mathrel{\hat{=}} \forall y : X_{\perp} \cdot x \, r \, y \Rightarrow (y \neq \perp \wedge q.y).$$

With the relational and transformer interpretations already given, that says: $wp.r.q$ holds at just those states from which termination is ensured, and in a state satisfying q.

Since wp is universally $(\cup, \geq)$-junctive (i.e. from $(Rel.X, \subseteq)$ to $(\mathcal{T}.X, \geq)$), it has an adjoint which we call the *relational projection*, rp. For $t : \mathcal{T}.X$, $rp.t$ is the binary relation on $X_{\perp}$ defined to be strict and to satisfy, for $x : X$ and $y : X_{\perp}$,

$$x \, (rp.t) \, y \mathrel{\hat{=}} \forall q : \mathbb{P}r.X \cdot t.q.x \Rightarrow q.y$$

(where, by definition, $\neg q . \perp$). Adjunction means

$$t \leq wp.r \equiv r \subseteq rp.t \tag{51}$$

so that the functions wp and rp form a Galois connection between the relational and transformer spaces with their orders reversed: from $(Rel.X, \subseteq)$ to $(\mathcal{T}.X, \geq)$. Standard theory [O44] shows that the Galois connection preserves much of the structure on the two semantics models, except for angelic nondeterminism. Gathering the (elementary) properties we need:

Theorem (Galois connection). The Galois connection (51) satisfies

$$wp.(r \mathbin{\stackrel{\circ}{\,}} s) = (wp.r) \circ (wp.s) \tag{52}$$

$$wp.(\mathrm{id}.X)_{\perp} = \mathrm{id}.(\mathcal{T}.X) \tag{53}$$

$$rp.(t \circ u) = (rp.t) \mathbin{\stackrel{\circ}{\,}} (rp.u) \tag{54}$$

$$rp.\,\mathrm{id}\,.(\mathcal{T}.X) \;=\; (\mathrm{id}\,.X)_{\perp} \tag{55}$$

$$\forall\, U \subseteq \mathcal{T}.X \,\cdot\, rp.\vee U \;=\; \cap\, rp.(\!|\,U\,|\!) \tag{56}$$

$$\forall\, U \subseteq \mathcal{T}.X \,\cdot\, rp.\wedge U \;=\; \cup\, rp.(\!|\,U\,|\!) \tag{57}$$

$$\forall\, S \subseteq Rel.X \,\cdot\, wp.(\cup S) \;=\; \wedge\, wp.(\!|\,S\,|\!) \tag{58}$$

$$\forall\, S \subseteq Rel.X \,\cdot\, wp.(\cap S) \;\geq\; \vee\, wp.(\!|\,S\,|\!)\,. \tag{59}$$

Trivial consequences, which it is helpful to have stated explicitly, are:

Corollary (Galois connection).   The Galois connection (51) also satisfies

$$r \subseteq s \;\Rightarrow\; wp.r \geq wp.s \tag{60}$$

$$t \geq u \;\Rightarrow\; rp.t \subseteq wp.u \tag{61}$$

$$rp \circ wp \;=\; \mathrm{id}\,.(Rel.X) \tag{62}$$

$$\mathrm{id}\,.(\mathcal{T}.X) \;\leq\; wp \circ rp \tag{63}$$

$$rp.true \;=\; \{\,\}_{\perp} \tag{64}$$

$$rp.false \;=\; X_{\perp} \times X_{\perp} \tag{65}$$

$$wp.\omega_{\perp} \;=\; false \tag{66}$$

$$wp.\{\,\}_{\perp} \;=\; true. \tag{67}$$

The fact that inequality (59) may be strict indicates why the embedding wp cannot be used to lift angelic nondeterminism from relations to transformers. (For example, with $r \;\widehat{=}\; [x := 0 \sqcap x := 1]_{\mathcal{R}}$ and $s \;\widehat{=}\; [x := 1 \sqcap x := 2]_{\mathcal{R}}$ we have $wp.(r \cap s) = [x := 1]_{\mathcal{R}} > (wp.r) \vee (wp.s)$, as can be seen by evaluating each side at the postcondition $x = 1$.) Otherwise, the transformer semantics is obtained from the relational semantics (Fig. 7) under the Galois embedding wp, as summarised in Fig. 8:

Theorem (Semantics).   With the exception of angelic choice, the denotations in Fig. 8 equal those obtained from lifting the $\mathcal{R}$-semantics from Fig. 7 via the Galois embedding wp: for every computation $A : Gcl.X$ not containing angelic choice,

$$[A]_{\mathcal{T}} = wp.[A]_{\mathcal{R}}\,.$$

In particular the semantics of code $A : gcl.X$ is given by that formula.

Proof.   To indicate the nature of the calculations involved, consider the cases of **skip** and demonic choice. For the former, we reason with postcondition q and (proper) state $x : X$,

$$wp.[\mathbf{skip}]_{\mathcal{R}}.q.x$$

$\equiv$ definition of wp

$$\forall\, y : X_{\perp} \,\cdot\, x\,[\mathbf{skip}]_{\mathcal{R}}\, y \Rightarrow (y \neq \perp \wedge q.y)$$

$\equiv$ Fig. 7

$$\forall\, y : X_{\perp} \,\cdot\, x = y \;\Rightarrow\; (y \neq \perp \wedge q.y)$$

$$\equiv \qquad\qquad\qquad\qquad\qquad\qquad\qquad\qquad\text{calculus}$$

$q.x$

$$\equiv \qquad\qquad\qquad\qquad\qquad\qquad\qquad\text{definition of id}$$

$\text{id} .(\mathbb{P}\text{r}.X).q.x \, ,$

as claimed in Fig. 8. For demonic choice, we reason:

$wp.[A \sqcap B]_{\mathcal{R}}$

$$\equiv \qquad\qquad\qquad\qquad\qquad\qquad\qquad\qquad\qquad\text{Fig. 7}$$

$wp.([A]_{\mathcal{R}} \cup [B]_{\mathcal{R}})$

$$\equiv \qquad\qquad\qquad\qquad\qquad\qquad\qquad\qquad\text{Law (58)}$$

$wp.[A]_{\mathcal{R}} \wedge wp.[B]_{\mathcal{R}}$

$$\equiv \qquad\qquad\qquad\qquad\qquad\qquad\qquad\qquad\text{induction}$$

$[A]_{\mathcal{T}} \wedge [B]_{\mathcal{T}} \, ,$

as required by Fig. 8. $\qquad\qquad\qquad\qquad\qquad\qquad\qquad\qquad\qquad\qquad\square$

4   Fibre Bundles

In this section we adapt the standard concept of a 'fibre bundle' [S51] to the context of computations. It enables us to consider a computation initial-state by initial-state without the need for a homogeneous model in which initial and final states have the same type.

4.1   Bundles and Sections

Definition (fibre bundle).   Let X and E be sets. Then E forms a *fibre bundle* over *base* X with *projection* $\pi : E \to X$ if π is a surjection. For each $x : X$ the inverse-image set $\pi^{\sim}.(\!|\, x \,|\!)$ is called the *fibre* at x. Note that the composition $\pi^{\sim}$ followed by π equals the identity relation on X whilst in the reverse order it equals the universal relation on each fibre:

$$\pi^{\sim} {}_{9}^{\circ} \pi \;=\; \text{id} .X \qquad\qquad\qquad (68)$$

$$e \,(\pi {}_{9}^{\circ} \pi^{\sim}) \, e' \;\equiv\; (\pi.e = \pi.e') \,. \qquad\qquad (69)$$

Definition (section).   By a (*relational*) *section* of a fibre bundle we mean a relation $s : X \leftrightarrow E$ that relates a base point x only to members of the fibre at x, namely

$$s {}_{9}^{\circ} \pi \subseteq \text{id} .X.$$

See Fig. 9. In the standard theory of fibre bundles s is required to be a function; we consider relations in order to represent nondeterministic computations from initial states in X to final states in the fibre at x. The standard definition also

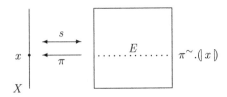

Fig. 9. The fibre bundle E over X with projection π and (relational) section s. The fibre at base element $x:X$ consists of the set of bundle elements $\pi^\sim.(\!|\,x\,|\!)$.

requires the previous inclusion to be an equality; we allow inclusion to permit unenabled computations: sections that are only partial. If the base and fibres are the same then the result is a *homogeneous* model of computation. One of the strengths of the bundle setting is its appropriateness also for nonhomogeneous models.

In this setting, the concept of fibre bundle is merely a notational convenience since with no structure on the base (traditionally either topological, differential or geometric) its local structure is identical to its global structure. Instead we derive from the spirit of the definition.

4.2   Examples

Examples of bundles E are obtained by instantiating fibres $\pi^\sim.(\!|\,x\,|\!)$ and imposing healthiness conditions on sections $s : X \leftrightarrow E$ as follows.

1. In the *relational* model (Sec. 3.1), the base and fibres both consist of the extended state space $X_\perp$ so that $\pi^\sim.(\!|\,x\,|\!) = X_\perp$. A section is interpreted as taking an element of the base—an initial state—to an element of its fibre—a final state; thus it is required to be strict and pointwise upclosed (unenabled initial states lie outside its domain). This example is homogeneous; the simpler nonhomogeneous alternative in which the base is simplified to just X is not viable because initial $\perp$ is required to ensure Law (29).
2. In the *predicate-transformer* model (Sec. 3.2), the base and fibres consist of predicates over state space X so that $\pi^\sim.(\!|\,x\,|\!) = \mathbb{P}r.X$. A section is interpreted as taking a postcondition [resp. precondition] to a weakest precondition [resp. strongest postcondition]. Thus a section representing a computation is required to be a monotone function, and a section representing a program is required also to be strict, positively conjunctive and continuous. This model is, like the relational model, homogeneous.
3. In the *probabilistic relational* model [HSM97, MM05], the base consists of state space X but fibres are sets of sub-distributions (i.e. probability distributions that sum to at most 1 over X) so that

$$\forall\, x:X \cdot \pi^\sim.(\!|\,x\,|\!) \;=\; \{f : X \to [0,1] \mid \textstyle\sum_X f \le 1\}.$$

Divergence from initial x is represented as $f.x = 0$ so that no virtual state is needed in this model. Sub-distributions are ordered pointwise (i.e. $f \leq g \ \hat{=} \ \forall x : X \cdot f.x \leq g.x$). A section is interpreted as taking an initial state to a demonic choice of sub-distributions. Thus a section s is required, pointwise, to be convex:

$$f, g \in s.(\!|\, x \,|\!) \wedge p \in [0, 1] \quad \Rightarrow \quad p \times f + (1-p) \times g \in s.(\!|\, x \,|\!) \,;$$

$\leq$-upclosed:

$$f \in s.(\!|\, x \,|\!) \wedge f \leq g \quad \Rightarrow \quad g \in s.(\!|\, x \,|\!) \,;$$

and, if state space X is infinite,[2] topologically closed: each $s.(\!|\, x \,|\!)$ is a topologically-closed subset of the product space $[0, 1]^{\#X}$. This model is not homogeneous (and requires the obvious (Kleisli/linearity) construction to make it so).

4. In the *multirelational* model [R06], the base consists of state space X and fibres are sets of states: $\pi^\sim.(\!|\, x \,|\!) = \mathbb{P}.X$. A section s is interpreted as relating an initial state to an angelic choice of sets of possible (demonically chosen) final states; it is thus required to be $\subseteq$-upclosed pointwise:

$$\forall x : X \cdot Q \in s.(\!|\, x \,|\!) \wedge Q \subseteq Q' \quad \Rightarrow \quad Q' \in s.(\!|\, x \,|\!).$$

Under its angelic refinement ordering, this model might be embedded in the previous model, by identifying a subset Q of a finite state space with the characteristic function of Q scaled by the number of states (which results in a uniform demonic choice of final state in Q). A Galois embedding ε is

$$\varepsilon.R.x \ \hat{=} \ \{(\#X)^{-1} \xi_Q \mid x \, R \, Q\}$$

where ξ_Q denotes the characteristic function of Q (upclosure is not required on the right-hand side since it follows from that of R).

5. The *expectation-transformer* model [MM05] is a simple extension of the predicate-transformer model in which predicate transformers are replaced by *expectations* (i.e. non-negative-real-valued functions on state space) with the lifted ordering. (A predicate, like a pre or post-condition, may be viewed as a $\{0, 1\}$-valued expectation.) Thus base and fibres consist of expectations over X, so that $\pi^\sim.(\!|\, x \,|\!) = X \to \mathbb{R}^{\geq 0}$. A section s is interpreted as taking a post-expectation (i.e. a state-dependent expected profit) to a greatest pre-expectation (i.e. state-dependent least expected profit) and is required to be *sublinear*: for all expectations f, g and all non-negative reals a, b, c,

$$(a(s.f) + b(s.g)) \ominus \mathbf{c} \ \leq \ s.((af + bg) \ominus \mathbf{c})$$

where $x \ominus y \ \hat{=} \ (x-y) \, \mathbf{max} \, 0$ and $\mathbf{c}$ is the constant function $\lambda x : X \cdot c$.

---

[2] If X is finite this property follows automatically; indeed $s.(\!|\, x \,|\!)$ is then the (closed) convex hull of its (finitely many) extreme points.

6. The *quantum* model [SZ00] is a simple restriction of the expectation transformer model in which state space consists of the space of quantum registers over the standard space X of the problem

$$\{\chi : X \to \mathbb{C} \mid \textstyle\sum_{x \in X} \mid \chi.x \mid^2 = 1\}.$$

A section is required to be the result of only unitary state transformations (which in particular maintain the sum-square invariant).

Further examples appear in Sec. 5.1; more may be drawn from the various computing paradigms, like process algebra, or from particular semantic formalisms, like [HH98].

4.3 Morphisms and Refinement

Definition (morphism). We define a *morphism* of fibre bundles $\pi : E \to X$ and $\pi' : E' \to X'$ to consist of a pair (χ, ε) in which χ is a relation from X to X' and ε is a relation from E to E' which together maintain consistency of fibre:

$$\chi : X \leftrightarrow X'$$
$$\varepsilon : E \leftrightarrow E'$$
$$\varepsilon \,\sfrac{9}{9}\, \pi' \subseteq \pi \,\sfrac{9}{9}\, \chi.$$

By (68) and (69), the fibre-consistency condition is equivalent to

$$\pi^\sim \,\sfrac{9}{9}\, \varepsilon \subseteq \chi \,\sfrac{9}{9}\, \pi'^\sim.$$

Definition (refinement). Given sections $s : X \leftrightarrow E$ and $s' : X' \leftrightarrow E'$ we define a *refinement* to be a morphism (χ, ε) of fibre bundles that acts as a simulation

$$\chi \,\sfrac{9}{9}\, s' \subseteq s \,\sfrac{9}{9}\, \varepsilon. \tag{70}$$

(The alternative condition, $s' \,\sfrac{9}{9}\, \varepsilon^\sim \subseteq \chi^\sim \,\sfrac{9}{9}\, s$, is equivalent to (70) if both χ and ε are injective, which is not often the case.) The advantage of this definition over the standard definitions of simulation [dRE98] is that it permits non-homogenous semantics of computations, of the kind demonstrated by the previous examples.

4.4 Examples

1. The special case of operational, or algorithmic, refinement in the relational model—in which the state space is unaltered but each result of the concrete computation is a result of the abstract—is captured by taking $X' = X$, $E' = E$, $\chi = \mathrm{id}.X$ and $\varepsilon = \mathrm{id}.E$ so that inclusion (70) becomes:

$$s' \subseteq s.$$

The same holds in the probabilistic relational model. In the predicate transformer model the rôles of s and s' are reversed (so that at each postcondition the refining computation has weaker precondition).

2. The important case of data refinement in the relational model—by a downwards simulation relation between abstract and concrete states—is captured by taking $\chi : X \leftrightarrow X'$ to be the simulation and $\varepsilon : E \leftrightarrow E'$ its fibre-wise lifting: $\varepsilon \mathrel{\widehat{=}} \pi \mathbin{\fatsemi} \chi \mathbin{\fatsemi} \pi'^{\sim}$. Then inclusion (70) becomes:

$$\chi \mathbin{\fatsemi} s' \subseteq s \mathbin{\fatsemi} \pi \mathbin{\fatsemi} \chi \mathbin{\fatsemi} \pi'^{\sim} \qquad \text{or} \qquad \chi \mathbin{\fatsemi} s' \mathbin{\fatsemi} \pi' \subseteq s \mathbin{\fatsemi} \pi \mathbin{\fatsemi} \chi$$

as expected.

5   Computations Fibrewise

Reasoning about computations is by tradition a curious blend of algebraic reasoning (using refinement laws) and semantic reasoning (using validity in a semantic model). In this section we promote to the level of algebra the kind of semantic reasoning that enables a computation to be investigated initial-state by initial-state or, as we shall say, fibre-wise.

We introduce another interpretation of fibre bundles as computations, focusing on the fibre as a collection of computations rather than a set of final states (the view of Sec. 4.2). It is this interpretation that we use in the remainder of the paper.

5.1   Computation Fibres

Motivated by that diversion into fibre bundles we now consider a computation A, an initial state x_0 and define the fibre [co-fibre] of A at x_0 to be the computation that is magic [aborts] off x_0, where it behaves like A. In other words the fibre [co-fibre] of A at x_0 consists of the coercion [assertion] $x = x_0$ followed by A.

Definition (fibre and co-fibre).   For $A : gcl.X$ and $E \subseteq X$, the fibre of A at E is

$$A * E \mathrel{\widehat{=}} (A \mathbin{\lhd} x \in E \mathbin{\rhd} \mathbf{magic})$$

and the co-fibre of A at E is

$$A \bullet E \mathrel{\widehat{=}} (A \mathbin{\lhd} x \in E \mathbin{\rhd} \mathbf{abort}).$$

For simplicity, $A * \{x_0\}$ and $A \bullet \{x_0\}$ are written $A * x_0$ and $A \bullet x_0$ respectively. Also, if b is a predicate on state space we abuse notation by writing $A * b$ and $A \bullet b$ for the fibre and co-fibre at the set $\{x : X \mid b.x\}$.

Fibres and co-fibres are distributed by the computation combinators. The proofs follow directly from the laws of Gcl.

Lemma (fibre distribution).   For $A, B : Gcl.X$, $x_0, x_1 : X$ and predicate b on X,

$$\mathbf{magic} * x_0 = \mathbf{magic} \tag{71}$$

$$(A \mathbin{\lhd} b \mathbin{\rhd} B) * x_0 = (A * x_0) \mathbin{\lhd} b \mathbin{\rhd} (B * x_0) \tag{72}$$

$$(A \, ; B) * x_0 \;=\; (A * x_0) \, ; B \tag{73}$$
$$(A \sqcap B) * x_0 \;=\; (A * x_0) \sqcap (B * x_0) \tag{74}$$
$$(A \sqcup B) * x_0 \;=\; (A * x_0) \sqcup (B * x_0) \tag{75}$$
$$(A * x_0) * x_1 \;=\; (A * x_0) \lhd x_0 = x_1 \rhd \mathbf{abort} \tag{76}$$
$$=\; (A * x_1) * x_0 \,. \tag{77}$$

Proof.

Law (71):

    $\mathbf{magic} * x_0$

    $=$ definition of $*$

    $\mathbf{magic} \lhd x = x_0 \rhd \mathbf{magic}$

    $=$ Law (18)

    $\mathbf{magic}.$

Law (72):

    $(A \lhd b \rhd B) * x_0$

    $=$ definition of $*$

    $(A \lhd b \rhd B) \lhd x = x_0 \rhd \mathbf{magic}$

    $=$ Laws (17) and (20)

    $(A \lhd x = x_0 \rhd \mathbf{magic}) \lhd b \rhd (B \lhd x = x_0 \rhd \mathbf{magic})$

    $=$ definition of $*$

    $(A * x_0) \lhd b \rhd (B * x_0) \,.$

Law (73):

    $(A \, ; B) * x_0$

    $=$ definition of $*$

    $(A \, ; B) \lhd x = x_0 \rhd \mathbf{magic}$

    $=$ Laws (31) and (21)

    $(A \lhd x = x_0 \rhd \mathbf{magic}) \, ; B$

    $=$ definition of $*$

    $(A * x_0) \, ; B \,.$

Law (74):

    $(A \sqcap B) * x_0$

    $=$ definition of $*$

    $(A \sqcap B) \lhd x = x_0 \rhd \mathbf{magic}$

    $=$ Laws (21), (12) and (31)

    $(\mathbf{skip} \lhd x = x_0 \rhd \mathbf{magic}) \, ; (A \sqcap B)$

$$=\qquad\qquad\text{Law (15)}$$

$(\textbf{skip} \triangleleft x = x_0 \triangleright \textbf{magic}) \mathbin{\text{\textcommabelow{9}}} A \quad\sqcap\quad (\textbf{skip} \triangleleft x = x_0 \triangleright \textbf{magic}) \mathbin{\text{\textcommabelow{9}}} B$

$$=\qquad\qquad\text{Laws (21), (12) and (31) again}$$

$(A \triangleleft x = x_0 \triangleright \textbf{magic}) \quad\sqcap\quad (B \triangleleft x = x_0 \triangleright \textbf{magic})$

$$=\qquad\qquad\text{definition of } *$$

$(A * x_0) \sqcap (B * x_0).$

Law (75):

$(A \sqcup B) * x_0$

$$=\qquad\qquad\text{definition of } *$$

$(A \sqcup B) \triangleleft x = x_0 \triangleright \textbf{magic}$

$$=\qquad\qquad\text{Laws (21), (12) and (31)}$$

$(\textbf{skip} \triangleleft x = x_0 \triangleright \textbf{magic}) \mathbin{\text{\textcommabelow{9}}} (A \sqcup B)$

$$=\qquad\qquad\text{Law (39)}$$

$(\textbf{skip} \triangleleft x = x_0 \triangleright \textbf{magic}) \mathbin{\text{\textcommabelow{9}}} A \quad\sqcup\quad (\textbf{skip} \triangleleft x = x_0 \triangleright \textbf{magic}) \mathbin{\text{\textcommabelow{9}}} B$

$$=\qquad\qquad\text{Laws (21), (12) and (31) yet again}$$

$(A \triangleleft x = x_0 \triangleright \textbf{magic}) \quad\sqcup\quad (B \triangleleft x = x_0 \triangleright \textbf{magic})$

$$=\qquad\qquad\text{definition of } *$$

$(A * x_0) \sqcup (B * x_0).$

Law (76):

$(A * x_0) * x_1$

$$=\qquad\qquad\text{definition of } *$$

$(A \triangleleft x = x_0 \triangleright \textbf{magic}) \triangleleft x = x_1 \triangleright \textbf{magic}$

$$=\qquad\qquad\text{Law (19)}$$

$A \triangleleft x = x_0 = x_1 \triangleright (\textbf{magic} \triangleleft x = x_1 \triangleright \textbf{magic})$

$$=\qquad\qquad\text{Law (18)}$$

$A \triangleleft x = x_0 = x_1 \triangleright \textbf{magic}$

$$=\qquad\qquad\text{Law (18)}$$

$A \triangleleft x = x_0 = x_1 \triangleright (\textbf{magic} \triangleleft x_0 = x_1 \triangleright \textbf{magic})$

$$=\qquad\qquad\text{Law (19)}$$

$(A \triangleleft x = x_0 \triangleright \textbf{magic}) \triangleleft x_0 = x_1 \triangleright \textbf{magic}$

$$=\qquad\qquad\text{definition of } *$$

$(A * x_0) \triangleleft x_0 = x_1 \triangleright \textbf{magic}.$

Law (77) is immediate from Law (76).   □

The 'co-fibre distribution' lemma is proved analogously.

Lemma (co-fibre distribution).   For $A, B : Gcl.X$, $x_0, x_1 : X$ and predicate b on X,

$$\mathbf{abort} \bullet x_0 = \mathbf{abort} \tag{78}$$

$$(A \lhd b \rhd B) \bullet x_0 = (A \bullet x_0) \lhd b \rhd (B \bullet x_0) \tag{79}$$

$$(A \mathbin{\substack{\circ \\ \circ}} B) \bullet x_0 = (A \bullet x_0) \mathbin{\substack{\circ \\ \circ}} B \tag{80}$$

$$(A \sqcap B) \bullet x_0 = (A \bullet x_0) \sqcap (B \bullet x_0) \tag{81}$$

$$(A \sqcup B) \bullet x_0 = (A \bullet x_0) \sqcup (B \bullet x_0) \tag{82}$$

$$(A \bullet x_0) \bullet x_1 = (A \bullet x_0) \lhd x_0 = x_1 \rhd \mathbf{abort} \tag{83}$$

$$= (A \bullet x_1) \bullet x_0 . \tag{84}$$

The means for constructing fibres and co-fibres from point fibres and point co-fibres are established similarly:

$$A * E = \sqcap \{A * e \mid e \in E\} \tag{85}$$

$$A \bullet E = \sqcup \{A \bullet e \mid e \in E\}. \tag{86}$$

In the transformer model with duality available, 'co-fibre distribution' follows from 'fibre distribution' by duality:

$$(A * E)^* = (A^*) \bullet E. \tag{87}$$

5.2   Fibre Representation

Theorem (fibre representation).   A computation is both the demonic choice of its fibres and the angelic choice of its co-fibres: for each $A : Gcl.X$,

$$A = \sqcap \{A * x_0 \mid x_0 \in X\} \tag{88}$$

$$= \sqcup \{A \bullet x_0 \mid x_0 \in X\} \tag{89}$$

(in spite of the fact that the right-hand side of (89) does not form a directed set).

In particular refinement is decided fibre-wise and co-fibre-wise:

$$A \sqsubseteq B = \forall x_0 : X \cdot A * x_0 \sqsubseteq B * x_0 \tag{90}$$

$$= \forall x_0 : X \cdot A \bullet x_0 \sqsubseteq B \bullet x_0 . \tag{91}$$

A powerful property of (singleton) fibres is this. If a predeterministic computation refines the point fibre of a demonic choice then it refines (at least) one of the point fibres: if D is predeterministic then

$$(A \sqcap B) \bullet x_0 \sqsubseteq D \;\Rightarrow\; (A \bullet x_0 \sqsubseteq D) \vee (B \bullet x_0 \sqsubseteq D). \tag{92}$$

Soundness is immediate (in either the relational or transformer model). Property (92) fails for larger fibres and for computations D that are not predeterministic, as simple examples demonstrate.

1. Suppose that x_0 and x_1 are distinct states. With

$$A \mathrel{\widehat{=}} (x := 0 \bullet x_0) \sqcap (x := 1 \bullet x_1) \quad \text{and} \quad B \mathrel{\widehat{=}} (x := 2 \bullet x_0) \sqcap (x := 3 \bullet x_1)$$

we find $A \sqcap B \sqsubseteq D$ where $D \mathrel{\widehat{=}} (x := 0 \bullet x_0) \sqcap (x := 3 \bullet x_1)$, although D is predeterministic but refines neither A nor B. Thus the result does not extend to larger fibres.

2. With

$$A \mathrel{\widehat{=}} (x := 0 \sqcap x := 1) \bullet x_0 \quad \text{and} \quad B \mathrel{\widehat{=}} (x := 1 \sqcap x := 2) \bullet x_0$$

we find $A \sqcap B \sqsubseteq D$ where $D \mathrel{\widehat{=}} (x := 0 \sqcap x := 2)$, although D refines neither A nor B. Thus the result does not extend to non-predeterministic D.

We also use (in Sec. 6) a special case of property (92): if the demonic choice of two computations at the same fibre aborts, then (at least) one of the fibres aborts.

$$(A \sqcap B) \bullet x_0 = \mathbf{abort} \quad \Rightarrow \quad (A \bullet x_0 = \mathbf{abort}) \vee (B \bullet x_0 = \mathbf{abort}) \tag{93}$$

5.3   Fibre Normal Forms

In the transformer model, the operational angel-versus-demon view of a computation results in a normal form in which any monotone predicate transformer is the sequential composition of an angelic choice (of 'intermediate postcondition' which the computation is certain to achieve from its initial state) followed by a demonic choice (of state satisfying that condition); see [BvW98], Theorem 13.10.

Since relational computations are embedded by wp as the positively conjunctive predicate transformers, that normal form specialises to relations, *loc. cit.*, Theorem 26.4, to show that any relational computation is the sequential composition of an assertion (with predicate equal to the domain of the relation) followed by a demonic choice (of related state). Thus relational computations are 'generated' by assertions, assignments, demonic choice and sequential composition. In our setting, with emphasis on point fibres, the relevant normal form is this (in which $\overline{E}$ denotes the complement of set E).

Theorem (fibre normal form).   In the relational model any computation is the demonic choice of point fibres, each of which is a demonic choice: for any computation A there is a partial function $\alpha : X \twoheadrightarrow \mathbb{P}.X$ whose domain consists of those states from which A does not diverge and for which

$$A \;=\; \sqcap\{(x :\in \alpha.x_0) * x_0 \mid x_0 \in \operatorname{dom}.\alpha\} \sqcap \mathbf{abort} * \overline{\operatorname{dom}.\alpha}. \tag{94}$$

Proof.   We forego a brief proof in the relational semantics (by explicit construction for an arbitrary healthy relational) because we wish the form to hold in any (relational) computation structure and to use the extra information provided by an algebraic proof by structural induction over computations. In each case the required partial functions are defined as follows.

$$\textbf{abort} \quad \{\,\} \tag{95}$$

$$\textbf{magic} \quad \lambda\, x : X \,\cdot\, \{\,\} \tag{96}$$

$$x := e \quad \lambda\, x : X \,\cdot\, e \tag{97}$$

$$A \sqcap B \quad \lambda\, x : \mathrm{dom}\,.\alpha_A \cap \mathrm{dom}\,.\alpha_B \,\cdot\, \alpha_A.x \cup \alpha_B.x \tag{98}$$

$$A \sqcup B \quad \lambda\, x : \mathrm{dom}\,.\alpha_A \cup \mathrm{dom}\,.\alpha_B \,\cdot\, \alpha_A.x \cap \alpha_B.x \tag{99}$$

$$A \,\mathring{,}\, B \quad \lambda\, x : \mathrm{dom}\,.\alpha_A \cap \mathrm{dom}\,.\alpha_B \,\cdot$$

$$\{t : X \mid \exists\, u{:}X \,\cdot\, u \in \alpha.x \,\wedge\, t \in \beta.u\} \tag{100}$$

$$\square$$

Observe that $\alpha.x_0 = \{\,\}$ for those states x_0 at which A is unenabled; and if A is a program then the range of α consists of only nonempty finite sets.

The dual of the 'fibre normal form' Theorem expresses each computation as the angelic choice of point co-fibres of a demonic assignment; it thus requires both forms of nondeterminism. However it does not require the 'correction' term for divergence.

Corollary (co-fibre normal form).   In the relational model any computation is the angelic choice of point co-fibres, each of which is a demonic choice:

$$A \;=\; \sqcup\{(x :\in \alpha.x_0) \bullet x_0 \mid x_0 \in \mathrm{dom}\,.\alpha\}. \tag{101}$$

Those normal forms refine Laws (10) and (28) and the 'fibre representation' Theorem by showing that all computations are 'generated' by demonically non-deterministic assignments and either point coercions and demonic choice, or point assertions and angelic choice.

Corollary (basis).   The space $(Gcl.X, \sqsubseteq)$ of relational computations is 'generated' by the set of (total deterministic) assignments, demonic choice and either point coercions, or point assertions and angelic choice.

6   Assertions Algebraically

In view of the importance placed on assertions and coercions by the fibre-wise approach to computations, the purpose of this section is to characterise them algebraically. But we wish to use this algebraic characterisation in any relational computational structure, so we prove it using the relational laws of $Gcl.X$, since a semantic proof would apply to just $Rel.X$. (For an elegant characterisation in the context of Kleene Algebra see [vW04].)

6.1   An Algebraic Property

We begin by formalising the observation that if $ass.b$ refines a computation B then B aborts off b and so the assertion followed by B equals B:

$$ass.b \,\mathring{,}\, B \;=\; B \lhd b \rhd B \;=\; B.$$

Theorem (assertions).   If $A : Gcl.X$ is an assertion then

$$\forall B : Gcl.X \ \cdot \ B \sqsubseteq A \ \Rightarrow \ A \ \mathbf{\mathring{,}} \ B = B \,. \tag{102}$$

Proof.   Suppose that for some predicate b on state space, $A \mathbin{\hat{=}} ass.b$, and that $B : Gcl.X$. We begin by proving that if the antecedent of (102) holds then B aborts off b.

$B \sqsubseteq A$
$\Rightarrow$ $\qquad\qquad\qquad\qquad$ Law (81) (as monotonicity of $\_ \bullet b$) and definition of A

$B \bullet \neg b \ \sqsubseteq \ (\mathbf{skip} \lhd b \rhd \mathbf{abort}) \bullet \neg b$
$\Rightarrow$ $\qquad\qquad\qquad\qquad$ Laws (79), (19), (16), (17) and (18)

$B \bullet \neg b \sqsubseteq \mathbf{abort}$
$\Rightarrow$ $\qquad\qquad\qquad\qquad$ Law (27)

$B \bullet \neg b \ = \ \mathbf{abort}$
$\Rightarrow$ $\qquad\qquad\qquad\qquad$ Leibniz

$(B \bullet b \sqcup B \bullet \neg b) \ = \ (B \bullet b \sqcup \mathbf{abort})$
$\Rightarrow$ $\qquad\qquad\qquad\qquad$ Laws (40) and (27)

$B \lhd b \rhd B \ = \ B \bullet b$
$\Rightarrow$ $\qquad\qquad\qquad\qquad$ Law (18) and definition of $\bullet$

$B \ = \ B \lhd b \rhd \mathbf{abort} \,.$

So we reason

$A \ \mathbf{\mathring{,}} \ B$
$=$ $\qquad\qquad\qquad\qquad$ definition of A

$(\mathbf{skip} \lhd b \rhd \mathbf{abort}) \ \mathbf{\mathring{,}} \ B$
$=$ $\qquad\qquad\qquad\qquad$ Law (21)

$(\mathbf{skip} \ \mathbf{\mathring{,}} \ B) \lhd b \rhd (\mathbf{abort} \ \mathbf{\mathring{,}} \ B)$
$=$ $\qquad\qquad\qquad\qquad$ Laws (12) and (13)

$B \lhd b \rhd \mathbf{abort}$
$=$ $\qquad\qquad\qquad\qquad$ above

$B \,.$

6.2   Algebraic Characterisation

The previous result is interesting because the condition it embodies is strong enough to characterise assertions algebraically.

Theorem (assertions algebraically).   If $A : Gcl.X$ satisfies condition (102) then A is an assertion.

Proof.   We define a predicate b on X pointwise: b holds at a point iff computation A skips there:

$b.x_0 \;\widehat{=}\; (A \bullet x_0 = \mathbf{skip} \bullet x_0).$

To show that A is the assertion $ass.b$, however, we must show that if it does not skip at a point then it aborts there:

$$A \bullet x_0 \neq \mathbf{skip} \bullet x_0 \quad \Rightarrow \quad A \bullet x_0 = \mathbf{abort}. \tag{103}$$

We use a succession of lemmas each of which inherits the assumptions to date. The first lemma shows that A is always enabled.

Lemma (A).   If $A : Gcl.X$ satisfies (102) then

$\forall x_0 : X \,\cdot\, A \bullet x_0 \neq \mathbf{magic} \bullet x_0 .$

Proof.

$\quad true$

$\quad =$ $\hspace{5cm}$ Implication (102) with $B \,\widehat{=}\, \mathbf{abort}$, by Law (27)

$\quad A \,\fatsemi\, \mathbf{abort} \;=\; \mathbf{abort}$

$\quad =$ $\hspace{7.5cm}$ Law (91)

$\quad \forall x_0 : X \,\cdot\, (A \bullet x_0) \,\fatsemi\, \mathbf{abort} \;=\; \mathbf{abort} \bullet x_0$

$\quad =$ $\hspace{7.5cm}$ Law (78)

$\quad \forall x_0 : X \,\cdot\, (A \bullet x_0) \,\fatsemi\, \mathbf{abort} \;=\; \mathbf{abort}$

$\quad \Rightarrow$ $\hspace{6cm}$ otherwise contradicts Law (31)

$\quad \forall x_0 : X \,\cdot\, A \bullet x_0 \neq \mathbf{magic} \bullet x_0 .$

$\hspace{13cm}\square$

The second lemma shows that there is some assignment which refines A at x_0 .

Lemma (B).   If $A : Gcl.X$ satisfies both (102) and the antecendent of implication (103) then

$\exists\, x_1 : X \setminus \{x_0\} \,\cdot\, A \bullet x_0 \;\sqsubseteq\; (x := x_1) \bullet x_0 .$

Proof.   Since A is enabled at x_0 (Lemma A) either A aborts from x_0 or it terminates, taking some value (perhaps demonically). We write $x :\in E$ for the demonic choice of x to a member of $E \subseteq X$ (i.e. $\sqcap \{x := e \mid e \in E\}$). Thus if E is empty it is $\mathbf{magic}$; if E is finite and nonempty it is a program; and if E is infinite it is a computation more general than code.

Writing A in co-fibre representation (with the dependence of E on x_0 made explicit)

$$A \;=\; \sqcup\{(x :\in E(x_0)) \bullet x_0 \mid x_0 \in X\} ,$$

we infer that the two cases of divergence and termination are, respectively,

$$A \bullet x_0 \;=\; \begin{cases} \mathbf{abort} \bullet x_0 \\ (x :\in E(x_0)) \bullet x_0 \end{cases}$$

where by assumption, $\{\} \neq E(x_0) \neq \{x_0\}$. But in the first case the result holds for any $x_1 \neq x_0$ whilst in the second it holds for any $x_1 \in E(x_0) \setminus \{x_0\}$, a choice possible by the antecedent of (103). □

Now we introduce an artifact,

$$Y \mathrel{\widehat{=}} \mathbf{abort} * x_1 \quad (= \mathbf{abort} \lhd x = x_1 \rhd \mathbf{magic}),$$

whose use will become apparent in the final step of the proof. For the moment we observe that Y does not abort at x_0:

$$Y \bullet x_0 \neq \mathbf{abort}. \tag{104}$$

That is indeed evident since

$Y \bullet x_0$
$=$ definition of $\bullet$

$Y \lhd x = x_0 \rhd \mathbf{abort}$
$=$ definition of Y

$(\mathbf{abort} \lhd x = x_1 \rhd \mathbf{magic}) \lhd x = x_0 \rhd \mathbf{abort}$
$=$ calculus, since $x_0 \neq x_1$ by Lemma B

$\mathbf{magic} \lhd x = x_0 \rhd \mathbf{abort}$
$\neq$

$\mathbf{abort}.$

What we really want of Y is this:

Lemma (C).

$$(A \bullet x_0) \mathbin{\raise.5ex\hbox{$\,_9^9\,$}} (A \sqcap Y) = \mathbf{abort}. \tag{105}$$

Proof. We reason

$(A \bullet x_0) \mathbin{\raise.5ex\hbox{$\,_9^9\,$}} (A \sqcap Y)$
$\sqsubseteq$ Law (35) (as monotonicity of $\mathbin{{}_9^9\!\_}$)

$(A \bullet x_0) \mathbin{\raise.5ex\hbox{$\,_9^9\,$}} Y$
$\sqsubseteq$ Lemma B and Law (34) (as monotonicity of $\_\mathbin{{}_9^9}$)

$(x := x_1) \bullet x_0 \mathbin{\raise.5ex\hbox{$\,_9^9\,$}} Y$
$=$ definition of Y

$(x := x_1) \bullet x_0 \mathbin{\raise.5ex\hbox{$\,_9^9\,$}} (\mathbf{abort} \lhd x = x_1 \rhd \mathbf{magic})$
$=$ definition of $\bullet$, Laws (21), (13), (26), (16) and (27)

$\mathbf{abort} \bullet x_0$
$=$ Law (78)

$\mathbf{abort},$

from which equality follows by Law (27). □

For the final step of the proof we recall that we are trying to establish the consequent of (103). To do so we set $B \mathrel{\widehat{=}} A \sqcap Y$ and observe that $A \sqcap Y \sqsubseteq A$. Hence by (102)

$$A \mathbin{\fatsemi} (A \sqcap Y) \;=\; A \sqcap Y. \tag{106}$$

Thus

    true

$=$ Lemma C

    $(A \mathbin{\fatsemi} (A \sqcap Y)) \bullet x_0 \;=\; \mathbf{abort}$

$\Rightarrow$ by (106)

    $(A \sqcap Y) \bullet x_0 \;=\; \mathbf{abort}$

$\Rightarrow$ Law (35)

    $(A \bullet x_0) \sqcap (Y \bullet x_0) \;=\; \mathbf{abort}$

$\Rightarrow$ Law (93)

    $(A \bullet x_0) = \mathbf{abort} \;\;\vee\;\; (Y \bullet x_0) = \mathbf{abort}$

$\Rightarrow$ Law (104)

    $A \bullet x_0 \;=\; \mathbf{abort}$

and so the proof of the theorem is complete. $\square$

6.3   Coercions

The analogous result for coercions is this; we omit the proof, which mimics that above.

Theorem (coercions).   Computation $C : Gcl.X$ is a coercion iff

$$\forall B : Gcl.X \;\cdot\; C \sqsubseteq B \;\Rightarrow\; C \mathbin{\fatsemi} B = B. \tag{107}$$

Note. A proof of the 'coercions' Theorem from the 'assertions Theorem' is available in the transformer model simply by taking duals, using the 'transformer duality' Theorem. It goes like this.

    $\forall B : Gcl.X \;\cdot\; C \sqsubseteq B \;\Rightarrow\; C \mathbin{\fatsemi} B = B$

$=$ $\mathcal{T}$ semantics and $*$ bijective ('transformer duality' Theorem)

    $\forall t : \mathcal{T}.X \;\cdot\; [C]_{\mathcal{T}} \leq t^* \;\Rightarrow\; [C]_{\mathcal{T}} \circ t^* = t^*$

$=$ Laws (42), (43) and (44)

    $\forall t : \mathcal{T}.X \;\cdot\; t \leq [C]_{\mathcal{T}}^* \;\Rightarrow\; [C]_{\mathcal{T}}^* \circ t = t$

$=$ 'assertions' Theorem

    $\exists b : \mathbb{P}r.X \;\cdot\; [C]_{\mathcal{T}}^* \;=\; [ass.b]_{\mathcal{T}}$

$=$ Law (44)

    $\exists b : \mathbb{P}r.X \;\cdot\; [C]_{\mathcal{T}} \;=\; [\mathbf{skip} \mathbin{\lhd} b \mathbin{\rhd} \mathbf{abort}]_{\mathcal{T}}^*$

$=$ $\mathcal{T}$ semantics and Law (49)

$$\exists\, b : \mathbb{P}\text{r}.X \ \cdot\ [C]_{\mathcal{T}} \ = \ [\textbf{skip}]^*_{\mathcal{T}} \lhd b \rhd [\textbf{abort}]^*_{\mathcal{T}}$$

$$= \hspace{5cm} \mathcal{T} \text{ semantics, Laws (45) and (46)}$$

$$\exists\, b : \mathbb{P}\text{r}.X \ \cdot\ [C]_{\mathcal{T}} \ = \ [\textbf{skip}]_{\mathcal{T}} \lhd b \rhd [\textbf{magic}]_{\mathcal{T}}$$

$$= \hspace{5cm} \text{definition of coercion}$$

$$\exists\, b : \mathbb{P}\text{r}.X \ \cdot\ [C]_{\mathcal{T}} \ = \ [coer.b]_{\mathcal{T}}\,.$$

$$\square$$

Unfortunately the discussion in Sec. 3.2 shows that a similar proof, using a relational dual in place of *, is not possible in the relational model.

7 Representing Computation Structures

The purpose of this section is to prove that a structure satisfying the relational laws of $Gcl.X$ is actually isomorphic to the space of relational computations over X, and similarly for the program substructure $gcl.X$ and relational programs.

But first we need the appropriate notion of isomorphism.

7.1 Morphisms of Computation Structures

Definition (computation morphism). If $\mathcal{X}$ and $\mathcal{X}'$ are (relational) computation structures (i.e. models of (relational) computation space $Gcl.X$) then a function T from the 'programs' of $\mathcal{X}$ to those of $\mathcal{X}'$ is a *computation morphism* iff it preserves sequential compositions, infima and suprema:

$$T(A \mathbin{\raise0.3ex\hbox{\tiny$\,^\circ_\circ$}} B) \ = \ (T.A) \mathbin{\raise0.3ex\hbox{\tiny$\,^\circ_\circ$}}{}' (T.B) \tag{108}$$

$$T.\wedge \mathcal{A} \ = \ \wedge' T.(\!|\mathcal{A}|\!) \tag{109}$$

$$T.\vee \mathcal{A} \ = \ \vee' T.(\!|\mathcal{A}|\!)\,. \tag{110}$$

Function T is a *program morphism* iff it preserves sequential compositions and finite nonempty infima. A *computation* [resp. *program*] *isomorphism* is a bijective computation [resp. program] morphism.

Recall that the Galois embedding wp is not $(\cap,\vee)$-junctive (Law (59)) and so wp does not preserve angelic choice (suprema). Indeed, in view of Fig. 1, the relational and transformer models of computation are not isomorphic.

Evaluation of the trivial cases shows:

Lemma (morphed structures). A program morphism preserves the identity for sequential composition ($T.1 = 1'$) and is monotone ($A \leq B \Rightarrow T.A \leq' T.B$), and a computation morphism also preserves minimum, maximum and atomic elements ($T.\top = \top'$ and $T.\bot = \bot'$) and is isotone ($A \leq B \Leftrightarrow T.A \leq' T.B$).

It is vital that the definition of isomorphism, although it focuses on only sequential composition and order, preserves also assertions and coercions as a result. For by the 'basis' Corollary, preservation of other computations follows.

Lemma (morphed assertions). A program isomorphism preserves assertions, point assertions, coercions and point coercions.

Proof. Suppose that $T : \mathcal{X} \to \mathcal{X}'$ is an isomorphism, where $\mathcal{X}$ and $\mathcal{X}'$ are as above. Since the proof of the 'assertions algebraically' Theorem used only properties of relational computation structures, we use it to reason:

$\quad a$ is assertion in $\mathcal{X}$

$=$ $\hfill$ 'assertions algebraically' Theorem

$\quad \forall\, y : Y \cdot y \leq a \;\Rightarrow\; a \,\mathring{,}\, y = y$

$\Rightarrow$ $\hfill$ 'morphed structures' Lemma

$\quad \forall\, y : Y \cdot T.y \leq' T.a \;\Rightarrow\; T.a \,\mathring{,}'\, T.y = T.y$

$=$ $\hfill$ T bijective

$\quad \forall\, y : Y \; y \leq' T.a \;\Rightarrow\; T.a \,\mathring{,}'\, y = y$

$=$ $\hfill$ 'assertions algebraically' Theorem

$\quad T.a$ is assertion in $\mathcal{X}'$.

Preservation of point assertions now follows (by the 'morphed structures' Lemma) since an assertion is a point assertion iff it is atomic, by the 'assertions' Lemma. The proof for coercions is similar. $\hfill \square$

7.2   Representing Computation Structures

As an axiom system, Gcl has models of each infinite cardinality, one for each choice of state space. Thus there are computation structures which are not isomorphic. Considering first models with a given state space X, we have:

Theorem (computation representation). Any (relational) computation structure with state space X is isomorphic to $Rel.X$.

Proof. Suppose that the computation structure is $R.X$ and define a putative isomorphism T fibrewise

$\quad T : R.X \to Rel.X$
$\quad T.A.(\!|\, x_0\, |\!) \;\mathrel{\widehat{=}}\; \{y : X \mid A \bullet x_0 \sqsubseteq x := y\}\,,$

provided $A \bullet x_0 \neq \mathbf{abort}$, in which case $T.A.(\!|\, x_0\, |\!) \;\mathrel{\widehat{=}}\; X_\perp$. (Observe that if the defining set is empty, for example A is not enabled at x_0, then $T.A$ is not enabled at x_0.) We must show that T is a bijective computation morphism. The definition follows two cases, depending on abortion or not. Verification of the former case is routine; we focus on the latter.

$\quad$Firstly T is injective, since

$\quad T.A = T.B$

$=$ $\hfill$ relational calculus

$\quad \forall\, x_0 : X \cdot T.A.(\!|\, x_0\, |\!) \;=\; T.B.(\!|\, x_0\, |\!)$

$=$ definition of T

$\forall\, x_0, y : X \;\cdot\; A \bullet x_0 \sqsubseteq x := y \;\Leftrightarrow\; B \bullet x_0 \sqsubseteq x := y$

$=$ definition of co-fibre and Law (28)

$\forall\, x_0 : X \;\cdot\; A \bullet x_0 \;=\; B \bullet x_0$

$=$ Law (91)

$A = B$.

Secondly T is surjective, since if $R : Rel.X$ with $R.(\!|\, x_0 \,|\!) \,=\, Y_{x_0}$, for some x_0-dependent $Y_{x_0} \subseteq X_\perp$, defining $A : R.X$ fibrewise

$$\forall\, x_0 : X \;\cdot\; A \bullet x_0 \;\widehat{=}\; (x :\in Y_{x_0}) \bullet x_0\,,$$

we find that by definition $T.A = R$.

    Thirdly T preserves sequential composition, since

$T.(A \,\S\, B).(\!|\, x_0 \,|\!)$

$=$ definition of T

$\{ y : X \mid (A \,\S\, B) \bullet x_0 \sqsubseteq x := y \}$

$=$ Law (73)

$\{ y : X \mid (A \bullet x_0) \,\S\, B \sqsubseteq x := y \}$

$=$ hypothesis, 'co-fibre normal form' Corollary and (100)

$T.B.(\!|\, \{ y : X \mid A \bullet x_0 \sqsubseteq x := y \} \,|\!)$

$=$ definition of T

$T.B.(\!|\, T.A.(\!|\, x_0 \,|\!) \,|\!)$

$=$ relational calculus

$((T.A) \,\S\, (T.B)).(\!|\, x_0 \,|\!)$.

    Fourthly T preserves demonic nondeterminism, since

$T.(A \sqcap B).(\!|\, x_0 \,|\!)$

$=$ definition of T

$\{ y : X \mid (A \sqcap B) \bullet x_0 \sqsubseteq x := y \}$

$=$ Law (92)

$\{ y : X \mid (A \bullet x_0 \sqsubseteq x := y) \vee (B \bullet x_0 \sqsubseteq x := y) \}$

$=$ set theory

$\{ y : X \mid A \bullet x_0 \sqsubseteq x := y \} \cup \{ y : X \mid B \bullet x_0 \sqsubseteq x := y \}$

$=$ definition of T

$(T.A \bullet x_0) \sqcap (T.A \bullet x_0)$

$=$ Law (81)

$(T.A \sqcap T.A) \bullet x_0$.

Finally T preserves angelic nondeterminism, since

$T.(A \sqcup B).(\!|\, x_0 \,|\!)$

$$=$$

definition of T

$$\{y:X \mid (A \sqcup B) \bullet x_0 \sqsubseteq x := y\}$$

$$=$$

Law (82)

$$\{y:X \mid (A \bullet x_0) \sqcup (B \bullet x_0) \sqsubseteq x := y\}$$

$$=$$

definition of supremum

$$\{y:X \mid (A \bullet x_0 \sqsubseteq x := y) \wedge (B \bullet x_0 \sqsubseteq x := y)\}$$

$$=$$

set theory

$$\{y:X \mid A \bullet x_0 \sqsubseteq x := y\} \cap \{y:X \mid B \bullet x_0 \sqsubseteq x := y\}$$

$$=$$

semantics, Fig. 7

$$(T.A \sqcup T.B).(\!| x_0 |\!) \, .$$

□

Corollary (program representation). Any (relational) program structure with state space X is isomorphic to $rel.X$.

Considering next models with different state spaces, we have:

Theorem (state-space representation). Two models of (relational) computations but with possibly different states spaces are isomorphic iff there is a bijection between their state spaces.

Proof. The proof establishes a bijective correspondence between isomorphisms $T : R.X \to S.Y$ and bijections $t : X \rightarrowtail\!\!\!\rightarrow Y$. For the forward direction, given T we define a relation $t : X \leftrightarrow Y$ by

$$x_0 \, t \, y_0 \; \hat{=} \; T.(ass.x_0) = ass.y_0 \, . \tag{111}$$

Relation t is well defined because, by the 'morphed assertions' Lemma, T preserves point assertions. It is surjective by similar reasoning and the fact that T is surjective. It is readily shown to be a function (because if two point assertions are equal then their points coincide) and injective (because T is injective).

Conversely, given bijection $t : X \rightarrowtail\!\!\!\rightarrow Y$, define T from $R.X$ to $S.Y$ firstly on assertions by (111), then to preserve $\sqcap$ and $\sqcup$ and finally on assignments by

$$T.(x := e) \; \hat{=} \; y := t.e$$

where t maps expressions (recall our assignments are total and deterministic) over X to expressions over Y by 'translation' (or trivial simulation):

$$t.e.y \; \hat{=} \; t.(\!| e.(t^{\sim}.y) |\!) \, .$$

Thus $T.(x :\in E) = y :\in t.(\!| E |\!)$ and in co-fibre normal form we find

$$T.(\sqcup\{(x :\in \alpha.x_0) \bullet x_0 \mid x_0 \in \mathrm{dom}.\alpha\}) \; =$$
$$\sqcup\{(y :\in t.\alpha.y_0) \bullet y_0 \mid y_0 \in \mathrm{dom}.(t.\alpha)\}$$

wherein t acts on fibre-form functions α as it does on expressions.

Then T is well defined and injective by existence and uniqueness of co-fibre normal form. It is surjective since if

$$B = \sqcup\{(y :\in t.\beta.y_0) \bullet y_0 \mid y_0 \in \text{dom}\,.\beta\}$$

then by the definition of T we find

$$T.(\sqcup\{(x :\in \alpha.x_0) \bullet x_0 \mid x_0 \in \text{dom}\,.\alpha\}) \;=\; B.$$

Finally, since T preserves $\sqcap$ and $\sqcup$ by definition, it remains to show that it preserves $\mathring{,}$. But that follows from a routine calculation using (100).   □

8   Conclusion

By taking the slightly unusual view that computations are sections of fibre bundles we have emphasised the fibre-wise nature of a computation. That has enabled us to treat assignment initial-state-by-initial state and to use powerful refinement laws like Law (92) that fail more generally. The result has been the fibre normal form for a (relational) computation and the isomorphism of any (relational) computation structure with the binary-relation model of computation. Although that isomorphism is based on the order and sequential composition combinators of the structure, by characterising assertions and coercions in those terms we have shown that the isomorphism also preserves them, and hence by the 'fibre normal form' Theorem truly is an isomorphism of computations. The laws of (relational) computation are categorical to within cardinality of state space.

Further work consists of capturing the transformer model similarly and of clarifying the extent to which, as remarked only in passing here, the transformer dual may be lifted to relations via the Galois connection between relations and transformers. It would also be of some interest to pursue the bundle approach to refinement, particularly in the non-homogeneous setting.

Acknowledgements

The author is grateful to the organisers of RelMiCS 2006 for the opportunity to explore, in this paper, the fibre-wise approach to computation and, at the conference, its connections with other approaches. He is grateful to Georg Struth and Renate Schmidt for super-editorial corrections and clarifications and for bringing to his attention several references.

This exposition has benefitted from drafts of joint work with Annabelle McIver and Carroll Morgan on the application of Galois connections to the study of various relational and transformer models of computation. Some of the work reported here, and some results mentioned in passing, have been supported by the University of Stellenbosch and the South African National Research Foundation under the auspices of Ingrid Rewitzky.

References

[BvW98]   R.-J. Back and J. von Wright. *Refinement Calculus: A Systematic Introduction.* Graduate Texts in Computer Science, Springer Verlag, 1998.

[D76]   E. W. Dijsktra. *A Discipline of Programming.* Prentice-Hall International, 1976.

[HSM97]   He, Jifeng, K. Seidel and A. K. McIver. Probabilistic models for the guarded command language. *Science of Computer Programming,* **28**:171–192, 1997.

[H92]   W. H. Hesselink. *Programs, Recursion and Unbounded Choice.* Cambridge University Press, 1992.

[H87]   C. A. R. Hoare *et al*, The laws of programming. *Communications of the ACM,* **30**:672–686, 1987.

[HH98]   C. A. R. Hoare and He, Jifeng. *Unifying Theories of Programming.* Prentice Hall, 1998.

[MM05]   A. K. McIver and C. C. Morgan. *Abstraction, Refinement and Proof for Probabilistic Systems.* Springer Monographs in Computer Science, 2005.

[N89]   G. Nelson. A generalisation of Dijkstra's calculus. ACM ToPLAS, **11**(4):517–561, 1989.

[O44]   O. Ore, Galois connexions. *Transactions of the American Mathematical Society,* **55**:494–513, 1944.

[R06]   I. M. Rewitzky. Monotone predicate transformers as up-closed multirelations. This volume, *Relations and Kleene Algebra in Computer Science (RelMics/AKA 2006)* Springer-Verlag, LNCS, 2006.

[dRE98]   W.-P. de Roever and K. Engelhardt, *Data Refinement: Model-Oriented Proof Methods and their Comparison.* Cambridge Tracts in Theoretical Computer Science, Cambridge University Press, 1998.

[SZ00]   J. W. Sanders and P. Zuliani. Quantum Programming. *Mathematics of Program Construction, 2000,* edited by J. N. Oliviera and R. Backhouse, Springer-Verlag LNCS **1837**:80–99, 2000.

[S51]   N. Steenrod. *The Topology of Fibre Bundles.* Princeton University Press, 1951.

[vW04]   J. von Wright. Towards a refinement algebra. *Science of Computer Programming,* **51**:23–45, 2004.

An Axiomatization of Arrays for Kleene Algebra with Tests

Kamal Aboul-Hosn

Department of Computer Science
Cornell University
Ithaca, NY 14853-7501, USA
kamal@cs.cornell.edu

Abstract. The formal analysis of programs with arrays is a notoriously difficult problem due largely to aliasing considerations. In this paper we augment the rules of Kleene algebra with tests (KAT) with rules for the equational manipulation of arrays in the style of schematic KAT. These rules capture and make explicit the essence of *subscript aliasing*, where two array accesses can be to the same element. We prove the soundness of our rules, as well as illustrate their usefulness with several examples, including a complete proof of the correctness of heapsort.

1 Introduction

Much work has been done in reasoning about programs with arrays. Arrays require more complex modeling than regular variables because of issues of *subscript aliasing*, where two array accesses can be to the same element, for example, $A(x)$ and $A(y)$ when $x = y$. Proving equivalence of programs with arrays often involves intricate read/write arguments based on program semantics or complex program transformations.

Reasoning about arrays dates back to seminal work of More [1] and Downey and Sethi [2]. Much research has also been based on early work by McCarthy on an extensional theory of arrays based on read/write operators [3]. A standard approach is to treat an array as a single variable that maps indices to values [4,5,6]. When an array entry is updated, say $A(i) := s$, a subsequent access $A(j)$ is treated as the program **if** $(i = j)$ **then** s **else** $A(j)$. Several other approaches of this nature are summarized in [7], where Bornat presents Hoare Logic rules for reasoning about programs with aliasing considerations.

More recently, there have been many attempts to find good theories of arrays in an effort to provide methods for the formal verification of programs with arrays. Recent work, including that of Stump et al. [8], focuses on decision procedures and NP-completeness outside the context of any formal system. Additionally, the theorem prover HOL has an applicable theory for finite maps [9].

In this paper we augment the rules of Kleene algebra with tests (KAT) with rules for the equational manipulation of arrays in the style of KAT. Introduced

R.A. Schmidt (Ed.): RelMiCS /AKA 2006, LNCS 4136, pp. 63–77, 2006.

in [10], KAT is an equational system for program verification that combines Kleene algebra (KA), the algebra of regular expressions, with Boolean algebra. KAT has been applied successfully in various low-level verification tasks involving communication protocols, basic safety analysis, source-to-source program transformation, concurrency control, compiler optimization, and dataflow analysis [10,11,12,13,14,15,16]. This system subsumes Hoare logic and is deductively complete for partial correctness over relational models [17].

Schematic KAT (SKAT), introduced in [11], is a specialization of KAT involving an augmented syntax to handle first-order constructs and restricted semantic actions. Rules for array manipulation in the context of SKAT were given in [12], but these rules were (admittedly) severely restricted; for instance, no nesting of array references in expressions was allowed. The paper [12] attempted to provide only enough structure to handle the application at hand, with no attempt to develop a more generally applicable system.

We extend the rules of [12] in two significant ways: (i) we provide commutativity and composition rules for sequences of array assignments; and (ii) we allow nested array references; that is, array references that can appear as subexpressions of array indices on both the left- and right-hand sides of assignments. The rules are *schematic* in the sense that they hold independent of the first-order interpretation.

In Section 2, we provide a brief introduction to KAT and SKAT. In Section 3, we give a set of rules for the equational manipulation of such expressions and illustrate their use with several interesting examples. These rules capture and make explicit the essence of subscript aliasing. Our main results are (i) a soundness theorem that generalizes the soundness theorem of [12] to this extended system; and (ii) a proof of the correctness of heapsort, presented in Section 5.

2   Preliminary Definitions

2.1   Kleene Algebra with Tests

Kleene algebra (KA) is the algebra of regular expressions [18,19]. The axiomatization used here is from [20]. A *Kleene algebra* is an algebraic structure $(K, +, \cdot, {}^{*}, 0, 1)$ that satisfies the following axioms:

$$
\begin{array}{rclcll}
(p+q)+r &=& p+(q+r) & & & (1) \\
p+q &=& q+p & & & (3) \\
p+0 &=& p+p &=& p & (5) \\
p(q+r) &=& pq+pr & & & (7) \\
1+pp^{*} &\leq& p^{*} & & & (9) \\
1+p^{*}p &\leq& p^{*} & & & (11)
\end{array}
$$

$$
\begin{array}{rclcll}
(pq)r &=& p(qr) & & & (2) \\
p1 &=& 1p &=& p & (4) \\
0p &=& p0 &=& 0 & (6) \\
(p+q)r &=& pr+qr & & & (8) \\
q+pr \leq r &\rightarrow& p^{*}q \leq r & & & (10) \\
q+rp \leq r &\rightarrow& qp^{*} \leq r & & & (12)
\end{array}
$$

This a universal Horn axiomatization. Axioms (1)–(8) say that K is an *idempotent semiring* under $+, \cdot, 0, 1$. The adjective *idempotent* refers to (5). Axioms (9)–(12) say that $p^{*}q$ is the $\leq$-least solution to $q + px \leq x$ and qp^{*} is the $\leq$-least solution to $q + xp \leq x$, where $\leq$ refers to the natural partial order on K defined by $p \leq q \overset{\text{def}}{\Longleftrightarrow} p + q = q$.

Standard models include the family of regular sets over a finite alphabet, the family of binary relations on a set, and the family of $n \times n$ matrices over another Kleene algebra. Other more unusual interpretations include the min,+ algebra, also known as the *tropical semiring*, used in shortest path algorithms, and models consisting of convex polyhedra used in computational geometry.

A *Kleene algebra with tests* (KAT) [10] is a Kleene algebra with an embedded Boolean subalgebra. That is, it is a two-sorted structure $(\mathcal{K}, \mathcal{B}, +, \cdot, {}^{*}, {}^{-}, 0, 1)$ such that

- $(\mathcal{K}, +, \cdot, {}^{*}, 0, 1)$ is a Kleene algebra,
- $(\mathcal{B}, +, \cdot, {}^{-}, 0, 1)$ is a Boolean algebra, and
- $\mathcal{B} \subseteq \mathcal{K}$.

Elements of $\mathcal{B}$ are called *tests*. The Boolean complementation operator $^{-}$ is defined only on tests.

The axioms of Boolean algebra are purely equational. In addition to the Kleene algebra axioms above, tests satisfy the equations

$$
\begin{aligned}
BC &= CB & BB &= B \\
B + CD &= (B+C)(B+D) & B+1 &= 1 \\
\overline{B+C} &= \overline{B}\,\overline{C} & \overline{BC} &= \overline{B} + \overline{C} \\
B + \overline{B} &= 1 & B\overline{B} &= 0 \\
\overline{\overline{B}} &= B
\end{aligned}
$$

2.2 Schematic KAT

Schematic KAT (SKAT) is a specialization of KAT involving an augmented syntax to handle first-order constructs and restricted semantic actions whose intended semantics coincides with the semantics of flowchart schemes over a ranked alphabet Σ [11]. Atomic propositions represent assignment operations, $x := t$, where x is a variable and t is a Σ-term.

Four identities are paramount in proofs using SKAT:

$$
\begin{aligned}
x := s; y := t &= y := t[x/s]; x := s \ (y \notin FV(s)) & (13) \\
x := s; y := t &= x := s; y := t[x/s] \ (x \notin FV(s)) & (14) \\
x := s; x := t &= x := t[x/s] & (15) \\
\varphi[x/t]; x := t &= x := t; \varphi & (16)
\end{aligned}
$$

where x and y are distinct variables and $FV(s)$ is the set of free variables occurring in s in (13) and (14). The notation $s[x/t]$ denotes the result of substituting t for all occurrences of x in s. Here φ is an atomic first order formula. When x is not a free variable in t or in φ, we get the commutativity conditions

$$
\begin{aligned}
x := s; y := t &= y := t; x := s \ (y \notin FV(s), x \notin FV(t)) & (17) \\
\varphi; x := t &= x := t; \varphi & (x \notin FV(\varphi)) & (18)
\end{aligned}
$$

Additional axioms include:

$$
\begin{aligned}
x := x &= 1 & (19) \\
x := s; x = s &= x = s[x/s] & (20) \\
s = t; x := s &= s = t; x := t & (21)
\end{aligned}
$$

Using these axioms, one can also reason about imperative programs by translating them to propositional formulas [21]. One can translate program constructs as follows:

$$x := s \equiv a$$
$$x = s \equiv A$$
$$\textbf{if } B \textbf{ then } p \textbf{ else } q \equiv Bp + \overline{B}q$$
$$\textbf{while } B \textbf{ do } p \equiv (Bp)^*\overline{B}$$

where a, is an atomic proposition and A is a Boolean test. With this translation, we can use propositional KAT to do most of the reasoning about a program independent of its meaning. We use the first order axioms only to verify premises we need at the propositional level.

3   Arrays in SKAT

Arrays have special properties that create problems when trying to reason about program equivalence. The axioms (13)-(21) do not hold without some preconditions. We want to identify the conditions under which we can apply these axioms to assignments with arrays.

Consider the statement

$$A(A(2)) := 3; A(4) := A(2)$$

We would like to use an array-equivalent version of (13) to show that

$$A(A(2)) := 3; A(4) := A(2) \ = \ A(4) := A(2); A(A(2)) := 3 \qquad (22)$$

With simple variables, this sort of equivalence holds. However, in (22), if $A(2) = 2$, the two sides are not equal. The left-hand side sets both $A(2)$ and $A(4)$ to 3, while the right-hand side sets $A(2)$ to 3 and $A(4)$ to 2. The problem is that $A(2) = A(A(2))$.

One solution is to limit array indices to simple expressions that contain no array symbols, the approach taken by Barth and Kozen [12]. Let i and j be expressions containing no array symbols. For an expression e, let e_x, e_y, and e_{xy} denote $e[x/A(i)]$, $e[y/A(j)]$, and $e[x/A(i), y/A(j)]$, respectively. The following axioms hold when expressions s and t contain no array symbols and $i \neq j$:

$$A(i) := s_x; A(j) := t_{xy} \ = \ A(j) := t_y[x/s_x]; A(i) := s_x \qquad (23)$$
$$A(i) := s_x; A(j) := t_{xy} \ = \ A(i) := s_x; A(j) := t_y[x/s_x] \qquad (24)$$
$$A(i) := s_x; A(i) := t_x \ = \ A(i) := t[x/s_x] \qquad (25)$$
$$\varphi[y/t_y]; A(j) := t_y \ = \ A(j) := t_y; \varphi \qquad (26)$$

where $y \notin FV(s)$ in (23) and $x \notin FV(s)$ in (24).

These rules place some strong limitations on the programs that one can reason about, although these limitations were acceptable in the context of reasoning in

Barth and Kozen's paper. These axioms allow no more than two array references (in most cases, only one) in a sequence of two assignment statements, which eliminates many simple program equivalences such as

$$A(3) := A(4); A(3) := A(5) \;=\; A(3) := A(5)$$

Our goal is to generalize these rules so we can have more than one array reference in a sequence of assignments and so we can allow nested array references.

In attempting to adapt (13) to arrays in a general way, we first note that an array index contains an expression that must be evaluated, which could contain another array variable. Therefore, we need to perform a substitution in that subterm as well:

$$A(i) := s; A(j) := t \;=\; A(j[A(i)/s]) := t[A(i)/s]; A(i) := s \qquad (27)$$

This rule poses several questions. First of all, what is meant by $t[A(i)/s]$? We want this to mean "replace all occurrences of $A(i)$ by s in the term t." However, this statement is somewhat ambiguous in a case such as

$$t = A(3) + A(2 + 1)$$

where i is 3. We could either replace $A(i)$ (i) *syntactically*, only substituting s for $A(3)$ in t, or (ii) *semantically*, replacing both $A(3)$ and $A(2 + 1)$. Besides being undecidable, (ii) is somewhat contrary to the sort of static analysis for which we use SKAT. Moreover, implementing these sorts of rules in a system such as KAT-ML [22] could be difficult and costly, requiring the system to perform evaluation.

However, (i) is unsound. For example,

$$A(2) := 4; A(3) := A(2) + A(1 + 1) \qquad (28)$$
$$= A(3) := 4 + A(1 + 1); A(2) := 4$$

is not true if $A(2) \neq 4$ before execution.

Our solution is to identify the preconditions that ensure that this sort of situation does not occur. The preconditions would appear as tests in the equation to which the axiom is being applied. While it is true that establishing these preconditions is as difficult as replacing occurrences of array references semantically, it is more true to the style of SKAT, separating out reasoning that requires interpreting expressions in the underlying domain.

Let $Arr(e)$ be the set of all array references (array variable and index) that appear in the term e and let $e' = e[A(i)/s]$. We also define

$$Arrs(e, A, i, s) \stackrel{def}{=} Arr(e') - ((Arr(s) - ((Arr(e) - \{A(i)\}) \cap Arr(s)) \cap Arr(e')$$

The appropriate precondition for (27) is

$$\forall k, A(k) \in (Arrs(j, A, i, s) \cup Arrs(t, A, i, s)) \Rightarrow k \neq i$$

The condition looks complex, but what it states is relatively straightforward: any array reference that occurs in j' or t' must either not be equal to $A(i)$ or it must have been introduced when the substitution of s for $A(i)$ occurred.

For example, the transformation in (28) would be illegal, because $Arrs(A(2) + A(1 + 1), A, 2, 4)$ is $\{A(1 + 1)\}$, and $1 + 1 = 2$. However,

$$A(2) := A(2) + 1; A(3) := A(2) + 4 \ = \ A(3) := A(2) + 1 + 4; A(2) := A(2) + 1$$

would be legal, since $Arrs(A(2) + 4, A, 2, A(2) + 1)$ is the empty set.

With this and a couple additional preconditions, we can use the syntactic notion of replacement as we do in all other axioms. The complete set of axioms corresponding to (13)-(16) is:

$$A(i) := s; A(j) := t \ = \ A(j') := t'; A(i) := s \tag{29}$$
$$\text{if} \ \ i \neq j'$$
$$\forall k, A(k) \in Arr(s) \cup Arr(i) \Rightarrow k \neq j'$$
$$\forall k, A(k) \in Arrs(j) \cup Arrs(t) \Rightarrow k \neq i$$

$$A(i) := s; A(j) := t \ = \ A(i) := s; A(j') := t' \tag{30}$$
$$\text{if} \ \ \forall k, A(k) \in Arr(s) \cup Arr(i) \Rightarrow k \neq i$$
$$\forall k, A(k) \in Arrs(j) \cup Arrs(t) \Rightarrow k \neq i$$

$$A(i) := s; A(j) := t \ = \ A(j') := t' \tag{31}$$
$$\text{if} \ \ i = j'$$
$$\forall k, A(k) \in Arrs(j) \cup Arrs(t) \Rightarrow k \neq i$$

$$\varphi'; A(i) := s \ = \ A(i) := s; \varphi \tag{32}$$
$$\text{if} \ \ \forall k, A(k) \in Arr(\varphi) \Rightarrow k \neq i$$

We also have axioms for the interaction between assignments to array variables and to regular variables.

$$x := s; A(j) := t \ = \ A(j[x/s]) := t[x/s]; x := s \tag{33}$$
$$\text{if} \ \ \forall k, A(k) \in Arr(s) \Rightarrow k \neq j[x/s]$$

$$A(i) := s; y := t \ = \ y := t'; A(i) := s \tag{34}$$
$$\text{if} \ \ y \notin FV(s) \cup FV(i)$$
$$\forall k, A(k) \in Arrs(t) \Rightarrow k \neq i$$

$$x := s; A(j) := t \ = \ x := s; A(j[x/s]) := t[x/s] \tag{35}$$
$$\text{if} \ \ x \notin FV(s)$$

$$A(i) := s; y := t \ = \ A(i) := s; y := t' \tag{36}$$
$$\text{if} \ \ \forall k, A(k) \in Arr(s) \cup Arr(i) \Rightarrow k \neq i$$
$$\forall k, A(k) \in Arrs(t) \Rightarrow k \neq i$$

In contrast to many other treatments of arrays, we prevent aliasing through preconditions instead of using updated arrays for subsequent accesses. In approaches such as those found in [3,6,7], a program $A(i) := s; A(j) := t$ is translated to $A(i) := s; [A(i)/s](j) := t$, where $[A(i)/s]$ represents the array A with element i assigned to the value of s. Additionally, all occurrences of A in j and t must be replaced by $[A(i)/s]$. The replacement amounts to changing all array accesses A into the program **if** $(i = j)$ **then** s **else** $A(j)$.

Such a translation is not well suited to SKAT, where we want assignment statements to be atomic propositions. Using the if-then-else construct still requires checking all of the preconditions we have; they are captured in the test for equality of i and j. However, our precondition approach allows one to test these conditions only when doing program transformations using the axioms. Array accesses outside these transformations need not be changed at all. Since considerations of subscript aliasing primarily come up in the context of reasoning about program equivalence, it makes sense to consider aliasing through preconditions within that reasoning.

These same axioms can be extended to multidimensional arrays. Consider an array B with n indices. Each condition requiring array references to be different in the one-dimensional array case must be true in the multi-dimensional case as well. In order for two array accesses of the same array to be different, they must differ on at least one of the indices. Formally, we can state the axiom corresponding to (29) as

$$B(i_1, \ldots, i_n) := s; B(j_1, \ldots, j_n) := t = B(j'_1, \ldots, j'_n) := t'; B(i_1 \ldots i_n) := s$$

if $i \neq j'$ and:

$$\forall k_1, \ldots k_n, A(k_1, \ldots, k_n) \in Arr(s) \cup \bigcup_{a=1}^{n} Arr(i_a) \Rightarrow \exists \ell.1 \leq \ell \leq n \wedge j'_\ell \neq k_\ell$$

$$\forall k_1, \ldots k_n, A(k_1, \ldots k_n) \in \bigcup_{a=1}^{n} Arrs(j_a) \Rightarrow \exists \ell.1 \leq \ell \leq n \wedge k'_\ell \neq i_\ell$$

$$\forall k_1, \ldots k_n, A(k_1, \ldots k_n) Arrs(t) \Rightarrow \exists \ell.1 \leq \ell \leq n \wedge k'_\ell \neq i_\ell$$

4   Soundness of Axioms

We have proven soundness for all these rules using a technique similar to the one used in [11]. We highlight the technique for the proof in this paper. For a more complete proof, see [23]. We consider interpretations over special Kripke frames called *Tarskian*, defined with respect to a first order structure D of signature Σ. States are *valuations*, assigning values in D to variables, denoted with Greek letters θ and η. For a valuation θ, $\theta[x/s]$ is the the state that agrees with θ on all variables except possibly x, which takes the value s. An array variable is interpreted as a map $D \to D$, as defined in [12]. We use $\theta(A(i))$ to represent $\theta(A)(\theta(i))$.

First, we need to relate substitution in the valuation and substitution in a term. This relation corresponds to the relation between the substitution model of evaluation and the environment model of evaluation. For simple terms, this is easy:

$$\theta(t[x/s]) = \theta[x/\theta(s)](t)$$

which was shown in [11]. For arrays, we have the same difficulties of aliasing we have in the definition of our rules. The corresponding lemma for array references requires a precondition:

Lemma 1.

$$\theta(t[A(i)/s]) = \theta\left[\frac{A(\theta(i))}{\theta(s)}\right](t)$$

if

$$\forall A(k) \in Arrs(t, A, i, s), i \neq k$$

where $\theta\left[\frac{A(i)}{s}\right]$ is the valuation that agrees with θ on all variables except possibly the array variable A, where $A(i)$ now maps to s. The proof is by induction on t.

With this lemma, we can prove the soundness of (29) - (36). We show the proofs for (29) - (32), as (33) - (36) are just special cases of these. For example, for the axiom, we prove

Theorem 1. $A(i) := s; A(j) := t = A(j[A(i)/s]) := t[A(i)/s]; A(i) := s$ *if*

$$i \neq j' \tag{37}$$
$$\forall k, A(k) \in Arr(s) \cup Arr(i) \Rightarrow k \neq j[A(i)/s] \tag{38}$$
$$\forall k, A(k) \in Arrs(j, A, i, s) \cup Arrs(t, A, i, s) \Rightarrow k \neq i \tag{39}$$

Proof. We need to show that for any Tarskian frame D,

$$[A(i) := s; A(j) := t]_D = [A(j[A(i)/s]) := t[A(i)/s]; A(i) := s]_D$$

From the left-hand side, we have

$$[A(i) := s; A(j) := t]_D$$
$$= [A(i) := s]_D \circ [A(j) := t]_D$$
$$= \left\{\theta, \theta\left[\frac{A(\theta(i))}{\theta(s)}\right] \mid \theta \in Val_D\right\} \circ \left\{\eta, \eta\left[\frac{A(\eta(j))}{\eta(t)}\right] \mid \eta \in Val_D\right\}$$
$$= \left\{\theta, \theta\left[\frac{A(\theta(i))}{\theta(s)}\right]\left[\frac{A(\theta\left[\frac{A(\theta(i))}{\theta(s)}\right](j))}{\theta\left[\frac{A(\theta(i))}{\theta(s)}\right](t)}\right] \mid \theta \in Val_D\right\}$$

Now consider the right-hand side.

$$[A(j[A(i)/s]) := t[A(i)/s]; A(i) := s]_D$$
$$= [A(j[A(i)/s]) := t[A(i)/s]]_D \circ [A(i) := s]_D$$
$$= \left\{\theta, \theta\left[\frac{A(\theta(j[A(i)/s]))}{\theta(t[A(i)/s])}\right] \mid \theta \in Val_D\right\} \circ \left\{\eta, \eta\left[\frac{A(\eta(i))}{\eta(s)}\right] \mid \eta \in Val_D\right\}$$
$$= \left\{\theta, \theta\left[\frac{A(\theta(j[A(i)/s]))}{\theta(t[A(i)/s])}\right]\left[\frac{A(\theta\left[\frac{A(\theta(j[A(i)/s]))}{\theta(t[A(i)/s])}\right](i))}{\theta\left[\frac{A(\theta(j[A(i)/s]))}{\theta(t[A(i)/s])}\right](s)}\right]\right\}$$

where $\theta \in Val_D$.

Therefore, it suffices to show for all $\theta \in Val_D$,

$$\theta\left[\frac{A(\theta(i))}{\theta(s)}\right]\left[\frac{A(\theta\left[\frac{A(\theta(i))}{\theta(s)}\right](j))}{\theta\left[\frac{A(\theta(i))}{\theta(s)}\right](t)}\right] = \theta\left[\frac{A(\theta(j[A(i)/s]))}{\theta(t[A(i)/s])}\right]\left[\frac{A(\theta\left[\frac{A(\theta(j[A(i)/s]))}{\theta(t[A(i)/s])}\right](i))}{\theta\left[\frac{A(\theta(j[A(i)/s]))}{\theta(t[A(i)/s])}\right](s)}\right]$$

We start with the right-hand side

$$\theta\left[\frac{A(\theta(j[A(i)/s]))}{\theta(t[A(i)/s])}\right]\left[\frac{A(\theta\left[\frac{A(\theta(j[A(i)/s]))}{\theta(t[A(i)/s])}\right](i))}{\theta\left[\frac{A(\theta(j[A(i)/s]))}{\theta(t[A(i)/s])}\right](s)}\right] = \theta\left[\frac{A(\theta(j[A(i)/s]))}{\theta(t[A(i)/s])}\right]\left[\frac{A(\theta(i))}{\theta(s)}\right] \quad \text{by (38)}$$

$$= \theta\left[\frac{A(\theta(i))}{\theta(s)}\right]\left[\frac{A(\theta(j[A(i)/s]))}{\theta(t[A(i)/s])}\right] \quad \text{by (37)}$$

$$= \theta\left[\frac{A(\theta(i))}{\theta(s)}\right]\left[\frac{A(\theta\left[\frac{A(\theta(i))}{\theta(s)}\right](j))}{\theta\left[\frac{A(\theta)(i)}{\theta(s)}\right](t)}\right] \quad \text{by Lemma 1,} \tag{39}$$

$\square$

The proofs for the remaining rules are similar.

With these new axioms, we can prove programs equivalent that contain arrays. In all examples, fragments of the statements that changed from one step to the next are in bold.

The following two programs for swapping array variables are equivalent, assuming that the domain of computation is the integers, $x \neq y$, and $\oplus$ is the bitwise xor operator.

$$\begin{array}{ll}
t := A(y); & A(x) := A(x) \oplus A(y); \\
A(y) := A(x); & A(y) := A(x) \oplus A(y); \\
A(x) := t; & A(x) := A(x) \oplus A(y); \\
t := 0 & t := 0
\end{array}$$

The program on the left uses a temporary variable to perform the swap while the program on the right uses properties of xor and the domain of computation to swap without a temporary variable. We set the variable t to 0 so that the two programs end in the same state, though we could set t to any value at the end. By (15), we know that the right-hand side is equivalent to

$$A(x) := A(x) \oplus A(y); A(y) := A(x) \oplus A(y);$$
$$A(x) := A(x) \oplus A(y); \mathbf{t := A(x); t := 0}$$

By (34), this is equivalent to

$$A(x) := A(x) \oplus A(y); A(y) := A(x) \oplus A(y);$$
$$\mathbf{t := A(x) \oplus A(y); A(x) := A(x) \oplus A(y)}; t := 0$$

We then use (35) to show that this is equal to

$$A(x) := A(x) \oplus A(y); A(y) := A(x) \oplus A(y);$$
$$t := A(x) \oplus A(y); \mathbf{A(x) := t}; t := 0$$

Using (34) and (35), this is equivalent to

$$A(x) := A(x) \oplus A(y); \boldsymbol{t := A(y); A(y) := A(x) \oplus t}; A(x) := t; t := 0$$

By (33), this is equivalent to

$$\boldsymbol{t := A(y); A(x) := A(x) \oplus t}; A(y) := A(x) \oplus t; A(x) := t; t := 0$$

By (29), where we need the condition that $x \neq y$, commutativity of xor, and the fact that $x \oplus x \oplus y = y$, this is equal to

$$t := A(y); \boldsymbol{A(y) := A(x); A(x) := A(x) \oplus t}; A(x) := t; t := 0$$

Finally, by (31), we end up with the left-hand side,

$$t := A(y); A(y) := A(x); \boldsymbol{A(x) := t}; t := 0$$

5 Proving Heapsort Correct

We can prove heapsort on an array correct using these new axioms and the axioms of SKAT to get some basic assumptions so that we can reason at the propositional level of KAT. The proof is completely formal, relying only on the axioms of KAT and some basic facts of number theory. Most proofs of this algorithm are somewhat informal, appealing to a general examination of the code. An exception is a formal proof of heapsort's correctness in Coq [24]. In this section, we provide an outline of the major steps of the proof. For the proof in its entirety, see [23].

We adapt the algorithm given in [25, Ch. 7]. Consider the function *heapify(A,i)*, which alters the array A such that the tree rooted at index i obeys the heap property: for every node i other than the root,

$$A(par(i)) \geq A(i)$$

where

$$par(i) = \lfloor i/2 \rfloor$$

We have the following property for these operators

$$i \geq 1 \Rightarrow (i = par(j) \Rightarrow rt(i) = j \vee lt(i) = j) \tag{40}$$

which states that node i is a child of its parent, where

$$lt(i) = 2i$$
$$rt(i) = 2i + 1$$

The code for the function is as follows, where the letters to the left represent the names given to the assignments and tests at the propositional level of KAT:

```
    heapify(A,root)
    {
a:      i := root;
B:      while(i != size(A) + 1)
        {
b:          l := lt(i);
c:          r := rt(i);
C:          if(l <= size(A) && A(l) > A(i))
d:              lgst := l
            else
e:              lgst := i
D:          if(r <= size(A) && A(r) > A(lgst))
f:              lgst := r
E:          if(lgst != i)
            {
g:              swap(A,i,lgst);
h:              i := lgst
            }
            else
j:              i := size(A) + 1
        }
    }
```

where

```
    swap(A,i,j)
    {
        t := A[i];
        A[i] := A[j];
        A[j] := t
    }
```

The variable $size(A)$ denotes the size of the heap rooted at $A(1)$ while $length(A)$ is the size of the entire array.

We wish to prove that the heapify function does in fact create the heap property for the tree rooted at index r. First, we express the property that a tree indexed at r is a heap, except for the trees under the node i and greater:

$$H'_{A,r,i} \stackrel{def}{\Leftrightarrow} 1 \leq r < i \Rightarrow$$
$$(lt(r) \leq size(A) \Rightarrow (A(r) \geq A(lt(r)) \wedge H'_{A,lt(r),i})) \wedge$$
$$(rt(r) \leq size(A) \Rightarrow (A(r) \geq A(rt(r)) \wedge H'_{A,rt(r),i}))$$

Now, we can easily define what it means to be a heap rooted at node r:

$$H_{A,r} \stackrel{def}{\Leftrightarrow} H'_{A,r,size(A)}$$

We also define the test

$$P_{A,r,i} \stackrel{def}{\Leftrightarrow} i \geq 1 \Rightarrow lt(i) \leq size(A) \Rightarrow A(par(i)) \geq A(lt(i)) \wedge$$
$$rt(i) \leq size(A) \Rightarrow A(par(i)) \geq A(rt(i))$$

We wish to prove that

$$root \geq 1; H_{A,lt(root)}; H_{A,rt(root)}; heapify(A, root) = heapify(A, root); H_{A,root}$$

First, we need a couple of lemmas. We show that swapping two values in an array reverses the relationship between them and that swapping two values maintains the heap property.

The majority of the proof is spent showing the loop invariant of the while loop in the heapify function.

Lemma 2.

$$(i \geq 1); P_{A,root,i}; H'_{A,root,i}; H_{A,lt(i)}; H_{A,rt(i)};$$
$$(B; b; c; (C; d + \overline{C}; e)(D; f + \overline{D})(E; g; h + \overline{E}; j)^*$$
$$= (B; b; c; (C; d + \overline{C}; e)(D; f + \overline{D})(E; g; h + \overline{E}; j)^*;$$
$$(i \geq 1); P_{A,root,i}; H'_{A,root,i}; H_{A,lt(i)}; H_{A,rt(i)}$$

Proof. The proof proceeds by using commuting our invariants through the program and citing distributivity and congruence. □

Now, we can prove the original theorem.

Theorem 2.

$$(root \geq 1); H_{A,lt(root)}; H_{A,rt(root)}; heapify(A, root) = heapify(A, root); H_{A,root}$$

Proof. The proof proceeds by commuting our the tests through the heapify function. □

Now that we have properties for the *heapify* function, we can show that the function *build-heap(A)*, which creates a heap from the array A, works correctly. The program is

```
build-heap(A)
     {
a:        size(A) = length(A);
b:        root := floor(size(A)/2);
B:        while(root >= 1)
          {
c:             heapify(A,root);
d:             root := root - 1
          }
     }
```

We show that the invariant of the loop $(B; c; d)^*$ is $\forall j > root, H_{A,j}$. It suffices to show that it is true for one iteration of the loop, i.e.

Lemma 3.

$$(\forall j > root, H_{A,j}); B; c; d = B; c; d; (\forall j > root, H_{A,j})$$

Proof. We define a predicate to represent the ancestor relationship:

$$ch(i, j) \Leftrightarrow i = par(j) \lor ch(i, par(j))$$

We then define our other properties in terms of this one and use Theorem 2, Boolean algebra rules, and (16) to prove the lemma. □

Now we need to show

Theorem 3.

$$a; b; (B; c; d)^*; \overline{B} = a; b; (B; c; d)^*; \overline{B}; H_{A,1}$$

Proof. We use the definition of $H_{A,root}$ and reflexivity. □

Finally, we can prove that the function heapsort works. The function is defined as:

```
   heapsort(A)
   {
a:      build-heap(A);
B:      while(size(A) != 1)
        {
b:          swap(A,1,size(A));
c:          size(A) := size(A) - 1;
d:          heapify(A,1);
        }
   }
```

Theorem 4.

$$heapsort(A) = heapsort(A); (\forall j, k, 1 \leq j < k \leq length(A) \Rightarrow A(j) \leq A(k))$$

Proof. To prove this, we prove that the invariant of the loop is

$$(size(A) \leq length(A) \Rightarrow A(size + 1) \geq A(size(A)));$$
$$(\forall j, k, size(A) < j < k \leq length(A) \Rightarrow A(k) \geq A(j)); H_{A,1}$$

We use commutativity and Theorem 2. □

Therefore, we know that the heapsort function sorts an array A.

6 Conclusions and Future Work

We have presented an axiomatization of arrays for use with KAT. Through the use of preconditions, we are able to capture the essence of aliasing considerations and consider them only where they are needed: when reasoning about program transformation. The axiomatization presented here applies to arrays. However, we believe it could be extended to pointers, since pointer analysis suffers from many of the same complications with aliasing as arrays. Providing a framework such as KAT for reasoning about pointers could be very valuable.

We would also like to implement these axioms in KAT-ML [26]. These extensions would be helpful in proving and verifying properties about everyday programs in an easily-transferable way. Arrays being so ubiquitous in programming today makes such an extension necessary to make the system useful. Inevitably, KAT and its implementation KAT-ML could provide an apparatus for verifying properties of programs written in a variety of languages.

Acknowledgments

This work was supported in part by NSF grant CCR-0105586 and ONR Grant N00014-01-1-0968. The views and conclusions contained herein are those of the authors and should not be interpreted as necessarily representing the official policies or endorsements, either expressed or implied, of these organizations or the US Government.

References

1. More, T.: Axioms and theorems for a theory of arrays. IBM J. Res. Dev. **17**(2) (1973) 135–175
2. Downey, P.J., Sethi, R.: Assignment commands with array references. J. ACM **25**(4) (1978) 652–666
3. McCarthy, J.: Towards a mathematical science of computation. In: IFIP Congress. (1962) 21–28
4. McCarthy, J., Painter, J.: Correctness of a compiler for arithmetic expressions. In Schwartz, J.T., ed.: Proceedings Symposium in Applied Mathematics, Vol. 19, Mathematical Aspects of Computer Science. American Mathematical Society, Providence, RI (1967) 33–41
5. Hoare, C.A.R., Wirth, N.: An axiomatic definition of the programming language PASCAL. Acta Informatica **2**(4) (1973) 335–355
6. Power, A.J., Shkaravska, O.: From comodels to coalgebras: State and arrays. Electr. Notes Theor. Comput. Sci. **106** (2004) 297–314
7. Bornat, R.: Proving pointer programs in Hoare logic. In: MPC '00: Proceedings of the 5th International Conference on Mathematics of Program Construction, London, UK, Springer-Verlag (2000) 102–126
8. Stump, A., Barrett, C.W., Dill, D.L., Levitt, J.R.: A decision procedure for an extensional theory of arrays. In: Logic in Computer Science. (2001) 29–37
9. Collins, G., Syme, D.: A theory of finite maps. In Schubert, E.T., Windley, P.J., Alves-Foss, J., eds.: Higher Order Logic Theorem Proving and Its Applications. Springer, Berlin, (1995) 122–137
10. Kozen, D.: Kleene algebra with tests. Transactions on Programming Languages and Systems **19**(3) (1997) 427–443
11. Angus, A., Kozen, D.: Kleene algebra with tests and program schematology. Technical Report 2001-1844, Computer Science Department, Cornell University (2001)
12. Barth, A., Kozen, D.: Equational verification of cache blocking in LU decomposition using Kleene algebra with tests. Technical Report 2002-1865, Computer Science Department, Cornell University (2002)

13. Cohen, E.: Lazy caching in Kleene algebra (1994) `http://citeseer.nj.nec.com/22581.html`.
14. Cohen, E.: Hypotheses in Kleene algebra. Technical Report TM-ARH-023814, Bellcore (1993)
15. Cohen, E.: Using Kleene algebra to reason about concurrency control. Technical report, Telcordia, Morristown, N.J. (1994)
16. Kozen, D., Patron, M.C.: Certification of compiler optimizations using Kleene algebra with tests. In Lloyd, J., Dahl, V., Furbach, U., Kerber, M., Lau, K.K., Palamidessi, C., Pereira, L.M., Sagiv, Y., Stuckey, P.J., eds.: Proc. 1st Int. Conf. Computational Logic (CL2000). Volume 1861 of Lecture Notes in Artificial Intelligence., London, Springer-Verlag (2000) 568–582
17. Kozen, D.: On Hoare logic and Kleene algebra with tests. Trans. Computational Logic **1**(1) (2000) 60–76
18. Kleene, S.C.: Representation of events in nerve nets and finite automata. In Shannon, C.E., McCarthy, J., eds.: Automata Studies. Princeton University Press, Princeton, N.J. (1956) 3–41
19. Conway, J.H.: Regular Algebra and Finite Machines. Chapman and Hall, London (1971)
20. Kozen, D.: A completeness theorem for Kleene algebras and the algebra of regular events. Infor. and Comput. **110**(2) (1994) 366–390
21. Fischer, M.J., Ladner, R.E.: Propositional modal logic of programs. In: Proc. 9th Symp. Theory of Comput., ACM (1977) 286–294
22. Aboul-Hosn, K., Kozen, D.: KAT-ML: An interactive theorem prover for Kleene algebra with tests. In: Proc. 4th Int. Workshop on the Implementation of Logics, University of Manchester (2003) 2–12
23. Aboul-Hosn, K.: An axiomatization of arrays for Kleene algebra with tests. Technical report, Cornell University (2006)
24. Filliâtre, J.C., Magaud, N.: Certification of sorting algorithms in the Coq system. In: Theorem Proving in Higher Order Logics: Emerging Trends. (1999)
25. Cormen, T.H., Leiserson, C.E., Rivest, R.L.: Introduction to Algorithms. The MIT Electrical Engineering and Computer Science Series. MIT Press/McGraw Hill (1990)
26. Aboul-Hosn, K., Kozen, D.: KAT-ML: An interactive theorem prover for Kleene algebra with tests. Journal of Applied Non-Classical Logics **16**(1) (2006)

Local Variable Scoping and Kleene Algebra with Tests

Kamal Aboul-Hosn and Dexter Kozen

Department of Computer Science
Cornell University
Ithaca, New York 14853-7501, USA
{kamal, kozen}@cs.cornell.edu

Abstract. Most previous work on the semantics of programs with local state involves complex storage modeling with pointers and memory cells, complicated categorical constructions, or reasoning in the presence of context. In this paper, we explore the extent to which relational semantics and axiomatic reasoning in the style of Kleene algebra can be used to avoid these complications. We provide (i) a fully compositional relational semantics for a first-order programming language with a construct for local variable scoping; and (ii) an equational proof system based on Kleene algebra with tests for proving equivalence of programs in this language. We show that the proof system is sound and complete relative to the underlying equational theory without local scoping. We illustrate the use of the system with several examples.

1 Introduction

Reasoning about programs with local state is an important and difficult problem that has attracted much attention over the years. Most previous work involves complex storage modeling with pointers and memory cells or complicated categorical constructions to capture the intricacies of programming with state. Reasoning about the equality of such programs typically involves the notion of *contextual* or *observable equivalence*, where two programs are considered equivalent if either can be put in the context of a larger program and yield the same value. Pitts [1] explains that these notions are difficult to define formally, because there is no clear agreement on the meaning of *program context* and *observable behavior*. A common goal is to design a semantics that is *fully abstract*, where observable equivalence implies semantic equivalence, although this notion makes the most sense in a purely functional context (see for example [2,3]).

Seminal work by Meyer and Sieber [4] introduced a framework for proving the equivalence of ALGOL procedures with no parameters. Much attention has focused on the use of denotational semantics to model a set of storage locations [5,6,7,8]. The inability to prove some simple program equivalences using traditional techniques led several researchers to take a categorical approach [9,10,11]. See [12] for more information regarding the history of these approaches.

R.A. Schmidt (Ed.): RelMiCS /AKA 2006, LNCS 4136, pp. 78–90, 2006.

More recently, several researchers have investigated the use of operational semantics to reason about ML programs with references. While operational semantics can be easier to understand, their use makes reasoning about programs more complex. Mason and Talcott [13,14,15] considered a λ-calculus extended with state operations. By defining axioms in the form of contextual assertions, Mason and Talcott were able to prove the equivalence in several examples of Meyer and Sieber. Pitts and Stark [1,16,17,18] also use operational semantics.

Others have looked at using game semantics to reason about programs with local state [19,20,21,22]. Several full abstraction results have come from using game semantics to represent languages with state.

In [23], we presented a fully compositional relational semantics for higher-order programs, and showed how it could be used to avoid intricate memory modeling and the explicit use of context in program equivalence proofs. We showed how to handle several examples of Meyer and Sieber [4] in our framework. However, in that paper, we did not attempt to formulate an equational axiomatization; all arguments were based on semantic reasoning.

In this paper we consider a restricted language without higher-order programs but with a *let* construct for declaring local variables with limited scope:

$$\text{let } x = t \text{ in } p \text{ end.} \tag{1}$$

In the presence of higher-order programs, this construct can be encoded as a λ-term $(\lambda x.p)t$, but here we take (1) as primitive. The standard relational semantics used in first-order KAT and Dynamic Logic involving valuations of program variables is extended to accommodate the *let* construct: instead of a valuation, a state consists of a stack of such valuations. The formal semantics captures the operational intuition that local variables declared in a *let* statement push a new valuation with finite domain, which is then popped upon exiting the scope.

This semantics is a restriction of the relational semantics of [23] for interpreting higher-order programs. There, instead of a stack, we used a more complicated tree-like structure called a *closure structure*. Nevertheless, it is worthwhile giving an explicit treatment of this important special case. The *let* construct interacts relatively seamlessly with the usual regular and Boolean operators of KAT, which have a well-defined and well-studied relational semantics and deductive theory. We are able to build on this theory to provide a deductive system for program equivalence in the presence of *let* that is complete relative to the underlying equational theory without *let*.

This paper is organized as follows. In Section 2, we define a compositional relational semantics of programs with *let*. In Section 3, we give a set of proof rules that allow *let* statements to be systematically eliminated. In Section 4, we show that the proof system is sound and complete relative to the underlying equational theory without local scoping, and provide a procedure for eliminating variable scoping expressions. By "eliminating variable scoping expressions," we do not mean that every program is equivalent to one without scoping expressions—that is not true, and a counterexample is given in Section 5—but rather that the equivalence of two programs with scoping expressions can be reduced to the equivalence of two programs without scoping expressions. We demonstrate the use of the proof system through several examples in Section 5.

2   Relational Semantics

The *domain of computation* is a first-order structure $\mathfrak{A}$ of some signature Σ. A *partial valuation* is a partial map $f : \mathsf{Var} \rightarrow |\mathfrak{A}|$, where Var is a set of program variables. The domain of f is denoted dom f. A stack of partial valuations is called an *environment*. Let $\sigma, \tau, \ldots$ denote environments. The notation $f :: \sigma$ denotes an environment with head f and tail σ; thus environments grow from right to left. The empty environment is denoted ε. The *shape* of an environment $f_1 :: \cdots :: f_n$ is dom $f_1 :: \cdots :: $ dom f_n. The *domain* of the environment $f_1 :: \cdots :: f_n$ is $\bigcup_{i=1}^{n}$ dom f_i. The shape of ε is ε and the domain of ε is $\varnothing$. The set of environments is denoted Env. A state of the computation is an environment, and programs will be interpreted as binary relations on environments.

In Dynamic Logic and KAT, programs are built inductively from atomic programs and tests using the regular program operators $+$, $;$, and *. In the first-order versions of these languages, atomic programs are simple assignments $x := t$, where x is a variable and t is a Σ-term. Atomic tests are atomic first-order formulas $R(t_1, \ldots, t_n)$ over the signature Σ.

To accommodate local variable scoping, we also include *let expressions* in the inductive definition of programs. A *let expression* is an expression

$$\mathsf{let}\ x_1 = t_1, \ldots, x_n = t_n\ \mathsf{in}\ p\ \mathsf{end} \tag{2}$$

where p is a program, the x_i are program variables, and the t_i are terms.

Operationally, when entering the scope (2), a new partial valuation is created and pushed onto the stack. The domain of this new partial valuation is $\{x_1, \ldots, x_n\}$, and the initial values of $x_1, \ldots, x_n$ are the values of $t_1, \ldots, t_n$, respectively, evaluated in the old environment. This partial valuation will be popped when leaving the scope. The locals in this partial valuation shadow any other occurrences of the same variables further down in the stack. When evaluating a variable in an environment, we search down through the stack for the first occurrence of the variable and take that value. When modifying a variable, we search down through the stack for the first occurrence of the variable and modify that occurrence. In reality, any attempt to evaluate or modify an undefined variable (one that is not in the domain of the current environment) would result in a runtime error. In the relational semantics, there would be no input-output pair corresponding to this computation.

To capture this formally in relational semantics, we use a *rebinding operator* $[x/a]$ defined on partial valuations and environments, where x is a variable and a is a value. For a partial valuation $f : \mathsf{Var} \rightarrow |\mathfrak{A}|$,

$$f[x/a](y) = \begin{cases} f(y), & \text{if } y \in \text{dom } f \text{ and } y \neq x, \\ a, & \text{if } y \in \text{dom } f \text{ and } y = x, \\ \text{undefined}, & \text{if } y \notin \text{dom } f. \end{cases}$$

For an environment σ,

$$\sigma[x/a] = \begin{cases} f[x/a] :: \tau, & \text{if } \sigma = f :: \tau \text{ and } x \in \text{dom } f, \\ f :: \tau[x/a], & \text{if } \sigma = f :: \tau \text{ and } x \notin \text{dom } f, \\ \varepsilon, & \text{if } \sigma = \varepsilon. \end{cases}$$

Note that rebinding does not change the shape of the environment. In particular, $\varepsilon[x/a] = \varepsilon$.

The value of a variable x in an environment σ is

$$\sigma(x) = \begin{cases} f(x), & \text{if } \sigma = f :: \tau \text{ and } x \in \text{dom } f, \\ \tau(x), & \text{if } \sigma = f :: \tau \text{ and } x \notin \text{dom } f, \\ \text{undefined}, & \text{if } \sigma = \varepsilon. \end{cases}$$

The value of a term t in an environment σ is defined inductively on t in the usual way. Note that $\sigma(t)$ is defined iff $x \in \text{dom } \sigma$ for all x occurring in t.

A program is interpreted as a binary relation on environments. The binary relation associated with p is denoted $[\![p]\!]$. The semantics of assignment is

$$[\![x := t]\!] = \{(\sigma, \sigma[x/\sigma(t)]) \mid \sigma(t) \text{ and } \sigma(x) \text{ are defined}\}.$$

Note that both x and t must be defined by σ for there to exist an input-output pair with first component σ.

The semantics of scoping is

$$[\![\text{let } x_1 = t_1, \ldots, x_n = t_n \text{ in } p \text{ end}]\!]$$
$$= \{(\sigma, \text{tail}(\tau)) \mid \sigma(t_i) \text{ is defined}, 1 \le i \le n, \text{ and } (f :: \sigma, \tau) \in [\![p]\!]\}, \quad (3)$$

where f is the environment such that $f(x_i) = \sigma(t_i)$, $1 \le i \le n$.

As usual with binary relation semantics, the semantics of the regular program operators $+$, ;, and * are union, relational composition, and reflexive transitive closure, respectively. For an atomic test $R(t_1, \ldots, t_n)$,

$$[\![R(t_1, \ldots, t_n)]\!]$$
$$= \{(\sigma, \sigma) \mid \sigma(t_i) \text{ is defined}, 1 \le i \le n, \text{ and } \mathfrak{A}, \sigma \vDash R(t_1, \ldots, t_n)\}.$$

where $\vDash$ is satisfaction in the usual sense of first-order logic. The Boolean operator ! (weak negation) is defined on atomic formulas by

$$[\![!R(t_1, \ldots, t_n)]\!]$$
$$= \{(\sigma, \sigma) \mid \sigma(t_i) \text{ is defined}, 1 \le i \le n, \text{ and } \mathfrak{A}, \sigma \vDash \neg R(t_1, \ldots, t_n)\}.$$

This is not the same as classical negation $\neg$, which we need in order to use the axioms of Kleene algebra with tests. However, in the presence of !, classical negation is tantamount to the ability to check whether a variable is undefined. That is, we must have a test $\mathsf{undefined}(x)$ with semantics

$$[\![\mathsf{undefined}(x)]\!] = \{(\sigma, \sigma) \mid \sigma(x) \text{ is undefined}\}.$$

Example 1. Consider the program

```
let  x = 1
in   x := y + z;
     let y = x + 2 in y := y + z; z := y + 1 end;
     y := x
end
```

Say we start in state $(y = 5, z = 20)$. Here are the successive states of the computation:

After...	the state is...
entering the outer scope	$(x = 1) :: (y = 5, z = 20)$
executing the first assignment	$(x = 25) :: (y = 5, z = 20)$
entering the inner scope	$(y = 27) :: (x = 25) :: (y = 5, z = 20)$
executing the next assignment	$(y = 47) :: (x = 25) :: (y = 5, z = 20)$
executing the next assignment	$(y = 47) :: (x = 25) :: (y = 5, z = 48)$
exiting the inner scope	$(x = 25) :: (y = 5, z = 48)$
executing the last assignment	$(x = 25) :: (y = 25, z = 48)$
exiting the outer scope	$(y = 25, z = 48)$

Lemma 1. *If $(\sigma, \tau) \in \llbracket p \rrbracket$, then σ and τ have the same shape.*

Proof. This is true of the assignment statement and preserved by all program operators. $\qquad\square$

The goal of presenting a semantics for a language with local state is to allow reasoning about programs without the need for context. A *context* $C[\text{-}]$ is just a program expression with a distinguished free program variable. Relational semantics captures all contextual information in the state, thus making contexts superfluous in program equivalence arguments. This is reflected in the following theorem.

Theorem 1. *For program expressions p and q, $\llbracket C[p] \rrbracket = \llbracket C[q] \rrbracket$ for all contexts $C[\text{-}]$ iff $\llbracket p \rrbracket = \llbracket q \rrbracket$.*

This is a special case of a result proved in more generality in [23]. The direction ($\rightarrow$) is immediate by taking $C[\text{-}]$ to be the trivial context consisting of a single program variable. The reverse direction follows from an inductive argument, observing that the semantics is fully compositional, the semantics of a compound expression being completely determined by the semantics of its subexpressions.

3   Axioms and Basic Properties

In this section we present a set of axioms that can be used to systematically eliminate all local scopes, allowing us to reduce the equivalence problem to equivalence in the traditional "flat" semantics in which all variables are global. Although the relational semantics presented in Section 2 is a special case of the semantics presented in [23] for higher-order programs, an axiomatization was not considered in that work.

Axioms

A. If the y_i are distinct and do not occur in p, $1 \le i \le n$, then the following two programs are equivalent:

$$\text{let } x_1 = t_1, \ldots, x_n = t_n \text{ in } p \text{ end}$$
$$\text{let } y_1 = t_1, \ldots, y_n = t_n \text{ in } p[x_i/y_i \mid 1 \le i \le n] \text{ end}$$

where $p[x_i/y_i \mid 1 \le i \le n]$ refers to the simultaneous substitution of y_i for all occurrences of x_i in p, $1 \le i \le n$, including bound occurrences and those on the left-hand sides of assignments. This transformation is known as α-conversion.

B. If y does not occur in s and y and x are distinct, then the following two programs are equivalent:

$$\text{let } x = s \text{ in let } y = t \text{ in } p \text{ end end}$$
$$\text{let } y = t[x/s] \text{ in let } x = s \text{ in } p \text{ end end}$$

In particular, the following two programs are equivalent, provided x and y are distinct, x does not occur in t, and y does not occur in s:

$$\text{let } x = s \text{ in let } y = t \text{ in } p \text{ end end}$$
$$\text{let } y = t \text{ in let } x = s \text{ in } p \text{ end end}$$

C. If x does not occur in s, then the following two programs are equivalent:

$$\text{let } x = s \text{ in let } y = t \text{ in } p \text{ end end}$$
$$\text{let } x = s \text{ in let } y = t[x/s] \text{ in } p \text{ end end}$$

This holds even if x and y are the same variable.

D. If x_1 does not occur in $t_2, \ldots, t_n$, then the following two programs are equivalent:

$$\text{let } x_1 = t_1, \ldots, x_n = t_n \text{ in } p \text{ end}$$
$$\text{let } x_1 = t_1 \text{ in let } x_2 = t_2, \ldots, x_n = t_n \text{ in } p \text{ end end}$$

E. If t is a closed term (no occurrences of variables), then the following two programs are equivalent:

$$\text{skip} \qquad \text{let } x = t \text{ in skip end}$$

where skip is the identity function on states.

F. If x does not occur in pr, then the following two programs are equivalent:

$$p; \text{let } x = t \text{ in } q \text{ end}; r \qquad \text{let } x = t \text{ in } pqr \text{ end}$$

G. If x does not occur in p and t is closed, then the following two programs are equivalent:

$$p + \mathsf{let}\ x = t\ \mathsf{in}\ q\ \mathsf{end} \qquad \mathsf{let}\ x = t\ \mathsf{in}\ p + q\ \mathsf{end}$$

The proviso "t is closed" is necessary: if value of t is initially undefined, then the program on the left may halt, whereas the program on the right never does.

H. If x does not occur in t, then the following two programs are equivalent:

$$(\mathsf{let}\ x = t\ \mathsf{in}\ p\ \mathsf{end})^* \qquad \mathsf{let}\ x = a\ \mathsf{in}\ (x := t; p)^*\ \mathsf{end}$$

where a is any closed term. The proviso that x not occur in t is necessary, as the following counterexample shows. Take $t = x$ and p the assignment $y := a$. The program on the right contains the pair $(y = b, y = a)$ for $b \neq a$, whereas the program on the left does not, since x must be defined in the environment in order for the starred program to be executed once.

I. If x does not occur in t and a is a closed term, then the following two programs are equivalent:

$$\mathsf{let}\ x = t\ \mathsf{in}\ p\ \mathsf{end} \qquad \mathsf{let}\ x = a\ \mathsf{in}\ x := t; p\ \mathsf{end}$$

J. If x does not occur in t, then the following two programs are equivalent:

$$\mathsf{let}\ x = s\ \mathsf{in}\ p\ \mathsf{end}; x := t \qquad x := s; p; x := t$$

Theorem 2. *Axioms A–J are sound with respect to the binary relation semantics of Section 2.*

Proof. Most of the arguments are straightforward relational reasoning. Perhaps the least obvious is Axiom H, which we argue explicitly. Suppose that x does not occur in t. Let a be any closed term. We wish to show that the following two programs are equivalent:

$$(\mathsf{let}\ x = t\ \mathsf{in}\ p\ \mathsf{end})^* \qquad \mathsf{let}\ x = a\ \mathsf{in}\ (x := t; p)^*\ \mathsf{end}$$

Extending the nondeterministic choice operator to infinite sets in the obvious way, we have

$$(\mathsf{let}\ x = t\ \mathsf{in}\ p\ \mathsf{end})^* = \sum_n (\mathsf{let}\ x = t\ \mathsf{in}\ p\ \mathsf{end})^n$$

$$\mathsf{let}\ x = a\ \mathsf{in}\ (x := t; p)^*\ \mathsf{end} = \mathsf{let}\ x = a\ \mathsf{in}\ \sum_n (x := t; p)^n\ \mathsf{end}$$

$$= \sum_n \mathsf{let}\ x = a\ \mathsf{in}\ (x := t; p)^n\ \mathsf{end}$$

the last by a straightforward infinitary generalization of Axiom G. It therefore suffices to prove that for any n,

$$(\mathsf{let}\ x = t\ \mathsf{in}\ p\ \mathsf{end})^n = \mathsf{let}\ x = a\ \mathsf{in}\ (x := t; p)^n\ \mathsf{end}$$

This is true for $n = 0$ by Axiom E. Now suppose it is true for n. Then

$$
\begin{aligned}
&(\text{let } x = t \text{ in } p \text{ end})^{n+1} \\
&= (\text{let } x = t \text{ in } p \text{ end})^n; \text{let } x = t \text{ in } p \text{ end} \\
&= \text{let } x = a \text{ in } (x := t; p)^n \text{ end}; \text{let } x = t \text{ in } p \text{ end} \quad (4) \\
&= \text{let } x = a \text{ in } (x := t; p)^n; x := t; p \text{ end} \quad (5) \\
&= \text{let } x = a \text{ in } (x := t; p)^{n+1} \text{ end}
\end{aligned}
$$

where (4) follows from the induction hypothesis and (5) follows from the identity

$$
\text{let } x = a \text{ in } q \text{ end}; \text{let } x = t \text{ in } p \text{ end} \quad = \quad \text{let } x = a \text{ in } q; x := t; p \text{ end} \quad (6)
$$

To justify (6), observe that since x does not occur in t by assumption, p is executed in exactly the same environment on both sides of the equation.

When proving programs equivalent, it is helpful to know we can permute local variable declarations and remove unnecessary ones.

Lemma 2.

(i) *For any permutation $\pi : \{1, \ldots, n\} \to \{1, \ldots, n\}$, the following two programs are equivalent:*

$$
\text{let } x_1 = t_1, \ldots, x_n = t_n \text{ in } p \text{ end}
$$
$$
\text{let } x_{\pi(1)} = t_{\pi(1)}, \ldots, x_{\pi(n)} = t_{\pi(n)} \text{ in } p \text{ end.}
$$

(ii) *If x does not occur in p, and if t is a closed term, then the following two programs are equivalent:*

$$
p \qquad \text{let } x = t \text{ in } p \text{ end.}
$$

The second part of Lemma 2 is similar to the first example of Meyer and Sieber [4] in which a local variable unused in a procedure call can be eliminated.

4 Flattening

To prove equivalence of two programs p, q with scoping, we transform the programs so as to remove all scoping expressions, then prove the equivalence of the two resulting programs. The transformed programs are equivalent to the original ones except for the last step. The two transformed programs are equivalent in the "flat" semantics iff the original ones were equivalent in the semantics of Section 2. Thus the process is complete modulo the theory of programs without scope. The transformations are applied in the following stages.

Step 1. Apply α-conversion (Axiom A) to both programs to make all bound variables unique. This is done from the innermost scopes outward. In particular, no bound variable in the first program appears in the second program and vice-versa. The resulting programs are equivalent to the originals.

Step 2. Let $x_1, \ldots, x_n$ be any list of variables containing all bound variables that occur in either program after Step 1. Use the transformation rules of Axioms A–J to convert the programs to the form let $x_1 = a, \ldots, x_n = a$ in p end and let $x_1 = a, \ldots, x_n = a$ in q end, where p and q do not have any scoping expressions and a is a closed term. The scoping expressions can be moved outward using Axioms F–H. Adjacent scoping expressions can be combined using Axioms C and D. Finally, all bindings can be put into the form $x = a$ using Axiom I.

Step 3. Now for p, q with no scoping and a a closed term, the two programs

$$\text{let } x_1 = a, \ldots, x_n = a \text{ in } p \text{ end}$$
$$\text{let } x_1 = a, \ldots, x_n = a \text{ in } q \text{ end}$$

are equivalent iff the two programs

$$x_1 := a; \cdots; x_n := a; p; x_1 := a; \cdots; x_n := a$$
$$x_1 := a; \cdots; x_n := a; q; x_1 := a; \cdots; x_n := a$$

are equivalent with respect to the "flat" binary relation semantics in which states are just partial valuations. We have shown

Theorem 3. *Axioms A–J of Section 3 are sound and complete for program equivalence relative to the underlying equational theory without local scoping.*

5   Examples

We demonstrate the use of the axiom system through several examples. The first example proves that two versions of a program to swap the values of two variables are equivalent when the domain of computation is the integers.

Example 2. The following two programs are equivalent:

$$
\begin{array}{ll}
\text{let } \; t = x & \quad x := x \oplus y; \\
\text{in} \quad x := y; & \quad y := x \oplus y; \\
\quad\quad y := t & \quad x := x \oplus y \\
\text{end} &
\end{array}
$$

where $\oplus$ is the bitwise xor operator. The first program uses a local variable to store the value of x temporarily. The second program does not need a temporary value; it uses xor to switch the bits in place. Without the ability to handle local variables, it would be impossible to prove these two programs equivalent, because the first program includes an additional variable t. In general, without specific information about the domain of computation and without an operator like $\oplus$, it would be impossible to prove the left-hand program equivalent to any let-free program.

Proof. We apply Lemma 2 to convert the second program to

$$
\begin{aligned}
&\textsf{let } \ t = a \\
&\textsf{in } \quad x := x \oplus y; \\
&\qquad\quad y := x \oplus y; \\
&\qquad\quad x := x \oplus y \\
&\textsf{end}
\end{aligned}
$$

where a is a closed term. Next, we apply Axiom I to the first program to get

$$
\begin{aligned}
&\textsf{let } \ t = a \\
&\textsf{in } \quad t := x; \\
&\qquad\quad x := y; \\
&\qquad\quad y := t \\
&\textsf{end}
\end{aligned}
$$

From Theorem 3, it suffices to show the following programs are equivalent:

$$
\begin{array}{ll}
t := a; & t := a; \\
t := x; & x := x \oplus y; \\
x := y; & y := x \oplus y; \\
y := t; & x := x \oplus y; \\
t := a & t := a
\end{array}
$$

We have reduced the problem to an equation between let-free programs. The remainder of the argument is a straightforward application of the axioms of schematic KAT [24] and the properties of the domain of computation.   □

The second example shows that a local variable in a loop need only be declared once if the variable's value is not changed by the body of the loop.

Example 3. If the final value of x after exectuing program p is always a, that is, if p is equivalent to $p; (x = a)$ for closed term a, then the following two programs are equivalent:

$$(\textsf{let } x = a \textsf{ in } p \textsf{ end})^* \qquad \textsf{let } x = a \textsf{ in } p^* \textsf{ end}.$$

Proof. First, we use Axiom H to convert the program on the left-hand side to

$$\textsf{let } x = a \textsf{ in } (x := a; p)^* \textsf{ end}.$$

It suffices to show the following flattened programs are equivalent:

$$x := a; (x := a; p)^*; x := a \qquad x := a; p^*; x := a.$$

The equivalence follows from basic theorems of KAT and our assumption $p = p; (x = a)$.   □

The next example is important in path-sensitive analysis for compilers. It shows that a program with multiple conditionals all guarded by the same test needs only one local variable for operations in both branches of the conditionals.

Example 4. If x and w do not occur in p and the program $(y = a); p$ is equivalent to the program $p; (y = a)$ (that is, the execution of p does not affect the truth of the test $y = a$), then the following two programs are equivalent:

```
let  x = 0, w = 0
in   (if y = a then x := 1 else w := 2); p; if y = a then y := x else y := w
end
```

```
let  x = 0
in   (if y = a then x := 1 else x := 2); p; y := x
end
```

Proof. First we note that it follows purely from reasoning in KAT that $(y = a); p$ is equivalent to $(y = a); p; (y = a)$ and that $(y \neq a); p$ is equivalent to $p; (y \neq a)$ and also to $(y \neq a); p; (y \neq a)$.

We use laws of distributivity and Boolean tests from KAT and our assumptions to transform the first program into

```
let  x = 0, w = 0
in   (y = a; x := 1; p; y = a; y := x) + (y ≠ a; w := 2; p; y ≠ a; y := w)
end
```

Axiom D allows us to transform this program into

```
let  x = 0
in   let  w = 0
     in   (y = a; x := 1; p; y = a; y := x) + (y ≠ a; w := 2; p; y ≠ a; y := w)
     end
end
```

By two applications of Axiom G, we get

$$
\left(\begin{array}{l} \text{let } x = 0 \\ \text{in } y = a; x := 1; p; y = a; y := x \\ \text{end} \end{array}\right) + \left(\begin{array}{l} \text{let } w = 0 \\ \text{in } y \neq a; w := 2; p; y \neq a; y := w \\ \text{end} \end{array}\right)
$$

Using α-conversion (Axiom A) to replace w with x, this becomes

$$
\left(\begin{array}{l} \text{let } x = 0 \\ \text{in } y = a; x := 1; p; y = a; y := x \\ \text{end} \end{array}\right) + \left(\begin{array}{l} \text{let } x = 0 \\ \text{in } y \neq a; x := 2; p; y \neq a; y := x \\ \text{end} \end{array}\right)
$$

This program is equivalent to

```
let  x = 0
in   (y = a; x := 1; p; y = a; y := x) + (y ≠ a; x := 2; p; y ≠ a; y := x)
end
```

by a simple identity

$$\text{let } x = a \text{ in } p + q \text{ end} \quad = \quad \text{let } x = a \text{ in } p \text{ end} + \text{let } x = a \text{ in } q \text{ end}$$

It is easy to see that this identity is true, as both p and q are executed in the same state on both sides of the equation. It can also be justified axiomatically using Axioms A, D, and G and a straightforward application of Theorem 3.

Finally, we use laws of distributivity and Booleans to get

$$
\begin{aligned}
&\mathsf{let}\quad x = 0 \\
&\mathsf{in}\quad (\mathsf{if}\ y = a\ \mathsf{then}\ x := 1\ \mathsf{else}\ x := 2); p; y := x \\
&\mathsf{end}
\end{aligned}
$$

which is what we wanted to prove.   □

6   Conclusion

We have presented a relational semantics for first-order programs with a *let* construct for local variable scoping and a set of equational axioms for reasoning about program equivalence in this language. The axiom system allows the *let* construct to be systematically eliminated, thereby reducing the equivalence arguments to the *let*-free case. This system admits algebraic equivalence proofs for programs with local variables in the equational style of schematic KAT. We have given several examples that illustrate that in many cases, it is possible to reason purely axiomatically about programs with local variables without resorting to semantic arguments involving heaps, pointers, or other complicated semantic constructs.

Acknowledgments

We would like to thank Matthew Fluet, Riccardo Pucella, Sigmund Cherem, and the anonymous referees for their valuable input.

References

1. Pitts, A.M.: Operational semantics and program equivalence. Technical report, INRIA Sophia Antipolis (2000) Lectures at the International Summer School On Applied Semantics, APPSEM 2000, Caminha, Minho, Portugal, September 2000.
2. Plotkin, G.: Full abstraction, totality and PCF (1997)
3. Cartwright, R., Felleisen, M.: Observable sequentiality and full abstraction. In: Conference Record of the Nineteenth Annual ACM SIGPLAN-SIGACT Symposium on Principles of Programming Languages, Albequerque, New Mexico (1992) 328–342
4. Meyer, A.R., Sieber, K.: Towards fully abstract semantics for local variables. In: Proc. 15th Symposium on Principles of Programming Languages (POPL'88), New York, NY, USA, ACM Press (1988) 191–203
5. Milne, R., Strachey, C.: A Theory of Programming Language Semantics. Halsted Press, New York, NY, USA (1977)
6. Scott, D.: Mathematical concepts in programmng language semantics. In: Proc. 1972 Spring Joint Computer Conferneces, Montvale, NJ, AFIPS Press (1972) 225–34

7. Stoy, J.E.: Denotational Semantics: The Scott-Strachey Approach to Programming Language Theory. MIT Press, Cambridge, MA, USA (1981)
8. Halpern, J.Y., Meyer, A.R., Trakhtenbrot, B.A.: The semantics of local storage, or what makes the free-list free?(preliminary report). In: POPL '84: Proceedings of the 11th ACM SIGACT-SIGPLAN symposium on Principles of programming languages, New York, NY, USA, ACM Press (1984) 245–257
9. Stark, I.: Categorical models for local names. LISP and Symbolic Computation **9**(1) (1996) 77–107
10. Reyonlds, J.: The essence of ALGOL. In de Bakker, J., van Vliet, J.C., eds.: Algorithmic Languages, North-Holland, Amsterdam (1981) 345–372
11. Oles, F.J.: A category-theoretic approach to the semantics of programming languages. PhD thesis, Syracuse University (1982)
12. O'Hearn, P.W., Tennent, R.D.: Semantics of local variables. In M. P. Fourman, P.T.J., Pitts, A.M., eds.: Applications of Categories in Computer Science. L.M.S. Lecture Note Series, Cambridge University Press (1992) 217–238
13. Mason, I.A., Talcott, C.L.: Axiomatizing operational equivalence in the presence of effects. In: Proc. 4th Symp. Logic in Computer Science (LICS'89), IEEE (1989) 284–293
14. Mason, I.A., Talcott, C.L.: Equivalence in functional languages with effects. Journal of Functional Programming **1** (1991) 287–327
15. Mason, I.A., Talcott, C.L.: References, local variables and operational reasoning. In: Seventh Annual Symposium on Logic in Computer Science, IEEE (1992) 186–197
16. Pitts, A.M., Stark, I.D.B.: Observable properties of higher order functions that dynamically create local names, or what's new? In Borzyszkowski, A.M., Sokolowski, S., eds.: MFCS. Volume 711 of Lecture Notes in Computer Science., Springer (1993) 122–141
17. Pitts, A.M.: Operationally-based theories of program equivalence. In Dybjer, P., Pitts, A.M., eds.: Semantics and Logics of Computation. Publications of the Newton Institute. Cambridge University Press (1997) 241–298
18. Pitts, A.M., Stark, I.D.B.: Operational reasoning in functions with local state. In Gordon, A.D., Pitts, A.M., eds.: Higher Order Operational Techniques in Semantics. Cambridge University Press (1998) 227–273
19. Abramsky, S., Honda, K., McCusker, G.: A fully abstract game semantics for general references. In: LICS '98: Proceedings of the 13th Annual IEEE Symposium on Logic in Computer Science, Washington, DC, USA, IEEE Computer Society (1998) 334–344
20. Laird, J.: A game semantics of local names and good variables. In Walukiewicz, I., ed.: FoSSaCS. Volume 2987 of Lecture Notes in Computer Science., Springer (2004) 289–303
21. Abramsky, S., McCusker, G.: Linearity, sharing and state: a fully abstract game semantics for idealized ALGOL with active expressions. Electr. Notes Theor. Comput. Sci. **3** (1996)
22. Abramsky, S., McCusker, G.: Call-by-value games. In Nielsen, M., Thomas, W., eds.: CSL. Volume 1414 of Lecture Notes in Computer Science., Springer (1997) 1–17
23. Aboul-Hosn, K., Kozen, D.: Relational semantics for higher-order programs. In: Proc. Mathematics of Program Construction (MPC06). (2006) To appear.
24. Angus, A., Kozen, D.: Kleene algebra with tests and program schematology. Technical Report 2001-1844, Computer Science Department, Cornell University (2001)

Computing and Visualizing Lattices of Subgroups Using Relation Algebra and RELVIEW

Rudolf Berghammer

Institut für Informatik und Praktische Mathematik
Christian-Albrechts-Universität Kiel
Olshausenstraße 40, D-24098 Kiel
rub@informatik.uni-kiel.de

Abstract. We model groups as relational systems and develop relation-algebraic specifications for direct products of groups, quotient groups, and the enumeration of all subgroups and normal subgroups. The latter two specifications immediately lead to specifications of the lattices of subgroups and normal subgroups, respectively. All specifications are algorithmic and can directly be translated into the language of the computer system RELVIEW. Hence, the system can be used for constructing groups and for computing and visualizing their lattices of subgroups and normal subgroups. This is demonstrated by some examples.

1   Introduction

A number of mathematical structures are generalizations of lattices. Especially this holds for relation algebra [16,14], which additionally possesses operations for forming complements, compositions, and transpositions. Its use in Computer Science is mainly due to the fact that many structures/datatypes can be modeled via relations, many problems on them can be specified naturally by relation-algebraic expressions and formulae, and, therefore, many solutions reduce to relation-algebraic reasoning and computations, respectively.

As demonstrated in [14], relation algebra is well suited for dealing with many problems concerning order relations in a component-free (also called point-free) manner. Taking ordered sets as a starting point for introducing lattices (instead of algebras having two binary operations $\sqcap$ and $\sqcup$), lattices are nothing else than partial order relations with the additional property that every pair x, y of elements has a greatest lower bound $x \sqcap y$ and a least upper bound $x \sqcup y$. This suggests to apply the formal apparatus of relation algebra and tools for its mechanization for lattice-theoretical problems, too. A first example for this approach is [4], where relation algebra and the computer system RELVIEW [1,3] are combined for computing and visualizing cut completions and concept lattices.

The material presented in this paper is a continuation of [4]. We combine again relation algebra and RELVIEW to compute and visualize the lattices of subgroups and normal subgroups of groups by means of appropriate algorithmic relation-algebraic specifications. These lattices are a powerful tool in group theory since many group theoretical properties are determined by the lattice of (normal)

R.A. Schmidt (Ed.): RelMiCS /AKA 2006, LNCS 4136, pp. 91–105, 2006.

subgroups and vice versa. As an example we mention that a finite group is cyclic iff its subgroup lattice is distributive. A lot of further results in this vein can be found in the monograph [15]. Fundamental for our approach is the modeling of groups as relational systems. As a consequence, construction principles on groups should be conducted within this framework, too, since frequently groups are presented as compositions of other ones. In this paper we treat two important principles, viz. the construction of direct products of groups and of quotient groups modulo normal subgroups. Again this is done by developing appropriate algorithmic relation-algebraic specifications.

2   Relational Preliminaries

We write $R : X \leftrightarrow Y$ if R is a relation with domain X and range Y, i.e., a subset of $X \times Y$. If the sets X and Y of R's type $X \leftrightarrow Y$ are finite and of size m and n, respectively, we may consider R as a Boolean matrix with m rows and n columns. Since this Boolean matrix interpretation is well suited for many purposes and also used by RELVIEW to depict relations, in the following we often use matrix terminology and matrix notation. Especially we speak about the rows and columns of R and write $R_{x,y}$ instead of $\langle x, y \rangle \in R$ or $x\,R\,y$.. We assume the reader to be familiar with the basic operations on relations, viz. R^{T} (transposition), $\overline{R}$ (complementation), $R \cup S$ (union), $R \cap S$ (intersection), and RS (composition), the predicate $R \subseteq S$ (inclusion), and the special relations O (empty relation), L (universal relation), and I (identity relation).

By $syq(R, S) := \overline{R^{\mathsf{T}}\overline{S}} \cap \overline{\overline{R}^{\mathsf{T}}S}$ the symmetric quotient $syq(R, S) : Y \leftrightarrow Z$ of two relations $R : X \leftrightarrow Y$ and $S : X \leftrightarrow Z$ is defined.

We also will use the pairing (or fork) $[R, S] : Z \leftrightarrow X \times Y$ of two relations $R : Z \leftrightarrow X$ and $S : Z \leftrightarrow Y$. Component-wisely it is defined by demanding for all $z \in Z$ and $u = \langle u_1, u_2 \rangle \in X \times Y$ that $[R, S]_{z,u}$ iff R_{z,u_1} and S_{z,u_2}. (Throughout this paper pairs u are assumed to be of the form $\langle u_1, u_2 \rangle$.) Using identity and universal relations of appropriate types, the pairing operation allows to define the two projection relations $\pi : X \times Y \leftrightarrow X$ and $\rho : X \times Y \leftrightarrow Y$ of the direct product $X \times Y$ as $\pi := [\mathsf{I}, \mathsf{L}]^{\mathsf{T}}$ and $\rho := [\mathsf{L}, \mathsf{I}]^{\mathsf{T}}$. Then the above definition implies for all $u \in X \times Y$, $x \in X$, and $y \in Y$ that $\pi_{u,x}$ iff $u_1 = x$ and $\rho_{u,y}$ iff $u_2 = y$. Also the parallel composition (or product) $R \otimes S : X \times X' \leftrightarrow Y \times Y'$ of two relations $R : X \leftrightarrow Y$ and $S : X' \leftrightarrow Y'$, such that $(R \otimes S)_{u,v}$ is equivalent to R_{u_1,v_1} and S_{u_2,v_2} for all $u \in X \times X'$ and $v \in Y \times Y'$, can be defined by means of pairing. We get the desired property if we define $R \otimes S := [\pi R, \rho S]$, where $\pi : X \times X' \leftrightarrow X$ and $\rho : X \times X' \leftrightarrow X'$ are the projection relations on $X \times X'$.

There are some relational possibilities to model sets. Our first modeling uses vectors, which are relations v with $v = v\mathsf{L}$. Since for a vector the range is irrelevant we consider in the following mostly vectors $v : X \leftrightarrow \mathbf{1}$ with a specific singleton set $\mathbf{1} = \{\perp\}$ as range and omit in such cases the second subscript, i.e., write v_x instead of $v_{x,\perp}$. Such a vector can be considered as a Boolean matrix with exactly one column, i.e., as a Boolean column vector, and represents the subset $\{x \in X \mid v_x\}$ of X. A non-empty vector v is said to be a point if $vv^{\mathsf{T}} \subseteq \mathsf{I}$, i.e., v is *injective*. This means that it represents a singleton subset of its domain or an element from it if we identify a singleton set with the only element it contains. In the Boolean matrix model a point $v : X \leftrightarrow \mathbf{1}$ is a Boolean column vector in which exactly one component is true.

As a second way to model sets we will apply the relation-level equivalents of the set-theoretic symbol $\in$, i.e., *membership-relations* $\mathsf{M} : X \leftrightarrow 2^X$ on X and its powerset 2^X. These specific relations are defined by demanding for all $x \in X$ and $Y \in 2^X$ that $\mathsf{M}_{x,Y}$ iff $x \in Y$. A Boolean matrix implementation of M requires exponential space. However, in [12,5] an implementation of M using reduced ordered binary decision diagrams (ROBDDs) is given, the number of nodes of which is linear in the size of X.

Finally, we will use injective functions for modeling sets. Given an injective function $\imath$ from Y to X, we may consider Y as a subset of X by identifying it with its image under $\imath$. If Y is actually a subset of X and $\imath$ is given as relation of type $Y \leftrightarrow X$ such that $\imath_{y,x}$ iff $y = x$ for all $y \in Y$ and $x \in X$, then the vector $\imath^\mathsf{T}\mathsf{L} : X \leftrightarrow \mathbf{1}$ represents Y as subset of X in the sense above. Clearly, the transition in the other direction is also possible, i.e., the generation of a relation $inj(v) : Y \leftrightarrow X$ from the vector representation $v : X \leftrightarrow \mathbf{1}$ of $Y \subseteq X$ such that for all $y \in Y$ and $x \in X$ we have $inj(v)_{y,x}$ iff $y = x$.

A combination of such relations with membership-relations allows a *column-wise enumeration* of sets of subsets. More specifically, if $v : 2^X \leftrightarrow \mathbf{1}$ represents a subset $\mathfrak{S}$ of the powerset 2^X in the sense defined above, then for all $x \in X$ and $Y \in \mathfrak{S}$ we get the equivalence of $(\mathsf{M}\,inj(v)^\mathsf{T})_{x,Y}$ and $x \in Y$. This means that $S := \mathsf{M}\,inj(v)^\mathsf{T} : X \leftrightarrow \mathfrak{S}$ is the relation-algebraic specification of membership on $\mathfrak{S}$, or, using matrix terminology, the elements of $\mathfrak{S}$ are represented precisely by the columns of S. Furthermore, a little reflection shows for all $Y, Z \in \mathfrak{S}$ the equivalence of $Y \subseteq Z$ and $\overline{S^\mathsf{T}\,\overline{S}}_{Y,Z}$. Therefore, $\overline{S^\mathsf{T}\,\overline{S}} : \mathfrak{S} \leftrightarrow \mathfrak{S}$ is the relation-algebraic specification of set inclusion on $\mathfrak{S}$.

3   Relational Modeling of Groups

Assuming that the reader is familiar with the fundamental notions of groups (otherwise see e.g., [11]), in this section we introduce our relational model of groups and show how to construct direct products and quotient groups within this framework. As usual, we denote a group $(G, \cdot, ^{-1}, 1)$ simply by its carrier set G and write xy instead of $x \cdot y$.

3.1   Basics of the Approach

Suppose G to be a group. Our relational modeling of G is rather simple. We use a *multiplication relation* $R : G{\times}G \leftrightarrow G$ for the binary operation $\cdot : G \times G \to G$, an *inversion relation* $I : G \leftrightarrow G$ for the unary operation $^{-1} : G \to G$, and a *neutral point* $e : G \leftrightarrow \mathbf{1}$ for the neutral element 1, i.e., demand for all $u \in G \times G$ and $x, y \in G$ the following equivalences to hold:

$$R_{u,x} \iff u_1 u_2 = x \qquad I_{x,y} \iff x^{-1} = y \qquad e_x \iff x = 1$$

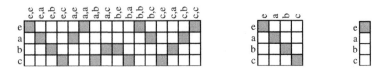

Fig. 1. Relational Model of the Kleinian group V_4

As a small example we consider the well-known *Kleinian group* V_4. Its carrier set consists of four elements e, a, b, c and the group structure is completely determined by demanding that e is the neutral element and that $aa = bb = e$. The pictures of Fig. 1 show the multiplication relation and inversion relation of V_4 as depicted by RELVIEW as Boolean matrices and the point for the neutral element e as depicted by RELVIEW as Boolean vector. For reasons of space, the multiplication relation is shown in transposed form.

Instead of the relational system (R, I, e), the multiplication relation R on its own can be used to model a group G. This is due to the fact that the inversion relation I and the neutral point e relation-algebraically can be specified using R. Let $\pi, \rho : G{\times}G \leftrightarrow G$ be the projection relations on $G \times G$. Then we have:

$$e = \overline{\rho^{\mathsf{T}}\, \overline{(\pi \cap R)\mathsf{L}}} \qquad I = \pi^{\mathsf{T}}(\rho \cap R e\mathsf{L})$$

We only prove the first equation. For all $x \in G$ we calculate as follows:

$$\overline{\rho^{\mathsf{T}}\, \overline{(\pi \cap R)\mathsf{L}}}_{\,x} \iff \neg \exists\, u : \rho^{\mathsf{T}}_{x,u} \wedge \overline{(\pi \cap R)\mathsf{L}}_{\,u}$$
$$\iff \forall\, u : \rho_{u,x} \to ((\pi \cap R)\mathsf{L})_u$$
$$\iff \forall\, u : u_2 = x \to \exists\, y : \pi_{u,y} \wedge R_{u,y} \wedge \mathsf{L}_y$$
$$\iff \forall\, u : u_2 = x \to \exists\, y : u_1 = y \wedge R_{u,y}$$
$$\iff \forall\, u : u_2 = x \to u_1 u_2 = u_1$$

This shows that $\overline{\rho^{\mathsf{T}}\, \overline{(\pi \cap R)\mathsf{L}}}$ represents the (right) neutral element of G.

Furthermore, it should be mentioned that the usual first-order group axioms can be translated into equivalent relational inclusions so that they can be checked by means of RELVIEW. But we do not want to go into details here.

3.2 Construction of Product Groups

Let G and G' be two groups. Then the direct product $G \times G'$ becomes a group if we define the binary operation component-wisely. The inverse of a pair is the pair of inverses and as neutral element we have the pair of neutral elements.

Now, suppose the relational systems (R, I, e) and (R', I', e') to be relational models of G and G', respectively, as introduced in Section 3.1. In the following we develop a relational model of the direct product of G and G'.

Due to the equations of Section 3.1 it suffices to develop a relation-algebraic specification of the multiplication relation of $G \times G'$. To this end we assume that

$\pi : G \times G' \leftrightarrow G$ and $\rho : G \times G' \leftrightarrow G'$ are the projection relations on $G \times G'$. Then we are able to calculate for all pairs $u, v, w \in G \times G'$ as follows:

$$w = \langle u_1 v_1, u_2 v_2 \rangle$$
$$\Longleftrightarrow\ w_1 = u_1 v_1 \wedge w_2 = u_2 v_2$$
$$\Longleftrightarrow\ (\exists z : u_1 = z_1 \wedge v_1 = z_2 \wedge z_1 z_2 = w_1) \wedge$$
$$\qquad (\exists z : u_2 = z_1 \wedge v_2 = z_2 \wedge z_1 z_2 = w_2)$$
$$\Longleftrightarrow\ (\exists z : \pi_{u,z_1} \wedge \pi_{v,z_2} \wedge R_{z,w_1}) \wedge (\exists z : \rho_{u,z_1} \wedge \rho_{v,z_2} \wedge R'_{z,w_2})$$
$$\Longleftrightarrow\ (\exists z : (\pi \otimes \pi)_{\langle u,v \rangle, z} \wedge R_{z,w_1}) \wedge (\exists z : (\rho \otimes \rho)_{\langle u,v \rangle, z} \wedge R'_{z,w_2})$$
$$\Longleftrightarrow\ ((\pi \otimes \pi)R)_{\langle u,v \rangle, w_1} \wedge ((\rho \otimes \rho)R')_{\langle u,v \rangle, w_2}$$
$$\Longleftrightarrow\ [(\pi \otimes \pi)R, (\rho \otimes \rho)R']_{\langle u,v \rangle, w}$$

If we remove the two subscripts $\langle u, v \rangle$ and w from the last expression of this calculation following the definition of relational equality, we arrive at the following relation-algebraic specification of the multiplication relation of $G \times G'$:

$$Pmrel(R, R') = [(\pi \otimes \pi)R, (\rho \otimes \rho)R'] : (G \times G') \times (G \times G') \leftrightarrow G \times G'$$

Likewise but more easily, we are able to develop the relation-algebraic specification $[\pi I, \rho I'] : G \times G' \leftrightarrow G \times G'$ of the inversion relation of $G \times G'$ from the fact that $v = u^{-1}$ iff $v_1 = u_1^{-1}$ and $v_2 = u_2^{-1}$ for all $u, v \in G \times G'$. Also the neutral point of the product group $G \times G'$ can be computed from the neutral points of G and G', respectively. Here we arrive at $\pi e \cap \rho e' : G \times G' \leftrightarrow \mathbb{1}$. Compared with the computations of the inversion relation and the neutral point of $G \times G'$ from the multiplication relation of $G \times G'$ as shown in Section 3.1, the computations using $[\pi I, \rho I']$ and $\pi e \cap \rho e'$ are much more efficient in that they use the projection relations π, ρ of the direct product $G \times G'$ instead of the much larger direct product $(G \times G') \times (G \times G')$ as the equations of Section 3.1 do.

3.3 Construction of Quotient Groups

Besides direct products, forming the quotient group (or factor group) G/N of a group G modulo a normal subgroup $N \subseteq G$ is a further important construction on groups. Here G/N denotes the set of all equivalence classes (called right cosets) of the equivalence relation $E : G \leftrightarrow G$, that is defined component-wisely by the equivalence of $E_{x,y}$ and $xy^{-1} \in N$ for all $x, y \in G$. The set G/N becomes a group if the binary operation is defined by means of the classes' representatives. Then the inverse of an equivalence class $[x]$ is the class $[x^{-1}]$ and the neutral equivalence class is the class $[1]$, which coincides with the normal subgroup N.

Again we assume that G is modeled by the relational system (R, I, e) and $\pi, \rho : G \times G \leftrightarrow G$ are the projection relations on $G \times G$. Furthermore, we suppose that the normal subgroup $N \subseteq G$ is represented by the vector $n : G \leftrightarrow \mathbb{1}$ in the sense of Section 2. Fundamental for obtaining a relational model of G/N is the following calculation, where $x, y \in G$ are arbitrarily chosen group elements:

$$
\begin{aligned}
xy^{-1} \in N \iff & \exists\, u : u_1 = x \wedge u_2 = y^{-1} \wedge u_1 u_2 \in N \\
\iff & \exists\, u : \pi_{u,x} \wedge (\rho I^\mathsf{T})_{u,y} \wedge \exists\, z : u_1 u_2 = z \wedge z \in N \\
\iff & \exists\, u : \pi_{u,x} \wedge (\rho I)_{u,y} \wedge \exists\, z : R_{u,z} \wedge n_z && \text{as } I = I^\mathsf{T} \\
\iff & \exists\, u : \pi^\mathsf{T}_{x,u} \wedge (\rho I)_{u,y} \wedge \exists\, z : R_{u,z} \wedge (nL)_{z,y} \\
\iff & \exists\, u : \pi^\mathsf{T}_{x,u} \wedge (\rho I \cap RnL)_{u,y} \\
\iff & (\pi^\mathsf{T}(\rho I \cap RnL))_{x,y}
\end{aligned}
$$

In combination with the definition of relational equality the last expression of this calculation yields the relation-algebraic description $E = \pi^\mathsf{T}(\rho I \cap RnL)$.

Next, we apply a result of [2] saying that the relation $\mathsf{M}\, inj(syq(\mathsf{M}, S)\mathsf{L})^\mathsf{T}$ is a column-wise enumeration of the set of all equivalence classes [1] of an equivalence relation S in the sense of Section 2. If we replace in this expression the relation S by $\pi^\mathsf{T}(\rho I \cap RnL)$ and use the corresponding membership-relation M of type $G \leftrightarrow 2^G$ we get a relation-algebraic specification of the canonical epimorphism $f : G \to G/N$, where $f(x) = [x]$, as follows:

$$
Cepi(R.I, n) = \mathsf{M}\, inj(syq(\mathsf{M}, \pi^\mathsf{T}(\rho I \cap RnL))\mathsf{L})^\mathsf{T} : G \leftrightarrow G/N
$$

Now, we are almost done. Define $C := Cepi(R.I, n)$. Then we have for all equivalence classes $a, b, c \in G/N$ the following property:

$$
\begin{aligned}
ab = c \iff & \exists\, v : [v_1] = a \wedge [v_2] = b \wedge c = [v_1 v_2] \\
\iff & \exists\, v : [v_1] = a \wedge [v_2] = b \wedge \exists\, x : [x] = c \wedge v_1 v_2 = x \\
\iff & \exists\, v : (\pi C)_{v,a} \wedge (\rho C)_{v,b} \wedge \exists\, x : C_{x,c} \wedge R_{v,x} \\
\iff & \exists\, v : [\pi C, \rho C]_{v,\langle a,b \rangle} \wedge (RC)_{v,c} \\
\iff & \exists\, v : [\pi C, \rho C]^\mathsf{T}_{\langle a,b \rangle, v} \wedge (RC)_{v,c} \\
\iff & ([\pi C, \rho C]^\mathsf{T} RC)_{\langle a,b \rangle, c}
\end{aligned}
$$

A consequence is the following relation-algebraic specification of the multiplication relation of the quotient group G/N:

$$
Qmrel(R, I, n) = [\pi C, \rho C]^\mathsf{T} RC : (G/N) \times (G/N) \leftrightarrow G/N
$$

As in the case of direct products, also the inversion relation and the neutral point of the quotient group G/N can be specified relation-algebraically using the relational model (R, I, e) of the group G only. Doing so, we obtain $C^\mathsf{T} I C :$ $G/N \leftrightarrow G/N$ for the inversion relation of G/N and $C^\mathsf{T} e : G/N \leftrightarrow \mathbb{1}$ for the neutral point of G/N. However, contrary to direct products, now these "direct" specifications are less efficient than the specifications based on the multiplication relation of G/N. The reason is that the direct product $(G/N) \times (G/N)$ usually is much smaller than the direct product $G \times G$.

---

[1] It should be mentioned that [2] also contains an efficient relational programm for the column-wise enumeration of the equivalence classes, that avoids the use of a membership-relation. We only use here $\mathsf{M}\, inj(syq(\mathsf{M}, S)\mathsf{L})^\mathsf{T}$ to simplify presentation.

4   Relation-Algebraic Specification of Subgroup Lattices

Having modeled groups and two construction principles on them via relation algebra, in this section we develop vector representations of the sets of subgroups and normal subgroups, respectively. They immediately lead to column-wise enumerations of these sets and to specifications of the corresponding lattices.

4.1   Lattice of Subgroups

In what follows, we assume a group G that it is modeled by the relational system (R, I, e). A non-empty subset of G is a subgroup if it is closed with respect to both group operations. If $\mathfrak{S}_G$ denotes the set of all subgroups of G, then the ordered set $(\mathfrak{S}_G, \subseteq)$ constitutes a lattice, called the *subgroup lattice* of G. In this lattice $Y \sqcap Z$ corresponds to the intersection $Y \cap Z$ and $Y \sqcup Z$ to $\langle Y \cup Z \rangle$, the least subgroup of G containing the union $Y \cup Z$. See e.g., [9,6].

Now, suppose G to be finite. Then $Y \subseteq G$ is closed with respect to both group operations iff it is closed with respect to the binary group operation only. The latter characterization is the starting point of the following calculation, where $M : G \leftrightarrow 2^G$ is a membership relation and $\pi, \rho : G{\times}G \leftrightarrow G$ are the projection relations on $G \times G$:

$$Y \text{ is a subgroup}$$
$$\Longleftrightarrow Y \neq \emptyset \wedge \forall u : u_1 \in Y \wedge u_2 \in Y \rightarrow u_1 u_2 \in Y$$
$$\Longleftrightarrow (\exists x : x \in Y) \wedge \forall u : u_1 \in Y \wedge u_2 \in Y \rightarrow \exists z : u_1 u_2 = z \wedge z \in Y$$
$$\Longleftrightarrow (\exists x : x \in Y) \wedge \forall u : u_1 \in Y \wedge u_2 \in Y \rightarrow \exists z : R_{u,z} \wedge z \in Y$$
$$\Longleftrightarrow (\exists x : \mathsf{M}_{x,Y} \wedge \mathsf{L}_x) \wedge \forall u : (\pi \mathsf{M})_{u,Y} \wedge (\rho \mathsf{M})_{u,Y} \rightarrow \exists z : R_{u,z} \wedge \mathsf{M}_{z,Y}$$
$$\Longleftrightarrow (\mathsf{M}^{\mathsf{T}}\mathsf{L})_Y \wedge \forall u : (\pi \mathsf{M})_{u,Y} \wedge (\rho \mathsf{M})_{u,Y} \rightarrow (R\mathsf{M})_{u,Y}$$
$$\Longleftrightarrow (\mathsf{M}^{\mathsf{T}}\mathsf{L})_Y \wedge \neg \exists u : (\pi \mathsf{M})_{u,Y} \wedge (\rho \mathsf{M})_{u,Y} \wedge \overline{R\mathsf{M}}_{u,Y}$$
$$\Longleftrightarrow (\mathsf{M}^{\mathsf{T}}\mathsf{L})_Y \wedge \neg \exists u : (\pi \mathsf{M} \cap \rho \mathsf{M} \cap \overline{R\mathsf{M}})^{\mathsf{T}}_{Y,u} \wedge \mathsf{L}_u$$
$$\Longleftrightarrow (\mathsf{M}^{\mathsf{T}}\mathsf{L} \cap \overline{(\pi \mathsf{M} \cap \rho \mathsf{M} \cap \overline{R\mathsf{M}})^{\mathsf{T}}\mathsf{L}})_Y$$

If we remove the subscript Y from the last expression of this calculation following the vector representation of sets as introduced in Section 2 and apply after that some well-known relation-algebraic rules to transpose[2] only a "row vector" instead of relations of types $G \leftrightarrow 2^G$ and $G{\times}G \leftrightarrow 2^G$, we get

$$Sg\,Vect(R) = (\mathsf{L}^{\mathsf{T}}\mathsf{M} \cap \overline{\mathsf{L}^{\mathsf{T}}(\pi \mathsf{M} \cap \rho \mathsf{M} \cap \overline{R\mathsf{M}})})^{\mathsf{T}} : 2^G \leftrightarrow \mathbf{1}$$

as relation-algebraic specification of the vector representing $\mathfrak{S}_G$ as subset of 2^G. The column-wise enumeration of the set $\mathfrak{S}_G$ by

$$SgList(R) = \mathsf{M}\,inj(Sg\,Vect(R))^{\mathsf{T}} : G \leftrightarrow \mathfrak{S}_G$$

---

[2] Using a ROBDD-implemention of relations, transposition of a relation with domain or range $\mathbf{1}$ only means to exchange domain and range; the ROBDD remains unchanged. See [13] for details.

and the relation-algebraic specification of set inclusion on the set $\mathfrak{S}_G$, i.e., the partial order relation of the subgroup lattice $(\mathfrak{S}_G, \subseteq)$, by

$$SgLat(R) = \overline{\overline{SgList(R)}^\mathsf{T}\ \overline{SgList(R)}} : \mathfrak{S}_G \leftrightarrow \mathfrak{S}_G$$

are immediate consequences of the technique shown at the end of Section 2.

Using $\overline{\mathsf{L}^\mathsf{T}(\mathsf{M} \cap \overline{I\mathsf{M}})}^\mathsf{T} : 2^G \leftrightarrow \mathbf{1}$ as vector representation of the subsets of G closed with respect to inversion, it is possible to refine $SgVect(R)$ to a relation-algebraic specifications that works for general groups. In regard to computability by a computer system like RELVIEW, the restriction to finite groups in the development of $SgVect(R)$ is irrelevant. Its only reason is to simplify the calculation and to obtain a more efficient solution.

4.2   Lattice of Normal Subgroups

A subgroup Y of a group G is a normal subgroup if for all $x \in G$ and $y \in Y$ we have $xyx^{-1} \in Y$. Also the set $\mathfrak{N}_G$ of normal subgroups of G can be made to a lattice. In this *lattice of normal subgroups* $(\mathfrak{N}_G, \subseteq)$ of G again greatest lower bounds $Y \sqcap Z$ correspond to intersections $Y \cap Z$. Compared with the subgroup lattice $(\mathfrak{S}_G, \subseteq)$ of G, however, the description of the least upper bound of two normal subgroups Y and Z is more simple. We have $Y \sqcup Z = \{yz \mid y \in Y, z \in Z\}$. The lattice of normal subgroups is modular; for more details see e.g., [9,6].

Making again the assumptions that the (finite) group G is modeled by the relational system (R, I, e), the above standard definition of normal subgroups is our starting point for developing relation-algebraic specifications of the vector description of $\mathfrak{N}_G$, the column-wise enumeration of $\mathfrak{N}_G$, and the partial order relation of the lattice $(\mathfrak{N}_G, \subseteq)$. Here is the decisive part of the development, where $Y \in \mathfrak{S}_G$ is an arbitrarily chosen subgroup of G:

$\qquad Y$ is a normal subgroup

$\Longleftrightarrow \forall\, u : u_2 \in Y \to u_1 u_2 u_1^{-1} \in Y$

$\Longleftrightarrow \forall\, u : (\rho\mathsf{M})_{u,Y} \to \exists\, v : u_1 u_2 = v_1 \wedge u_1^{-1} = v_2 \wedge v_1 v_2 \in Y$

$\Longleftrightarrow \forall\, u : (\rho\mathsf{M})_{u,Y} \to \exists\, v : R_{u,v_1} \wedge (\pi I)_{u,v_2} \wedge v_1 v_2 \in Y$

$\Longleftrightarrow \forall\, u : (\rho\mathsf{M})_{u,Y} \to \exists\, v : R_{u,v_1} \wedge (\pi I)_{u,v_2} \wedge \exists\, z : v_1 v_2 = z \wedge z \in Y$

$\Longleftrightarrow \forall\, u : (\rho\mathsf{M})_{u,Y} \to \exists\, v : R_{u,v_1} \wedge (\pi I)_{u,v_2} \wedge \exists\, z : R_{v,z} \wedge \mathsf{M}_{z,Y}$

$\Longleftrightarrow \forall\, u : (\rho\mathsf{M})_{u,Y} \to \exists\, v : [R, \pi I]_{u,v} \wedge (R\mathsf{M})_{v,Y}$

$\Longleftrightarrow \forall\, u : (\rho\mathsf{M})_{u,Y} \to ([R, \pi I]R\mathsf{M})_{u,Y}$

$\Longleftrightarrow \neg\exists\, u : (\rho\mathsf{M})^\mathsf{T}_{Y,u} \wedge \overline{[R, \pi I]R\mathsf{M}}^\mathsf{T}_{Y,u}$

$\Longleftrightarrow \neg\exists\, u : (\rho\mathsf{M} \cap \overline{[R, \pi I]R\mathsf{M}})^\mathsf{T}_{Y,u} \wedge \mathsf{L}_u$

$\Longleftrightarrow \overline{(\rho\mathsf{M} \cap \overline{[R, \pi I]R\mathsf{M}})^\mathsf{T}\mathsf{L}}_Y.$

If we remove the subscript Y from the last expression, apply after that, as in Section 4.1, some simple transformations to improve efficiency in view of an implementation of relations via ROBDDs, and intersect the result with the vector representation of the set of all subgroups of G, this leads to

$$NsgVect(R, I) = SgVect(R) \cap \overline{\mathsf{L}^\mathsf{T}(\rho\mathsf{M} \cap \overline{[R, \pi I]R\mathsf{M}})}^\mathsf{T} : 2^G \leftrightarrow 1$$

as vector representing $\mathfrak{N}_G$ as subset of 2^G. Immediate consequences are

$$NsgList(R, I) = \mathsf{M}\, inj(NsgVect(R, I))^\mathsf{T} : G \leftrightarrow \mathfrak{N}_G$$

as relation-algebraic specification of the column-wise enumeration of $\mathfrak{N}_G$ and

$$NsgLat(R, I) = \overline{\overline{NsgList(R, I)^\mathsf{T}}\, \overline{NsgList(R, I)}} : \mathfrak{N}_G \leftrightarrow \mathfrak{N}_G$$

as relation-algebraic specification of the partial order relation of the lattice of normal subgroups $(\mathfrak{N}_G, \subseteq)$.

5   Implementation and Examples

In order to illustrate our approach, in the following we give a short description of the RELVIEW system and show afterwards by means of three simple examples some possibilites and special features of this tool.

5.1   The RELVIEW-System

Relational algebra has a fixed and surprisingly small set of constants and operations which – in the case of finite carrier sets – can be implemented very efficiently. At Kiel University we have developed a visual computer system for the visualization and manipulation of relations and for relational prototyping and programming, called RELVIEW. The tool is written in the C programming language, uses ROBDDs for very efficiently implementing relations [12,13], and makes full use of the X-windows graphical user interface. Details and applications can be found, for instance, in [1,2,3,4,5].

One of the main purposes of the RELVIEW tool is the evaluation of relation-algebraic expressions. These are constructed from the relations of its workspace using pre-defined operations (including the basic ones, symmetric quotient, and pairing) and tests and user-defined relational functions and programs.

A relational program in RELVIEW is much like a function procedure in the programming languages Pascal or Modula 2, except that it only uses relations as datatype. It starts with a head line containing the program name and the formal parameters. Then the declaration of the local relational domains, functions, and variables follows. Domain declarations can be used to introduce projection relations. The third part of a RELVIEW-program is the body, a while-program over relations. Recursive calls of programs are allowed. As a RELVIEW-program computes a value, finally, its last part consists of a RETURN-clause, which essentially is a relation-algebraic expression whose value after the execution of the body is the result.

All relation-algebraic specifications we have developed so far immediately can be translated into RELVIEW-code. For example, the following relational program SgVect is a RELVIEW-implementation of the relation-algebraic specification $SgVect(R)$ of Section 4.1:

```
SgVect(R)
   DECL PP = PROD(R^*R,R^*R);
        pi, rho, M, L1, L2
   BEG  pi = p-1(PP);
        rho = p-2(PP);
        M = epsi(O1n(R)^);
        L1 = L1n(R);
        L2 = Ln1(R)^
        RETURN (L1*M & -(L2 * (pi*M & rho*M & -(R*M))))^
   END.
```

The first declaration introduces PP as a name for the direct product $G \times G$. Using PP, the projection relations are then computed by the first two assignments of the body and stored as pi and rho. The next three assignments compute the membership relation M : $G \leftrightarrow 2^G$ and two universal relations of type $1 \leftrightarrow G$ and $G \times G \leftrightarrow 1$, respectively. Finally, the RETURN-clause consists of a direct translation of the right-hand side of $SgVect(R)$ into RELVIEW-syntax.

5.2 A First Example

As already mentioned, the lattices of normal subgroups are modular. The same is not true for subgroup lattices. This is e.g., shown in [9] by the subgroup lattice of the *alternating group* A_4 of the even permutations $p_1, \ldots, p_{12}$ on the set $\{1, 2, 3, 4\}$. We have verified this example with the aid of RELVIEW.

In doing so, we started with the following multiplication tabls of A_4, where the number i, $1 \le i \le 12$, abbreviates the permutation p_i on the set $\{1, 2, 3, 4\}$. Using cycle-notation these are specified as $p_1 = ()$, $p_2 = (1\ 2)(3\ 4)$, $p_3 = (1\ 3)(2\ 4)$, $p_4 = (1\ 4)(2\ 3)$, $p_5 = (1\ 2\ 3)$, $p_6 = (1\ 3\ 2)$, $p_7 = (1\ 2\ 4)$, $p_8 = (1\ 4\ 2)$, $p_9 = (1\ 3\ 4)$, $p_{10} = (1\ 4\ 3)$, $p_{11} = (2\ 3\ 4)$, and $p_{12} = (2\ 4\ 3)$.

	1	2	3	4	5	6	7	8	9	10	11	12
1	1	2	3	4	5	6	7	8	9	10	11	12
2	2	1	4	3	12	10	11	9	8	6	7	5
3	3	4	1	2	8	11	10	5	12	7	6	9
4	4	3	2	1	9	7	6	12	5	11	10	8
5	5	9	12	8	6	1	3	10	11	4	2	7
6	6	11	7	10	1	5	12	4	2	8	9	3
7	7	10	6	11	4	9	8	1	3	12	5	2
8	8	12	9	5	11	3	1	7	6	2	4	10
9	9	5	8	12	7	4	2	11	10	1	3	6
10	10	7	11	6	2	12	5	3	1	9	8	4
11	11	6	10	7	3	8	9	2	4	5	12	1
12	12	8	5	9	10	2	4	6	7	3	1	11

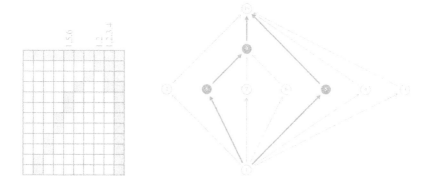

Fig. 2. Enumeration of Subgroups and Subgroup Lattice of A_4

In the next step we transformed the above table into the relational model of A_4. Loading the relations of this model (which are too large to be presented here) from an ASCII-file into RELVIEW we then computed the subgroups of A_4 and the partial order relation of the subgroup lattice. In the two pictures of Fig. 2 the results of these computations are shown. The 12×10 Boolean matrix on the left-hand side column-wisely enumerates the 10 subgroups of A_4 and the directed graph on the right-hand side depicts the inclusion order on them by means of the Hasse diagram. Additionally we have labeled three columns of the enumeration matrix, where the labels indicate the permutations forming the respective subgroup, drawn the corresponding nodes of the graph as black circles, and emphasized the subgraph generated by the black nodes and the nodes 1, 10 by boldface arrows. From the relationships drawn as boldface arrows we immediately see that the subgraph lattice of A_4 contains a "pentagon sublattice" N_5. Hence, the so-called M_3-N_5-theorem [6] implies that it is not modular.

We also have used RELVIEW to compute the three normal subgroups of A_4 and the corresponding lattice. The latter forms a chain of length two, with a normal subgroup N isomorphic to the Kleinian group V_4 as element in the middle. The three pictures of Fig. 3 concern the quotient group of A_4 modulo this specific normal subgroup N, i.e., the quotient group A_4/V_4. On the left-hand side the equivalence relation E of Section 3.3 is shown, the columns of the matrix in the

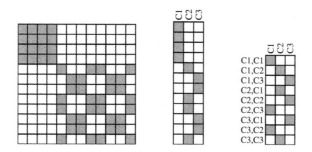

Fig. 3. Quotient Construction A_4/V_4

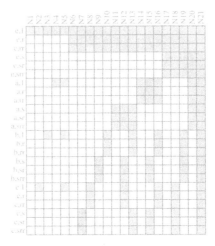

Fig. 4. Column-wise Enumeration of the Normal Subgroups of $V_4 \times D_3$

middle enumerate the set A_4/V_4, and the matrix on the right-hand side is the multiplication relation of the quotient group A_4/V_4 (which, obviously, coincides with the multiplication relation of the cyclic group $\mathbb{Z}_3$).

5.3   A Second Example

In the following, we treat a product construction. Assume D_3 to be the *dihedral group* of order 6. This group is generated by two elements, r and s say, such that the equations $r^3 = 1 = s^2$ and $rsr = s$ hold. From this description we obtain $D_3 = \{1, r, rr, s, sr, srr\}$ and the following multiplication table:

	1	r	rr	s	sr	srr
1	1	r	rr	s	sr	srr
r	r	rr	1	srr	s	sr
rr	rr	1	r	sr	srr	s
s	s	sr	srr	1	r	rr
sr	sr	srr	s	rr	1	r
srr	srr	s	sr	r	rr	1

We have computed a relational model of the dihedral group D_3 and combined it with that of the Kleinian group V_4 to get a relational model of the product group $V_4 \times D_3$ with RELVIEW's help. Again the later model is too large to be presented here; its multiplication relation has 576 rows and 24 columns.

Then we used the system to compute all normal subgroups of this product group and the lattice of normal subgroups, too. The picture of Fig. 4 shows the column-wise enumeration of the 21 normal subgroups of the group $V_4 \times D_3$ by a 24×21 Boolean matrix. A graphical representation of the lattice of normal subgroups of $V_4 \times D_3$ by means of the Hasse diagram is depicted in Fig 5. From it and again the M_3-N_5-theorem we obtain that the lattice of normal subgroups of $V_4 \times D_3$ is not distributive since it contains a "diamond sublattice" M_3.

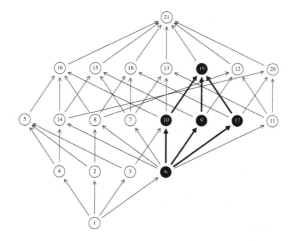

Fig. 5. The Lattice of Normal Subgroups of $V_4 \times D_3$

To give an impression of the potential of RELVIEW and the positive effects of the ROBDD-implementation of relations, we want to mention that (on a Sun-Fire 880 workstation running Solaris 9 at 750 MHz) the system required 0.5 sec. to filter out from the 15 777 216 subsets of $V_4 \times D_3$ those 21 subsets which form normal subgroups and to compute the partial order relation of the lattice of normal subgroups. We also have determined the 54 subgroups of $V_4 \times D_3$ and the relation/graph of the subgroup lattice. Here RELVIEW required 0.4 sec. For reasons of space we renounce the corresponding RELVIEW pictures.

5.4   A Third Example

When interested on the generation of a group G, the *Frattini subgroup* of G [15] plays an important role. It is defined as the intersection of all maximal (proper) subgroups of G and consists of the non-generating elements of G, i.e, of all elements $x \in G$ such that whenever $G = \langle X \rangle$ then $G = \langle X \setminus \{x\} \rangle$.

Fig. 6. Maximal Subgroups and Frattini Subgroup of $\mathbb{Z}_4 \times \mathbb{Z}_4$

The RELVIEW-picture of Fig. 6 visualizes the computation of the Frattini subgroup for the product group $\mathbb{Z}_4 \times \mathbb{Z}_4$. It shows the Hasse diagram of the subgroup lattice of $\mathbb{Z}_4 \times \mathbb{Z}_4$, where the three maximal subgroups are drawn as squares, the Frattini subgroup is drawn as a black circle, and the intersection relationships are emphasized as boldface arrows. The vector representing the maximal subgroups is $m := max(Q, \overline{gre(Q, \mathsf{L})})$ and the point representing the Frattini subgroup is $glb(Q, m)$, where Q is the partial order relation of the subgroup lattice and the relational functions max, gre, and glb specify maximal elements, greatest elements, and greatest lower bounds following [14].

We have RELVIEW also used column-wisely to enumerate the 15 subgroups of $\mathbb{Z}_4 \times \mathbb{Z}_4$ as a 16×15 Boolean matrix. From its 10th column we obtained $\langle 0, 0 \rangle$, $\langle 0, 2 \rangle$, $\langle 2, 0 \rangle$, and $\langle 2, 2 \rangle$ as the elements of the Frattini subgroup of $\mathbb{Z}_4 \times \mathbb{Z}_4$. This is what we have expected, as 0 and 2 are the non-generating elements of $\mathbb{Z}_4$. In contrast with this example, the Frattini subgroups of all our former examples V_4, A_4, A_4/V_4, D_3, and $V_4 \times D_3$ are trivial.

6   Conclusion

Besides the examples of Section 5 we have applied the RELVIEW-programs resulting from the relational specifications of Section 4 to other groups. Most of them have been constructed as products of small groups (using RELVIEW-programs obtained from the specifications of Section 3.2), like $D_3 \times D_3$ (26 subgroups; 10 are normal subgroups) and $V_4 \times D_3 \times \mathbb{Z}_2$ (236 subgroups; 83 are normal subgroups). Due to the use of membership-relations, for groups with $|G| \geq 50$ the computations can take very long time or be complete unfeasible – despite of the very efficient ROBDD-implementation of relations in RELVIEW. Therefore, we can not compete with algorithms specifically tailored to the problems we have treated (cf. e.g., the times of [10] for computing all normal subgroups).

Nowadays, systematic experiments are accepted as a way for obtaining new mathematical insights. Hence, tools for experimental computations and visualizations become increasingly important in many areas as one proceeds in the investigation of new and more complex notions. We believe that the real attraction of RELVIEW in respect thereof lies in its flexibility, its manifold visualization possibilities, and the concise form of its programs. RELVIEW proved to be an ideal tool for experimenting while avoiding unnecessary overhead. Programs are built very quickly and their correctness is guaranteed by the completely formal developments. For example, using $Inf(R) = [R, R]^{\mathsf{T}} \cap \overline{[R, R]^{\mathsf{T}} \overline{R}}$: $L \times L \leftrightarrow L$ as relation-algebraic specification of the lattice operation $\sqcap$ in terms of the lattice's partial order relation $R : L \leftrightarrow L$, the program SgVect of Section 5.1 immediately can be used for computing all sublattices of L. Thus, we have been able to find out that, e.g., the 32-element Boolean lattice possesses exactly 12 084 sublattices, whereas only 52 of them are Boolean sublattices.

At this place, also the advantages of the system when using it in teaching should be mentioned. We found it very attractive to use RELVIEW for producing good examples. These frequently have been proven for students to be the key of fully understanding an advanced concept. We have further recognized that it is sometimes very helpful to demonstrate how a certain algorithm works. In RELVIEW this is possible by executing computations in a stepwise fashion.

Relation algebra easily allows to specify the conjugacy-relation on a group G with multiplication relation $R : G{\times}G \leftrightarrow G$ as $\rho^{\mathsf{T}}(RR^{\mathsf{T}} \cap \pi\rho^{\mathsf{T}})\pi : G \leftrightarrow G$, where $\pi, \rho : G{\times}G \leftrightarrow G$ are the projection relations on $G \times G$. Based on this fact, we presently investigate a procedure to get the normal subgroups of a group without using a membership-relation, viz. as normal closures of the conjugacy classes. We also combine relation algebra and RELVIEW to solve many other problems on orders and lattices and to visualize their solutions, like the test of order and lattice properties, the enumeration of specific elements and subsets, questions concerning concept lattices [8], and the construction of free lattices from partial order relations on the generator set following [7]. Our ultimate aim is a library of RELVIEW-programs for order- and lattice-theoretical tasks that hides most of the relational notions and notation and, therefore, facilitates the use of the tool and its manifold possibilities for people not being specialists in relation algebra.

References

1. Behnke R., et al.: RELVIEW – A system for calculation with relations and relational programming. In: Astesiano E. (ed.): Proc. 1st Conf. *Fundamental Approaches to Software Engineering*, LNCS 1382, Springer, 318-321 (1998).
2. Berghammer R., Hoffmann T.: Modelling sequences within the RELVIEW system. J. Universal Comput. Sci. 7, 107-13 (2001).
3. Berghammer R., Neumann F.: RELVIEW– An OBDD-based Computer Algebra system for relations. In: Gansha V.G. et al. (eds.): Proc. 8th Int. Workshop *Computer Algebra in Scientific Computing*, LNCS 3718, Springer, 40-51 (2005)
4. Berghammer R.: Computation of cut completions and concept lattices using relational algebra and RELVIEW. J. Rel. Meth. in Comput. Sci. 1, 50-72 (2004).
5. Berghammer R., Leoniuk B., Milanese U.: Implementation of relation algebra using binary decision diagrams. In: de Swart H. (ed.): Proc. 6th Int. Workshop *Relational Methods in Computer Science*, LNCS 2561, Springer, 241-257 (2002).
6. Davey B.A., Priestley H.A.: Introduction to lattices and order. Cambridge Univ. Press (1990).
7. Freese R., Jezek J., Nation J.B.: Free lattices. Mathematical Surveys and Monographs, Vol. 42, American Math. Society (1995).
8. Ganter B., Wille R.: Formal concept analysis. Springer, (1999).
9. Hermes H.: Introduction to lattice theory (in German). Springer, 2nd ed. (1967).
10. Hulpke A.: Computing normal subgroups. In: Proc. Int. Symposium on Symbolic and Algebraic Computation, ACM Press, 194-198 (1998).
11. Lang S.: Algebra. Springer, rev. 3rd ed. (2002).
12. Leoniuk B.: ROBDD-based implementation of relational algebra with applications (in German). Ph.D. thesis, Inst. für Informatik und Prak. Math., Univ. Kiel (2001).
13. Milanese U.: On the implementation of a ROBDD-based tool for the manipulation and visualization of relations (in German). Ph.D. thesis, Inst. für Informatik und Prak. Math., Univ. Kiel (2003).
14. Schmidt G., Ströhlein T.: Relations and graphs. Springer (1993).
15. Schmidt R.: Subgroup lattices of groups, de Gruyter (1994).
16. Tarski A.: On the calculus of relations. J. Symb. Logic 6, 73-89 (1941).

On the Complexity of the Equational Theory of Relational Action Algebras

Wojciech Buszkowski

Faculty of Mathematics and Computer Science, Adam Mickiewicz University in
Poznań
Faculty of Mathematics and Computer Science, University of Warmia and Mazury in
Olsztyn
buszko@amu.edu.pl

Abstract. Pratt [22] defines action algebras as Kleene algebras with residuals. In [9] it is shown that the equational theory of *-continuous action algebras (lattices) is Π_1^0−complete. Here we show that the equational theory of relational action algebras (lattices) is Π_1^0−hard, and some its fragments are Π_1^0−complete. We also show that the equational theory of action algebras (lattices) of regular languages is Π_1^0−complete.

1 Introduction

A *Kleene algebra* is an algebra $\mathcal{A} = (A, \vee, \cdot,^*, 0, 1)$ such that $(A, \vee, 0)$ is a join semilattice with the least element 0, $(A, \cdot, 1)$ is a monoid, product $\cdot$ distributes over join $\vee$, 0 is an annihilator for product, and * is a unary operation on A, fulfilling the conditions:

$$1 \vee aa^* \leq a^* \ , \ 1 \vee a^*a \leq a^* \ , \tag{1}$$

$$ab \leq b \Rightarrow a^*b \leq b \ , \ ba \leq b \Rightarrow ba^* \leq b \ , \tag{2}$$

for all $a, b \in A$. One defines: $a \leq b$ iff $a \vee b = b$. The notion of a Kleene algebra has been introduced by Kozen [15,16] to provide an algebraic axiomatization of the algebra of regular expressions. Regular expressions on an alphabet Σ can be defined as terms of the first-order language of Kleene algebras whose variables are replaced by symbols from Σ (treated as individual constants). Each regular expression α on Σ denotes a regular language $L(\alpha) \subseteq \Sigma^*$. The Kozen completeness theorem states that $L(\alpha) = L(\beta)$ if and only if $\alpha = \beta$ is valid in Kleene algebras.

The class of Kleene algebras is a quasi-variety, but not a variety. Redko [23] shows that the equations true for regular expressions cannot be axiomatized by any finite set of equations. Pratt [22] shows that the situation is different for Kleene algebras with residuals, called action algebras. *An action algebra*

R.A. Schmidt (Ed.): RelMiCS /AKA 2006, LNCS 4136, pp. 106–119, 2006.

is a Kleene algebra $\mathcal{A}$ supplied with two binary operations $/, \backslash$, fulfilling the equivalences:

$$ab \leq c \Leftrightarrow a \leq c/b \Leftrightarrow b \leq a\backslash c , \qquad (3)$$

for all $a, b, c \in A$. Operations $/, \backslash$ are called *the left* and *right residual*, respectively, with respect to product. Pratt writes $a \to b$ for $a\backslash b$ and $a \leftarrow b$ for a/b; we use the slash notation of Lambek [18]. Pratt [22] proves that the class of action algebras is a finitely based variety. Furthermore, in the language without residuals, the equations true in all action algebras are the same as those true in all Kleene algebras. Consequently, in the language with residuals, one obtains a finite, equational axiomatization of the algebra of regular expressions.

On the other hand, the logic of action algebras differs in many essential aspects from the logic of Kleene algebras. Although regular languages are (effectively) closed under residuals, the Kozen completeness theorem is not true for terms with residuals. For instance, since $L(a) = \{a\}$, then $L(a/a) = \{\epsilon\}$, while $a/a = 1$ is not true in action algebras (one only gets $1 \leq a/a$). It is known that $L(\alpha) = L(\beta)$ iff $\alpha = \beta$ is valid in relational algebras (α, β do not contain residuals). Consequently, the equational theory of Kleene algebras equals the equational theory of relational Kleene algebras. This is not true for action algebras (see below).

A Kleene algebra is said to be **-continuous*, if $xa^*y = \sup\{xa^ny : n \in \omega\}$, for all elements x, a, y. Relational algebras with operations defined in the standard way and algebras of (regular) languages are *-continuous. The equational theory of Kleene algebras equals the equational theory of *-continuous Kleene algebras. Again, it is not the case for action algebras. The equational theory of all action algebras is recursively enumerable (it is not known if it is decidable), while the equational theory of *-continuous action algebras is Π_1^0−complete [9], and consequently, it possesses no recursive axiomatization.

In this paper we study the complexity of relational action algebras and lattices. *An action lattice* is an action algebra $\mathcal{A}$ supplied with meet $\wedge$ such that $(A, \vee, \wedge)$ is a lattice; Kleene lattices are defined in a similar way. If $\mathcal{K}$ is a class of algebras, then $\mathrm{Eq}(\mathcal{K})$ denotes the equational theory of $\mathcal{K}$, this means, the set of all equations valid in $\mathcal{K}$. KA, KL, ACTA, ACTL will denote the classes of Kleene algebras, Kleene lattices, action algebras, and action lattices, respectively. KA* denotes the class of *-continuous Kleene algebras, and similarly for the other classes.

Let U be a set. $P(U^2)$ (the powerset of U^2) is the set of all binary relations on U. For $R, S \subseteq U^2$, one defines: $R \vee S = R \cup S$, $R \wedge S = R \cap S$, $R \cdot S = R \circ S$, $1 = I_U = \{(x, x) : x \in U\}$, $0 = \emptyset$, $R^0 = I_U$, $R^{n+1} = R^n \circ R$, $R^* = \bigcup_{n \in \omega} R^n$, and:

$$R/S = \{(x, y) \in U^2 : \{(x, y)\} \circ S \subseteq R\} , \qquad (4)$$

$$R\backslash S = \{(x, y) \in U^2 : R \circ \{(x, y)\} \subseteq S\} , \qquad (5)$$

$P(U^2)$ with so-defined operations and designated elements is an action lattice (it is a complete lattice). Algebras of this form will be called *relational action lattices*; without meet, they will be called *relational action algebras*. Omitting residuals, one gets relational Kleene lattices and algebras, respectively. RKA, RKL, RACTA, RACTL will denote the classes of relational Kleene algebras, relational Kleene lattices, relational action algebras and relational action lattices, respectively.

All relational algebras and lattices, mentioned above, are *-continuous. Consequently, Eq(KA)⊆Eq(KA*)⊆Eq(RKA), and similar inclusions are true for classes KL, KL*, RKL, classes ACTA, ACTA*, RACTA, and classes ACTL, ACTL*, RACTL. It is known that Eq(KA)=Eq(KA*)=Eq(RKA) (this follows from the Kozen completeness theorem and the fact mentioned in the third paragraph of this section). We do not know if Eq(KL)=Eq(KL*). All relational Kleene lattices are distributive lattices, but there exist nondistributive *-continuous Kleene lattices, which yields Eq(KL*)≠Eq(RKL). Since Eq(ACTA) is Σ_1^0, and Eq(ACTA*) is Π_1^0−complete, then Eq(ACTA)≠Eq(ACTA*); also, Eq(ACTL)≠Eq(ACTL*), for similar reasons [9].

It is easy to show that Eq(ACTA*) (resp. Eq(ACTL*)) is strictly contained in Eq(RACTA) (resp. Eq(RACTL)); see section 2. Then, Π_1^0−completeness of the former theory does not directly provide any information on the complexity of the latter. In section 3, we prove that Eq(RACTA) and Eq(RACTL) are Π_1^0−hard. The argument is similar to that in [9] which yields Π_1^0−hardness of Eq(ACTA*) and Eq(ACTL*): we show that the total language problem for context-free grammars is reducible to Eq(RACTL) and Eq(RACTA).

We do not know whether Eq(RACTA) and Eq(RACTL) are Π_1^0. In [21], it has been shown that Eq(ACTA*) and Eq(ACTL*) are Π_1^0, using an infinitary logic for ACTL* [9], which satisfies the cut-elimination theorem and a theorem on elimination of negative occurrences of *. The elimination procedure replaces each negative occurrence of α^* by the disjunction $1 \vee \alpha \vee \ldots \vee \alpha^n$, for some $n \in \omega$. As a result, one obtains expressions which contain 1. Unfortunately, the exact complexity of Eq(RACTA), Eq(RACTL) is not known; without * they are Σ_1^0. Andréka and Mikulás [1] prove a representation theorem for residuated meet semilattices which implies that, in language with $\wedge, \cdot, /, \backslash$ only, order formulas $\alpha \leq \beta$ valid in RACTL possess a cut-free, finitary axiomatization (the Lambek calculus admitting meet and empty antecedents of sequents), and consequently, the validity problem for such formulas is decidable (other results and proofs can be found in [10,8]). We use this fact in section 3 (lemma 2) to prove the results mentioned in the above paragraph and to prove Π_1^0−completeness of some fragments of Eq(RACTL).

In section 4, we consider analogous questions for the equational theory of action algebras (lattices) of regular languages, and we show that this theory is Π_1^0−complete. We use the completeness of the product-free fragment L (with $\wedge$) with respect to algebras of regular languages; the proof is a modification of the proof of finite model property for this system [5,7].

Our results show that there exists no finitary dynamic logic (like PDL), complete with respect to standard relational frames, which handles programs formed by residuals and regular operations. Programs with residuals can express the weakest prespecification and postspecification of a program and related conditions; see Hoare and Jifeng [13].

2   Sequent Systems

To provide a cut-free axiom system for the logic of *-continuous action algebras (lattices) it is expedient to consider *sequents* of the form $\Gamma \Rightarrow \alpha$ such that Γ is a finite sequence of terms (of the first-order language of these algebras), and α is a term. (Terms are often called *formulas*.) Given an algebra $\mathcal{A}$, an *assignment* is a homomorphism f from the term algebra to $\mathcal{A}$; one defines $f(\Gamma)$ by setting: $f(\epsilon) = 1$, $f(\alpha_1, \ldots, \alpha_n) = f(\alpha_1) \cdot \ldots \cdot f(\alpha_n)$. One says that $\Gamma \Rightarrow \alpha$ is *true* in $\mathcal{A}$ under f, if $f(\Gamma) \leq f(\alpha)$. Clearly, $f(\alpha) = f(\beta)$ iff both $f(\alpha) \leq f(\beta)$ and $f(\beta) \leq f(\alpha)$. A sequent is said to be *true* in $\mathcal{A}$, if it is true in $\mathcal{A}$ under any assignment, and *valid* in a class $\mathcal{K}$, if it is true in all algebras from $\mathcal{K}$. Since $\mathrm{Eq}(\mathcal{K})$ and the set of sequents valid in $\mathcal{K}$ are simply interpretable in each other, then the complexity of one of these sets equals the complexity of the other.

The sequents valid in ACTL* can be axiomatized by the following system. The axioms are:

$$\text{(I)}\ \alpha \Rightarrow \alpha \ , \text{(1)}\ \Rightarrow 1 \ , \text{(0)}\ \alpha, 0, \beta \Rightarrow \gamma \ , \tag{6}$$

and the inference rules are:

$$\frac{\Gamma, \alpha, \Delta \Rightarrow \gamma;\ \Gamma, \beta, \Delta \Rightarrow \gamma}{\Gamma, \alpha \vee \beta, \Delta \Rightarrow \gamma} \quad \frac{\Gamma \Rightarrow \alpha_i}{\Gamma \Rightarrow \alpha_1 \vee \alpha_2}, \tag{7}$$

$$\frac{\Gamma, \alpha_i, \Delta \Rightarrow \gamma}{\Gamma, \alpha_1 \wedge \alpha_2, \Delta \Rightarrow \gamma}, \quad \frac{\Gamma \Rightarrow \alpha;\ \Gamma \Rightarrow \beta}{\Gamma \Rightarrow \alpha \wedge \beta}, \tag{8}$$

$$\frac{\Gamma, \alpha, \beta, \Delta \Rightarrow \gamma}{\Gamma, \alpha \cdot \beta, \Delta \Rightarrow \gamma}, \quad \frac{\Gamma \Rightarrow \alpha;\ \Delta \Rightarrow \beta}{\Gamma, \Delta \Rightarrow \alpha \cdot \beta}, \tag{9}$$

$$\frac{\Gamma, \alpha, \Delta \Rightarrow \gamma;\ \Phi \Rightarrow \beta}{\Gamma, \alpha/\beta, \Phi, \Delta \Rightarrow \gamma}, \quad \frac{\Gamma, \beta \Rightarrow \alpha}{\Gamma \Rightarrow \alpha/\beta}, \tag{10}$$

$$\frac{\Gamma, \beta, \Delta \Rightarrow \gamma;\ \Phi \Rightarrow \alpha}{\Gamma, \Phi, \alpha \backslash \beta, \Delta \Rightarrow \gamma}, \quad \frac{\alpha, \Gamma \Rightarrow \beta}{\Gamma \Rightarrow \alpha \backslash \beta}, \tag{11}$$

$$\frac{\Gamma, \Delta \Rightarrow \alpha}{\Gamma, 1, \Delta \Rightarrow \alpha}, \tag{12}$$

$$\frac{(\Gamma, \alpha^n, \Delta \Rightarrow \beta)_{n \in \omega}}{\Gamma, \alpha^*, \Delta \Rightarrow \beta};\ \frac{\Gamma_1 \Rightarrow \alpha; \ldots; \Gamma_n \Rightarrow \alpha}{\Gamma_1, \ldots, \Gamma_n \Rightarrow \alpha^*}. \tag{13}$$

These rules are typical left- and right-introduction rules for Gentzen-style sequent systems. For each pair of rules, the left-hand rule will be denoted by (operation-L), and the right-hand rule by (operation-R). For instance, rules (7) will be denoted $(\vee-L)$ and $(\vee-R)$, respectively. Rule (12) will be denoted (1-L). Rule (*-L) is an infinitary rule (a kind of ω-rule); here α^n stands for the sequence of n copies of α. (*-R) denotes an infinite set of finitary rules: one for any fixed $n \in \omega$. For $n = 0$, (*-R) has the empty set of premises, so it is, actually, an axiom $\Rightarrow \alpha^*$; this yields $1 \Rightarrow \alpha^*$, by (1-L).

Without * and rules (13), the system is known as Full Lambek Calculus (FL); see Ono [19], Jipsen [14]. The rule (CUT):

$$\frac{\Gamma, \alpha, \Delta \Rightarrow \beta; \ \Phi \Rightarrow \alpha}{\Gamma, \Phi, \Delta \Rightarrow \beta} \tag{14}$$

is admissible in FL, this means: if both premises are provable in FL, then the conclusion is provable in FL [19]. The $(\cdot, /, \backslash)$-fragment of FL is the Lambek calculus L (admitting empty antecedents of sequents), introduced by Lambek [18] (in a form not admitting empty antecedents) who has proved the cut-elimination theorem for L.

A *residuated lattice* is an algebra $\mathcal{A} = (A, \vee, \wedge, \cdot, /, \backslash, 0, 1)$ such that $(A, \vee, \wedge)$ is a lattice with the least element 0, $(A, \cdot, 1)$ is a monoid, and $/, \backslash$ are residuals for product (they fulfill (3)). It is known that FL is complete with respect to residuated lattices: a sequent is provable in FL iff it is valid in the class of residuated lattices. A *residuated monoid* is a structure $\mathcal{A} = (A, \leq, \cdot, /, \backslash, 1)$ such that $(A, \leq)$ is a poset, $(A, \cdot, 1)$ is a monoid, and $/, \backslash$ are residuals for product. L is complete with respect to residuated monoids. These completeness theorems can be proved in a standard way: soundness is obvious, and completeness can be shown by the construction of a Lindenbaum algebra. Residuated monoids and lattices are applied in different areas of logic and computer science; see e.g. [19,20,6].

The following monotonicity conditions are true in all residuated monoids: if $a \leq c$ and $b \leq d$, then $ab \leq cd$, $a/d \leq c/b$, $d \backslash a \leq b \backslash c$ (in lattices also: $a \vee b \leq c \vee d$, $a \wedge b \leq c \wedge d$, in action algebras also: $a^* \leq c^*$).

FL with * and rules (13) has been introduced in [9] and denoted by ACTω. The set of provable sequents can be defined in the following way. For a set X, of sequents, $C(X)$ is defined as the set of all sequents derivable from sequents from X by a single application of some inference rule (axioms are treated as inference rules with the empty set of premises). Then, $C(\emptyset)$ is the set of all axioms. One defines a transfinite chain C_ζ, for ordinals ζ, by setting: $C_0 = \emptyset$, $C_{\zeta+1} = C(C_\zeta)$, $C_\lambda = \bigcup_{\zeta < \lambda} C_\zeta$. Since C is a monotone operator and $C_0 \subseteq C_1$, then $C_\zeta \subseteq C_{\zeta+1}$, for all ζ, and consequently, $C_\zeta \subseteq C_\eta$ whenever $\zeta < \eta$. The join of this chain equals the set of sequents provable in ACTω. *The rank* of a provable sequent equals the least ζ such that this sequent belongs to C_ζ.

The cut-elimination theorem for ACTω is proved in [21] by a triple induction: (1) on the complexity of formula α in (CUT), (2) on the rank of $\Gamma, \alpha, \Delta \Rightarrow \beta$, (3) on the rank of $\Phi \Rightarrow \alpha$ (following an analogous proof for L in [4]). Let us show one case of induction (1): $\alpha = \gamma^*$. Assume that $\Gamma, \alpha, \Delta \Rightarrow \beta$ and $\Phi \Rightarrow \alpha$ are provable. We start induction (2). If the left premise is an axiom (I), then the conclusion of (CUT) is the right premise. If the left premise is an axiom (0), then the conclusion of (CUT) is also an axiom (0). Assume that the left premise gets its rank on the basis of an inference rule R; then, each premise of R is of a smaller rank. If R is any rule, not introducing the designated occurrence of α, then we directly apply the hypothesis of induction (2). If R introduces the designated occurrence of α, then R is (*-L) with premises $\Gamma, \gamma^n, \Delta \Rightarrow \beta$, for all $n \in \omega$. We start induction (3). If $\Phi \Rightarrow \alpha$ is an axiom (I), then the conclusion of (CUT) is the left premise of (CUT). If $\Phi \Rightarrow \alpha$ is an axiom (0), then the conclusion of (CUT) is also an axiom (0). If $\Phi \Rightarrow \alpha$ is a conclusion of (*-R), then the premises are $\Phi_1 \Rightarrow \gamma, \ldots, \Phi_n \Rightarrow \gamma$, for some $n \in \omega$. For $n = 0$, we get $\Phi = \epsilon$, and the conclusion of (CUT) is the premise of (*-L) for $n = 0$. For $n > 0$, one of the premises of (*-L) is $\Gamma, \gamma^n, \Delta \Rightarrow \beta$, and we use n times the hypothesis of induction (1). If $\Phi \Rightarrow \alpha$ is a conclusion of a rule different from (*-R), then we directly apply the hypothesis of induction (3).

Since the rule (CUT) is admissible in ACTω, then a standard argument yields the completeness of ACTω with respect to *-continuous action lattices [21]. Soundness is obvious, and completeness can be shown by the construction of a Lindenbaum algebra. Using (1-L),(*-L) and (*-R), one easily proves $1 \Rightarrow \alpha^*$, $\alpha, \alpha^* \Rightarrow \alpha^*$ and, using (CUT), derives the following rules:

$$\frac{\alpha, \beta \Rightarrow \beta}{\alpha^*, \beta \Rightarrow \beta}, \frac{\beta, \alpha \Rightarrow \beta}{\beta, \alpha^* \Rightarrow \beta}, \tag{15}$$

and consequently, the Lindenbaum algebra is an action lattice. By (*-L), it is *-continuous.

Since ACTω is cut-free, then it possesses the subformula property: every provable sequent admits a proof in which all sequents consist of subformulas of formulas appearing in this sequent. In particular, ACTω is a conservative extension of all its fragments, obtained by a restriction of the language, e.g. L, FL, the $\vee$−free fragment, the $\wedge$−free fragment, and so on. All *-free fragments are finitary cut-free systems, admitting a standard proof-search decision procedure. So, they are decidable.

Now, we show that Eq(ACTA*)$\neq$Eq(RACTA). In relational algebras, for $R, S \subseteq I_U$, we have $R \circ S = R \cap S$. Fix a variable p. In L, from $p \Rightarrow p$, one infers $\Rightarrow p/p$, by (/-R). Then, $1 \Rightarrow 1$ yields $1/(p/p) \Rightarrow 1$, by (/-L). So, the sequent $1/(p/p) \Rightarrow (1/(p/p)) \cdot (1/(p/p))$ is valid in RACTA. It is not valid in ACTA*, since it is not provable in L. (Use the proof-search procedure; notice that $\Rightarrow p$, $p/p \Rightarrow 1$, $\Rightarrow 1/(p/p)$ are not provable.) The same example shows Eq(ACTL*)$\neq$Eq(RACTL) (another proof: the distribution of $\wedge$ over $\vee$ is not valid in ACTL*, since it is not provable in FL).

We define positive and negative occurrences of subterms in terms: α is positive in α; if γ is positive (resp. negative) in α or β, then it is positive (resp. negative) in $\alpha \vee \beta$, $\alpha \wedge \beta$, $\alpha \cdot \beta$, α^*; if γ is positive (resp. negative) in β, then it is positive (resp. negative) in β/α, $\alpha\backslash\beta$; if γ is positive (resp. negative) in α, then it is negative (resp. positive) in β/α, $\alpha\backslash\beta$.

For $n \in \omega$, let $\alpha^{\leq n}$ denote $\alpha^0 \vee \ldots \vee \alpha^n$; here α^i stands for the product of i copies of α and α^0 is the constant 1. We define two term transformations P_n, N_n, for any $n \in \omega$ [9]. Roughly, $P_n(\gamma)$ (resp. $N_n(\gamma)$) arises from γ by replacing any positive (resp. negative) subterm of the form α^* by $\alpha^{\leq n}$.

$$P_n(\alpha) = N_n(\alpha) = \alpha \ , \text{ if } \alpha \text{ is a variable or a constant,} \tag{16}$$

$$P_n(\alpha \circ \beta) = P_n(\alpha) \circ P_n(\beta) \ , \text{ for } \circ = \vee, \wedge, \cdot \ , \tag{17}$$

$$N_n(\alpha \circ \beta) = N_n(\alpha) \circ N_n(\beta) \ , \text{ for } \circ = \vee, \wedge, \cdot \ , \tag{18}$$

$$P_n(\alpha/\beta) = P_n(\alpha)/N_n(\beta) \ , P_n(\alpha\backslash\beta) = N_n(\alpha)\backslash P_n(\beta) \ , \tag{19}$$

$$N_n(\alpha/\beta) = N_n(\alpha)/P_n(\beta) \ , N_n(\alpha\backslash\beta) = P_n(\alpha)\backslash N_n(\beta) \ , \tag{20}$$

$$P_n(\alpha^*) = (P_n(\alpha))^{\leq n} \ , N_n(\alpha^*) = (N_n(\alpha))^* \ . \tag{21}$$

For a sequent $\Gamma \Rightarrow \alpha$, we set $N_n(\Gamma \Rightarrow \alpha) = P_n(\Gamma) \Rightarrow N_n(\alpha)$, where:

$$P_n(\epsilon) = \epsilon \ , P_n(\alpha_1, \ldots, \alpha_k) = P_n(\alpha_1), \ldots, P_n(\alpha_k) \ . \tag{22}$$

A term occurs positively (resp. negatively) in $\Gamma \Rightarrow \alpha$ if it occurs positively (resp. negatively) in α or negatively (resp. positively) in Γ.

Palka [21] proves the following theorem on elimination of negative occurrences of $*$: for any sequent $\Gamma \Rightarrow \alpha$, this sequent is provable in ACTω iff, for all $n \in \omega$, the sequent $N_n(\Gamma \Rightarrow \alpha)$ is provable in ACTω.

As a consequence of this theorem, the set of sequents provable in ACTω is Π_1^0. Indeed, the condition

$$N_n(\Gamma \Rightarrow \alpha) \text{ is provable in ACT}\omega \tag{23}$$

is recursive, since $N_n(\Gamma \Rightarrow \alpha)$ contains no negative occurrences of $*$, whence it is provable in ACTω iff it is provable in ACTω^-, i.e. ACTω without rule (*-L), and the latter system is finitary and admits an effective proof-search procedure. Actually, no result of the present paper relies upon Palka's theorem except for some remark at the end of section 3.

3   Eq(RACTL) and Eq(RACTA) Are Π_1^0−Hard

A *context-free grammar* is a quadruple $G = (\Sigma, N, s, P)$ such that Σ, N are disjoint, finite alphabets, $s \in N$, and P is a finite set of production rules of the form $p \mapsto x$ such that $p \in N$, $x \in (\Sigma \cup N)^*$. Symbols in Σ are called *terminal* symbols and symbols in N are called *nonterminal* symbols. The relation $\Rightarrow_G$ is defined as follows: $x \Rightarrow_G y$ iff, for some $z, u, v \in (\Sigma \cup N)^*$, $p \in N$, we have $x = upv$, $y = uxv$ and $(p \mapsto x) \in P$. The relation $\Rightarrow_g^*$ is the reflexive and transitive closure of $\Rightarrow_G$. *The language* of G is the set:

$$L(G) = \{x \in \Sigma^* : s \Rightarrow_G x\} \ . \tag{24}$$

A context-free grammar G is said to be ϵ−free, if $x \neq \epsilon$, for any rule $p \mapsto x$ in P. If G is ϵ−free, then $\epsilon \notin L(G)$. The following problem is Π_1^0−complete [12]: for any context-free grammar G, decide if $L(G) = \Sigma^*$. Since the problem if $\epsilon \in L(G)$ is decidable, and every grammar G can be effectively transformed into an ϵ−free grammar G' such that $L(G') = L(G) - \{\epsilon\}$, then also the following problem is Π_1^0-complete: for any $\epsilon-$free context-free grammar G, decide if $L(G) = \Sigma^+$ [9].

Types will be identified with $(/)-$terms of the language of ACTω, this means, terms formed out of variables by means of $/$ only. A *Lambek categorial grammar* is a tuple $G = (\Sigma, I, s)$ such that Σ is a finite alphabet, I is a finite relation between symbols from Σ and types, and s is a designated variable. For $a \in \Sigma$, $I(a)$ denotes the set of all types α such that $aI\alpha$. (The relation I is called the initial type assignment of G.) For a string $a_1 \ldots a_n \in \Sigma^+$, $a_i \in \Sigma$, and a type α, we write $a_1 \ldots a_n \to_G \alpha$ if there are $\alpha_1 \in I(a_1)$, ..., $\alpha_n \in I(a_n)$ such that $\alpha_1, \ldots, \alpha_n \Rightarrow \alpha$ is provable in L. We define the language of G as the set of all $x \in \Sigma^+$ such that $x \to_G s$. (Notice that we omit commas between symbols in strings on Σ, but we write them in sequences of terms appearing in sequents.) In general, Lambek categorial grammars admit types containing $\cdot, \backslash$ and, possibly, other operations [6], but we do not employ such grammars in this paper.

It is well-known that, for any $\epsilon-$free context-free grammar G, one can effectively construct a Lambek categorial grammar G' with the same alphabet Σ and such that $L(G) = L(G')$; furthermore, the relation I of G' employs very restricted types only: of the form p, p/q, $(p/q)/r$, where p, q, r are variables. This fact has been proved in [2] for classical categorial grammars and extended to Lambek categorial grammars by several authors; see e.g. [4,9]. One uses the fact that, for sequents $\Gamma \Rightarrow s$ such that Γ is a finite sequence of types of the above form and s is a variable, Γ reduces to s in the sense of classical categorial grammars iff $\Gamma \Rightarrow s$ is provable in L.

Consequently, the problem if $L(G) = \Sigma^+$, for Lambek categorial grammars G, is Π_1^0-complete. In [9] it is shown that this problem is reducible to the decision problem for ACTω. Then, Eq(ACTL*) is Π_1^0-hard, and the same holds for Eq(ACTA*). Below we show that this reduction also yields the Π_1^0-hardness of Eq(RACTL) and Eq(RACTA).

Let $G = (\Sigma, I, s)$ be a Lambek categorial grammar. We can assume $I_G(a) \neq \emptyset$, for any $a \in \Sigma$; otherwise $L(G) \neq \Sigma^+$ immediately. We can also assume that all types involved in I are of one of the forms: p, p/q, $(p/q)/r$, where p, q, r are variables. Fix $\Sigma = \{a_1, \ldots, a_k\}$, where $a_i \neq a_j$ for $i \neq j$. Let $\alpha_1^i, \ldots, \alpha_{n_i}^i$ be all distinct types $\alpha \in I(a_i)$. For any $i = 1, \ldots, k$, we form a term $\beta_i = \alpha_1^i \wedge \ldots \wedge \alpha_{n_i}^i$. We also define a term $\gamma(G) = \beta_1 \vee \ldots \vee \beta_k$. The following lemma has been proved in [9].

Lemma 1. $L(G) = \Sigma^+$ *iff* $(\gamma(G))^*, \gamma(G) \Rightarrow s$ *is provable in* ACTω.

Proof. For the sake of completeness, we sketch the proof. $L(G) = \Sigma^+$ iff, for all $n \geq 1$ and all sequences $(i_1, \ldots, i_n)$ of integers from the set $[k] = \{1, \ldots, k\}$, $a_{i_1} \ldots a_{i_n} \to_G s$. The latter condition is equivalent to the following: for any $j = 1, \ldots, n$, there exists $\alpha_{l_j}^{i_j} \in I(a_{i_j})$ such that $\alpha_{l_1}^{i_1}, \ldots, \alpha_{l_n}^{i_n} \Rightarrow s$ is provable in L. The latter condition is equivalent to the following: $\beta_{i_1}, \ldots, \beta_{i_n} \Rightarrow s$ is provable in FL. One uses the following fact: if $\Gamma \Rightarrow \alpha$ is a $(\wedge, /)-$sequent in which all occurrences of $\wedge$ are negative, and $\gamma_1 \wedge \gamma_2$ occurs in this sequent (as a subterm of a term), then $\Gamma \Rightarrow \alpha$ is provable in FL iff both $\Gamma' \Rightarrow \alpha'$ and $\Gamma'' \Rightarrow \alpha''$ are provable in FL, where $\Gamma' \Rightarrow \alpha'$ (resp. $\Gamma'' \Rightarrow \alpha''$) arises from $\Gamma \Rightarrow \alpha$ by replacing the designated occurrence of $\gamma_1 \wedge \gamma_2$ by γ_1 (resp. γ_2). Now, for $n \geq 1$, $\beta_{i_1}, \ldots, \beta_{i_n} \Rightarrow s$ is provable in FL, for all sequents $(i_1, \ldots, i_n) \in [k]^n$, iff $(\gamma(G))^n \Rightarrow s$ is provable in FL (here we use the distribution of product over join). By (*-L), (*-R), the latter condition is equivalent to the following: $(\gamma(G))^*, \gamma(G) \Rightarrow s$ is provable in ACTω. $\square$

Andréka and Mikulás [1] prove that every residuated meet semilattice is embeddable into a relational algebra. The embedding h does not preserve 1; one only gets: $1 \leq a$ iff $I_U \subseteq h(a)$. It follows that the $(\wedge, \cdot, /, \backslash)$−fragment of FL is (even strongly) complete with respect to relational algebras, which is precisely stated by the following lemma (from [1]; also see [10,8] for different proofs).

Lemma 2. *Let $\Gamma \Rightarrow \alpha$ be a $(\wedge, \cdot, /, \backslash)$−sequent of the language of FL. Then, $\Gamma \Rightarrow \alpha$ is provable in FL iff it is valid in RACTL.*

We use lemmas 1 and 2 to prove the following theorem.

Theorem 1. *Eq(RACTL) is Π_1^0−hard.*

Proof. We show that $(\gamma(G))^*, \gamma(G) \Rightarrow s$ is provable in ACTω iff the sequent is valid in RACTL. The implication ($\Rightarrow$) is obvious. Assume that $(\gamma(G))^*, \gamma(G) \Rightarrow s$ is not provable in ACTω. Then, for some $n \in \omega$, $(\gamma(G))^n, \gamma(G) \Rightarrow s$ is not provable in ACTω, whence it is not provable in FL (it is *-free). By (CUT) and ($\cdot$-R), $(\gamma(G))^n \cdot \gamma(G) \Rightarrow s$ is not provable in FL The term $(\gamma(G))^n \cdot \gamma(G)$ is equivalent in FL to the disjunction of all terms $\beta_{i_1} \cdot \cdots \cdot \beta_{i_{n+1}}$ such that $(i_1, \ldots, i_{n+1}) \in [k]^{n+1}$. By ($\vee$−L), ($\vee$−R) and (CUT), a sequent $\gamma_1 \vee \ldots \vee \gamma_m \Rightarrow \gamma$ is provable in FL iff all sequents $\gamma_i \Rightarrow \gamma$, for $i = 1, \ldots, m$, are provable in FL. Consequently, there exists a sequence $(i_1, \ldots, i_{n+1}) \in [k]^{n+1}$ such that $\beta_{i_1} \cdot \cdots \cdot \beta_{i_{n+1}} \Rightarrow s$ is not provable in FL. The latter sequent does not contain operation symbols other than $\wedge, \cdot, /, \backslash$, so it is not valid in RACTL, by lemma 2. Consequently, $(\gamma(G))^n, \gamma(G) \Rightarrow s$ is not valid in RACTL. Then, $(\gamma(G))^*, \gamma(G) \Rightarrow s$ is not valid in RACTL (we use the fact that $f(\alpha^n) \subseteq f(\alpha^*)$, for any assigment f and any formula α). Using lemma 1, we obtain: $L(G) = \Sigma^+$ iff $(\gamma(G))^*, \gamma(G) \Rightarrow s$ is valid in RACTL. $\square$

For RACTA, we need a modified reduction. We use a lemma, first proved in [4].

Lemma 3. *Let $\alpha_1, \ldots, \alpha_n$ be types, and let s be a variable. Then, $\alpha_1, \ldots, \alpha_n \Rightarrow s$ is provable in L iff $s/(s/\alpha_1), \ldots, s/(s/\alpha_n) \Rightarrow s$ is provable in L.*

Proof. We outline the proof. A type α/Γ is defined by induction on the length of Γ: $\alpha/\epsilon = \alpha$, $\alpha/(\Gamma\beta) = (\alpha/\beta)/\Gamma$. So, $p/(qr) = (p/r)/q$. We consider the $(/)$−fragment of L. One shows: if $(s/\beta_1 \ldots \beta_k), \Delta \Rightarrow s$ is provable in this system, then there exist $\Delta_1, \ldots, \Delta_k$ such that $\Delta = \Delta_1 \ldots \Delta_k$ and, for each $i = 1, \ldots, k$, $\Delta_i \Rightarrow \alpha_i$ is provable (use induction on cut-free proofs; the converse implication also holds, by (I), (/−L)).

The 'only if' part of the lemma holds, by applying (/−R), (I), (/−L) n times. Now, assume that the right-hand sequent is provable. Denote $\beta_i = s/(s/\alpha_i)$.

By the above paragraph, $\beta_2, \ldots, \beta_n \Rightarrow s/\alpha_1$ is provable, so $\beta_2, \ldots, \beta_1, \alpha_1 \Rightarrow s$ is provable (the rule $(/-R)$ is reversible, by (CUT) and the provable sequent $\alpha/\beta, \beta \Rightarrow \alpha$). Repeat this step $n-1$ times. $\qquad\qquad\qquad\square$

Let $G = (\Sigma, I, s)$ be a Lambek categorial grammar. We construct a Lambek categorial grammars $G' = (\Sigma, I', s)$ such that I' assigns $s/(s/\alpha)$ to $a_i \in \Sigma$ iff G assigns α to a_i. By lemma 3, $L(G') = L(G)$. For G', we construct terms $(\alpha_j^i)'$, $(\beta_i)'$ and $\gamma(G')$ in a way fully analogous to the construction of α_j^i, β_i and $\gamma(G)$. Now, the term $(\beta_i)'$ is of the form:

$$(s/(s/\alpha_1^i)) \wedge \ldots \wedge (s/(s/\alpha_{n_i}^i)) . \tag{25}$$

Using the equation $(a/b) \wedge (a/c) = a/(b \vee c)$, valid in residuated lattices, we can transform the above term into an equivalent (in FL) $\wedge$−free term:

$$s/[(s/\alpha_1^i) \vee \ldots \vee (s/\alpha_{n_i}^i)]. \tag{26}$$

Let $\delta(G')$ be the term arising from $\gamma(G')$ by transforming each constituent $(\beta_i)'$ as above. Then, $f(\delta(G')) = f(\gamma(G'))$, for any assignment f.

Theorem 2. *Eq(RACTA) is Π_1^0−hard.*

Proof. $L(G) = \Sigma^+$ iff $L(G') = \Sigma^+$. As in the proofs of lemma 1 and theorem 1, one shows that the second condition is equivalent to: $(\gamma(G'))^*, \gamma(G') \Rightarrow s$ is valid in RACTL. The latter condition is equivalent to: $(\delta(G'))^*, \delta(G') \Rightarrow s$ is valid in RACTL. But the latter sequent is $\wedge$−free, whence it is valid in RACTL iff it is valid in RACTA. $\qquad\qquad\qquad\square$

We can also eliminate $\vee$ (preserving $\wedge$). Using the equation $(a \vee b)^* = (a^*b)^*a^*$, valid in all Kleene algebras, we can transform $(\gamma(G))^*$ into an equivalent (in ACTω) term $\phi(G)$, containing $*, \wedge, \cdot, /$ only. Then, $(\gamma(G))^*, \gamma(G) \Rightarrow s$ is valid in RACTL iff $\phi(G), \gamma(G) \Rightarrow s$ is valid in RACTL iff $\phi(G) \Rightarrow s/\gamma(G)$ is valid in RACTL, and $s/\gamma(G)$ is equivalent to a $\vee$−free term (see the equation between (25) and (26)). Since $a \leq b$ iff $a \wedge b = a$, then we can reduce $L(G) = \Sigma^+$ to a $\vee$−free equation.

Corollary 1. *The $\vee$−free fragment of Eq(RACTL) is Π_1^0−hard.*

We have found a lower bound for the complexity of Eq(RACTL): it is at least Π_1^0. We did not succeed in determining the upper bound. Both 1 and $\vee$ cause troubles. In section 2, we have shown a sequent with 1 which is valid in RACTL, but not valid in ACTL*. According to the author's knowledge, the precise complexity of the equational theory of relational residuated lattices (upper semilattices) is not known; it must be Σ_1^0, since valid equations can be faithfully interpreted as valid formulas of first-order logic.

We can show some Π_1^0−complete fragments of Eq(RACTL). For instance, the set of all sequents of the form $\alpha, \gamma^*, \beta \Rightarrow p$, with α, β, γ being finite disjunctions of $(/, \backslash, \wedge)$−terms, valid in RACTL is Π_1^0−complete. This sequent is valid iff, for all $n \in \omega$, $\alpha, \gamma^n, \beta \Rightarrow \delta$ is valid, and the latter sequents are valid iff they are provable in FL (see the proof of theorem 1). Consequently, this set of sequents is Π_1^0. It is $\Pi_1^{0)}$−hard, again by the proof of theorem 1. This set can be extended as follows.

A term is said to be *good* if it is formed out of $(\wedge, /, \backslash)$−terms by $\cdot$ and $*$ only. A sequent $\Gamma \Rightarrow \alpha$ is said to be *nice* if it is a $(\wedge, \cdot, ^*, /, \backslash)$−sequent, and any negatively occurring term of the form β^* occurs in this sequent within a good term γ, which appears either as an element of Γ, or in a context δ/γ or $\gamma\backslash\delta$. Using the *-elimination theorem [21], one can prove that the set of nice sequents valid in RACTL is Π_1^0−complete.

4   Algebras of Regular Languages

A language on Σ is a set $L \subseteq \Sigma^*$. $P(\Sigma^*)$ is the set of all languages on Σ; it is a complete action lattice with operations and designated elements, defined as follows: $L_1 \vee L_2 = L_1 \cup L_2$, $L_1 \wedge L_2 = L_1 \cap L_2$, $L_1 \cdot L_2 = \{xy : x \in L_1, y \in L_2\}$, $1 = \{\epsilon\}$, $0 = \emptyset$, $L^0 = \{\epsilon\}$, $L^{n+1} = L^n \cdot L$, $L^* = \bigcup_{n \in \omega} L^n$, and:

$$L_1/L_2 = \{x \in \Sigma^* : \{x\} \cdot L_2 \subseteq L_1\} \ , \tag{27}$$

$$L_1\backslash L_2 = \{x \in \Sigma^* : L_1 \cdot \{x\} \subseteq L_2\} \ . \tag{28}$$

By LAN we denote the class of all action lattices of the form $P(\Sigma^*)$, for finite alphabets Σ. We add symbols from Σ to the language of ACTω as new individual constants. Regular expressions on Σ can be defined as variable-free terms without meet and residuals. An assignment $L(a) = \{a\}$, for $a \in \Sigma$, is uniquely extended to all regular expressions; it is a homomorphism from the term algebra to $P(\Sigma^*)$. Languages of the form $L(\alpha)$, α is a regular expression on Σ, are called *regular languages* on Σ. By REGL(Σ) we denote the set of all regular languages on Σ. It is well-known that REGL(Σ) is a subalgebra of the action lattice $P(\Sigma^*)$, whence it is a *-continuous action lattice. By REGLAN we denote the class of all action lattices REGL(Σ), for finite alphabets Σ.

We will show that Eq(REGLAN) is Π_1^0−complete. It is quite easy to show that Eq(REGLAN) is Π_1^0. Since regular languages are effectively closed under meet and residuals, $L(\alpha)$ can be computed for all variable-free terms α with individual constants from Σ. An equation $\alpha = \beta$ is valid in REGLAN iff $L(\sigma(\alpha))) = L(\sigma(\beta))$, for all finite alphabets Σ and all substitutions σ assigning regular expressions on Σ to variables.

We note that Eq(RACTL) is different from Eq(REGLAN) and Eq(LAN). The sequent $p, 1/p \Rightarrow 1$ is valid in LAN, and consequently, in REGLAN. Let f be an assignment of terms in $P(\Sigma^*)$. If $f(p) = \emptyset$, then $f(p, 1/p) = \emptyset$. If $f(p) = \{\epsilon\}$, then $f(1/p) = \{\epsilon\}$ and $f(p, 1/p) = \{\epsilon\} = f(1)$. Otherwise $f(1/p) = \emptyset$ and $f(p, 1/p) = \emptyset$. This sequent is not valid in RACTL. Let $U = \{a, b\}$, $a \neq b$, and $f(p) = \{(a, b)\}$. Then, $f(1/p) = \{(a, b), (b, a), (b, b)\}$, so $f(p, 1/p) = \{(a, a), (a, b)\}$ is not contained in I_U.

In [5], it has been shown that the $(\wedge, /, \backslash)$−fragment of FL possesses finite model property. The proof yields, actually, the completeness of this fragment with respect to so-called co-finite models $(P(\Sigma^*), f)$ such that $f(p)$ is a co-finite subset of Σ^*, for any variable p. Then, $f(p)$ is a regular language on Σ. We obtain the following lemma. The proof is a modification of the proof of finite model property of this fragment, given in [7].

Lemma 4. *Let $\Gamma \Rightarrow \alpha$ be a $(\wedge, /, \backslash)-sequent$. Then, $\Gamma \Rightarrow \alpha$ is provable in FL iff it is valid in REGLAN.*

Proof. The 'only if' part is obvious. For the 'if' part, assume that $\Gamma \Rightarrow \alpha$ is not provable. Let T be the set of all subterms appearing in this sequent. We consider languages on the alphabet T. An assignment f_n, $n \in \omega$, is defined as follows: for any variable p, $f_n(p)$ equals the set of all $\Delta \in T^*$ such that either $v(\Delta) > n$, or $\Delta \Rightarrow p$ is provable ($v(\Delta)$ denotes the total number of occurrences of variables in Δ). As usual, f_n is extended to a homomorphism from the term algebra to $P(T^*)$. Since all languages $f_n(p)$ are co-finite, then all languages $f_n(\beta)$ are regular. If $\Delta \in T^*$, $v(\Delta) > n$, then $\Delta \in f_n(\beta)$, for all terms β (easy induction on β).

By induction on $\beta \in T$, we prove: (i) if $v(\Delta) \leq n - v(\beta)$ and $\Delta \in f_n(\beta)$, then $\Delta \Rightarrow \beta$ is provable, (ii) if $v(\beta) \leq v(\Delta)$ and $\Delta \Rightarrow \beta$ is provable, then $\Delta \in f_n(\beta)$. For $\beta = p$, (i) and (ii) follow from the definition of f_n.

Let $\beta = \gamma/\delta$. Assume $v(\Delta) \leq n - v(\beta)$ and $\Delta \in f_n(\beta)$. Since $v(\delta) \leq v(\delta)$, then $\delta \in f_n(\delta)$, by (I) and the induction hypothesis (use (ii)). So, $(\Delta\delta) \in f_n(\gamma)$, by the definition of residuals in $P(T^*)$. Since $v(\Delta\delta) \leq n - v(\gamma)$, then $\Delta, \delta \Rightarrow \gamma$ is provable (use (i)). By $(/-R)$, $\Delta \Rightarrow \beta$ is provable. Assume that $v(\beta) \leq v(\Delta)$ and $\Delta \Rightarrow \beta$ is provable. By the reversibility of $(/-R)$, $\Delta, \delta \Rightarrow \gamma$ is provable. Let $\Phi \in f_n(\delta)$. Case 1: $v(\Phi) > n - v(\delta)$. Then, $v(\Delta\Phi) > n$, whence $(\Delta\Phi) \in f_n(\gamma)$. Case 2: $v(\Phi) \leq n - v(\delta)$. Then, $\Phi \Rightarrow \delta$ is provable, by the induction hypothesis (use (i)), and consequently, $\Delta, \Phi \Rightarrow \gamma$ is provable, by (CUT). Since $v(\gamma) \leq v(\Delta\Phi)$, then $(\Delta\Phi) \in f_n(\gamma)$, by the induction hypothesis (use (ii)). So, $\Delta \in f_n(\beta)$. The case $\beta = \delta\backslash\gamma$ is dual.

Let $\beta = \gamma \wedge \delta$. Assume $v(\Delta) \leq n - v(\beta)$ and $\Delta \in f_n(\beta)$. Then, $v(\Delta) \leq n - v(\gamma)$ and $\Delta \in f_n(\gamma)$. Also $v(\Delta) \leq n - v(\delta)$ and $\Delta \in f_n(\delta)$. By the induction hypothesis, $\Delta \Rightarrow \gamma$ and $\Delta \Rightarrow \delta$ are provable, and consequently, $\Delta \Rightarrow \beta$ is provable, by $(\wedge-R)$. Assume that $v(\beta) \leq v(\Delta)$ and $\Delta \Rightarrow \beta$ is provable. Since $\beta \Rightarrow \gamma$ and $\beta \Rightarrow \delta$ are provable, by (I) and $(\wedge-L)$, then $\Delta \Rightarrow \gamma$ and $\Delta \Rightarrow \delta$ are provable, by (CUT). We have $v(\gamma) \leq v(\Delta)$ and $v(\delta) \leq v(\Delta)$, and consequently, $\Delta \in f_n(\gamma)$ and $\Delta \in f_n(\delta)$, by the induction hypothesis, which yields $\Delta \in f_n(\beta)$.

Take $n = v(\Gamma \Rightarrow \alpha)$. Let $\Gamma = \alpha_1 \ldots \alpha_k$. Since $v(\alpha_i) \leq v(\alpha_i)$, then $\alpha_i \in f_n(\alpha_i)$, by (I) and (ii). Consequently, $\Gamma \in f_n(\Gamma)$. Since $v(\Gamma) = n - v(\alpha)$, then $\Gamma \notin f_n(\alpha)$, by the assumption and (i) (this also holds for $\Gamma = \epsilon$). Consequently, $\Gamma \Rightarrow \alpha$ is not valid in REGLAN. $\qquad\square$

Theorem 3. *Eq(REGLAN) is $\Pi_1^0-complete$.*

Proof. We know that this set is Π_1^0. We show that it is Π_1^0-hard. We return to lemma 1 in section 3. We show that $(\gamma(G))^*, \gamma(G) \Rightarrow s$ is provable in ACTω iff this sequent is valid in REGLAN. The implication $(\Rightarrow)$ is obvious. To prove

($\Leftarrow$) assume that $(\gamma(G))^*, \gamma(G) \Rightarrow s$ is not provable in ACTω. As in the proof of theorem 1, we show that there exists a sequence $(i_1, \ldots, i_n) \in [k]^n$, $n \geq 1$, such that $\beta_{i_1} \cdots \beta_{i_n} \Rightarrow s$ is not provable in FL. By ($\cdot$-L), $\beta_{i_1}, \ldots, \beta_{i_n} \Rightarrow s$ is not provable in FL. By lemma 4, the latter sequent is not valid in REGLAN. As in the proof of theorem 1, we show that $(\gamma(G))^*, \gamma(G) \Rightarrow s$ is not valid in REGLAN. So, $L(G) = \Sigma^+$ iff $(\gamma(G))^*, \gamma(G) \Rightarrow s$ is valid in REGLAN. $\square$

We note that Eq(LAN) belongs to a higher complexity class. The Horn formulas valid in LAN can be expressed by equations valid in LAN. Notice that $\alpha \leq \beta$ is true iff $1 \leq \beta/\alpha$ is true. Also the conjunction of formulas $1 \leq \alpha_i$, $i = 1, \ldots, n$, is true iff $1 \leq \alpha_1 \wedge \cdots \wedge \alpha_n$ is true. Finally, the implication 'if $1 \leq \alpha$ then $1 \leq \beta$'x is true iff $1 \wedge \alpha \leq \beta$ is true.

The Horn theory of LAN, restricted to $(/, \backslash)$−terms, is Σ_1^0−complete [3]. The proof of theorem 3 yields the Π_1^0−hardness of Eq(LAN); so, it is not Σ_1^0. If it were Π_1^0, then this restricted Horn theory of LAN would be recursive. So, Eq(LAN) is neither Π_1^0, nor Σ_1^0.

In [17,11] the Horn theory of KA* and the Horn theory of RKA are shown to be Π_1^1−complete. This yields a lower bound for the complexity of Horn theories of ACTA* and RACTA (every *-continuous Kleene algebra is embeddable into a complete, whence *-continuous, action lattice [9]).

References

1. H. Andréka and S. Mikulaś, Lambek calculus and its relational semantics: completeness and incompleteness, *Journal of Logic, Language and Information* 3 (1994), 1-37.
2. Y. Bar-Hillel, C. Gaifman and E. Shamir, On categorial and phrase structure grammars, *Bulletin Res. Council Israel* F9 (1960), 155-166.
3. W. Buszkowski, Some decision problems in the theory of syntactic categories, *Zeitschrift f. math. Logik und Grundlagen der Mathematik* 28 (1982), 539-548.
4. W. Buszkowski, The equivalence of unidirectional Lambek categorial grammars and context-free grammars, *Zeitschrift f. math. Logik und Grundlagen der Mathematik* 31 (1985), 369-384.
5. W. Buszkowski, The finite model property for BCI and related systems, *Studia Logica* 57 (1996), 303-323.
6. W. Buszkowski, Mathematical Linguistics and Proof Theory, in [24], 683-736.
7. W. Buszkowski, Finite models of some substructural logics, *Mathematical Logic Quarterly* 48 (2002), 63-72.
8. W. Buszkowski, Relational models of Lambek logics, in: *Theory and Applications of Relational Structures as Knowledge Instruments*, Lecture Notes in Comp. Science 2929, 2003, 196-213.
9. W. Buszkowski, On action logic: Equational theories of action algebras, to appear in *Journal of Logic and Computation*.
10. W. Buszkowski and M. Kołowska-Gawiejnowicz, Representation of residuated semigroups in some algebras of relations. (The method of canonical models.), *Fundamenta Informaticae* 31 (1997), 1-12.
11. C. Hardin and D. Kozen, On the complexity of the Horn theory of REL, manuscript, 2003.

12. J.E. Hopcroft and J.D. Ullman, *Introduction to Automata Theory, Languages and Computation*, Addison-Wesley, Reading, 1979.

13. C. Hoare and H. Jifeng, The weakest prespecification, *Fundamenta Informaticae* 9 (1986), 51-84, 217-252.

14. P. Jipsen, From semirings to residuated Kleene algebras, *Studia Logica* 76 (2004), 291-303.

15. D. Kozen, On Kleene algebras and closed semirings, in: *Proc. MFCS 1990, Lecture Notes in Comp. Science* 452, 1990, 26-47.

16. D. Kozen, A completeness theorem for Kleene algebras and the algebra of regular events, *Information and Computation* 110:2 (1994), 366-390.

17. D. Kozen, On the complexity of reasoning in Kleene algebras, *Information and Computation* 179 (2002), 152-162.

18. J. Lambek, The mathematics of sentence structure, *American Mathematical Monthly* 65 (1958), 154-170.

19. H. Ono, Semantics for Substructural Logics, in: *Substructural Logics*, (P. Schroeder-Heister and K. Dosen, eds.), Clarendon Press, Oxford, 1993, 259-291.

20. E. Orłowska and A.M. Radzikowska, Double residuated lattices and their applications, in: *Relational Methods in Computer Science, Lecture Notes in Comp. Science* 2561, 2002, 171-189.

21. E. Palka, An infinitary sequent system for the equational theory of *-continuous action lattices, to appear in *Fundamenta Informaticae*.

22. V. Pratt, Action logic and pure induction, in: *Logics in AI. Proc. JELIA'90, Lecture Notes in Artif. Intelligence* 478, 1990, 97-120.

23. V.N. Redko, On defining relations for the algebra of regular events, *Ukrain. Mat. Z.* 16 (1964), 120-126. In Russian.

24. J. van Benthem and A. ter Meulen (eds.), *Handbook of Logic and Language*, Elsevier, Amsterdam, The MIT Press, Cambridge Mass., 1997.

Demonic Algebra with Domain

Jean-Lou De Carufel and Jules Desharnais

Département d'informatique et de génie logiciel
Université Laval, Québec, QC, G1K 7P4, Canada
jldec1@ift.ulaval.ca, Jules.Desharnais@ift.ulaval.ca

Abstract. We first recall the concept of Kleene algebra with domain (KAD). Then we explain how to use the operators of KAD to define a demonic refinement ordering and demonic operators (many of these definitions come from the literature). Then, taking the properties of the KAD-based demonic operators as a guideline, we axiomatise an algebra that we call *Demonic algebra with domain* (DAD). The laws of DAD not concerning the domain operator agree with those given in the 1987 CACM paper *Laws of programming* by Hoare et al. Finally, we investigate the relationship between demonic algebras with domain and KAD-based demonic algebras. The question is whether every DAD is isomorphic to a KAD-based demonic algebra. We show that it is not the case in general. However, if a DAD $\mathcal{D}$ is isomorphic to a demonic algebra based on a KAD $\mathcal{K}$, then it is possible to construct a KAD isomorphic to $\mathcal{K}$ using the operators of $\mathcal{D}$. We also describe a few open problems.

1 Introduction

The basic operators of Kleene algebra (KA) or relation algebra (RA) can directly be used to give an abstract angelic semantics of while programs. For instance, $a + b$ corresponds to an angelic non-deterministic choice between programs a and b, and $(t \cdot b)^* \cdot \neg t$ is the angelic semantics of a loop with condition t and body b. One way to express demonic semantics in KA or RA is to define demonic operators in terms of the basic operators; these demonic operators can then be used in the semantic definitions. In RA, this has been done frequently (see for instance [1,2,6,7,16,19,23]); in KA, much less [11,12].

In the recent years, various algebras for program refinement have seen the day [3,13,14,15,21,22,24]. The refinement algebra of von Wright is an abstraction of predicate transformers, while the laws of programming of Hoare et al. have an underlying relational model. Möller's lazy Kleene algebra has weaker axioms than von Wright's and can handle systems in which infinite sequences of states may occur.

Our goal is also to design a refinement algebra, that we call a *Demonic algebra* (DA). Rather than designing it with a concrete model in mind, our first goal is to come as close as possible to the kind of algebras that one gets by defining demonic operators in *KA with domain* (KAD) [8,9,10], as is done in [11,12], and then forgetting the basic angelic operators of KAD. Starting from KAD

R.A. Schmidt (Ed.): RelMiCS /AKA 2006, LNCS 4136, pp. 120–134, 2006.

means that DA abstracts many concrete models, just like KA does. We hope that the closeness to KA will eventually lead to decision procedures like those of KA. A second longer term goal, not pursued here, is to precisely determine the relationship of DA with the other refinement algebras; we will say a few words about that in the conclusion.

In Section 2, we recall the definitions of Kleene algebra and its extensions, *Kleene algebra with tests* (KAT) and *Kleene algebra with domain* (KAD). This section also contains the definitions of demonic operators in terms of the KAD operators. Section 3 presents the axiomatisation of DA and its extensions, *DA with tests* (DAT) and *DA with domain* (DAD), as well as derived laws. It turns out that the laws of DAT closely correspond to the laws of programming of [13,14]. In Section 4, we begin to investigate the relationship between KAD and DAD by first defining angelic operators in terms of the demonic operators (call this transformation $\mathcal{G}$). Then we investigate whether the angelic operators thus defined by $\mathcal{G}$ induce a KAD. Not all answers are known there and we state a conjecture that we believe holds and from which the conditions that force $\mathcal{G}$ to induce a KAD can be determined. It is stated in Section 5 that the conjecture holds in those DADs obtained from a KAD by defining demonic operators in terms of the angelic operators (call this transformation $\mathcal{F}$). The good thing is that $\mathcal{F}$ followed by $\mathcal{G}$ is the identity. Section 6 simply describes the main unsolved problem. We conclude in Section 7 with a description of future research and a quick comparison with other refinement algebras.

Due to restricted space, we cannot include proofs. They are published in a research report available on the web [5].

2   Kleene Algebra with Domain and KAD-Based Demonic Operators

In this section, we recall basic definitions about KA and its extensions, KAT and KAD. Then we present the KAD-based definition of the demonic operators.

Definition 1 (Kleene algebra). *A* Kleene algebra *(KA) [4,17] is a structure* $(K, +, \cdot, {}^*, 0, 1)$ *such that the following properties hold for all* $x, y, z \in K$.

$$(x + y) + z = x + (y + z) \tag{1}$$
$$x + y = y + x \tag{2}$$
$$x + x = x \tag{3}$$
$$0 + x = x \tag{4}$$
$$(x \cdot y) \cdot z = x \cdot (y \cdot z) \tag{5}$$
$$0 \cdot x = x \cdot 0 = 0 \tag{6}$$
$$1 \cdot x = x \cdot 1 = x \tag{7}$$
$$x \cdot (y + z) = x \cdot y + x \cdot z \tag{8}$$
$$(x + y) \cdot z = x \cdot z + y \cdot z \tag{9}$$

$$x^* = x \cdot x^* + 1 \tag{10}$$
$$x^* = x^* \cdot x + 1 \tag{11}$$

Addition induces a partial order $\leq$ such that, for all $x, y \in K$,

$$x \leq y \iff x + y = y . \tag{12}$$

Finally, the following properties must be satisfied for all $x, y, z \in K$.

$$z \cdot x + y \leq z \implies y \cdot x^* \leq z \tag{13}$$
$$x \cdot z + y \leq z \implies x^* \cdot y \leq z \tag{14}$$

To reason about programs, it is useful to have a concept of condition, or test. It is provided by Kleene algebra with tests, which is further extended by Kleene algebra with domain.

Definition 2 (Kleene algebra with tests). *A KA with tests (KAT) [18] is a structure $(K, \text{test}(K), +, \cdot, {}^*, 0, 1, \neg)$ such that $\text{test}(K) \subseteq \{t \mid t \in K \wedge t \leq 1\}$, $(K, +, \cdot, {}^*, 0, 1)$ is a KA and $(\text{test}(K), +, \cdot, \neg, 0, 1)$ is a Boolean algebra.*

In the sequel, we use the letters s, t, u, v for tests and w, x, y, z for programs.

Definition 3 (Kleene algebra with domain). *A KA with domain (KAD) [8,9,12,10] is a structure $(K, \text{test}(K), +, \cdot, {}^*, 0, 1, \neg, {}^\ulcorner)$ such that $(K, \text{test}(K), +, \cdot, {}^*, 0, 1, \neg)$ is a KAT and, for all $x \in K$ and $t \in \text{test}(K)$,*

$$x \leq {}^\ulcorner x \cdot x , \tag{15}$$
$${}^\ulcorner(t \cdot x) \leq t , \tag{16}$$
$${}^\ulcorner(x \cdot {}^\ulcorner y) \leq {}^\ulcorner(x \cdot y) . \tag{17}$$

These axioms force the test algebra $\text{test}(K)$ to be the maximal Boolean algebra included in $\{x \mid x \leq 1\}$ [10]. Property (17) is called *locality*. There are many other properties about KAT and KAD and we present some of the most important ones concerning the domain operator. See [8,10,12] for proofs.

Proposition 4. *The following hold for all $t \in \text{test}(K)$ and all $x, y \in K$.*

1. $t \cdot t = t$
2. ${}^\ulcorner x = \min_{\leq}\{t \mid t \in \text{test}(K) \wedge t \cdot x = x\}$
3. ${}^\ulcorner x \cdot x = x$
4. ${}^\ulcorner t = t$
5. ${}^\ulcorner(t \cdot x) = t \cdot {}^\ulcorner x$
6. ${}^\ulcorner(x + y) = {}^\ulcorner x + {}^\ulcorner y$

The following operator characterises the set of points from which no computation as described by x may lead outside the domain of y.

Definition 5 (KA-Implication). *Let x and y be two elements of a KAD. The KA-implication $x \rightarrow y$ is defined by $x \rightarrow y = \neg{}^\ulcorner(x \cdot \neg{}^\ulcorner y)$.*

We are now ready to introduce the demonic operators. Most of the proofs can be found in [12].

Definition 6 (Demonic refinement). *Let x and y be two elements of a KAD. We say that x refines y, noted $x \sqsubseteq_A y$, when $\ulcorner y \leq \ulcorner x$ and $\ulcorner y \cdot x \leq y$.*

The subscript A in $\sqsubseteq_A$ indicates that the demonic refinement is defined with the operators of the angelic world. An analogous notation will be introduced when we define angelic operators in the demonic world. It is easy to show that $\sqsubseteq_A$ is a partial order. Note that for all tests s and t, $s \sqsubseteq_A t \iff t \leq s$. This definition can be simply illustrated with relations. Let $Q = \{(1,2),(2,4)\}$ and $R = \{(1,2),(1,3)\}$. Then $\ulcorner R = \{(1,1)\} \subseteq \{(1,1),(2,2)\} = \ulcorner Q$. Since in addition $\ulcorner R; Q = \{(1,2)\} \subseteq R$, we have $Q \sqsubseteq_A R$ (";" is the usual relational composition).

Proposition 7 (Demonic upper semilattice).

1. *The partial order $\sqsubseteq_A$ induces an upper semilattice with demonic join $\sqcup_A$:*

$$x \sqsubseteq_A y \iff x \sqcup_A y = y.$$

2. *Demonic join satisfies the following two properties.*

$$x \sqcup_A y = \ulcorner x \cdot \ulcorner y \cdot (x + y)$$
$$\ulcorner(x \sqcup_A y) = \ulcorner x \sqcup_A \ulcorner y = \ulcorner x \cdot \ulcorner y$$

Definition 8 (Demonic composition). *The demonic composition of two elements x and y of a KAD, written $x \mathbin{\square_A} y$, is defined by $x \mathbin{\square_A} y = (x \to y) \cdot x \cdot y$.*

Definition 9 (Demonic star). *Let $x \in K$, where K is a KAD. The unary iteration operator $^{\times_A}$ is defined by $x^{\times_A} = x^* \mathbin{\square_A} \ulcorner x$.*

Definition 10 (Conditional). *For each $t \in \mathsf{test}(K)$ and $x, y \in K$, the t-conditional is defined by $x \mathbin{\sqcap_{A t}} y = t \cdot x + \neg t \cdot y$. The family of t-conditionals corresponds to a single ternary operator $\sqcap_{A\bullet}$ taking as arguments a test t and two arbitrary elements x and y.*

The demonic join operator $\sqcup_A$ is used to give the semantics of demonic non-deterministic choices and $\square_A$ is used for sequences. Among the interesting properties of $\square_A$, we cite $t \mathbin{\square_A} x = t \cdot x$, which says that composing a test t with an arbitrary element x is the same in the angelic and demonic worlds, and $x \mathbin{\square_A} y = x \cdot y$ if $\ulcorner y = 1$, which says that if the second element of a composition is total, then again the angelic and demonic compositions coincide. The ternary operator $\sqcap_{A\bullet}$ is similar to the conditional choice operator $\_ \triangleleft \_ \triangleright \_$ of Hoare et al. [13,14]. It corresponds to a guarded choice with disjoint alternatives. The iteration operator $^{\times_A}$ rejects the finite computations that go through a state from which it is possible to reach a state where no computation is defined (e.g., due to blocking or abnormal termination).

As usual, unary operators have the highest precedence, and demonic composition $\square_A$ binds stronger than $\sqcup_A$ and $\sqcap_{A\bullet}$, which have the same precedence.

Proposition 11 (KA-based demonic operators). *The demonic operators $\sqcup_A, \square_A, \sqcap_{A\bullet}, ^{\times_A}$ and $\ulcorner$ satisfy the axioms of demonic algebra with domain presented in Section 3 (Definitions 12, 13, 16).*

3   Axiomatisation of Demonic Algebra with Domain

The demonic operators introduced at the end of the last section satisfy many properties. We choose some of them to become axioms of a new structure called demonic algebra with domain. For this definition, we follow the same path as for the definition of KAD. That is, we first define demonic algebra, then demonic algebra with tests and, finally, demonic algebra with domain.

3.1   Demonic Algebra

Demonic algebra, like KA, has a sum, a composition and an iteration operator. Here is its definition.

Definition 12 (Demonic algebra). *A demonic algebra (DA) is a structure* $(A_\mathcal{D}, \sqcup, \square, {}^\times, 0, 1)$ *such that the following properties are satisfied for* $x, y, z \in A_\mathcal{D}$.

$$x \sqcup (y \sqcup z) = (x \sqcup y) \sqcup z \tag{18}$$

$$x \sqcup y = y \sqcup x \tag{19}$$

$$x \sqcup x = x \tag{20}$$

$$0 \sqcup x = 0 \tag{21}$$

$$x \square (y \square z) = (x \square y) \square z \tag{22}$$

$$0 \square x = x \square 0 = 0 \tag{23}$$

$$1 \square x = x \square 1 = x \tag{24}$$

$$x \square (y \sqcup z) = x \square y \sqcup x \square z \tag{25}$$

$$(x \sqcup y) \square z = x \square z \sqcup y \square z \tag{26}$$

$$x^\times = x \square x^\times \sqcup 1 \tag{27}$$

$$x^\times = x^\times \square x \sqcup 1 \tag{28}$$

There is a partial order $\sqsubseteq$ *induced by* $\sqcup$ *such that for all* $x, y \in A_\mathcal{D}$,

$$x \sqsubseteq y \iff x \sqcup y = y \ . \tag{29}$$

The next two properties are also satisfied for all $x, y, z \in A_\mathcal{D}$.

$$z \square x \sqcup y \sqsubseteq z \implies y \square x^\times \sqsubseteq z \tag{30}$$

$$x \square z \sqcup y \sqsubseteq z \implies x^\times \square y \sqsubseteq z \tag{31}$$

When comparing Definitions 1 and 12, one observes the obvious correspondences $+ \leftrightarrow \sqcup, \cdot \leftrightarrow \square, {}^* \leftrightarrow {}^\times, 0 \leftrightarrow 0, 1 \leftrightarrow 1$. The only difference in the axiomatisation between KA and DA is that 0 is the left and right identity of addition in KA ($+$), while it is a left and right zero of addition in DA ($\sqcup$). However, this minor difference has a rather important impact. While KAs and DAs are upper semilattices with $+$ as the join operator for KAs and $\sqcup$ for DAs, the element 0 is the bottom of the semilattice for KAs and the top of the semilattice for DAs. Indeed, by (21) and (29), $x \sqsubseteq 0$ for all $x \in A_\mathcal{D}$.

All operators are isotone with respect to the refinement ordering $\sqsubseteq$. That is, for all $x, y, z \in A_{\mathcal{D}}$,

$$x \sqsubseteq y \implies z \sqcup x \sqsubseteq z \sqcup y \ \wedge\ z \mathbin{\square} x \sqsubseteq z \mathbin{\square} y \ \wedge\ x \mathbin{\square} z \sqsubseteq y \mathbin{\square} z \ \wedge\ x^{\times} \sqsubseteq y^{\times} \ .$$

This can easily be derived from (19), (20), (24), (25), (26), (27), (29) and (31).

3.2   Demonic Algebra with Tests

Now comes the first extension of DA, demonic algebra with tests. This extension has a concept of tests like the one in KAT and it also adds the conditional operator $\sqcap_t$. In KAT, $+$ and $\cdot$ are respectively the join and meet operators of the Boolean lattice of tests. But in DAT, it will turn out that for any tests s and t, $s \sqcup t = x \mathbin{\square} t$, and that $\sqcup$ and $\square$ both act as the join operator on tests (this is also the case for the KAD-based definition of these operators given in Section 2, as can be checked). Introducing $\sqcap_t$ provides a way to express the meet of tests, as will be shown below. Here is how we deal with tests in a demonic world.

Definition 13 (Demonic algebra with tests). *A demonic algebra with tests (DAT) is a structure $(A_{\mathcal{D}}, B_{\mathcal{D}}, \sqcup, \square, {}^{\times}, 0, 1, \sqcap_{\bullet})$ such that*

1. $(A_{\mathcal{D}}, \sqcup, \square, {}^{\times}, 0, 1)$ *is a DA;*
2. $\{1, 0\} \subseteq B_{\mathcal{D}} \subseteq A_{\mathcal{D}};$
3. *for all $t \in B_{\mathcal{D}}$, $1 \sqsubseteq t;$*
4. $\sqcap_{\bullet}$ *is a ternary operator of type $B_{\mathcal{D}} \times A_{\mathcal{D}} \times A_{\mathcal{D}} \to A_{\mathcal{D}}$ that can be thought of as a family of binary operators. For each $t \in B_{\mathcal{D}}$, $\sqcap_t$ is an operator of type $A_{\mathcal{D}} \times A_{\mathcal{D}} \to A_{\mathcal{D}}$, and of type $B_{\mathcal{D}} \times B_{\mathcal{D}} \to B_{\mathcal{D}}$ if its two arguments belong to $B_{\mathcal{D}};$*
5. $\sqcap_{\bullet}$ *satisfies the following properties for all $s, t \in B_{\mathcal{D}}$ and all $x, y, z \in A_{\mathcal{D}}$. In these axioms, we use the negation operator $\neg$, defined by*

$$\neg t = 0 \sqcap_t 1. \tag{32}$$

$$x \sqcap_t y = y \sqcap_{\neg t} x \tag{33}$$

$$t \mathbin{\square} x \sqcap_t y = x \sqcap_t y \tag{34}$$

$$x \sqcap_t x = x \tag{35}$$

$$x \sqcap_t 0 = t \mathbin{\square} x \tag{36}$$

$$(x \sqcap_t y) \mathbin{\square} z = x \mathbin{\square} z \sqcap_t y \mathbin{\square} z \tag{37}$$

$$s \mathbin{\square} (x \sqcap_t y) = (s \mathbin{\square} x) \sqcap_t (s \mathbin{\square} y) \tag{38}$$

$$x \sqcap_t (y \sqcup z) = (x \sqcap_t y) \sqcup (x \sqcap_t z) \tag{39}$$

$$x \sqcup (y \sqcap_t z) = (x \sqcup y) \sqcap_t (x \sqcup z) \tag{40}$$

$$t \sqcup \neg t = 0 \tag{41}$$

$$\neg(1 \sqcap_t s) = \neg t \sqcup \neg s \tag{42}$$

The elements in $B_{\mathcal{D}}$ are called (demonic) tests.

The axioms for $\sqcap_t$ given in the definition of DAT are all satisfied by the choice operator $\_\vartriangleleft t\vartriangleright\_$ of Hoare et al. [13,14]. The conditional operator satisfies a lot of additional laws, as shown by the following proposition, and more can be found in the precursor paper [20] (with a different syntax).

Proposition 14. *The following properties are true for all* $s, t \in B_{\mathcal{D}}$ *and all* $x, x_1, x_2, y, y_1, y_2, z \in A_{\mathcal{D}}$.

1. $\neg\neg t = t$
2. $x \sqsubseteq y \implies x \sqcap_t z \sqsubseteq y \sqcap_t z$
3. $0 \sqcap_t x = \neg t \square x$
4. $x \sqcap_t \neg t \square y = x \sqcap_t y$
5. $t \square t = t$
6. $s \sqcup t = s \square t$
7. $t \square \neg t = 0$
8. $s \square t = t \square s$
9. $\neg 1 = 0$
10. $\neg 0 = 1$
11. $t \square x \sqsubseteq x \iff 0 \sqsubseteq \neg t \square x$
12. $s \sqsubseteq t \implies \neg t \sqsubseteq \neg s$
13. $x \sqsubseteq y \iff t \square x \sqsubseteq t \square y \wedge \neg t \square x \sqsubseteq \neg t \square y$
14. $x = y \iff t \square x = t \square y \wedge \neg t \square x = \neg t \square y$
15. $t \square (x \sqcap_t y) = t \square x$
16. $x \sqsubseteq y \sqcap_t z \iff x \sqsubseteq t \square y \wedge x \sqsubseteq \neg t \square z$
17. $x \sqcap_t y \sqsubseteq z \iff x \sqsubseteq t \square z \wedge y \sqsubseteq \neg t \square z$
18. $(x_1 \sqcap_s x_2) \sqcap_t (y_1 \sqcap_s y_2) = (x_1 \sqcap_t y_1) \sqcap_s (x_2 \sqcap_t y_2)$

As a direct consequence, one can deduce the next corollary.

Corollary 15. *The set* $B_{\mathcal{D}}$ *of demonic tests forms a Boolean algebra with bottom* 1 *and top* 0. *The supremum of* s *and* t *is* $s \sqcup t$, *their infimum is* $1 \sqcap_s t$ *and the negation of* t *is* $\neg t = 0 \sqcap_t 1$ *(see (32)).*

Thus, tests have quite similar properties in KAT and DAT. But there are important differences. The first one is that $\sqcup$ and $\square$ behave the same way on tests (Proposition 14-6). The second one concerns Laws 13 and 14 of Proposition 14, which show how a proof of refinement or equality can be done by case analysis by decomposing it with cases t and $\neg t$. The same is true in KAT. However, in KAT, this decomposition can also be done on the right side, since for instance the law $x = y \iff x \cdot t = y \cdot t \wedge x \cdot \neg t = y \cdot \neg t$ holds, while the corresponding law does not hold in DAT. For example, $\{(0,0), (0,1), (1,0), (1,1)\} \square \{(0,0)\} = \{(0,0), (0,1), (1,0), (1,1)\} \square \{(1,1)\} = \{\}$, while $\{(0,0), (0,1), (1,0), (1,1)\} \neq \{\}$. In DAT, there is an asymmetry between left and right that can be traced back to laws (37) and (38). In (37), right distributivity holds for arbitrary elements, while left distributivity in (38) holds only for tests. Another law worth noting is Proposition 14-11. On the left of the equivalence, t acts as a *left preserver* of x and on the right, $\neg t$ acts as a *left annihilator*.

3.3   Demonic Algebra with Domain

The next extension consists in adding a domain operator to DAT. It is denoted by the symbol $^\ulcorner$.

Definition 16 (Demonic algebra with domain). *A demonic algebra with domain (DAD) is a structure* $(A_{\mathcal{D}}, B_{\mathcal{D}}, \sqcup, \circ, {}^\times, 0, 1, \sqcap_\bullet, {}^\ulcorner)$, *where* $(A_{\mathcal{D}}, B_{\mathcal{D}}, \sqcup, \circ, {}^\times, 0, 1, \sqcap_\bullet)$ *is a DAT, and the following properties hold for all* $t \in B_{\mathcal{D}}$ *and all* $x, y \in A_{\mathcal{D}}$.

$$^\ulcorner(x \circ t) \circ x = x \circ t \tag{43}$$
$$^\ulcorner(x \sqcup y) = {}^\ulcorner x \sqcup {}^\ulcorner y \tag{44}$$
$$^\ulcorner(x \circ y) = {}^\ulcorner(x \circ {}^\ulcorner y) \tag{45}$$

As noted above, the axiomatisation of DA is very similar to that of KA, so one might expect the resemblance to continue between DAD and KAD. In particular, looking at the angelic version of Definition 16, namely Definition 3, one might expect to find axioms like $^\ulcorner x \circ x \sqsubseteq x$ and $t \sqsubseteq {}^\ulcorner(t \circ x)$, or equivalently, $t \sqsubseteq {}^\ulcorner x \Longleftrightarrow t \circ x \sqsubseteq x$. These three properties can be derived from the chosen axioms (see Propositions 17-2, 17-3 and 17-4) but (43) cannot be derived from them, even when assuming (44) and (45). But (43) holds in KAD-based demonic algebras. Since our goal is to come as close as possible to these, we include (43) as an axiom.

In KAD, it is not necessary to have an axiom like (44), because additivity of $^\ulcorner$ (Proposition 4-6) follows from the axioms of KAD (Definition 3) and the laws of KAT. The proof that works for KAD does not work here. In fact, (43), (44) and (45) are independent.

Law (45) is locality in a demonic world.

By Proposition 17-2 below, $^\ulcorner x$ is a left preserver of x. By Proposition 17-4, it is the greatest left preserver. Similarly, by Proposition 17-6, $\neg^\ulcorner x$ is a left annihilator of x. By Proposition 17-5, it is the least left annihilator (since Proposition 17-5 can be rewritten as $\neg^\ulcorner x \sqsubseteq t \Longleftrightarrow 0 \sqsubseteq t \circ x$).

Proposition 17. *In a DAD, the demonic domain operator satisfies the following properties. Take* $x, y \in A_{\mathcal{D}}$ *and* $t \in B_{\mathcal{D}}$.

1. $^\ulcorner x = \max_\sqsubseteq \{t \mid t \in B_{\mathcal{D}} \wedge t \circ x = x\}$
2. $^\ulcorner x \circ x = x$
3. $t \sqsubseteq {}^\ulcorner(t \circ x)$
4. $t \sqsubseteq {}^\ulcorner x \Longleftrightarrow t \circ x \sqsubseteq x$
5. $t \sqsubseteq {}^\ulcorner x \Longleftrightarrow 0 \sqsubseteq \neg t \circ x$
6. $\neg^\ulcorner x \circ x = 0$
7. $x \sqsubseteq y \Longrightarrow {}^\ulcorner x \sqsubseteq {}^\ulcorner y$
8. $^\ulcorner x \sqsubseteq {}^\ulcorner(x \circ y)$
9. $^\ulcorner t = t$
10. $^\ulcorner(t \circ x) = t \circ {}^\ulcorner x$

11. $\ulcorner x = 0 \Longleftrightarrow x = 0$

12. $\ulcorner (x \sqcap_t y) = \ulcorner x \sqcap_t \ulcorner y$

All the above laws except 12 are identical to laws of $\ulcorner$, after compensating for the reverse ordering of the Boolean lattice (on tests, $\sqsubseteq$ corresponds to $\geq$).

To simplify the notation when possible, we will use the abbreviation

$$x \sqcap y = x \sqcap_{\ulcorner x} y \ . \tag{46}$$

Under special conditions, $\sqcap$ has easy to use properties, as shown by the next corollary.

Corollary 18. *Let x, y, z be arbitrary elements and s, t be tests of a DAD. Then $s \sqcap t$ is the meet of s and t in the Boolean lattice of tests. Furthermore, the following properties hold.*

$$\ulcorner x \diamond y = \ulcorner y \diamond x \implies x \sqcap y = y \sqcap x \tag{47}$$

$$\ulcorner x \diamond \ulcorner y = 0 \implies \ulcorner x \diamond y = \ulcorner y \diamond x \tag{48}$$

$$0 \sqcap x = x \sqcap 0 = x \tag{49}$$

$$x \sqcap x = x \tag{50}$$

$$t \diamond (x \sqcap y) = t \diamond x \sqcap t \diamond y \tag{51}$$

$$(x \sqcap y) \sqcap z = x \sqcap (y \sqcap z) \tag{52}$$

$$x \sqcup (y \sqcap z) = (x \sqcup y) \sqcap (x \sqcup z) \tag{53}$$

$$x \sqcap y \sqsubseteq x \tag{54}$$

$$\ulcorner (x \sqcap y) = \ulcorner x \sqcap \ulcorner y \tag{55}$$

The two most useful cases of the previous corollary are when $\sqcap$ is used on tests and when $\ulcorner x \diamond \ulcorner y = 0$.

4   Definition of Angelic Operators in DAD

Our goal in this section is to define angelic operators from demonic ones, as was done when going from the angelic to the demonic universe (Section 2). This is done in order to study transformations between KAD and DAD (Sections 5 and 6). We add a subscript D to the angelic operators defined here, to denote that they are defined by demonic expressions. We start with the angelic partial order $\leq_D$.

Definition 19 (Angelic refinement). *Let x, y be elements of a DAD. We say that $x \leq_D y$ when the following two properties are satisfied.*

$$\ulcorner y \sqsubseteq \ulcorner x \tag{56}$$

$$x \sqsubseteq \ulcorner x \diamond y \tag{57}$$

Theorem 21 below states that $\leq_D$ is a partial order. Moreover, it gives a formula using demonic operators for the angelic supremum with respect to this partial order. In order to prove this theorem, we need the following lemma.

Lemma 20. *The function*

$$f : A_{\mathcal{D}} \times A_{\mathcal{D}} \to A_{\mathcal{D}}$$
$$(x, y) \mapsto (x \sqcup y) \sqcap \neg\ulcorner y\urcorner\square x \sqcap \neg\ulcorner x\urcorner\square y$$

satisfies the following four properties for all $x, y, z \in A_{\mathcal{D}}$.

1. $\ulcorner f(x, y)\urcorner = \ulcorner x\urcorner \sqcap \ulcorner y\urcorner$
2. $f(x, x) = x$
3. $f(x, y) = f(y, x)$
4. $f(x, f(y, z)) = f(f(x, y), z)$

Theorem 21 (Angelic choice). *The angelic refinement of Definition 19 satisfies the following three properties.*

1. *For all* x, $0 \leq_D x$.
2. *For all* x, y,

$$x \leq_D y \iff f(x, y) = y \ ,$$

   where f *is the function defined in Lemma 20.*
3. $\leq_D$ *is a partial order. Letting* $x +_D y$ *denote the supremum of* x *and* y *with respect to* $\leq_D$, *we have*

$$x +_D y = f(x, y) \ .$$

The following expected properties are a direct consequence of Lemma 20 and Theorem 21.

$$(x +_D y) +_D z = x +_D (y +_D z) \tag{58}$$
$$x +_D y = y +_D x \tag{59}$$
$$x +_D x = x \tag{60}$$
$$0 +_D x = x \tag{61}$$

We now turn to the definition of angelic composition. But things are not as simple as for $\leq_D$ or $+_D$. The difficulty is due to the asymmetry between left and right caused by the difference between axioms (37) and (38), and by the absence of a codomain operator for "testing" the right-hand side of elements as can be done with the domain operator on the left. Consider the two relations

$$Q = \{(0,0), (0,1), (1,2), (2,3)\} \quad \text{and} \quad R = \{(0,0), (2,2)\} \ .$$

The angelic composition of Q and R is $Q \cdot R = \{(0,0), (1,2)\}$, while their demonic composition is $Q \square R = \{(1,2)\}$. There is no way to express $Q \cdot R$ only in terms of $Q \square R$. What we could try to do is to decompose Q as follows using the demonic meet

$$Q = Q \square \ulcorner R\urcorner \sqcap Q \square \neg\ulcorner R\urcorner \sqcap (Q_1 \sqcup Q_2) \ ,$$

where $Q_1 = \{(0,0)\}$ and $Q_2 = \{(0,1)\}$. Note that $Q \square \ulcorner R\urcorner = \{(1,2)\}$ and $Q \square \neg\ulcorner R\urcorner = \{(2,3)\}$ so that the domains of the three operands of $\sqcap$ are disjoint. The effect of $\sqcap$ is then just union. With these relations, it is possible to express the angelic composition as $Q \cdot R = Q \square R \sqcap Q_1 \square R$. Now, it is possible to extract $Q_1 \sqcup Q_2$ from Q, since $Q_1 \sqcup Q_2 = \neg^\ulcorner (Q \square^\ulcorner R) \square \neg^\ulcorner (Q \square \neg^\ulcorner R) \square Q$. The problem is that it is not possible to extract Q_1 from $Q_1 \sqcup Q_2$. On the one hand, Q_1 and Q_2 have the same domain; on the other hand, there is no test t such that $Q_1 = (Q_1 \sqcup Q_2) \square t$. Note that $Q_1 \square^\ulcorner R = Q_1$ and $Q_2 \square^\ulcorner R = \neg Q_2$. This is what leads us to the following definition.

Definition 22. *Let t be a test. An element x of a DAD is said to be t-decomposable iff there are unique elements x_t and $x_{\neg t}$ such that*

$$x = x_\square t \sqcap x_\square \neg t \sqcap (x_t \sqcup x_{\neg t}) \ ,$$

$$^\ulcorner x_t = {}^\ulcorner x_{\neg t} = \neg^\ulcorner (x \square t) \square \neg^\ulcorner (x \square \neg t) \square^\ulcorner x \ ,$$

$$x_t = x_t \square t \ ,$$

$$x_{\neg t} = x_{\neg t} \square \neg t \ .$$

And x is said to be decomposable *iff it is t-decomposable for all tests t.*

It is easy to see that all tests are decomposable. Indeed, the (unique) t-decomposition of a test s is

$$s = s \square t \sqcap s \square \neg t \sqcap (0 \sqcup 0) \ .$$

One may wonder whether there exists a DAD with non-decomposable elements. The answer is yes. The following nine relations constitute such a DAD, with the operations given (they are the standard demonic operations on relations), omitting $\sqcap_\bullet$. The set of tests is $\{0, s, t, 1\}$.

$$0 = \begin{pmatrix} 0 & 0 \\ 0 & 0 \end{pmatrix} \quad s = \begin{pmatrix} 1 & 0 \\ 0 & 0 \end{pmatrix} \quad t = \begin{pmatrix} 0 & 0 \\ 0 & 1 \end{pmatrix} \quad 1 = \begin{pmatrix} 1 & 0 \\ 0 & 1 \end{pmatrix}$$

$$a = \begin{pmatrix} 1 & 0 \\ 1 & 1 \end{pmatrix} \quad b = \begin{pmatrix} 1 & 1 \\ 0 & 1 \end{pmatrix} \quad c = \begin{pmatrix} 1 & 1 \\ 1 & 1 \end{pmatrix} \quad d = \begin{pmatrix} 1 & 1 \\ 0 & 0 \end{pmatrix} \quad e = \begin{pmatrix} 0 & 0 \\ 1 & 1 \end{pmatrix}$$

$\sqcup$	0	s	t	1	a	b	c	d	e
0	0	0	0	0	0	0	0	0	0
s	0	s	0	s	s	d	d	d	0
t	0	0	t	t	e	t	e	0	e
1	0	s	t	1	a	b	c	d	e
a	0	s	e	a	a	c	c	d	e
b	0	d	t	b	c	b	c	d	e
c	0	d	e	c	c	c	c	d	e
d	0	d	0	d	d	d	d	d	0
e	0	0	e	e	e	e	e	0	e

$\square$	0	s	t	1	a	b	c	d	e
0	0	0	0	0	0	0	0	0	0
s	0	s	0	s	s	d	d	d	0
t	0	0	t	t	e	t	e	0	e
1	0	s	t	1	a	b	c	d	e
a	0	s	0	a	a	c	c	d	0
b	0	0	t	b	c	b	c	0	e
c	0	0	0	c	c	c	c	0	0
d	0	0	0	d	d	d	d	0	0
e	0	0	0	e	e	e	e	0	0

$\times$	
0	0
s	s
t	t
1	1
a	a
b	b
c	c
d	0
e	0

$^\ulcorner$	
0	0
s	s
t	t
1	1
a	1
b	1
c	1
d	s
e	t

$\neg$	
0	1
s	t
t	s
1	0
a	1
b	1
c	1
d	s
e	t

The elements a, b, c, d and e are not decomposable. For instance, to decompose c with respect to s would require the existence of relations

$$\begin{pmatrix} 1 & 0 \\ 1 & 0 \end{pmatrix} \quad \text{and} \quad \begin{pmatrix} 0 & 1 \\ 0 & 1 \end{pmatrix} \ ,$$

which are not there.

Definition 23 (Angelic composition). *Let x and y be elements of a DAD such that x is decomposable. Then the angelic composition $\cdot_D$ is defined by*

$$x \cdot_D y = x \square y \sqcap x_{\ulcorner y} \square y .$$

Proposition 24. *Let x, y, z be decomposable elements of a KAD. Then,*

1. $1 \cdot_D x = x \cdot_D 1 = x$,
2. $0 \cdot_D x = x \cdot_D 0 = 0$,
3. $\ulcorner(x \cdot_D (y \cdot_D z)) = \ulcorner((x \cdot_D y) \cdot_D z)$.

We have not yet been able to show the associativity of $\cdot_D$ nor its distributivity over $+_D$.

The last angelic operator that we define here is the iteration operator that corresponds to the Kleene star.

Definition 25 (Angelic iteration). *Let x be an element of a DAD. The angelic finite iteration operator ^{*_D} is defined by*

$$x^{*_D} = (x \sqcap 1)^\times \sqcup 1 .$$

Although we are still struggling to ascertain the properties of $\cdot_D$ (and, as a side effect, those of ^{*_D}), we have a conjecture that most probably holds. At least, it holds for a very important case (see Section 5).

Conjecture 26 (Subalgebra of decomposable elements).

1. The set of decomposable elements of a DAD A_D is a subalgebra of A_D.
2. For the subalgebra of decomposable elements of A_D, the composition $\cdot_D$ is associative and distributes over $+_D$ (properties (5), (8) and (9)).
3. For the subalgebra of decomposable elements of A_D, the iteration operator ^{*_D} satisfies the unfolding and induction laws of the Kleene star (properties (10), (11), (13) and (14)).

5   From KAD to DAD and Back

In this section, we introduce two transformations between the angelic and demonic worlds. The ultimate goal is to show how KAD and DAD are related one to the other.

Definition 27. *Let $(K, \mathsf{test}(K), +, \cdot, {}^*, 0, 1, \neg, {}^{\ulcorner}\,)$ be a KAD. Let $\mathcal{F}$ denote the transformation that sends it to*

$$(K, \mathsf{test}(K), \sqcup_A, \square_A, {}^{\times_A}, 0, 1, \sqcap_{A_\bullet}, {}^{\ulcorner}\,) ,$$

where $\sqcup_A, \square_A, {}^{\times_A}$ and $\sqcap_{A_\bullet}$ are the operators defined in Proposition 7 and Definitions 8, 9 and 10, respectively.

Similarly, let $(A_\mathcal{D}, B_\mathcal{D}, \sqcup, \square, \times, 0, 1, \sqcap_\bullet, \ulcorner)$ *be a DAD. Let* $\mathcal{G}$ *denote the transformation that sends it to*

$$(A_\mathcal{D}, B_\mathcal{D}, +_\mathcal{D}, \cdot_\mathcal{D}, {}^{*\mathcal{D}}, 0, 1, \neg_\mathcal{D}, \ulcorner) \ ,$$

where $+_\mathcal{D}, \cdot_\mathcal{D}, {}^{*\mathcal{D}}$ *and* $\neg_\mathcal{D}$ *are the operators defined in Theorem 21, Definitions 23 and 25, and (32), respectively (since no special notation was introduced in Definition 13 to distinguish DAT negation from KAT negation, we have added a subscript D to* $\neg$ *in order to avoid confusion in Theorem 28).*

By this definition, the transformations $\mathcal{F}$ and $\mathcal{G}$ transport the domain operator unchanged between the angelic and demonic worlds. Indeed, it turns out that $\ulcorner x = \ulcorner x$ is the right transformation.

Having defined $\mathcal{F}$ and $\mathcal{G}$, we can now state the following theorem.

Theorem 28. *Let* $\mathcal{K} = (K, \mathrm{test}(K), +, \cdot, {}^*, 0, 1, \neg, \ulcorner)$ *be a KAD and let* $\mathcal{F}$ *and* $\mathcal{G}$ *be the transformations introduced in Definition 27.*

1. $\mathcal{F}(\mathcal{K})$ *is a DAD.*
2. *All elements of* $\mathcal{F}(\mathcal{K})$ *are decomposable.*
3. $\mathcal{G} \circ \mathcal{F}$ *is the identity on* K *and* $\mathrm{test}(K)$, *i.e., the algebra* $(K, \mathrm{test}(K), +_\mathcal{D}, \cdot_\mathcal{D}, {}^{*\mathcal{D}}, 0, 1, \neg_\mathcal{D}, \ulcorner)$ *derived from the DAD* $\mathcal{F}(\mathcal{K})$ *is isomorphic to* $\mathcal{K}$ *(only the symbols denoting the operators differ).*
4. *If a DAD* $\mathcal{D}$ *is isomorphic to* $\mathcal{F}(\mathcal{K})$, *then* $\mathcal{K}$ *is isomorphic to* $\mathcal{G}(\mathcal{D})$.

Saying that $\mathcal{F}(\mathcal{K})$ is a DAD is just a compact restatement of Proposition 11.

Due to this theorem, the conjecture stated in the previous section holds for the DAD $\mathcal{F}(\mathcal{K})$. This is a very important case. Since the elements of $\mathcal{F}(\mathcal{K})$ are decomposable, this result gives much weight to the conjecture.

6   From DAD to KAD and Back

Let $\mathcal{D} = (A_\mathcal{D}, B_\mathcal{D}, \sqcup, \square, \times, 0, 1, \sqcap_\bullet, \ulcorner)$ be a DAD. If $A_\mathcal{D}$ has non-decomposable elements, then $\mathcal{D}$ cannot be the image $\mathcal{F}(\mathcal{K})$ of a KAD $\mathcal{K}$, by Theorem 28-2. The question that is still not settled is whether the subalgebra $\mathcal{D}_d$ of decomposable elements of $\mathcal{D}$ is the image $\mathcal{F}(\mathcal{K})$ of some KAD $\mathcal{K}$. If Conjecture 26 holds, then this is the case and the composition of transformations $\mathcal{F} \circ \mathcal{G}$ is the identity on $\mathcal{D}_d$. This problem will be the subject of our future research.

7   Conclusion

The work on demonic algebra presented in this paper is just a beginning. Many avenues for future research are open. First and foremost, Conjecture 26 must be solved. In relation to this conjecture, the properties of non-decomposable elements are also intriguing. Are there concrete models useful for Computer Science where these elements play a rôle?

Another line of research is the precise relationship of DAD with the other refinement algebras and most particularly those of [15,21,22,24]. DAD has stronger axioms than these algebras, and thus these contain a DAD as a substructure. Some basic comparisons can already be done. For instance, DADs can be related to the *command algebras* of [15] as follows. Suppose a KAD $\mathcal{K} = (K, \text{test}(K), +, \cdot, ^*, 0, 1, \neg, ^\ulcorner)$. A *command* on $\mathcal{K}$ is an ordered pair (x, s), where $x \in K$ and $s \in \text{test}(K)$. The test s denotes the "domain of termination" of x. If $s \leq \ulcorner x$, the command (x, s) is said to be *feasible*; otherwise, it is *miraculous*. The set of non-miraculous commands of the form $(x, \ulcorner x)$, with the appropriate definition of the operators, is isomorphic to the KAD-based demonic algebra $\mathcal{D}$ obtained from $\mathcal{K}$. If K is the set of all relations over a set S, then $\mathcal{D}$ is isomorphic to the non-miraculous conjunctive predicate transformers on S; this establishes a relationship with the refinement algebras of [22,24], which have predicate transformers as their main models. The algebras in [22,24] have two kinds of tests, *guards* and *assertions*. Assertions correspond to the tests of DAD and the *termination* operator τ of [22] corresponds to the domain operator of DAD.

Finally, let us mention the problem of infinite iteration. In DAD, there is no infinite iteration operator. One cannot be added by simply requiring it to be the greatest fixed point of $\lambda(z :: x \,\square_A\, z \sqcup_A 1)$, since this greatest fixed point is always 0. In [12], tests denoting the starting points of infinite iterations for an element x are obtained by using the greatest fixed point (in a KAD) of $\lambda(t :: \ulcorner(x \cdot t))$. We intend to determine whether a similar technique can be used in DAD.

Acknowledgements

The authors thank Bernhard Möller and the anonymous referees for helpful comments. This research was partially supported by NSERC (Natural Sciences and Engineering Research Council of Canada) and FQRNT (Fond québécois de la recherche sur la nature et les technologies).

References

1. Backhouse, R.C., van der Woude, J.: Demonic operators and monotype factors. Mathematical Structures in Computer Science **3** (1993) 417–433
2. Berghammer, R., Zierer, H.: Relational algebraic semantics of deterministic and nondeterministic programs. Theoretical Computer Science **43** (1986) 123–147
3. Cohen, E.: Separation and reduction. In: Mathematics of Program Construction. Volume 1837 of Lecture Notes in Computer Science, Springer (2000) 45–59
4. Conway, J.: Regular Algebra and Finite Machines. Chapman and Hall, London (1971)
5. De Carufel, J.L., Desharnais, J.: Demonic algebra with domain. Research report DIUL-RR-0601, Département d'informatique et de génie logiciel, Université Laval, Canada (2006). Available at http://www.ift.ulaval.ca/~Desharnais/Recherche/RR/DIUL-RR-0601.pdf

6. Desharnais, J., Belkhiter, N., Sghaier, S., Tchier, F., Jaoua, A., Mili, A., Zaguia, N.: Embedding a demonic semilattice in a relation algebra. Theoretical Computer Science **149** (1995) 333–360

7. Desharnais, J., Mili, A., Nguyen, T.: Refinement and demonic semantics. In Brink, C., Kahl, W., Schmidt, G., eds.: Relational Methods in Computer Science, Springer (1997) 166–183

8. Desharnais, J., Möller, B., Struth, G.: Kleene algebra with domain. Technical Report 2003-7, Institut für Informatik, Augsburg, Germany (2003)

9. Desharnais, J., Möller, B., Struth, G.: Modal Kleene algebra and applications — a survey. JoRMiCS — Journal on Relational Methods in Computer Science **1** (2004) 93–131

10. Desharnais, J., Möller, B., Struth, G.: Kleene algebra with domain. To appear in ACM Transactions on Computational Logic (2006)

11. Desharnais, J., Möller, B., Tchier, F.: Kleene under a demonic star. In: AMAST 2000. Volume 1816 of Lecture Notes in Computer Science, Springer (2000) 355–370

12. Desharnais, J., Möller, B., Tchier, F.: Kleene under a modal demonic star. Journal of Logic and Algebraic Programming, Special issue on Relation Algebra and Kleene Algebra **66** (2006) 127–160

13. Hoare, C.A.R., Hayes, I.J., Jifeng, H., Morgan, C.C., Roscoe, A.W., Sanders, J.W., Sorensen, I.H., Spivey, J.M., Sufrin, B.A.: Laws of programming. Communications of the ACM **30** (1987) 672–686

14. Hoare, C.A.R., Jifeng, H.: Unifying Theories of Programming. International Series in Computer Science. Prentice Hall (1998)

15. Höfner, P., Möller, B., Solin, K.: Omega algebra, demonic refinement algebra and commands. These proceedings

16. Kahl, W.: Parallel composition and decomposition of specifications. Information Sciences **139** (2001) 197–220

17. Kozen, D.: A completeness theorem for Kleene algebras and the algebra of regular events. Information and Computation **110** (1994) 366–390

18. Kozen, D.: Kleene algebra with tests. ACM Transactions on Programming Languages and Systems **19** (1997) 427–443

19. Maddux, R.: Relation-algebraic semantics. Theoretical Computer Science **160** (1996) 1–85

20. McCarthy, J.: A basis for a mathematical theory of computation. In Braffort, P., Hirschberg, D., eds.: Computer Programming and Formal Systems, North-Holland, Amsterdam (1963) 33–70. Available at `http://www-formal.stanford.edu/jmc/basis/basis.html`

21. Möller, B.: Lazy Kleene algebra. In Kozen, D., Shankland, C., eds.: Mathematics of Program Construction. Volume 3125 of Lecture Notes in Computer Science, Springer (2004) 252–273

22. Solin, K., von Wright, J.: Refinement algebra with operators for enabledness and termination. In: Mathematics of Program Construction. Lecture Note in Computer Science, Springer-Verlag (2006). In press

23. Tchier, F., Desharnais, J.: Applying a generalisation of a theorem of Mills to generalised looping structures. In: Colloquium on Science and Engineering for Software Development, organised in the memory of Dr. Harlan D. Mills, and affiliated to the 21st International Conference on Software Engineering, Los Angeles (1999) 31–38

24. von Wright, J.: Towards a refinement algebra. Science of Computer Programming **51** (2004) 23–45

Topological Representation of Contact Lattices

Ivo Düntsch[1,*], Wendy MacCaull[2,*],
Dimiter Vakarelov[3,**], and Michael Winter[1,*]

[1] Department of Computer Science
Brock University
St. Catharines, ON, Canada
{duentsch, mwinter}@brocku.ca
[2] Department of Mathematics, Statistics and Computer Science
St. Francis Xavier University
Antigonish, NS, Canada
wmaccaul@stfx.ca
[3] Department of Mathematical Logic
Sofia University
Sofia, Bulgaria
dvak@fmi.uni-sofia.bg

Abstract. The theory of Boolean contact algebras has been used to represent a region based theory of space. Some of the primitives of Boolean algebras are not well motivated in that context. One possible generalization is to drop the notion of complement, thereby weakening the algebraic structure from Boolean algebra to distributive lattice. The main goal of this paper is to investigate the representation theory of that weaker notion, i.e., whether it is still possible to represent each abstract algebra by a substructure of the regular closed sets of a suitable topological space with the standard Whiteheadean contact relation.

1 Introduction

In the classical approach to space the basic primitive is the notion of a point, and geometric figures are considered to be sets of points. Contrary to this, the region-based approach to space adopts as its primitives more realistic spatial notions. In this theory, regions, as abstractions of "solid" spatial bodies, and several basic relations and operations between regions are considered. Some of the relations have their origin in mereology, e.g. "part-of" ($x \leq y$), "overlap" (xOy), its dual "underlap" (xUy) and others definable by them. A region based theory of space extends classical mereology by considering some new relations between

---

[*] The author gratefully acknowledges support from the Natural Sciences and Engineering Research Council of Canada.
[**] This author was supported by the project NIP-1510 by the Bulgarian Ministry of Science and Education

R.A. Schmidt (Ed.): RelMiCS /AKA 2006, LNCS 4136, pp. 135–147, 2006.

regions, topological in nature, such as "contact" (xCy), "non-tangential part-of" ($x \ll y$) and many others definable mainly by means of the contact and the part-of relations. This motivates some authors to call the new direction "mereo-topology". The most simple algebraic counterparts of mereotopology are contact algebras. They appear in different papers under various names (see for instance [4,5,6,7,8,9,13,14]). Contact algebras are Boolean algebras extended with the contact relation C satisfying some axioms. The elements of the Boolean algebra represent regions and the Boolean operations – join $x + y$, meet $x \cdot y$ and the Boolean complement x^* – allow the construction of new regions from given ones (such as the "universal" region 1, having all other regions as its parts, and the zero region 0, which is part of all regions). Part-of, underlap and overlap are definable by the Boolean part of the algebra, i.e., $x \leq y$ is the lattice ordering, $xOy \leftrightarrow x \cdot y \neq 0$ (motivating the name "overlap") and $xUy \leftrightarrow x + y \neq 1$ (motivating the name "underlap"). So, the Boolean part of a contact algebra incorporates the mereological aspect of the theory, while the contact relation C corresponds to the "topological" part of the formalism. The word "topology" in this context is well motivated, because standard models of contact algebras are the Boolean algebras of regular closed (or regular open) sets of a given topological space. We recall that a is regular closed if $a = Cl(Int(a))$, and, dually, c is regular open if $c = Int(Cl(c))$, where Cl and Int are the topological operations of closure and interior in the topological space. Note that in either case the Boolean operations are not the standard set-theoretic operations, i.e., we have $a \cdot b = Cl(Int(a \cap b))$, $a + b = a \cup b$, and $a^* = Cl(-a)$ for regular closed sets a and b, and $c \cdot d = c \cap d$, $c + d = Int(Cl(c \cup d))$, and $c^* = Int(-c)$ for regular open sets c and d. The contact relation between regular closed regions is given by $aCb \leftrightarrow a \cap b \neq \varnothing$ and for open regions by $cCd \leftrightarrow Cl(c) \cap Cl(d) \neq \varnothing$. Topological contact algebras correspond to the point-based aspect of the theory.

A major current stream in the theory of contact algebras is their topological representation theory, i.e., the construction of an embedding of the contact algebra into a suitable topological space. In addition, one usually requires that the embedding generates/determines the topology. For Boolean contact algebras it is natural to require that the image of the embedding is a basis for the topology (similar to the Stone representation theory for Boolean algebras). In the case of distributive lattices this approach works if and only if underlap satisfies an additional property (Corollary 2 and Theorem 5). By relaxing that strict interpretation of "generate the topology" we get a general representation theorem for distributive contact algebras (Theorem 7). Once such a theorem is established, it shows that, although the notion of point is not a primitive notion of the theory, it can be defined within the algebra via its isomorphic copy in the topological contact algebra provided by the representation theorem. This shows that the "pointless" abstraction is adequate and corresponds to the classical point-based approach. Recent results in this direction can be found in [4,5,8,14]. For instance, in [4] several versions of abstract points for contact algebras are discussed such as clans, E-filters etc. All versions generalize the notion of an ultrafilter, which is the abstract point suitable for representing Boolean algebras as ordinary sets.

Let us note that ordinary sets can also be considered as regions in a topological space endowed with the discrete topology, but such regions are "bad" regions in the sense that they do not have nice topological properties such as a boundary, a non-tangential part etc.

One of the main goals of this paper is to generalize the notion of contact algebras by weakening the algebraic structure from Boolean algebras to distributive lattices, but to keep the intended semantics of regions to be regular sets in topological spaces. In other words, we simply remove the operation of Boolean complement *. From a philosophical point of view, the complement of a region is not well motivated. If the region a represents a solid body what is then represented by a^*? (One can formulate similar criticisms for (certain aspects) of some of the other Boolean primitives, which are not discussed here.) Notice that the definitions of the mereological relations part-of, overlap and underlap do not depend on the existence of complements. Moreover, in all known definitions and variations of Boolean contact algebras, the axioms for contact do not rely on complements. So, studying a theory based on weaker assumptions will reveal more deeply the nature of the mereological and mereotopological relations. The mereo-topological relations usually considered such as "non-tangential part" $x \ll y \leftrightarrow a\overline{C}b^*$ and dual contact $a\check{C}b \leftrightarrow a^*Cb^*$ are definable from contact using complements. In the case of distributive lattices these relations must be primitives. Using this approach a deeper insight into the separate roles of the different mereological relations and their interactions may be achieved. For instance, in the Boolean case a certain mereological relation may possess some properties, which must be postulated separately for distributive lattices. An example is the property, "U is extensional", which implies that part-of is definable by U in the sense that $a \leq b$ if and only if $(\forall c)(bUc \rightarrow aUc)$. It turns out (Corollary 2) that this property is exactly the necessary and sufficient property for representing the contact structure in the strict sense. On the other hand, for Boolean contact algebras such a representation is always possible, because in that case underlap is extensional.

The paper is organized as follows. Section 2 introduces the algebraic notions such as distributive lattices, overlap and underlap, distributive contact algebras, filters and clans. Section 3 provides the topological background and a pure topological theorem relating extensionality of underlap to the generation of the topology by means of regular closed sets. It also shows the necessity of the distributivity of the lattice. In Section 4 we show how to represent U-extensional distributive contact lattices in a lattice of regular closed sets of some topological space. As a side result we obtain Cornish's theorem [3] for U-extensional distributive lattices. Here we prove also that every distributive contact lattice can be embedded in a lattice of regular closed sets, so that the image of the lattice generates the topology in a weaker sense. This is done in two steps. First we show that every distributive contact lattice can be embedded into a U-extensional contact lattice; then, we simply apply the representation theorem for U-extensional contact lattices. The last section contains some conclusions and future work.

A standard reference for distributive lattices is [2] and for topology [10].

2   Notation and First Observations

For any set X and $Y \subseteq X$ we denote by $-Y$ the complement of Y in X, if no confusion can arise. If R is a binary relation on D, and $x \in D$, we let $R(x) = \{y : xRy\}$, which is called the *range of x with respect to R*.

Distributive Lattices

Throughout this paper, $(D, 0, 1, +, \cdot)$ is a bounded distributive lattice; we usually denote algebras by their base set. The dual lattice D^{op} of D is the lattice $(D, 1, 0, \cdot, +)$ based on the reversed order of D. A sublattice D' of D is called dense (in D) iff for each element $0 \neq a \in D$ there is an element $0 \neq b \in D'$ with $b \leq a$. A dually dense sublattice of D is a dense sublattice of D^{op}. We call an embedding $h : D' \to D$ dense iff the image $h(D') = \{h(a) : a \in D'\}$ of D' is dense in D. Finally, an element $d \in D$ is called meet(join)-irreducible if $d = a \cdot b$ $(d = a + b)$ implies $d = a$ or $d = b$ for all $a, b \in D$.

We define two relations on D, which are of importance in the sequel:

$$(2.1) \qquad\qquad xOy \iff x \cdot y \neq 0, \qquad\qquad \text{``overlap''},$$
$$(2.2) \qquad\qquad xUy \iff x + y \neq 1, \qquad\qquad \text{``underlap''}.$$

The proof of the following is straightforward:

Lemma 1. *If $a \leq b$, then $O(a) \subseteq O(b)$ and $U(b) \subseteq U(a)$.*

O is called *extensional* if

$$(2.3) \qquad\qquad\qquad (\forall x, y)[O(x) = O(y) \Rightarrow x = y].$$

Analogously, we say that U is *extensional* if

$$(2.4) \qquad\qquad\qquad (\forall a, b)[U(a) = U(b) \Rightarrow a = b].$$

In [15] distributive lattices which satisfy (2.3) are called *disjunctive lattices*. If D is a Boolean algebra, then, clearly, both O and U are extensional. Extensionality of O and U has been considered earlier in the literature, and these results show that such extensionalities can influence the underlying algebraic structure; in particular, the following holds for a bounded distributive pseudocomplemented lattice (i.e., a bounded distributive lattice equipped with an operation $*$ satisfying $a \leq b^*$ iff $a \cdot b = 0$):

Theorem 1. *1. Suppose that D is a bounded distributive pseudocomplemented lattice. Then, D is a Boolean algebra if and only if O is extensional.*
2. Suppose that D is a bounded distributive dually pseudocomplemented lattice. Then, D is a Boolean algebra if and only if U is extensional.

Proof. 1. was shown in [9], and 2. follows by duality.   □

In particular, if D is finite and not a Boolean algebra, or if D is a chain, then neither O nor U is extensional. Furthermore, if 0 is meet irreducible then O is not extensional, and if 1 is join irreducible, then U is not extensional. For example, the lattice of all cofinite subsets of ω together with $\emptyset$ is U-extensional, but not O-extensional; dually, the set of all finite subsets of ω together with ω is O-extensional, but not U-extensional.

In [3] further characterizations of disjunctive and dually disjunctive lattices were given; these relate to dense sublattices of Boolean algebras:

Theorem 2. *1. O is extensional if and only if D is isomorphic to a dense sublattice of a complete Boolean algebra.*
  2. U is extensional if and only if D is isomorphic to a dually dense sublattice of a complete Boolean algebra.
  3. O and U are both extensional if and only if the Dedekind completion of D is a complete Boolean algebra.

Below, we will give additional conditions for O, respectively U, to be extensional. It is worthy of mention that each of these is strictly stronger than extensionality in the case that D is not distributive [9].

Lemma 2. *1. O is extensional if and only if $(\forall a, b)[O(a) \subseteq O(b) \Rightarrow a \leq b]$.*
  2. U is extensional if and only if $(\forall a, b)[U(b) \subseteq U(a) \Rightarrow a \leq b]$.

Proof. 1. "$\Rightarrow$": Suppose that $O(a) \subseteq O(b)$. Then, $O(b) = O(a) \cup O(b) = O(a + b)$. Extensionality of O implies that $b = a + b$, i.e. $a \leq b$.
  "$\Leftarrow$": Let $O(a) = O(b)$; then $O(a) \subseteq O(b)$ and $O(b) \subseteq O(a)$, and, by the hypothesis, $a \leq b$ and $b \leq a$, i.e., $a = b$.
2. is proved dually. $\square$

Later, we use extensionality in the equivalent form given by Lemma 2. If, for instance, U is extensional then we will say that the lattice is U-extensional.

A subset F of a lattice D is called a filter if $x, y \in F$ and $z \in D$ implies $x \cdot y \in F$ and $x + z \in F$. We call a filter F of D *prime* if $x + y \in F$ implies $x \in F$ or $y \in F$. Prime(D) is the set of prime filters of D. For each $x \in D$ we denote by $h_{\mathrm{Prime}}(x) = \{F \in \mathrm{Prime}(D) : x \in F\}$, the set of all prime filters containing x. Stone's well known representation theorem now states:

Theorem 3. *[2,12]*

  1. The mapping h_{Prime} is a lattice embedding of D into the lattice $2^{Prime(D)}$ of all subsets of $Prime(D)$.
  2. The collection $\{h_{Prime}(a) : a \in D\}$ forms a basis for the closed sets of a compact T_0 topology τ on $Prime(D)$ for which each set $(Prime(D) \setminus h(a))$ is compact open. Furthermore, τ is a T_1 topology if and only if D is relatively complemented, and a T_2 topology if and only if D is a Boolean algebra.

For later use we observe that $h_{\mathrm{Prime}}(a)$ is not necessarily regular closed.

Contact Relations and Distributive Contact Lattices
A binary relation C on D is called a *contact relation* (CR) if it satisfies:

C0. $(\forall a)0(-C)a$;
C1. $(\forall a)[a \neq 0 \Rightarrow aCa]$;
C2. $(\forall a)(\forall b)[aCb \Rightarrow bCa]$;
C3. $(\forall a)(\forall b)(\forall c)[aCb \text{ and } b \leq c \Rightarrow aCc]$;
C4. $(\forall a)(\forall b)(\forall c)[aC(b+c) \Rightarrow (aCb \text{ or } aCc)]$.

The pair $\langle D, C \rangle$ is called a *distributive contact lattice* (DCL). If D is a Boolean algebra, then $\langle D, C \rangle$ is a *Boolean contact algebra* (BCA). Let $\mathcal{C}$ denote the set of contact relations on D. The next lemma shows that $\mathcal{C}$ is not empty.

Lemma 3. *O is the smallest contact relation on D.*

Proof. Suppose that $C \in \mathcal{C}$. If $x \cdot y \neq 0$, then $(x \cdot y)C(x \cdot y)$ by C1, and C3 now implies that xCy. □

An extensive investigation of lattices of contact relations on a Boolean algebra is provided by [7].

In the next lemma we relate a contact relation to products of prime filters:

Lemma 4. *If $C \in \mathcal{C}$, then $C = \bigcup\{F \times G : F, G \in Prime(D), \ F \times G \subseteq C\}$.*

Proof. This was proved in [6] for Boolean contact lattices, and an analysis of the proof shows that it also holds for distributive contact lattices. □

A *clan* is a nonempty subset Γ of D which satisfies:

CL1. If $x, y \in \Gamma$ then xCy;
CL2. If $x + y \in \Gamma$ then $x \in \Gamma$ or $y \in \Gamma$;
CL3. If $x \in \Gamma$ and $x \leq y$, then $y \in \Gamma$.

Note that each proper prime filter is a clan. The set of all clans of D will be denoted by $\mathrm{Clan}(D)$.

Corollary 1. *aCb iff there exists a clan Γ such that $\{a, b\} \subseteq \Gamma$.*

Proof. Suppose that aCb; by the previous Lemma, there are $F, G \in Prime(D)$ such that $a \in F$, $b \in G$, and $F \times G \subseteq C$. Clearly, $F \cup G$ is a clan containing both a and b. The converse follows from the definition of clan. □

3   Topological Models

First we want to recall some notions from topology. By a topological space $(X, C(X))$ we mean a set X provided with a family $C(X)$ of subsets, called closed sets, which contains the empty set $\varnothing$ and the whole set X, and is closed with respect to finite unions and arbitrary intersections. The system $(C(X), \varnothing, X, \cap, \cup)$

is a distributive lattice, called the lattice of closed sets of X: $\varnothing$ is the zero element and X is the unit element of the lattice and the set inclusion is the lattice ordering. Fixing $C(X)$ we say that X is endowed with a topology. A subset $a \subseteq X$ is called *open* if it is the complement of a closed set. The family $Op(X)$ of open sets of X is also a lattice with respect to the same operations. A family of closed sets $\mathbf{B}(X)$ is called a *closed basis* of the topology if every closed set can be represented as an intersection of sets from $\mathbf{B}(X)$. Consequently, $X \in \mathbf{B}(X)$ and $B(X)$ is closed under finite unions; hence, $(\mathbf{B}(X), X, \cup)$ is an upper semi-lattice. Finally, a family of closed sets B is called a *(closed) sub-basis* of the topology if the set of finite unions of elements of B is a closed basis.

In every topological space one can define the following operations on subsets $a \subseteq X$:

1. $Cl(a) = \bigcap\{c \in C(X) : a \subseteq c\}$ (the closure of a), i.e., the intersection of all closed sets containing a.
2. $Int(a) = \bigcup\{o \in Op(X) : a \subseteq o\}$ (the interior of a), i.e., the union of all open sets contained in a.

Cl and Int are interdefinable, i.e. $Cl(a) = -Int(-a)$ and $Int(a) = -Cl(-a)$. If $\mathbf{B}(X)$ is a closed base of X, then obviously:

$$Cl(a) = \bigcap\{b \in \mathbf{B}(X) : a \subseteq b\}.$$

The next two facts follow from above:

$$x \in Cl(a) \text{ iff } (\forall b \in \mathbf{B}(X))(a \subseteq b \rightarrow x \in b);$$
$$x \in Int(a) \text{ iff } (\exists b \in \mathbf{B}(X))(a \subseteq b \text{ and } x \notin b).$$

A subset a of X is called *regular closed* if $Cl(Int(a)) = a$, and, dually, *regular open* if $Int(Cl(a)) = a$ (in this paper we will mainly work with regular closed sets). We denote by $RC(X)$ the family of regular closed sets of X. It is a well known fact that $RC(X)$ is a Boolean algebra with respect to the following operations and constants:

$$0 = \varnothing, 1 = X, a + b = a \cup b \text{ and } a \cdot b = Cl(Int(a \cap b)).$$

$RC(X)$ naturally provides a contact relation C defined by aCb iff $a \cap b \neq \varnothing$. C is called the standard (or Whiteheadean) contact relation on $RC(X)$.

A topological space is called *semi-regular* if it has a closed base of regular closed sets.

Every topological space X can be made semi-regular by defining a new topology taking the set $RC(X)$ as a base. It is a well known fact that this new topology generates the same set of regular closed sets.

The following topological theorem gives necessary and sufficient conditions for a closed base of a topology to be semi-regular.

Theorem 4. [Characterization theorem for semi-regularity]
Let X be a topological space and $\mathbf{B}(X)$ be a closed base for X. Suppose that $\cdot$ is a binary operation defined on the set $\mathbf{B}(X)$ so that $(\mathbf{B}(X), \varnothing, X, \cup, \cdot)$ is a lattice (not necessarily distributive). Then we have:

1. *The following conditions are equivalent:*
   (a) $\mathbf{B}(X)$ *is* U-*extensional.*
   (b) $\mathbf{B}(X) \subseteq RC(X)$.
   (c) *For all* $a, b \in \mathbf{B}(X)$, $a \cdot b = Cl(Int(a \cap b))$.
   (d) $(\mathbf{B}(X), \varnothing, X, \cup, \cdot)$ *is a dually dense sublattice of the Boolean algebra* $RC(X)$.
2. *If any of the (equivalent) conditions (a),(b),(c) or (d) of 1. is fulfilled then:*
   (a) $(\mathbf{B}(X), \varnothing, X, \cup, \cdot)$ *is a* U-*extensional distributive lattice.*
   (b) X *is a semi-regular space.*

Proof. 1. **(a)** $\to$ **(b)**. Let $\mathbf{B}(X)$ be U-extensional, i.e., for all $a, b \in \mathbf{B}(X)$ the following holds:

$$(\forall c \in \mathbf{B}(X))(a \cup c = X \to b \cup c = X) \to a \subseteq b.$$

We must show that for every $a \in \mathbf{B}(X)$, $a = Cl(Int(a))$. This follows from the following chain of equivalences:

$x \in Cl(Int(a))$
$\iff (\forall b \in \mathbf{B}(X))(Int(a) \subseteq b \to x \in b)$
$\iff (\forall b \in \mathbf{B}(X))((\forall y)(y \in Int(a) \to y \in b) \to x \in a)$
$\iff (\forall b \in \mathbf{B}(X))((\forall y)((\exists c \in \mathbf{B}(X))(a \cup c = X \wedge y \notin c) \to y \in b) \to x \in b)$
$\iff (\forall b \in \mathbf{B}(X))((\forall y)(\forall c \in \mathbf{B}(X)(a \cup c = X \to y \in c \vee y \in b)) \to x \in b)$
$\iff (\forall b \in \mathbf{B}(X))((\forall c \in \mathbf{B}(X))(a \cup c = X \to (\forall y)(y \in c \vee y \in b) \to x \in b))$
$\iff (\forall b \in \mathbf{B}(X))((\forall c \in \mathbf{B}(X))(a \cup c = X \to b \cup c = X) \to x \in b)$
$\iff (\forall b \in \mathbf{B}(X))(a \subseteq b \to x \in b)$
$\iff x \in Cl(a) = a.$

(b) $\to$ **(a)**. Let $\mathbf{B}(X) \subseteq RC(X)$. In order to show that $\mathbf{B}(X)$ is U-extensional let $a, b \in \mathbf{B}(X)$ with $a \nsubseteq b$ and $a \cup c = X$. We must show that $b \cup c \neq X$. The assumption (b) shows $Cl(Int(a)) \nsubseteq b$, which implies that there is an $x \in Cl(Int(a))$ with $x \notin b$. We obtain $Int(a) \subseteq c$ implies $x \in c$ for all $c \in \mathbf{B}(X)$, and, hence, $Int(a) \nsubseteq b$. This implies the existence of a $y \in X$ such that $y \in Int(a)$ and $y \notin b$. Again, we obtain that there is $c \in \mathbf{B}(X)$ such that $a \cup c = X$ and $y \notin c$, and, hence, $b \cup c \neq X$.
(b) $\to$ **(c)**. Let $\mathbf{B}(X) \subseteq RC(X)$. Then for any $a \cdot b \in \mathbf{B}(X)$ we have $a \cdot b = Cl(Int(a \cdot b))$. Since $\cdot$ is a lattice meet we obtain that $a \cdot b \subseteq a$, $a \cdot b \subseteq b$, and, hence, $a \cdot b \subseteq a \cap b$. We conclude $a \cdot b = Cl(Int(a \cdot b)) \subseteq Cl(Int(a \cap b))$. For the converse inclusion, we have $Cl(Int(a \cap b)) \subseteq Cl(Int(a)) = a$ and $Cl(Int(a \cap b)) \subseteq Cl(Int(b)) = b$, and, hence, $Cl(Int(a \cap b)) \subseteq a \cdot b$.
(c) $\to$ **(b)**. Let $Cl(Int(a \cap b)) = a \cdot b$. Then $a = a \cdot a = Cl(Int(a \cap a)) = Cl(Int(a))$, which shows that $\mathbf{B}(X) \subseteq RC(X)$.
(b) $\to$ **(d)**. Since (b) implies (c) we conclude that $(\mathbf{B}(X), \varnothing, X, \cup, \cdot)$ is in fact a sublattice of the Boolean algebra $RC(X)$. In order to show that $\mathbf{B}(X)$ is dually dense in $RC(X)$, let $a \in RC(X)$ where $a \neq X$. Since $a = Cl(Int(a))$

and $\mathbf{B}(X)$ is a basis of the closed sets, there exists $c \in \mathbf{B}(X)$ such that $Int(a) \subseteq c$. Furthermore, $a \neq X$ implies that there is an $x \notin Cl(Int(a))$, and, hence, $x \notin c$, which implies $c \neq X$. We conclude $a = Cl(Int(a)) \subseteq Cl(c) = c$, which proves the assertion.

(d) $\rightarrow$ (b). Obvious.

2. This follows immediately since all properties in 1. are equivalent and imply (a) and (b). □

We get the following corollary.

Corollary 2. *Let X be a topological space, $L = (L, 0, 1, +, \cdot)$ be a lattice and let h be an embedding of the upper semi-lattice $(L, 0, 1, +)$ into the lattice $C(X)$ of closed sets of X. Suppose that the set $\mathbf{B} = \{h(a) : a \in L\}$ forms a closed base for the topology of X. Then we have:*

1. *The following conditions are equivalent:*
   (a) L is U-extensional.
   (b) $\mathbf{B} \subseteq RC(X)$.
   (c) For all $a, b \in L$, $h(a \cdot b) = Cl(Int(h(a) \cap h(b)))$.
   (d) h is a dually dense embedding of L into the Boolean algebra $RC(X)$.
2. *If any of the (equivalent) conditions (a),(b),(c) or (d) of 1. is fulfilled then:*
   (a) L is a U-extensional distributive lattice.
   (b) X is a semi-regular space.

This corollary shows that if we require that a lattice L be embeddable into the Boolean algebra $RC(X)$ of some topological space X with the properties of Corollary 2, then the lattice must be both distributive and U-extensional. In the next section we will show that this can be extended to U-extensional distributive contact lattices.

4   Topological Representation of Distributive Contact Lattices

The next theorem is the first main result of the paper.

Theorem 5. [Topological representation theorem for U-extensional distributive contact lattices]
Let $D = (D, 0, 1, +, \cdot, C)$ be an U-extensional distributive contact lattice. Then there exists a semi-regular T0-space and a dually dense embedding h of D into the Boolean contact algebra $RC(X)$ of the regular closed sets of X.

Proof. Let $X = \text{Clan}(D)$ be the set of all clans of D and for $a \in D$, suppose $h(a) = \{\Gamma \in X : a \in \Gamma\}$. Using the properties of clans one can easily check that $h(0) = \varnothing$, $h(1) = X$ and that $h(a + b) = h(a) \cup h(b)$. This shows that the set $\mathbf{B}(X) = \{h(a) : a \in D\}$ is closed under finite unions and can be taken as a closed basis for a topology of X.

In order to show that h is an embedding we must show that $a \leq b$ iff $h(a) \subseteq h(b)$. From the left to the right this follows directly by the properties of clans. Suppose that $a \not\leq b$. Then there exists a prime filter F such that $a \in F$ and $b \notin F$. Since prime filters are clans this shows that $h(a) \not\subseteq h(b)$. Consequently, h is an embedding of the upper semi-lattice $(L, 0, 1, +)$ into the lattice of closed sets $C(X)$ of the space X. By Corollary 2, X is a semi-regular space and h is a dually dense embedding of D into the Boolean algebra $RC(X)$.

Now, we want to show that X is a T0-space. Let $\Gamma \neq \Delta$ be two different points (clans) of X; we will show that there exists a closed set A containing exactly one of them. Suppose $\Gamma \not\subseteq \Delta$. Then there exists $a \in \Gamma$ with $a \notin \Delta$, and, hence, $\Gamma \in h(a)$ and $\Delta \notin h(a)$ so that the $A = h(a)$ will work. In the case $\Delta \not\subseteq \Gamma$ the assertion is shown analogously.

It remains to show that h preserves the contact relation C. But this is a direct consequence of Corollary 1.   □

Notice that Theorem 5 generalizes Theorem 5.1 from [4] to the distributive case. As a consequence of Theorem 5 we obtain the following corollary, which has Theorem 2(2) as a special case. Recall that this theorem was already proved in [3].

Corollary 3. [Topological representation theorem for U-extensional distributive lattices]
Let $D = (D, 0, 1, +, \cdot)$ be a U-extensional distributive lattice. Then there exists a semi-regular T0-space and a dually dense embedding h of D into the Boolean contact algebra $RC(X)$ of the regular closed sets of X.

Proof. Since the overlap O is a contact relation on D the assertion follows immediately from Theorem 5.   □

Due to Corollary 2 we already know that a representation in the sense of Theorem 5 for distributive contact lattices that are not U-extensional is not possible. As mentioned in the introduction we have to use a weaker version of the property that the image $h(D)$ of the embedding h generates (or determines) the topology.

In order to prove such a representation theorem we consider "discrete" Boolean contact algebras defined in [6] as follows. Let (W, R) be a relational system where $W \neq \varnothing$, and R is a reflexive and symmetric relation in W. Subsets of W are considered as (discrete) regions and contact between two subsets $a, b \subseteq W$ is defined by $aĈb$ iff there is $x \in a$ and $y \in b$ such that xRy. Let $\mathbf{D}(W, R)$ denote the distributive lattice of all subsets of W (which is, in fact, a Boolean algebra) with a contact $Ĉ$ defined by R. It was shown in [6] that $\mathbf{D}(W, R)$ is indeed a Boolean and, hence, a distributive contact lattice. Since Boolean algebras are always U-extensional (and in addition, O-extensional) $\mathbf{D}(W, R)$ is a U-extensional distributive contact lattice. It is proved in [6] (using another terminology) that every Boolean contact algebra can be isomorphically embedded into an algebra of the above type. Inspecting the proof given in [6] one can see that it can be transferred easily to the distributive case.

Theorem 6. *Each distributive contact lattice $D = (D, 0, 1, +, \cdot, C)$ can be iso-morphically embedded into a Boolean contact algebra of the form $\mathbf{D}(W, R)$.*

Proof. Let $W = \mathrm{Prime}(D)$ be the set of prime filters of D, let $F, G \in \mathrm{Prime}(D)$ and define R as FRG iff $F \times G \subseteq C$. Consider the Stone embedding $h_{\mathrm{Prime}} : D \to \mathrm{Prime}(D)$. It remains to show that h preserves the contact relation. We observe that

$$h_{\mathrm{Prime}}(x) \hat{C} h_{\mathrm{Prime}}(y)$$
$$\Longleftrightarrow (\exists F, G \in \mathrm{Prime}(D))[x \in F, \ y \in G, \text{ and } F \times G \subseteq C]$$
$$\Longleftrightarrow xCy. \qquad\qquad\qquad\qquad\qquad \text{Lemma 4}$$

This completes the proof.   □

The following corollary is a direct consequence of the last theorem.

Corollary 4. [Extension lemma for distributive contact lattices]
Each distributive contact lattice can be embedded into a (U-extensional) Boolean contact algebra.

Now, we are ready to prove the second main result of this paper.

Theorem 7. [Topological representation theorem for distributive contact lattices]
Let $D = (D, 0, 1, +, \cdot, C)$ be distributive contact lattice. Then there exists a semi-regular T0-space, an embedding h of D into the Boolean contact algebra $RC(X)$ of the regular closed sets of X and an embedding k of D into the Boolean algebra $RC(X)^{op}$ so that $\{h(a) : a \in D\} \cup \{k(a) : a \in D\}$ is a sub-basis of the regular closed sets of X.

Proof. The proof can be realized in two steps. First, by Corollary 4, D can be embedded into a (U-extensional) Boolean contact algebra B. Let e_1 be the corresponding embedding. In the second step, we apply Theorem 5. Consequently, we get an embedding e_2 from B into a semi-regular T0-space X. Now, let $h = e_2 \circ e_1$, i.e. $h(a) = e_2(e_1(a))$ and $k(a) = e_2(e_1(a)^*)$ ($(e_1(a)^*$ is the complement (in B) of embedding of a). Then h is as required. Since the set $\{e_1(a) : a \in D\} \cup \{e_1(a)^* : a \in D\}$ generates the Boolean algebra B we get the last assertion.   □

Next, we want to discuss the two representation theorems proved in this paper in more detail.

Discussion. 1. Notice that there is a difference in the usage of topologies in the topological representation Theorems 5 and 7, and in the Stone topological representation theorems for distributive lattices and Boolean algebras. In Stone's theorem, topology is used to describe the image of the representation up to isomorphism. In our case, the topology is used to obtain good images of the elements of the lattice as regions, e.g., they should have a boundary, etc. For that reason Theorems 5 and 7 are just embedding theorems. In this respect they correspond much more to the embedding theorems for distributive lattices and Boolean algebras in algebras of sets. In our case, sets are replaced by regular closed sets.

2. If we consider contact structures as abstract "pointless" geometries, the question is which notion of points is suitable. In distributive contact lattices we may define two different kinds of points, i.e., prime filters and clans. Prime filters are in some sense "bad" points with respect to the contact structure. They correspond to the lattice part of the structure and can provide a representation by ordinary sets. It is possible to define a contact relation between those sets by means of the contact relation between points. Such a representation is constructed, for instance, in Theorem 6. Clans are "good points" with respect to the contact structure. They guarantee that the image $h(a)$ of each element of the lattice is a region, i.e., has a boundary, interior part, etc. The representation constructed in the proof of Theorem 7 can be interpreted as follows. In a first step we use "bad" points (prime filters) to represent the lattice as a lattice of sets ("bad" regions) and lift the contact relation to that structure. As a positive side-effect we end up with the property of U-extensionality. In the second step, the "good points" (clans) and U-extensionality are used to construct a representation with the intended topological properties. Since prime filters are clans they are among the "good points" of the second step, but they just appear in the interior part of the regions.

These informal explanations are reminiscent of considering prime filters and clans as atoms and molecules – the real points of the real spatial bodies. Similar ideas have been used in [5] for obtaining topological representation theorems for discrete versions of region-based theories of space.

5   Conclusion and Outlook

In this paper we have generalized the notion of Boolean contact algebras by weakening the algebraic part to distributive lattices. This provided a deeper insight into the interaction of several notions used in the representation theory. As a result we obtained a characterization theorem for semi-regularity in topological spaces, which appeared as one of the main tools in the representation theory. We have given two representation theorems of such lattices in algebras of regular closed sets of some topological spaces, considered as standard models for a region-based theory of space. These theorems are direct generalizations of some results from [6] and [4]. Because of the full duality of Boolean algebras, representations in algebras of regular closed sets and regular open sets are dual in that case. In fact, one can construct one representation in terms of the other by duality. In the distributive case duality is preserved only for the lattice part. Consequently, representations in algebras of regular open sets will need different techniques, which we plan to investigate in the future. Kripke semantics and associated reasoning mechanisms may also be developed (as in [1,11]). Another direction of research is to extend the vocabulary of distributive contact lattices with other mereotopological relations such as the non-tangential part-of, $\ll$, and dual contact, $\check{C}$. Last but not least, an open problem is the representation theory of a further generalization to non-distributive contact structures. First results of this direction can be found in [9]. Some non-topological representation theorems for non-distributive lattices may be found in [11]. The main problem here is that it is not obvious what kind of structure we want to consider as a standard model of a non-distributive contact lattice. Obviously, this question has to be resolved before the corresponding representation theory can be developed.

References

1. Allwein, G. and MacCaull, W. (2001). A Kripke semantics for the logic of Gelfand quantales. *Studia Logica*, 61:1-56.
2. Balbes, R. and Dwinger, P. (1974). *Distributive Lattices*. University of Missouri Press, Columbia.
3. Cornish, W. H. (1974). Crawley's completion of a conditionally upper continuous lattice. *Pac J Math*, 51(2):397-405.
4. Dimov, G. and Vakarelov, D. (2006). Contact algebras and region–based theory of space: A proximity approach. *Fundamenta Informaticae*. To appear.
5. Dimov, G. and Vakarelov, D. (2006). Topological Representation of Precontact algebras. In: W. MacCaull, M. Winter and I. Duentsch (Eds.), Relational Methods in Computer Science, LNCS No 3929, To appear.
6. Düntsch, I. and Vakarelov, D. (2006). Region–based theory of discrete spaces: A proximity approach. *Discrete Applied Mathematics*. To appear.
7. Düntsch, I. and Winter , M. (2005). Lattices of contact relations. Preprint.
8. Düntsch, I. and Winter, M. (2005). A representation theorem for Boolean contact algebras. *Theoretical Computer Science (B)*, 347:498-512.
9. Düntsch, I. and Winter, M. (2006). Weak contact structures. In: W. MacCaull, M. Winter and I. Duentsch (Eds.), Relational Methods in Computer Science, LNCS No 3929:73-82.
10. Engelking, R., General topology, PWN, 1977.
11. MacCaull, W. and Vakarelov, D. (2001). Lattice-based Paraconsistent Logic. In: W. MacCaull, M. Winter and I. Duentsch (Eds.), Relational Methods in Computer Science, LNCS No 3929:178-189.
12. Stone, M. (1937). Topological representations of distributive lattices and Brouwerian logics. *Časopis Pěst. Mat.*, 67:1-25.
13. Vakarelov, D., Düntsch, I., and Bennett, B. (2001). A note on proximity spaces and connection based mereology. In Welty, C. and Smith, B., editors, *Proceedings of the 2nd International Conference on Formal Ontology in Information Systems (FOIS'01)*, pages 139-150. ACM.
14. Vakarelov, D., Dimov, G.,Düntsch, I. & Bennett, B. A proximity approach to some region-based theory of space. *Journal of applied non-classical logics*, vol. 12, No3-4 (2002), 527-559
15. Wallman, H. (1938). Lattices and topological spaces. *Math. Ann.*, 39:112-136.

Betweenness and Comparability Obtained from Binary Relations

Ivo Düntsch[1,*] and Alasdair Urquhart[2,*]

[1] Department of Computer Science,
Brock University,
St. Catharines, Ontario, Canada, L2S 3A1
duentsch@brocku.ca
[2] Department of Philosophy,
University of Toronto,
Toronto, Ontario, Canada, M5S 1A2
urquhart@cs.toronto.edu

Abstract. We give a brief overview of the axiomatic development of betweenness relations, and investigate the connections between these and comparability graphs. Furthermore, we characterize betweenness relations induced by reflexive and antisymmetric binary relations, thus generalizing earlier results on partial orders. We conclude with a sketch of the algorithmic aspects of recognizing induced betweenness relations.

1 Introduction

The study of betweenness relations goes back to at least 1917, when Huntington and Kline [10] published "Sets of independent postulates for betweenness." The concept of betweenness can have rather different meanings – we quote from [10]:

- K is the class of points on a line; AXB means that point X lies between the points A and B.
- K is the class of natural numbers, AXB means that number X is the product of the numbers A and B.
- K is the class of human beings; AXB means that X is a descendant of A and an ancestor of B.
- K is the class of points on the circumference of a circle; AXB means that the arc $A - X - B$ is less than $180°$.

In the sequel they concentrate on the geometric case. Throughout, B is a ternary relation on a suitable set, and $B(x, y, z)$ is read as "y lies between x and z." Quantifier free axioms are assumed to be universally quantified. The notation $\#M$ means that all elements of M are different.

---

* Both authors gratefully acknowledge support from the Natural Sciences and Engineering Research Council of Canada.

R.A. Schmidt (Ed.): RelMiCS /AKA 2006, LNCS 4136, pp. 148–161, 2006.

Their first set of four postulates is concerned with three elements:

HK A. $B(a,b,c) \implies B(c,b,a)$.
HK B. $\#\{a,b,c\} \implies B(b,a,c) \vee B(c,a,b) \vee B(a,b,c) \vee B(c,b,a) \vee B(a,c,b) \vee B(b,c,a)$.
HK C. $\#\{a,b,c\} \implies \neg(B(a,b,c) \wedge B(a,c,b))$.
HK D. $B(a,b,c) \implies \#\{a,b,c\}$.

They proceed by adding another eight universal postulates which describe the configurations with four distinct elements, and state "If we think of a and b as two given points on a line, the hypotheses of these postulates state all the possible relations in which two other distinct points x and y of the line can stand in regard to a and b." For later use, we mention

HK 1. $\#\{b,c\} \wedge B(a,b,c) \wedge B(b,c,d) \implies B(a,b,d)$.
HK 2. $B(a,b,c) \wedge B(b,d,c) \implies B(a,b,d)$.

While HK A is widely accepted in many contexts (unless one requires one-way streets), the other postulates make very strong assumptions: HK B says that for any three different elements, one is between the other two, and HK D rules out what one might call degenerate triples. Postulate HK C prohibits "nesting." Their set of postulates completely axiomatizes betweenness if we restrict the domain to linear orders. The postulates HK 1 and HK 2 subsequently became known as "outer transitivity" and "inner transitivity", respectively. Some years later, Huntington [9] proposed a ninth postulate,

H 9. $\#\{a,b,c,x\} \wedge B(a,b,c) \implies B(a,b,x) \vee B(x,b,c)$.

and showed that the axiom system $\{$HK A $-$ HK D, H 9$\}$ is equivalent to the one given in [10].

Betweenness in metric spaces was investigated by Karl Menger [12], and, in a further development, Pitcher and Smiley [14] direct their interest to betweenness relations in lattices, and define $B(a,b,c) \iff a \cdot b + b \cdot c = b = (a+b) \cdot (b+c)$., see Figure 1. Observe that $a \leq b \leq c$ implies $B(a,b,c)$.

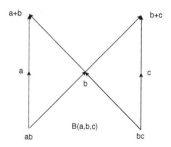

Fig. 1. Betweenness in lattices

The main difference from the system of [9] is the omission of HK B, which is geared to linear orders, and the introduction of "degenerate triples" which contain at most two distinct variables. Thus, their basic system consists only of the symmetry axiom HK A, and

PS β. $B(a,b,c) \wedge B(a,c,b) \Longleftrightarrow b = c.$.

They continue to explore various transitivity conditions and their connection to lattice properties. For example,

Theorem 1. *[14] A lattice L is modular if and only if its betweenness relation satisfies HK 2*

In a parallel development, Tarski proposed an axiom system for first order Euclidean plane geometry based on two relations: equidistance and betweenness. An overview of the system and its history can be found in [16], and axiom numbers below refer to this exposition. His axioms for betweenness are of course very much tailored for the purpose that he had in mind, and many of these are specific to the geometric context. We mention those which are of a more general nature:

Ax 6. $B(a,b,a) \Rightarrow a = b.$	(Identity)
Ax 12. $B(a,b,b).$	(Reflexivity)
Ax 13. $a = b \Rightarrow B(a,b,a).$	(Equality)
Ax 14. $B(a,b,c) \Rightarrow B(c,b,a).$	(Symmetry)
Ax 15. $B(a,b,c)$ and $B(b,d,c) \Rightarrow B(a,b,d).$	(Inner transitivity)
Ax 16. $B(a,b,c)$ and $B(b,c,d)$ and $b \neq c \Rightarrow B(a,b,d).$	(Outer transitivity)

In a further generalization, Birkhoff [2] defines a betweenness relation on U by the condition $B(a,b,c) \Longleftrightarrow a \leq b \leq c$ or $c \leq b \leq a$, where $\leq$ is a partial order. The reader is then invited to prove the following:

Birk 0. $B(a,b,c) \Rightarrow B(c,b,a).$
Birk 1. $B(a,b,c)$ and $B(a,c,b) \Rightarrow b = c.$
Birk 2. $B(a,b,c)$ and $B(a,d,b) \Rightarrow B(a,d,c).$
Birk 3. $B(a,b,c)$ and $B(b,c,d)$ and $b \neq c \Rightarrow B(a,b,d).$
Birk 4. $B(a,b,c)$ and $B(a,c,d) \Rightarrow B(b,c,d).$

It turns out that Birk 1 and Birk 2 follow from Birk 0, Birk 3, and Birk 4. Furthermore, these properties are not sufficient to characterize those betweenness relations which are induced by a partial order. In a fundamental article in 1950, Martin Altwegg [1] obtained an axiom system for such betweenness relations:

Z_1. $B(a,a,a).$
Z_2. $B(a,b,c) \Rightarrow B(c,b,a).$
Z_3. $B(a,b,c) \Rightarrow B(a,a,b).$
Z_4. $B(a,b,a) \Rightarrow a = b.$
Z_5. $B(a,b,c)$ and $B(b,c,d)$ and $b \neq c \Rightarrow B(a,b,d).$
Z_6. Suppose that $\langle a_0, a_1, \ldots, a_{2n}, a_{2n+1} \rangle$ and $a_0 = a_{2n+1}$. If $B(a_{i-1}, a_{i-1}, a_i)$ for all $0 < i \leq 2n+1$, and not $B(a_{i-1}, a_i, a_{i+1})$ for all $0 < i < 2n+1$, then $B(a_{2n}, a_0, a_1).$

Shortly after Altwegg's paper, Sholander [15] investigated betweenness for various kind of orderings, and derived Altwegg's characterization as a Corollary. His system, however, is somewhat shorter. In addition to Z_6, he only supposes

Sho B. $B(a,b,a) \Longleftrightarrow a = b$.
Sho C. $B(a,b,c) \wedge B(b,d,e) \implies B(c,b,d) \vee B(e,b,a)$.

Altwegg's work seems to have been largely forgotten – a notable exception being [3] –, and a search on the Science Citation Index reveals only three citations since 1965. A case in point is the widely studied area of comparability graphs that are closely connected to betweenness relations; as far as we know, researchers in this area were not aware of the earlier results. It is one of the aims of this paper to draw attention to Altwegg's work, and point out some connections between betweenness relations and comparability graphs.

2   Notation

The universe of our relations is a non-empty set U. The identity relation on U is denoted by $1'_U$, or just $1'$, if U is understood. For each $n \in \omega$, $\mathbf{n}$ denotes the set of all $k < n$. For $M \subseteq U$, we abbreviate by $\#M$ the statement that all elements of M are different.

A partial order $\leq$ is called *connected* if there is a path in $\leq \cup \geq$ from a to b for all $a, b \in U$. A *component of* $\leq$ is a maximally connected subset of $<$.

Graphs are assumed to be undirected, without loops or multiple edges. In other words, a graph is just a symmetric irreflexive binary relation on U. A *cycle C in G of length n* is a sequence of elements $a_0, \ldots, a_{n-1}$ of U such that $a_0 G a_1 \ldots a_{n-2} G a_{n-1} G a_0$; repetitions are allowed. A cycle is sometimes called a *closed path in G*. A cycle is *strict*, if $\#\{a_0, \ldots, a_{n-1}\}$, and we denote by C_n the strict cycle of length n. A *triangular chord of the cycle C* is an edge $\{a_i, a_{i+2}\}$ of G; here, addition is modulo n. For example, the graph of Figure 2 contains the cycle $d, a, b, e, b, c, f, c, a, d$ of length 9, which has no triangular chords [5].

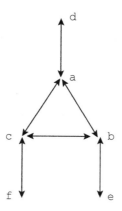

Fig. 2. A graph with a 9–cycle without triangular chords

3   Comparability Graphs

If P is a partial order on U, its *comparability graph* G_P is the set of all comparable proper pairs, i.e. $G_P = (P \cup P^{\smile}) \setminus 1'$; here, $P^{\smile}$ is the relational converse of P. A graph G

is called a *comparability graph*, if $G = G_P$ for some partial order P. We denote the class of comparability graphs by $\mathbb{G}_\leq$.

Two partial orders P, Q on the same set U are called *equivalent* if for all components M_P of P, $P \upharpoonright M_P = Q \upharpoonright M_P$ or $P \upharpoonright M_P = Q^{\smile} \upharpoonright M_P$.

Example 1. *Consider the partial orderings P, Q, shown in Figure 3; obviously, these are not equivalent. In both cases, the only non–comparable elements are b and c, so that $G_P = G_Q$.* □

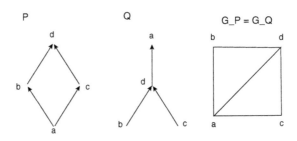

Fig. 3. Non-equivalent partial orders with the same comparability graph

Comparability graphs have been investigated since the early 1960s, and we invite the reader to consult the overview by Kelly [11] for more information. A characterization of comparability graphs is as follows:

Theorem 2. *[4,5]*

GH. *G is a comparability graph if and only if every odd cycle of G contains a triangular chord.*

For example, the graph of Figure 2 is not a comparability graph. It is instructive, and will be useful later on, to consider the strict partial orders $<$ obtained from strict cycles of even length n. As these cycles have no triangles, each path in $<$ has length 2, and, consequently, $a_0 < a_1 > a_2 < a_3, > \dots, a_{n-2} < a_{n-1} > a_0$, or its converse. see Figure 4. Conversely, each crown induces a cycle of even length.

In the sequel, let J be the set of all odd natural numbers greater than 3.

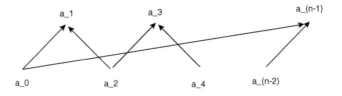

Fig. 4. A crown ordering induced by a strict cycle of even length

If G is a graph and $n \in J$, let σ_n express that each cycle of G of length n has a triangular chord:

$$\sigma_n : (\forall x_0, \ldots, x_{n-1})[x_0 G x_1 G \ldots G x_{n-1} G x_0 \implies x_0 G x_2 \vee x_1 G x_3 \vee \ldots \vee x_{n-2} G x_0 \vee x_{n-1} G x_1].$$

By Theorem 2, G is a comparability graph if and only if it satisfies σ_n for each $n \in J$, and thus, $\mathbb{G}_\leq$ has a universal first order axiomatization. Hence, $\mathbb{G}_\leq$ is closed with respect to substructures.

The following comes as no surprise:

Theorem 3. $\mathbb{G}_\leq$ *is not axiomatizable with a finite number of variables..*

Proof. Assume that Σ is a set of sentences with altogether n variables which axiomatizes $\mathbb{G}_\leq$; we can assume w.l.o.g. that $n = 2r \geq 4$. Let $U = \mathbf{n} + \mathbf{1}$, and G be the cycle on U of length $n + 1$, say, $0, 1, 2, 3, \ldots n, 0$. Then, since G is an odd cycle without triangular chord, it is not in $\mathbb{G}_\leq$.

Suppose that $U' \subseteq U$ with $|U'| = n$, w.l.o.g. $U' = \{0, 1, \ldots, n-1\}$, and H is the restriction of G to U'. Then, $H = G \setminus \{\langle n-1, n \rangle, \langle n, 0 \rangle\}$, and H is the comparability graph of the crown of Figure 4 with $\langle a_0, a_{n-1} \rangle$ removed.

Now, since the satisfaction in $\langle U, G \rangle$ of sentences with at most n variables depends only on its satisfaction in the n - generated substructures of $\langle U, G \rangle$, we have $\langle U, G \rangle \models \Sigma$. This contradicts the fact that $G \notin \mathbb{G}_\leq$.

4   Betweenness Relations

Considering the plethora of proposed axiomatizations of betweenness relations, we need to decide which axioms to use. Our strategy will be to start with those postulates that are most common, present the least restrictions and still let us obtain a sensible theory. We will allow "degenerate triples", as it enables us to go from triples to pairs and vice versa.

A ternary relation on a set U is called a *betweenness relation* if it satisfies

BT 0. $B(a, a, a)$.
BT 1. $B(a, b, c) \Rightarrow B(c, b, a)$.
BT 2. $B(a, b, c) \Rightarrow B(a, a, b)$
BT 3. $B(a, b, c)$ and $B(a, c, b) \Rightarrow b = c$.

We denote the class of all betweenness relations by $\mathbb{B}$. Observe that at this stage we do not include any transitivity conditions. Since $\mathbb{B}$ is a universal Horn class, it is closed under substructures, and thus, under unions of chains, and also under direct products; note that for any set M of triples consistent with the axioms, there is a smallest betweenness relation B containing M, and that B is finite just when M is finite.

With BT 1, BT 2, and BT 3 one can easily prove

Lemma 1.   *1. $B(a, b, c)$ implies $B(a, b, b)$, $B(b, b, c)$, and $B(b, c, c)$.*
2. $B(a, b, a) \Rightarrow a = b$.

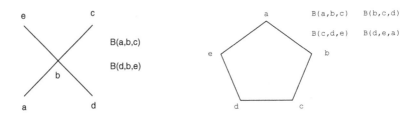

Fig. 5. A betweenness relation not induced by a binary relation

Fig. 6. A betweenness relation based on a pentagon

However, in the absence of transitivity axioms, $B(a,a,c)$ does not follow.

A triple $\langle a,b,c \rangle$ is called *proper*, if $\#\{a,b,c\}$. We say that a,b are *comparable*, if $B(a,a,b)$, and let C_B be the set of all comparable pairs. By BT 0, C_B is reflexive, and by Lemma 1 it is symmetric. If aC_Bb, and $a \neq b$ we call a and b *strictly comparable*, and denote the graph of all strictly comparable pairs by C_B^+. Note that C_B^+ does not necessarily determine B, as Example 1 shows.

Conversely, if R is a reflexive antisymmetric binary relation on U we let $B_R = \{\langle a,b,c \rangle : aRbRc \text{ or } cRbRa\}$, and say that B *is induced by* R, if $B = B_R$; it is straightforward to see that $B_R \in \mathbb{B}$. The question arises whether every betweenness relation is induced by a binary relation. The following two examples show that the answer is "no."

Example 2. *Let $U = \{a,\ldots,e\}$ and B be the smallest betweenness relation on U containing $\langle a,b,c \rangle$ and $\langle d,b,e \rangle$, see Figure 5 ; inspection shows that these are the only nontrivial triples of B.*

Assume that B is induced by the binary relation R. Then, $B(a,b,c)$ implies $aRbRc$ or $cRbRa$, and $B(d,b,e)$ implies $dRbRe$ or $eRbRd$. If, for example, $aRbRc$ and $dRbRe$, then $dRbRc$, implying $B(d,b,c)$, which is not the case. The other cases are similar. □

Another instructive example is the pentagon shown in Figure 6. We let B be the smallest betweenness relation containing $\langle a,a,b \rangle, \langle b,b,c \rangle, \langle c,c,d \rangle, \langle d,d,e \rangle$, and $\langle e,e,a \rangle$; then, C_B is the pentagon, and each triple in B contains at most two different entries..

Assume that $B = B_R$ for some reflexive and antisymmetric relation R. Then, aRb or bRa; suppose w.l.o.g. that aRb. Since B contains no proper triples, we then must have cRb, cRd, eRd, eRa. But then, $B(e,a,b)$, a contradiction. On the other hand, if B is generated by $\{\langle a,b,c \rangle, \langle b,c,d \rangle, \ldots, \langle e,a,b \rangle\}$, then B is induced by the reflexive and antisymmetric relation $aRbRcRdReRa$ enhanced by $1'$. This shows that C_B can have cycles of odd length without a triangular chord.

5   Axiomatizing Betweenness Relations

In this section, we generalize Altwegg's theorem characterizing betweenness relations arising from partially ordered sets. It turns out that the basic ideas of his proof go through even in the absence of transitivity. Our main theorem in this section characterizes the betweenness relations arising from reflexive, antisymmetric orderings, from which we obtain Altwegg's result as a corollary.

The construction $R \longmapsto B_R$, as defined in §4, produces a structure satisfying the axioms of a betweenness relation, defined on the same universe, if R is reflexive and antisymmetric. The main idea of the present section is to show that there is an inverse map $B \longmapsto R_B$; however, the map is not unique, rather it depends on an arbitrary choice of orientation for each component of the strict comparability graph C_B^+. If we produce a new relation R', from a reflexive, antisymmetric relation R by reversing the direction of all pairs in R within a fixed component of its comparability graph, then R and R' generate the same betweenness relation. However, if we regard two such relations as equivalent if they differ from one another only with respect to an arbitrary choice of orientation for a set of components, then the map $B \longmapsto R_B$ determines a relation that is unique up to equivalence. This notion generalizes the concept of equivalence for partially ordered sets defined in §3.

We begin by recalling some terminology from Altwegg's paper [1]. Suppose that B is a betweenness relation on U. A sequence $a_0, a_1, a_2, \ldots, a_{n-1}, a_n$ is called a *C–path* in B, if $a_0 C_B a_1 C_B \ldots C_B a_n$, i.e. every two consecutive entries are comparable. It is called a *B–path*, if $B(a_i, a_{i+1}, a_{i+2})$ for all $i \leq n-2$. Every C–path can be made into a B–path by doubling a_i for each $0 < i < n$.

Having derived a B–path $a_0, a_1, \ldots, a_n$ from a C–path, it can be reduced in the following way:

1. If $a_{i_0} = a_{i_1} = \ldots = a_{i_k}$, then remove $a_{i_2}, \ldots, a_{i_k}$.
2. If $a_i, a_{i+1}, a_{i+2}, a_{i+3}$, $a_{i+1} = a_{i+2}$, and $B(a_i, a_{i+1}, a_{i+3})$, then remove a_{i+2}.

A completely reduced path is called a *chain*. Note that by the construction of a chain $a_0, a_1, \ldots, a_n$ from a B–path, for $0 \leq i, i+1, i+2, i+3 \leq n$,

$$(5.1) \qquad \#\{a_i, a_{i+1}, a_{i+2}\} \Rightarrow B(a_i, a_{i+1}, a_{i+2}),$$

$$(5.2) \qquad a_{i+1} = a_{i+2} \Rightarrow \neg B(a_i, a_{i+1}, a_{i+3}).$$

A chain is called *simple*, if consecutive entries are different. We also call a, b a simple chain, if $a \neq b$ and $B(a, a, b)$. Clearly, for each $0 \leq k < m \leq n$, $a_k, a_{k+1}, \ldots, a_m$ is again a simple chain, and the inverse $a_n, \ldots, a_0$ of a (simple) chain $a_0 \ldots, a_n$ is also a (simple) chain.

Definition 1. *We define the notion of a B-walk of size n by induction on n:*

1. *A simple chain is a B-walk of size 1;*
2. *If $W = a, \ldots, p, q$ is a B-walk of size n, and $C = q, r \ldots, z$ a simple chain where $\neg B(p, q, r)$, then the sequence $a, \ldots, p, q, r, \ldots, z$ obtained by identifying the last element of W with the first element of C is a B-walk of size $n + 1$.*

In other words, a *B*-walk consists of a sequence obtained by gluing together simple chains; the gluing consists of identifying their endpoints. If $W = a, b, \ldots, c, d$ is a *B*-walk, then we say that it is a *B-walk from a, b to c, d*. A *B*-walk is even or odd depending on whether its size is even or odd. The *length* of a *B*-walk is its length, considered as a sequence, so, for example, the *B*-walk a, b, a, b has length 4. Note that length and size may differ; indeed, it is the definition of *size* in the various scenarios that causes GH, Z_6, and BT 4 below to look so similar. A *B-cycle* is a *B*-walk $a_0, a_1 \ldots, a_{n-1}, a_n$, in which the first and last two elements are the same ($a_0 = a_n$), and $\neg B(a_{n-1}, a_0, a_1)$.

Lemma 2. *Let R be a reflexive, antisymmetric relation, and B_R the betweenness relation generated by R. Assume that aRb and that $W = a, b, \ldots, c, d$ is a B_R-walk from a, b to c, d. If W is odd, then cRd, while if W is even, then dRc.*

Proof. The proof is by induction on the length of W. For a B_R-walk of length 2, it holds by assumption. Assuming for walks of length $n > 1$, let $a, b, \ldots, c, d, e$ be an odd B_R-walk of length $n + 1$. If $a, b, \ldots, c, d$ is also odd, then $B_R(c, d, e)$, and cRd, hence dRe. On the other hand, if $a, b, \ldots, c, d$ is even, then $\neg B_R(c, d, e)$, and dRc by inductive assumption, showing that dRe. The proof for even walks is symmetrical.   □

The main theorem of this section is:

Theorem 4. *The theory $\mathbb{B}_R$ of betweenness relations generated by a reflexive, antisymmetric relation is axiomatized by the following postulates:*

> *BT 0.* $B(a, a, a)$.
> *BT 1.* $B(a, b, c) \Rightarrow B(c, b, a)$.
> *BT 2.* $B(a, b, c) \Rightarrow B(a, a, b)$.
> *BT 3.* $B(a, b, c)$ *and* $B(a, c, b) \Rightarrow b = c$.
> *BT 4.* *There are no odd B-cycles.*

The fact that BT 4 holds for a betweenness relation B_R generated by a reflexive, antisymmetric relation R follows easily from Lemma 2. Note that Altwegg's postulate Z_6 is a special case of our BT 4. The more general formulation is needed here because the transitivity axioms are not available. To illustrate the axiom BT 4, let us consider two of the betweenness relations from the previous section. In Example 2, the sequence a, b, e, b, d, b, a is an odd B-cycle. The five simple chains making up the cycle are

$$a, b \mid b, e \mid e, b, d \mid d, b \mid b, a.$$

In the next example (the pentagon of Figure 6), the sequence a, b, c, d, e, a is an odd B-cycle. For any betweenness relation that is not generated by a reflexive, antisymmetric relation, there is an odd B-cycle that is a witness to this fact.

Lemma 3. *Let B be a betweenness relation satisfying the axiom BT 4 whose strict comparability graph C_B^+ is connected. If $\{a, b\}$, $\{c, d\}$ are distinct edges in this strict graph, then there is an odd B-walk from a, b to c, d or an odd B-walk from a, b to d, c, but not both.*

Proof. Since the strict comparability graph of B is connected, there is a C-path, and hence a B-path joining the edges $\{a, b\}$ and $\{c, d\}$. This path (or its inverse) must have one of the four forms: $a, b, \ldots, c, d$, $b, a, \ldots, c, d$, $b, a, \ldots d, c$ or $a, b, \ldots d, c$. By successive reductions, we can assume that this B-path is in fact a chain. Simplify this chain by removing immediate repetitions from it. Then the result is a B-walk from a, b to c, d, or from b, a to c, d, or from b, a to d, c, or from a, b to d, c.

For any $e, f, g, h \in U$, there is an even B-walk from e, f to g, h. if and only if there is an odd B-walk from e, f to h, g, since if $e, f, \ldots, g, h$ is an even B-walk from e, f to g, h, then $e, f, \ldots, g, h, g$ is an odd B-walk from e, f to h, g, and conversely. If the B-walk joining the edges $\{a,b\}$ and $\{c,d\}$ starts with a,b, then we are through. If on the other hand, it starts with b,a, then there is a B-walk of opposite parity starting with a,b, by the same argument as above. Hence, there is an odd B-walk from a,b to c,d or an odd B-walk from a,b to d,c.

It remains to be shown that there cannot be odd walks from a,b to both e,f and f,e, for any distinct comparable elements e,f. Suppose that $W_1 = a,b,\ldots,e,f$ and $W_2 = a,b,\ldots f,e$ are both odd B-walks. Then the inverse of W_2, $W_3 = e,f,\ldots b,a$ is odd. Let W_4 be the walk $a,b,\ldots,e,f,\ldots b,a$ resulting from the identification of the last two elements of W_1 and the first two elements of W_3. Then W_4 is an odd B-cycle, contradicting BT 4.   □

If U is a fixed universe, and R a reflexive, antisymmetric relation defined on U, then we write $\mathscr{B}(R) = \langle U, B_R \rangle$ for the betweenness relation defined from R. In the next definition, we describe the inverse construction.

Definition 2. *Let B be a betweenness relation defined on the set U, satisfying the axiom BT 4, and whose strict comparability graph C_B^+ is connected. In addition, let $\{a,b\}$ be an edge in C_B^+. Then $\mathscr{R}(B,a,b)$ is the relational structure $\langle U,R \rangle$ defined on U as follows. For $c,d \in U$, cRd holds if and only if $c = d$, or $\{c,d\}$ is an edge in C_B^+, and there is an odd B-walk from a,b to c,d.*

It follows from Lemma 3 that $\mathscr{R}(B,a,b)$ is reflexive and antisymmetric.

Theorem 5. *Let R be a reflexive, antisymmetric relation on U, and let aRb, where $a \neq b$. In addition, assume that the strict comparability graph of $\mathscr{B}(R)$ is connected. Then $\mathscr{R}(\mathscr{B}(R),a,b) = \langle U,R \rangle$.*

Proof. This follows from Lemmas 2 and 3.   □

Theorem 6. *Let B be a betweenness relation defined on the set U, satisfying the axiom BT 4, and whose strict comparability graph C_B^+ is connected. In addition, let $\{a,b\}$ be any edge in C_B^+. Then $\mathscr{B}(\mathscr{R}(B,a,b)) = \langle U,B \rangle$*

Proof. Let R be the relation defined from B in $\mathscr{R}(B,a,b)$. We need to show for any c,d,e in U, that $B(c,d,e)$ holds if and only if $B_R(c,d,e)$.

First, let us assume that $B(c,c,d)$, $c \neq d$. Then the edge $\{c,d\}$ belongs to the comparability graph C_B^+, so that cRd or dRc holds by Lemma 3, hence $B_R(c,c,d)$. Second, assume that $B(c,d,e)$ holds, where $\#\{c,d,e\}$. Then $\{c,d\}$ and $\{d,e\}$ are edges in C_B^+, and so cRd or dRc holds, and similarly dRe or eRd. Let us suppose that $\{c,d\}$ and $\{d,e\}$ are not consistently oriented, so that (say) cRd, but eRd. By construction, there are odd B-walks $W_1 = a,b,\ldots,c,d$ and $W_2 = a,b,\ldots,e,d$. Now consider the B-walk $a,b,\ldots,c,d,e,\ldots,b,a$ obtained by identifying the last element of W_1 with the first element of the inverse of W_2. This walk is an odd B-cycle, contradicting BT 4. It follows that $B_R(c,d,e)$.

For the converse, let us assume that $B_R(c,d,e)$, where $\#(c,d,e)$, but not $B(c,d,e)$. By construction, there are odd B-walks $W_1 = a,b,\ldots,c,d$ and $W_2 = a,b,\ldots,d,e$, hence there is an even B-walk $W_3 = a,b,\ldots,e,d$. The B-cycle $a,b,\ldots,c,d,e,\ldots,b,a$ obtained by identifying the last element of W_1 with the first element of W_3 is odd, contradicting BT 4. Hence, $B(c,d,e)$ must hold, completing the proof.   □

We can now use the previous results to prove the main theorem.

Proof of Theorem 4: We have already observed that Lemma 2 implies that betweenness relations generated from reflexive, antisymmetric relations satisfy BT 4. Conversely, let B be a betweenness relation satisfying BT 4. Then $B = \bigcup_{i \in I} B_i$, where each B_i is the restriction of B to one of the connected components C_i of C_B^+. Each such B_i also satisfies BT 4. If C_i contains no edges, then the universe U_i of this component is a unit set, and we can set R_i to be the identity relation on U_i. If C_i contains at least one edge $\{a_i, b_i\}$, then choose an orientation a_i, b_i for this edge. By Theorem 6, $\mathscr{B}_i(\mathscr{R}(B_i, a_i, b_i)) = \langle U_i, B_i \rangle$. Hence, setting $\mathscr{R}(B) = \bigcup_{i \in I} \mathscr{R}(B_i, a_i, b_i)$, $\mathscr{B}(\mathscr{R}(B)) = \langle U, B \rangle$, showing that the class of betweenness structures satisfying BT 4 is identical with those betweenness structures arising from reflexive, antisymmetric relations.   □

Theorem 4 is quite powerful, and we can deduce results for restricted classes of relations with its help. The next theorem is equivalent to Altwegg's result of 1950; it shows that it is sufficient to add the outer transitivity axiom to our basic set of postulates.

Theorem 7. *The theory* $\mathbb{B}_\leq$ *of betweenness relations generated by a partial order is axiomatized by the following postulates:*

BT 0. $B(a,a,a)$.
BT 1. $B(a,b,c) \Rightarrow B(c,b,a)$.
BT 2. $B(a,b,c) \Rightarrow B(a,a,b)$
BT 3. $B(a,b,a) \Rightarrow a = b$.
BT 4. *There are no odd B-cycles.*
BT 5. $B(a,b,c)$ *and* $B(b,c,d)$ *and* $b \neq c \Rightarrow B(a,b,d)$.

Proof. In view of Theorem 6, it is sufficient to prove that if a betweenness relation B satisfies BT 5, that the relation $\mathscr{R}(B)$ is transitive. Suppose that aRb and bRc hold in $\mathscr{R}(B)$, where $\#\{a,b,c\}$. Then by Theorem 6, $B(a,b,c)$ holds. By BT 2 and BT 5, we have $B(a,a,c)$, so that a and c are comparable, hence aRc or cRa. Now if cRa holds, then we have $B(b,c,a)$, so by BT 2, $B(a,b,a)$, a contradiction. Hence, aRc, showing that R is transitive.

To prove Altwegg's theorem, that his axiom system Z_1 to Z_6 also characterizes this set of betweenness relations, it is sufficient to show that BT 4 can be deduced from BT 0, BT 1, BT 2, BT 3 and BT 5, together with Z_6. Now Altwegg's postulate Z_6 asserts, using our earlier terminology, that there is no odd B-cycle in which the simple chains that compose it are of length 2 (that is to say, of minimum length). However, in the presence of the outer transitivity axiom, it is not hard to show that if $a, b, \ldots, c, d, e, \ldots, f, g$ is a simple chain, then so is $a, b, \ldots, c, e, \ldots, f, g$; that is to say, the intermediate elements in a simple chain can be removed, and the result is still a simple chain. Consequently, if we postulate outer transitivity, then Z_6 implies the more general version BT 4.   □

The following can be proved using basically the same construction as in Theorem 3:

Theorem 8. *The theories* $\mathbb{B}_\leq$ *and* $\mathbb{B}_R$ *are not axiomatizable with a finite number of variables.*

6   Algorithmic Aspects

In this section, we give a brief sketch of the algorithmic aspects of betweenness relations. In the case of comparability graphs arising from partially ordered sets, very
efficient algorithms are known for both the recognition problem and colouring problems. The reader is referred to the work of Golumbic [6,7,8] for descriptions of these
algorithms, and to the article by Möhring [13] for an informative survey of this area.

The characterization given in §5 rests on the fact that if we have assigned orientations
to some edges in the comparability graph of a betweenness relation, then other orientations are forced by the betweenness structure. If we use the notation $a \to b$, $a \leftarrow b$ to
symbolize the fact that we have assigned the orientation (a,b) (respectively (b,a)) to
the unordered edge $\{a,b\}$, then the following implications hold:

Imp 0. $[B(a,b,c) \wedge (a \neq b) \wedge (b \neq c) \wedge (a \to b)] \Rightarrow (b \to c)$;
Imp 1. $[B(a,b,c) \wedge (a \neq b) \wedge (b \neq c) \wedge (a \leftarrow b)] \Rightarrow (b \leftarrow c)$;
Imp 2. $[\neg B(a,b,c) \wedge (a \neq b) \wedge (b \neq c) \wedge (a \to b)] \Rightarrow (b \leftarrow c)$;
Imp 3. $[\neg B(a,b,c) \wedge (a \neq b) \wedge (b \neq c) \wedge (a \leftarrow b)] \Rightarrow (b \to c)$.

Let us say that a set S of ordered pairs (a,b), $a \neq b$, where a,b belong to the universe
of a betweenness relation $\langle U, B \rangle$, is *implicationally closed* if it is closed under these
implications (interpreting "$a \to b$" as "$(a,b) \in S$" and "$a \leftarrow b$" as "$(b,a) \in S$") and that
it is an *implicational class of B* if it is a minimal non-empty implicationally closed
subset of U. If A is an implicational class, then $A^{\smile}$ is the implicational class representing
the result of reversing the orientation of all edges in A. Using this terminology, we
can give an alternative characterization of betweenness relations arising from reflexive,
antisymmetric relations; this is the analogue of a corresponding theorem of Golumbic
for comparability graphs [6].

Theorem 9. *A betweenness relation B is generated by a reflexive antisymmetric relation if and only $A \cap A^{\smile} = \emptyset$ for all implicational classes of B.*

Proof. If B is a betweenness relation that is *not* generated by such a relation, then by
Theorem 4, there must be an odd B-cycle. Choose any edge (a,b) in this cycle, and
consider the smallest implicational class A containing it. By the argument of Lemma 2,
the oriented edge (b,a) must also belong to this class, showing that $A = A^{\smile}$.

For the converse, let us assume that $A \cap A^{\smile} \neq \emptyset$ for some implicational class A, and let
$(a,b),(b,a) \in A$. Since A is the smallest implicational class containing (a,b), it follows
that there is a sequence of elements of U, $a_0, \ldots, a_k$, and a sequence of statements
$S_1, \ldots, S_i, \ldots, S_k$, where each statement S_i is of the form $(a_{i-1} \to a_i)$ or $(a_{i-1} \leftarrow a_i)$,
$S_1 = (a_0 \to a_1) = (a \to b)$, $S_k = (b \to a)$, and every statement in the sequence, except
for the first, is derived from the preceding statement by one of the implicational rules
Imp 0 – Imp 3. In Example 2 of §4, such a sequence of statements is given by: $(a \to
b),(b \leftarrow e),(e \to b),(b \to d),(d \leftarrow b),(b \to a)$. Then it is straightforward to check that
the sequence of elements $a_0, \ldots, a_k$ is an odd B-cycle.   □

Theorem 9 immediately suggests an algorithm to determine whether or not a betweenness relation is generated by a reflexive, antisymmetric relation. The algorithm consists

of generating all of the implicational classes generated by directed edges in the comparability graph of the relation, while checking to see whether any overlap ever occurs between an implication class A and its converse $A^{\vee}$. If we succeed in generating all such classes without an overlap, then they can be used to orient the edges appropriately, while if an overlap occurs, then Theorem 9 tells us that the betweenness relation cannot be generated by a reflexive, antisymmetric relation.

If $\langle U, B \rangle$ is a betweenness relation, and $b \in U$, then the *betweenness degree* of b is the number of proper triples (a,b,c) in B; the *betweenness degree* $\Delta(B)$ of the relation B is the maximum betweenness degree of any element in U. The *comparability degree* $\delta(B)$ of the relation B is the maximum degree of any vertex of the comparability graph of B.

Theorem 10. *There is an algorithm to determine whether a given betweenness relation B is generated by a reflexive, antisymmetric relation that runs in $O((\Delta(B) + \delta(B))|B|)$ time and $O(|B|)$ space.*

Proof. We provide only a brief sketch of this result. The basic ideas of the algorithm are all to be found in the original paper of Golumbic [6], and the reader can consult this paper for the details of the implementation.

We initialize the data structures for the algorithm by setting up two arrays of linked lists, one for the proper triples in B, the other for the edges in the comparability graph. This takes space $O(|B|)$. Then we start from an arbitrarily selected edge $\{a,b\}$ in the comparability graph, and generate the smallest implicational class A containing (a,b), simultaneously with its converse $A^{\vee}$. The time complexity of the algorithm can be estimated through an upper bound on the time taken to look up the appropriate implication, when extending the implication classes. Suppose that (a,b) belongs to our class, and that we wish to see if there is an edge (b,c) or (c,b) that should be added because of some implication. First, we search for such an edge in the array representing the comparability graph; this takes time $O(\delta(B))$. Second, if we have found such an edge, we look for an appropriate proper triple with the middle element b; this takes time $O(\Delta(B))$, assuming that we have indexed such triples by their middle elements. Consequently, the entire procedure takes time $O((\Delta(B) + \delta(B))|B|)$.   □

7 Summary and Outlook

We have given an outline of the history of axiomatizations of betweenness relations, and have shown that the class of betweenness relations generated by a reflexive and antisymmetric binary relation is first order axiomatizable, albeit with an infinite number of variables. Furthermore, we have pointed out the connection of betweenness relations to comparability graphs. Such a graph may be generated by essentially different partial orders; in contrast, betweenness relations carry, in some sense, total information: If B is generated by a reflexive and antisymmetric binary relation, then this relation is determined up to taking converse on its components. In further work, we plan to investigate more deeply the relation between comparability graphs and betweenness relations, and also to give characterizations of induced betweenness relations in terms of forbidden substructures.

References

1. Martin Altwegg. Zur Axiomatik der teilweise geordneten Mengen. *Commentarii Mathematici Helvetici*, 24:149–155, 1950.
2. Garrett Birkhoff. *Lattice Theory*. American Mathematical Society, 1948. Second revised edition.
3. Nico Düvelmeyer and Walter Wenzel. A characterization of ordered sets and lattices via betweenness relations. *Resultate der Mathematik*, 46:237–250, 2004.
4. Alain Ghouila-Houri. Caractérisation des graphes non orientés dont on peut orienter les arêtes de manière à obtenir le graphe d'une relation d'ordre. *C.R. Acad. Sci. Paris*, pages 1370–1371, 1962.
5. Paul C. Gilmore and Alan J.Hoffman. A characterization of comparability graphs and of interval graphs. *Canadian Journal of Mathematics*, 16:539–548, 1964.
6. Martin Charles Golumbic. Comparability graphs and a new matroid. *Journal of Combinatorial Theory*, 22:68–90, 1977.
7. Martin Charles Golumbic. The complexity of comparability graph recognition and coloring. *Computing*, 18:199–208, 1977.
8. Martin Charles Golumbic. *Algorithmic graph theory and perfect graphs*. Academic Press, New York, 1980.
9. Edward V. Huntington. A new set of postulates for betweenness, with proof of complete independence. *Trans Am Math Soc*, 26:257–282, 1924.
10. Edward V. Huntington and J. Robert Kline. Set of independent postulates for betweenness. *Trans. Am. Math. Soc.*, 18:301–325, 1917.
11. David Kelly. Comparability graphs. In Ivan Rival, editor, *Graphs and Order*, pages 3–40. D. Reidel Publishing Company, 1985.
12. Karl Menger. Untersuchungen über die allgemeine Metrik. *Mathematische Annalen*, 100:75–163, 1928.
13. Rolf H. Möhring. Algorithmic aspects of comparability graphs and interval graphs. In Ivan Rival, editor, *Graphs and Order*, pages 41–101. D. Reidel Publishing Company, 1985.
14. Everett Pitcher and M.F. Smiley. Transitivities of betweenness. *Trans. Am. Math. Soc.*, 52:95–114, 1942.
15. Marlow Sholander. Trees, lattices, order and betweenness. *Proc. Am. Math. Soc.*, 3(3):369–381, 1952.
16. Alfred Tarski and Steven Givant. Tarski's system of geometry. *The Bulletin of Symbolic Logic*, 5(2):175–214, 1998.

Relational Representation Theorems for General Lattices with Negations[*]

Wojciech Dzik[1], Ewa Orlowska[2], and Clint van Alten[3]

[1] Institute of Mathematics, Silesian University
Bankowa 12, 40–007 Katowice, Poland
dzikw@silesia.top.pl
[2] National Institute of Telecommunications
Szachowa 1, 04–894 Warsaw, Poland [**]
orlowska@itl.waw.pl
[3] School of Mathematics, University of the Witwatersrand
Private Bag 3, Wits 2050, Johannesburg, South Africa
cvalten@maths.wits.ac.za

Abstract. Relational representation theorems are presented for general (i.e., non-distributive) lattices with the following types of negations: De Morgan, ortho, Heyting and pseudo-complement. The representation is built on Urquhart's representation for lattices where the associated relational structures are doubly ordered sets and the canonical frame of a lattice consists of its maximal disjoint filter-ideal pairs. For lattices with negation, the relational structures require an additional binary relation satisfying certain conditions which derive from the properties of the negation. In each case, these conditions are sufficient to ensure that the lattice with negation is embeddable into the complex algebra of its canonical frame.

1 Introduction

A relational semantics for the class of lattices was developed by Urquhart in [9]. These semantics were extended by Allwein and Dunn in [1] to include other operations on lattices such as negation, fusion and implication. In particular, they obtained a relational semantics for lattices with a De Morgan negation. In this paper we shall develop relational semantics for lattices with other negations, namely, Heyting negation, pseudo-complement and ortho-negation. The relational structures are extensions of Urquhart's relational structures for lattices, which are of the type $\langle X, \leqslant_1, \leqslant_2 \rangle$ where $\leqslant_1$ and $\leqslant_2$ are quasi-orders on X satisfying: $x \leqslant_1 y$ and $x \leqslant_2 y \Rightarrow x = y$. The *complex algebra* of such a relational structure is a bounded lattice with universe consisting of 'ℓ-closed' subsets of X

[*] Supported by the NRF-funded bilateral Poland/RSA research project GUN 2068034: Logical and Algebraic Methods in Formal Information Systems.
[**] E.O. acknowledges a partial support from the INTAS project 04-77-7080 Algebraic and Deduction Methods in Non-classical Logic and their Applications to Computer Science.

R.A. Schmidt (Ed.): RelMiCS /AKA 2006, LNCS 4136, pp. 162–176, 2006.

ordered by inclusion. To extend this to De Morgan negations, Allwein and Dunn introduce a relation N on X which is a function with some order-preserving properties. The complex algebra of such a relational structure is a bounded lattice with a definable negation that is De Morgan. We find suitable conditions on N that further ensure that the definable negation is, in fact, an ortho-negation (i.e., also satisfies the Heyting property: $a \wedge (\neg a) = 0$).

The other negations we consider are derived from Dunn's 'kite of negations'. We consider a 'quasi-minimal' Heyting negation. In this case, a binary relation C is added to Urquhart's relational structures, together with some conditions on C, $\leqslant_1$ and $\leqslant_2$ that ensure that the complex algebra has a definable negation that is also quasi-minimal and Heyting. A 'pseudo-complemented' lattice is a lattice with negation satisfying: $a \wedge b = 0 \ \Leftrightarrow \ a \leq \neg b$. Such algebras form a subclass of the class just considered and their relational structures require an additional condition to ensure that that a complex algebra is a pseudo-complemented lattice.

For each type of lattice with negation considered, a *canonical frame* may be constructed that is a relational structure of the correct type and satisfies the required conditions. Thus, we may consider the complex algebra of the canonical frame of each lattice with negation. The main result stated in each case is that the lattice with negation may be embedded into the complex algebra of its canonical frame.

The framework described above serves, on the one hand, as a tool for investigation of classes of lattices with negation operations and, on the other hand, as a means for developing Kripke-style semantics for the logics whose algebraic semantics is given. The representation theorems play an essential role in proving completeness of the logics with respect to a Kripke-style semantics determined by a class of frames associated with a given class of algebras. In this paper we deal with the algebraic aspects of lattices with negation. The framework presented above has been used in [8] and [5] in the context of lattice-based modal logics. It has been applied to lattice-based relation algebras in [4] and to double residuated lattices in [6] and [7].

2   Negations

We follow J. M. Dunn's analysis of negations, also known as "Dunn's Kite of Negations". Dunn's study of negation in non-classical logics as a negative modal operator is an application of his gaggle theory, cf. [3], which is a generalization of the Jonsson-Tarski Theorem. In gaggle theory, negation $\neg$ is treated as a Galois connective on an underlying poset or bounded lattice. This treatment requires the condition:

(N2) $a \leq \neg b \Leftrightarrow b \leq \neg a$ (Quasi-minimal)

and leads to the following conditions for $\neg$:

(N1) $a \leq b \Rightarrow \neg b \leq \neg a$ (Preminimal)

(N2') $a \leq \neg\neg a$
(Int) $a \leq b$ and $a \leq \neg b \Rightarrow a \leq x$ for all x (Intuitionistic)
(DeM) $\neg\neg a \leq a$ (De Morgan).

Note that from (N2) one may derive (N2') and (N1) as follows. From $\neg a \leq \neg a$ we get $a \leq \neg\neg a$. If $a \leq b$, then $a \leq \neg\neg b$ hence $\neg b \leq \neg a$. Conversely, (N2) is derivable from (N1) and (N2') as follows. If $a \leq \neg b$, then $\neg\neg b \leq \neg a$ hence $b \leq \neg a$. Thus, in any partially ordered set with $\neg$ the following implications hold:

$$(\text{N1}) \ \& \ (\text{N2'}) \ \Leftrightarrow \ (\text{N2})$$

In the presence of a lattice meet operation $\wedge$, (Int) can be restated as

(Int) $a \wedge (\neg a) = 0$.

We shall consider bounded, not necessarily distributive, lattices with negation. We assume that 1 is the greatest element and 0 the smallest. From $0 \leq \neg 1$ and (N2) we get $1 \leq \neg 0$ hence $\neg 0 = 1$. In the presence of the (Int) identity, we also have $\neg 1 = 0$ since then $0 = 1 \wedge \neg 1 = \neg 1$. One may also derive this from (DeM) and (N2') since then $\neg 1 = \neg\neg 0 = 0$.

In particular, we consider the following four classes of bounded (not necessarily distributive) lattices with negations: $\mathcal{M}$, $\mathcal{W}$, $\mathcal{O}$ and $\mathcal{P}$.

1) $\mathcal{M}$ denotes the variety of all *De Morgan lattices*, i.e., bounded lattices with De Morgan negation, that is, negation satisfying (N2) and (DeM).
2) $\mathcal{W}$ denotes the variety of all *weakly pseudo-complemented lattices*, that is, bounded lattices with Heyting negation, that is, satisfying (N1), (N2') and (Int) (weak intuitionistic negation). The quasi-identity

$$x \leq \neg y \ \Rightarrow \ x \wedge y = 0$$

is satisfied but the converse is not; see the Example 2.3 below.
3) $\mathcal{O}$ denotes the variety of all *ortholattices*, i.e., bounded lattices with ortho-negation, that is, negation satisfying (N2), (DeM) and (Int);
4) $\mathcal{P}$ denotes the class of all *pseudo-complemented lattices*, i.e., bounded lattices with pseudo-complement (intuitionistic negation), that is, satisfying the quasi-identities

$$x \wedge y = 0 \ \Leftrightarrow \ x \leq \neg y.$$

Both (N2) and (Int) are derivable from the above quasi-identities. Thus, each of the four classes we consider satisfies (N1), (N2), (N2'), $\neg 0 = 1$ and $\neg 1 = 0$.

Corollary 2.1. *The following connections hold between $\mathcal{M}$, $\mathcal{W}$, $\mathcal{O}$ and $\mathcal{P}$:*

(a) $\mathcal{O} \subset \mathcal{M}$
(b) $\mathcal{O} \subset \mathcal{W}$
(c) $\mathcal{O} = \mathcal{M} \cap \mathcal{W} \neq \mathcal{P}$
(d) $\mathcal{P} \subset \mathcal{W}, \ \mathcal{P} \not\subset \mathcal{M}$.

The proper inclusions are shown by examples below. In each case, 1 is the top element and 0 is the bottom element with $\neg 1 = 0$ and $\neg 0 = 1$.

Example 2.1. Let L_3 be a lattice of 3-valued Łukasiewicz logic with $1, 0, a$, where $\neg a = a$. Then L_3 is in $\mathcal{M}$, but (Int) is false: $\neg a \wedge a = a \neq 0$, so L_3 is neither in $\mathcal{W}$ nor in $\mathcal{P}$ nor in $\mathcal{O}$.

Example 2.2. Let $\mathcal{N}_5$ be the "pentagon", a lattice with 5 elements $1, 0, a, b, c$, where $a < b$ and c is incomparable with a, b. Let $\neg a = \neg b = c$ and $\neg c = b$. Then $\mathcal{N}_5$ is in $\mathcal{W}$ and in $\mathcal{P}$, but, since (DeM) is false: $a < b = \neg\neg a$, it is neither in $\mathcal{M}$ nor in $\mathcal{O}$.

Example 2.3. Let $\mathcal{M}_{02}$ be a lattice with 6 elements $1, 0, a, b, c, d,$, where a, b, c, d are incomparable. Let $\neg a = b$, $\neg b = a$, $\neg c = d$ and $\neg d = c$. Then $\mathcal{M}_{02}$ is in $\mathcal{M}$, in $\mathcal{O}$ and in $\mathcal{W}$ but not in $\mathcal{P}$ as the quasi-identity: $x \wedge y = 0 \Rightarrow x \leq \neg y$ fails. This shows that a weakly pseudo-complemented lattice need not be pseudo-complemented.

We shall need the following lemma. We use $(X]$ to denote the downward closure of a subset X of a lattice and $[X)$ for the upward closure.

Lemma 2.1. *Let* $\mathbf{W} = \langle W, \wedge, \vee, \neg, 0, 1 \rangle$ *be a lattice with negation satisfying (N2) (equivalently, (N1) and (N2')) and let F be a proper filter of* $\mathbf{W}$. *Then*

(a) $(\neg F]$ *is an ideal of* $\mathbf{W}$,
(b) $\neg a \in F$ *iff* $a \in (\neg F]$, *for all* $a \in W$.

If, in addition, $\mathbf{W}$ *satisfies (Int), then*

(c) $F \cap (\neg F] = \emptyset$.

Proof. (a) By definition, $(\neg F]$ is downward closed. Suppose that $a, b \in (\neg F]$. Then $a \leq \neg c$ and $b \leq \neg d$ for some $c, d \in F$. Since F is a filter, $c \wedge d \in F$ so $\neg(c \wedge d) \in \neg F$. From (N1) one easily derives $(\neg c) \vee (\neg d) \leq \neg(c \wedge d)$ hence $a \vee b \leq \neg(c \wedge d)$ so $a \vee b \in (\neg F]$. Thus, $(\neg F]$ is an ideal.
   (b) If $\neg a \in F$ then $\neg\neg a \in (\neg F]$ hence $a \in (\neg F]$ by (N2'). If $a \in (\neg F]$ then $a \leq \neg b$ for some $b \in F$, so $b \leq \neg a$ by (N2) hence $\neg a \in F$.
   (c) Suppose there is some $a \in F \cap (\neg F]$. Then $a \leq \neg b$ for some $b \in F$, so $b \leq \neg a$. Thus, $\neg a \in F$ hence $0 = a \wedge (\neg a) \in F$ which is a contradiction.

3   Preliminaries

We give here the necessary background on the relational representation of non-distributive lattices in the style of Urquhart [9]. (see also [4] and [8]). The representations of non-distributive lattices with negations is built on top of this framework following the methods of Allwein and Dunn [1].
   Let X be a non-empty set and let $\leqslant_1$ and $\leqslant_2$ be two quasi-orders on X. The structure $\langle X, \leqslant_1, \leqslant_2 \rangle$ is called a *doubly ordered set* iff it satisfies:

$$(\forall x, y)((x \leqslant_1 y \text{ and } x \leqslant_2 y) \Rightarrow x = y). \tag{1}$$

For a doubly ordered set $\boldsymbol{X} = \langle X, \leqslant_1, \leqslant_2 \rangle$, $A \subseteq X$ is $\leqslant_1$–*increasing* (resp., $\leqslant_2$–*increasing*) iff, for all $x, y \in X$, $x \in A$ and $x \leqslant_1 y$ (resp., $x \leqslant_2 y$) imply $y \in A$. We define two mappings $\ell, r : 2^X \to 2^X$ by

$$\ell(A) = \{x \in X : (\forall y \in X)\, x \leqslant_1 y \Rightarrow y \notin A\} \tag{2}$$
$$r(A) = \{x \in X : (\forall y \in X)\, x \leqslant_2 y \Rightarrow y \notin A\}. \tag{3}$$

Then $A \subseteq X$ is called ℓ–*stable* (resp., r–*stable*) iff $\ell(r(A)) = A$ (resp., $r(\ell(A)) = A$). The family of all ℓ-stable subsets of X will be denoted by $L(\boldsymbol{X})$.

Lemma 3.1. [4],[8] *Let* $\langle X, \leqslant_1, \leqslant_2 \rangle$ *be a doubly ordered set. Then, for* $A \subseteq X$,

(a) $l(A)$ *is* $\leqslant_1$–*increasing and* $r(A)$ *is* $\leqslant_2$–*increasing,*
(b) *if* A *is* $\leqslant_1$–*increasing, then* $A \subseteq l(r(A))$,
(c) *if* A *is* $\leqslant_2$–*increasing, then* $A \subseteq r(l(A))$.

Lemma 3.2. [9] *Let* $\langle X, \leqslant_1, \leqslant_2 \rangle$ *be a doubly ordered set. Then the mappings* l *and* r *form a Galois connection between the lattice of* $\leqslant_1$–*increasing subsets of* X *and the lattice of* $\leqslant_2$–*increasing subsets of* X. *In particular, for every* $\leqslant_1$–*increasing set* A *and* $\leqslant_2$–*increasing set* B,

$$A \subseteq l(B) \text{ iff } B \subseteq r(A).$$

Let $\boldsymbol{X} = \langle X, \leqslant_1, \leqslant_2 \rangle$ be a doubly ordered set. Define two binary operations $\wedge^C$ and $\vee^C$ on 2^X and two constants 0^C and 1^C as follows: for all $A, B \subseteq X$,

$$A \wedge^C B = A \cap B \tag{4}$$
$$A \vee^C B = l(r(A) \cap r(B)) \tag{5}$$
$$0^C = \emptyset \tag{6}$$
$$1^C = X. \tag{7}$$

Observe that the definition of $\vee^C$ in terms of $\wedge^C$ resembles a De Morgan law with two different negations. In [9] it was shown that $L(\boldsymbol{X}) = \langle L(\boldsymbol{X}), \wedge^C, \vee^C, 0^C, 1^C \rangle$ is a bounded lattice; it is called the **complex algebra** of $\boldsymbol{X}$.

Let $\boldsymbol{W} = \langle W, \wedge, \vee, 0, 1 \rangle$ be a bounded lattice. By a *filter-ideal pair* of $\boldsymbol{W}$ we mean a pair (x_1, x_2) such that x_1 is a filter of $\boldsymbol{W}$, x_2 is an ideal of $\boldsymbol{W}$ and $x_1 \cap x_2 = \emptyset$. Define the following three quasi-ordering relations on filter-ideal pairs:

$$(x_1, x_2) \leqslant_1 (y_1, y_2) \text{ iff } x_1 \subseteq y_1$$
$$(x_1, x_2) \leqslant_2 (y_1, y_2) \text{ iff } x_2 \subseteq y_2$$
$$(x_1, x_2) \leqslant (y_1, y_2) \text{ iff } (x_1, x_2) \leqslant_1 (y_1, y_2) \text{ and } (x_1, x_2) \leqslant_2 (y_1, y_2).$$

We say that a filter-ideal pair (x_1, x_2) is *maximal* iff it is maximal with respect to $\leqslant$. The set of all maximal filter-ideal pairs of $\boldsymbol{W}$ will be denoted by $X(\boldsymbol{W})$. We shall use x, y, z, etc. to denote maximal disjoint filter-ideal pairs and, if $x \in X(\boldsymbol{W})$, then we use the convention that the filter part of x is x_1 and the ideal part of x is x_2, so that $x = (x_1, x_2)$. The same convention holds for y, z, etc.

Note that $X(\boldsymbol{W})$ is a binary relation on 2^W. It was shown in [9] that for any filter-ideal pair (x_1, x_2) there exists $(y_1, y_2) \in X(\boldsymbol{W})$ such that $(x_1, x_2) \leqslant (y_1, y_2)$; in this case, we say that (x_1, x_2) has been *extended* to (y_1, y_2).

If $\boldsymbol{W} = \langle W, \wedge, \vee, 0, 1 \rangle$ is a bounded lattice then the **canonical frame** of $\boldsymbol{W}$ is defined as the relational structure $\boldsymbol{X}(\boldsymbol{W}) = \langle X(\boldsymbol{W}), \leqslant_1, \leqslant_2 \rangle$.

Consider the complex algebra $\boldsymbol{L}(\boldsymbol{X}(\boldsymbol{W}))$ of the canonical frame of a bounded lattice $\boldsymbol{W}$. Note that $\boldsymbol{L}(\boldsymbol{X}(\boldsymbol{W}))$ is an algebra of subrelations of $X(\boldsymbol{W})$. Define a mapping $h : W \to 2^{X(\boldsymbol{W})}$ by

$$h(a) = \{x \in X(\boldsymbol{W}) : a \in x_1\}.$$

Then h is in fact a map from W to $\boldsymbol{L}(\boldsymbol{X}(\boldsymbol{W}))$ and, moreover, we have the following result.

Proposition 3.1. [9] *For every bounded lattice $\boldsymbol{W}$, h is a lattice embedding of $\boldsymbol{W}$ into $\boldsymbol{L}(\boldsymbol{X}(\boldsymbol{W}))$.*

The following theorem is a weak version of the Urquhart result.

Theorem 3.1 (Representation theorem for lattices). *Every bounded lattice is embeddable into the complex algebra of its canonical frame.*

4   Lattices with De Morgan Negation

Recall that $\mathcal{M}$ denotes the variety of all De Morgan lattices, which are bounded lattices $\boldsymbol{W} = \langle W, \wedge, \vee, \neg, 0, 1 \rangle$ with a unary operation $\neg$ satisfying (N2) and (DeM). Recall that from (N2) and (DeM) one may derive (N1), (N2'), $\neg 1 = 0$ and $\neg 0 = 1$. The following are also derivable in $\mathcal{M}$:

$\neg\neg a = a$
$\neg(a \vee b) = (\neg a) \wedge (\neg b)$
$\neg(a \wedge b) = (\neg a) \vee (\neg b)$
$\neg a = \neg b \Rightarrow a = b.$

We will denote by $\mathcal{R}_M$ the class of all relational structures of type $\boldsymbol{X} = \langle X, \leqslant_1, \leqslant_2, N \rangle$, where $\langle X, \leqslant_1, \leqslant_2 \rangle$ is a doubly ordered set (i.e., $\leqslant_1$ and $\leqslant_2$ are quasi-orders satisfying (1)), $N : X \to X$ is a function and the following hold:

(M1) $(\forall x)(N(N(x)) = x)$,
(M2) $(\forall x, y)(x \leqslant_1 y \Rightarrow N(x) \leqslant_2 N(y))$,
(M3) $(\forall x, y)(x \leqslant_2 y \Rightarrow N(x) \leqslant_1 N(y))$.

The representation in this section essentially comes from [1], where the function N is called a 'generalized Routley-Meyer star operator'. We give full details here and in the next section extend the method to ortholattices.

For each $\boldsymbol{W} \in \mathcal{M}$, the **canonical frame** of $\boldsymbol{W}$ is defined as the relational structure $\boldsymbol{X}(\boldsymbol{W}) = \langle X(\boldsymbol{W}), \leqslant_1, \leqslant_2, N \rangle$, where $X(\boldsymbol{W})$ is the set of all maximal disjoint filter-ideal pairs of $\boldsymbol{W}$ and, for $x = (x_1, x_2)$, $y = (y_1, y_2) \in X(\boldsymbol{W})$,

$x \leqslant_1 y$ iff $x_1 \subseteq y_1$,
$x \leqslant_2 y$ iff $x_2 \subseteq y_2$,
$N(x) = (\neg x_2, \neg x_1)$, where $\neg A = \{\neg a : a \in A\}$ for any $A \subseteq W$.

Lemma 4.1. *If $\boldsymbol{W} \in \mathcal{M}$ then $\boldsymbol{X}(\boldsymbol{W}) \in \mathcal{R}_M$.*

Proof. We have already observed that $\langle X(\boldsymbol{W}), \leqslant_1, \leqslant_2 \rangle$ is a doubly ordered set. Condition (M1) follows from (DeM) and conditions (M2) and (M3) are immediate. Thus, we need only show that N is a function from $X(\boldsymbol{W})$ to $X(\boldsymbol{W})$. That is, if $x = (x_1, x_2) \in X(\boldsymbol{W})$, we must show that $N(x) = (\neg x_2, \neg x_1)$ is a maximal disjoint filter-ideal pair. Let $a_1, a_2 \in x_2$ hence $\neg a_1, \neg a_2 \in \neg x_2$. Then $(\neg a_1) \wedge (\neg a_2) = \neg(a_1 \vee a_2)$ and $a_1 \vee a_2 \in x_2$, hence $\neg x_2$ is closed under $\wedge$. If $\neg a_1 \leq b$ then $\neg b \leq \neg\neg a_1 = a_1$, so $\neg b \in x_2$. Then $b = \neg\neg b \in \neg x_2$, so $\neg x_2$ is upward closed. Thus, $\neg x_2$ is a filter. Similarly, $\neg x_1$ is an ideal. Also, $\neg x_1$ and $\neg x_2$ can be shown disjoint using the implication: $\neg b = \neg c \Rightarrow b = c$ and the fact that x_1 and x_2 are disjoint. To show maximality, suppose $y \in X(\boldsymbol{W})$ and $\neg x_1 \subseteq y_2$ and $\neg x_2 \subseteq y_1$. Then $\neg\neg x_1 \subseteq \neg y_2$, i.e., $x_1 \subseteq \neg y_2$ and also $x_2 \subseteq \neg y_1$. Since $(\neg y_2, \neg y_1)$ is a disjoint filter-ideal pair, the maximality of x implies $x_1 = \neg y_2$ and $x_2 = \neg y_1$. Thus, $\neg x_1 = y_2$ and $\neg x_2 = y_1$ so $N(x)$ is maximal.

If $\boldsymbol{X} = \langle X, \leqslant_1, \leqslant_2, N \rangle \in \mathcal{R}_M$, then $\langle X, \leqslant_1, \leqslant_2 \rangle$ is a doubly ordered set, so we may consider its complex algebra $\langle L(X), \wedge^C, \vee^C, 0^C, 1^C \rangle$, where $L(X)$ is the set of ℓ-stable sets and the operations are as in (4–7). We extend this definition to define the ***complex algebra*** of $\boldsymbol{X}$ as $\boldsymbol{L}(\boldsymbol{X}) = \langle L(X), \wedge^C, \vee^C, \neg^C, 0^C, 1^C \rangle$, where for $A \in L(X)$,

$$\neg^C A = \{x \in X : N(x) \in r(A)\}.$$

Lemma 4.2. *If $\boldsymbol{X} \in \mathcal{R}_M$ then $\boldsymbol{L}(\boldsymbol{X}) \in \mathcal{M}$.*

Proof. We need to show that $\neg^C A$ is ℓ-stable, i.e., $\ell r(\neg^C A) = \neg^C A$, and that $\boldsymbol{L}(\boldsymbol{X})$ satisfies (N2) and (DeM). Since ℓ and r form a Galois connection, by Lemma 3.2, we have $\neg^C A \subseteq \ell r(\neg^C A)$ iff $r(\neg^C A) \subseteq r(\neg^C A)$. For the converse, suppose that for every y, if $x \leqslant_1 y$ then $y \notin r(\neg^C A)$ and assume, to the contrary, that $x \notin \neg^C A$. Then $N(x) \notin r(A)$ and there is z such that $N(x) \leqslant_2 z$ and $z \in A$. It follows by (M3) and (M1) that $x \leqslant_1 N(z)$ and hence, by the above assumption, $N(z) \notin r(\neg^C A)$. Thus, there is t such that $N(z) \leqslant_2 t$ and $t \in \neg^C A$. By application of N and (M3) and (M1), we have that $z \leqslant_1 N(t)$ and $N(t) \in r(A)$, in particular $N(t) \notin A$. But $z \in A$ and A is $\leqslant_1$–increasing, as $A = \ell r(A)$, hence $N(t) \in A$, a contradiction.

To prove (N2), suppose that $A \subseteq \neg^C B$. Then, for every x, if $x \in A$ then $N(x) \in r(B)$. Suppose that $x \in B$ and, to the contrary, that $x \notin \neg^C A$, i.e., $N(x) \notin r(A)$, in which case $N(x) \leqslant_2 y$ and $y \in A$, for some y. By (M3) and (M1), $x \leqslant_1 N(y)$ hence $N(y) \in B$ since $B = \ell r(B)$ is $\leqslant_1$–increasing. But also $y \in \neg^C B$, by the assumption, and $N(y) \in r(B)$, a contradiction since $B \cap r(B) = \emptyset$.

To prove (DeM), let $x \in \neg^C \neg^C A$, hence $N(x) \in r(\neg^C A)$. We show that $x \in \ell(r(A))$ which equals A since A is ℓ-closed. Let $x \leqslant_1 w$. Then $N(x) \leqslant_2 N(w)$, by (M2), hence $N(w) \in r(\neg^C A)$ since $r(\neg^C A)$ is $\leqslant_2$–increasing. Thus, $N(w) \notin \neg^C A$, i.e., $w = N(N(w)) \notin r(A)$. Thus, $x \in \ell(r(A)) = A$.

The above lemmas imply that if $W \in \mathcal{M}$, then the complex algebra of the canonical frame of W, namely $L(X(W))$, is in $\mathcal{M}$ as well.

Theorem 4.1. *Each $W \in \mathcal{M}$ is embeddable into $L(X(W))$.*

Proof. Recall that the function $h : W \to L(X(W))$ defined by

$$h(a) = \{x \in L(X(W)) : a \in x_1\}$$

is an embedding of the lattice part of W into $L(X(W))$. We need only show that $h(\neg a) = \neg^C h(a)$ for all $a \in W$, where

$$h(\neg a) = \{x : \neg a \in x_1\}$$

and

$$
\begin{aligned}
\neg^C h(a) &= \{x : N(x) \in r(h(a))\} \\
&= \{x : \neg x_1 \subseteq y_2 \Rightarrow a \notin y_1, \text{ for all } y\}.
\end{aligned}
$$

First, let $x \in h(\neg a)$. Then $\neg a \in x_1$, hence $a = \neg\neg a \in \neg x_1$. Suppose that $\neg x_1 \subseteq y_2$. Then $a \notin y_1$, since y_1 and y_2 are disjoint.

Next, let $x \in \neg^C h(a)$. Suppose, to the contrary, that $\neg a \notin x_1$. By Lemma 2.1 it follows that $a \notin (\neg x_1]$ and hence that so $\langle [a), (\neg x_1] \rangle$ is a disjoint filter-ideal pair, which may be extended to a maximal one, say y. Thus, $\neg x_1 \subseteq y_2$, so $a \notin y_1$, but $[a) \subseteq y_1$, a contradiction.

5   Lattices with Ortho-negation (Ortholattices)

Recall that $\mathcal{O}$ denotes the variety of all ortholattices, which are bounded lattices $W = \langle W, \wedge, \vee, \neg, 0, 1 \rangle$ with a unary operation $\neg$ which satisfies (N2), (DeM) and (Int) (hence also (N1) and (N2')). That is, the negation in an ortholattice is both De Morgan and Heyting. We extend the relational representation for De Morgan lattices to ortholattices

We will denote by $\mathcal{R}_O$ the class of all relational structures of type $X = \langle X, \leqslant_1, \leqslant_2, N \rangle$, where $\langle X, \leqslant_1, \leqslant_2 \rangle$ is a doubly ordered set and $N : X \to X$ is a function such that (M1), (M2) and (M3) hold, as well as

(O) $(\forall x)(\exists y)(x \leqslant_1 y \ \& \ N(x) \leqslant_2 y)$.

That is, $\mathcal{R}_O$ is the subclass of $\mathcal{R}_M$ defined by (O).

If $W \in \mathcal{O}$, then $W \in \mathcal{M}$ hence its canonical frame is the relational structure $X(W) = \langle X(W), \leqslant_1, \leqslant_2, N \rangle$, where $X(W)$ is the set of all maximal disjoint filter-ideal pairs of W and, for $x, y \in X(W)$,

$x \leqslant_1 y$ iff $x_1 \subseteq y_1$
$x \leqslant_2 y$ iff $x_2 \subseteq y_2$
$N(x) = (\neg x_2, \neg x_1)$, where $\neg A = \{\neg a : a \in A\}$.

Lemma 5.1. *If $W \in \mathcal{O}$ then $X(W) \in \mathcal{R}_O$.*

Proof. We must show that $X(W)$ satisfies (O). Let $x \in X(W)$. By Lemma 2.1, $(x_1, \neg x_1)$ is a disjoint filter-ideal pair, so we may extend it to a maximal disjoint filter-ideal pair, say y. Then $x_1 \subseteq y_1$ and $\neg x_1 \subseteq y_2$, so we have found a y that satisfies the required conditions of (O).

If $X = \langle X, \leqslant_1, \leqslant_2, N \rangle \in \mathcal{R}_O$, then $X \in \mathcal{R}_M$ so it has a canonical algebra $L(X) = \langle L(X), \wedge^C, \vee^C, \neg^C, 0^C, 1^C \rangle$ defined as in the De Morgan negation case.

Lemma 5.2. *If $X \in \mathcal{R}_O$ then $L(X) \in \mathcal{O}$.*

Proof. We need only show that $L(X)$ satisfies $A \wedge^C (\neg^C A) = 0^C$. Suppose, to the contrary, that there exists $A \in L(X)$ such that $A \cap (\neg^C A) \neq \emptyset$, and let $x \in A \cap (\neg^C A)$. By (O), there exists y such that $x \leqslant_1 y$ and $N(x) \leqslant_2 y$. Since A is $\leqslant_1$-increasing, $y \in A$. Since $x \in \neg^C A$, $N(x) \in r(A)$. But then $N(x) \leqslant_2 y$ implies $y \notin A$, a contradiction.

Thus, the above lemmas imply that if $W \in \mathcal{O}$, then $L(X(W)) \in \mathcal{O}$ as well. Since the map h is an embedding of De Morgan lattices, we have the following result.

Theorem 5.1. *Each $W \in \mathcal{O}$ is embeddable into $L(X(W))$.*

6   Lattices with Heyting Negation

Recall that $\mathcal{W}$ denotes the variety of all weakly pseudo-complemented lattices, which are bounded lattices $W = \langle W, \wedge, \vee, \neg, 0, 1 \rangle$ with a unary operation $\neg$ satisfying (N1), (N2') and (Int) (hence also (N2), $\neg 0 = 1$ and $\neg 1 = 0$).

We will denote by $\mathcal{R}_W$ the class of all relational structures of type $X = \langle X, \leqslant_1, \leqslant_2, C \rangle$, where $\langle X, \leqslant_1, \leqslant_2 \rangle$ is a doubly ordered set and C is a binary relation on X such that the following hold:

(FC1) $(\forall x, y, z)((xCy \text{ and } z \leqslant_1 x) \Rightarrow zCy)$
(FC2) $(\forall x, y, z)((xCy \text{ and } y \leqslant_2 z) \Rightarrow xCz)$
(FC3) $(\forall x)(\exists y)(xCy \text{ and } x \leqslant_1 y)$
(FC4) $(\forall x, y)(xCy \Rightarrow (\exists z)(yCz \text{ and } x \leqslant_1 z))$
(FC5) $(\forall s, t, y)[(yCs \text{ and } s \leqslant_2 t) \Rightarrow (\exists z)(y \leqslant_1 z \text{ and } (\forall u)(z \leqslant_2 u \Rightarrow tCu))]$.

The relation C is intended to capture the negation in the relational structure in a similar manner that N was used in the De Morgan negation case.

For each $W \in \mathcal{W}$ we define the **canonical frame** of W as the relational structure $X(W) = \langle X(W), \leqslant_1, \leqslant_2, C \rangle$, where $X(W)$ is the set of all maximal disjoint filter-ideal pairs of W and, for all $x, y \in X(W)$,

$$x \leqslant_1 y \text{ iff } x_1 \subseteq y_1$$
$$x \leqslant_2 y \text{ iff } x_2 \subseteq y_2$$
$$xCy \text{ iff } (\forall a)(\neg a \in x_1 \Rightarrow a \in y_2).$$

Lemma 6.1. *If $W \in \mathcal{W}$ then $X(W) \in \mathcal{R}_W$.*

Proof. We know that $\langle X(W), \leqslant_1, \leqslant_2 \rangle$ is a doubly ordered set. Properties (FC1) and (FC2) are straightforward to prove. For (FC3), suppose $x \in X(W)$. By Lemma 2.1, $\langle x_1, (\neg x_1] \rangle$ is a disjoint filter-ideal pair, so we can extend it to a maximal one, say y. If $\neg a \in x_1$ then $a \in (\neg x_1]$ (by Lemma 2.1(b)) hence $a \in y_2$. Thus, xCy. Also, $x_1 \subseteq y_1$, i.e., $x \leqslant_1 y$, so we have found the required y.

For (FC4), suppose $x, y \in X(W)$ and xCy. By Lemma 2.1(a), $(\neg y_1]$ is an ideal. If $a \in x_1 \cap (\neg y_1]$ then $a \in x_1$ implies $\neg\neg a \in x_1$, which implies $\neg a \in y_2$. But $a \in (\neg y_1]$ implies $\neg a \in y_1$ (by Lemma 2.1(b)), which contradicts the fact that $y_1 \cap y_2 = \emptyset$. Thus, $x_1 \cap (\neg y_1] = \emptyset$. Thus, we can extend $\langle x_1, (\neg y_1] \rangle$ to a maximal disjoint filter-ideal pair, say z. If $\neg a \in y_1$ then $a \in (\neg y_1]$ hence $a \in z_2$, so yCz. Also, $x \leqslant_1 z$, so we have proved (FC4).

For (FC5), suppose that $s, t, y \in X(W)$ such that yCs and $s \leqslant_2 t$. First, we show that $y_1 \cap (\neg t_1] = \emptyset$. Suppose $a \in y_1 \cap (\neg t_1]$. Then, $\neg\neg a \in y_1$ hence $\neg a \in s_2$. Since $s \leqslant_2 t$ we have $\neg a \in t_2$. Also, $a \leq \neg b$ for some $b \in t_1$, so $\neg a \geq \neg\neg b \geq b$ hence $\neg a \in t_1$. This contradicts the fact that t_1 and t_2 are disjoint.

We therefore have that $\langle y_1, (\neg t_1] \rangle$ is a disjoint filter-ideal pair, so we may extend it to a maximal one, say z. Then, $y_1 \subseteq z_1$, i.e., $y \leqslant_1 z$. Suppose $z \leqslant_2 w$ and $\neg a \in t_1$. Then $\neg\neg a \in \neg t_1$ so $a \in (\neg t_1] \subseteq z_2 \subseteq w_2$ hence $a \in w_2$. Thus, we have proved (FC5).

Let $X = \langle X, \leqslant_1, \leqslant_2, C \rangle \in \mathcal{R}_W$. Then $\langle X, \leqslant_1, \leqslant_2 \rangle$ is a doubly ordered set hence we may consider its complex algebra $\langle L(X), \wedge^C, \vee^C, 0^C, 1^C \rangle$, where $L(X)$ is the set of ℓ-stable sets and the operations are defined as in (4–7). The **complex algebra** of X is defined as $L(X) = \langle L(X), \wedge^C, \vee^C, \neg^C, 0^C, 1^C \rangle$, where

$$\neg^C A = \{x \in X : \forall y (xCy \Rightarrow y \notin A)\}.$$

Lemma 6.2. *If A is ℓ-stable then so is $\neg^C A$.*

Proof. We have $\neg^C A = \{x : \forall y (xCy \Rightarrow y \notin A)\}$ and

$$\ell r(\neg^C A) = \{x : x \leqslant_1 s \Rightarrow (\exists t)(s \leqslant_2 t \text{ and } (\forall u)(tCu \Rightarrow u \notin A))\}.$$

Let $x \in \neg^C A$ and suppose that $x \leqslant_1 s$ for some s. We claim that $t = s$ satisfies the required properties. Clearly, $s \leqslant_2 s$. If sCu, then xCu since $x \leqslant_1 s$, by (FC1) hence $u \notin A$. Thus, $x \in \ell r(\neg^C A)$ so $\neg^C A \subseteq \ell r(\neg^C A)$.

For the reverse inclusion, note that, since A is ℓ-stable, we have

$$\neg^C A = \neg^C \ell r(A) = \{x : xCy \Rightarrow (\exists z)(y \leqslant_1 z \text{ and } (\forall u)(z \leqslant_2 u \Rightarrow u \notin A))\}.$$

Let $x \in \ell r(\neg^C A)$ and suppose that xCy for some y. By (FC4), there exists s such that

$$x \leqslant_1 s \text{ and } yCs.$$

Then, since $x \in \ell r(\neg^C A)$ and $x \leqslant_1 s$, there exists t such that

$$s \leqslant_2 t \text{ and } (\forall u)(tCu \Rightarrow u \notin A).$$

Since yCs and $s \leqslant_2 t$, by (FC5) there exists z such that

$$y \leqslant_1 z \text{ and } (\forall u)(z \leqslant_2 u \Rightarrow tCu)).$$

Thus, $(\forall u)(z \leqslant_2 u \Rightarrow u \notin A))$, so we have found the required z, so $x \in \neg^C lr(A) = \neg^C A$.

Lemma 6.3. *If $X \in \mathcal{R}_W$ then $L(X) \in \mathcal{W}$.*

Proof. To see that (N1) holds, suppose A, B are ℓ-stable sets and $A \subseteq B$. Let $x \in \neg^C B$. Then, xCy implies $y \notin B$ hence also $y \notin A$, so $x \in \neg^C A$.

To see that (N2') holds, note that

$$\neg^C \neg^C A = \{x : xCy \Rightarrow (\exists z)(yCz \text{ and } z \in A)\}.$$

Let $x \in A$ and suppose that xCy for some y. By (FC4), there exists z such that yCz and $x \leqslant_1 z$. Since A is $\leqslant_1$–increasing and $x \in A$, we have $z \in A$. Thus, the required z exists, showing that $x \in \neg^C \neg^C A$.

To see that (Int) holds, let A be an ℓ-stable set and suppose there exists $x \in A \cap \neg^C A$. By (FC3), there exists a y such that xCy and $x \leqslant_1 y$. Since $x \in \neg^C A$ and xCy we have $y \notin A$. But $x \in A$ and A is ℓ-stable, hence $\leqslant_1$–increasing, so $x \leqslant_1 y$ implies $y \in A$, a contradiction.

The above lemmas show that if $W \in \mathcal{W}$ then so is $L(X(W))$.

Theorem 6.1. *Each $W \in \mathcal{W}$ is embeddable into $L(X(W))$.*

Proof. Recall that the function $h : W \to L(X(W))$ defined by

$$h(a) = \{x \in L(X(W)) : a \in x_1\}$$

is an embedding of the lattice part of W into $L(X(W))$. We need only show that $h(\neg a) = \neg^C h(a)$ for all $a \in W$, where

$$h(\neg a) = \{x : \neg a \in x_1\}$$

and

$$\neg^C h(a) = \{x : xCy \Rightarrow a \notin y_1\}.$$

First, let $x \in h(\neg a)$ and suppose that xCy. Then, $\neg a \in x_1$ so $a \in y_2$ hence $a \notin y_1$, as required.

Next, let $x \in \neg^C h(a)$ and suppose that $\neg a \notin x_1$. Then $a \notin (\neg x_1]$ (by Lemma 2.1(b)) so $\langle [a), (\neg x_1] \rangle$ forms a disjoint filter-ideal pair which we can extend to a maximal one, say y. If $\neg c \in x_1$ then $c \in (\neg x_1]$ so xCy hence $a \notin y_1$, a contradiction since $[a) \subseteq y_1$.

Remark 6.1. The class of ortholattices is the intersection of De Morgan lattices and weakly pseudo-complemented lattices. In Section 5 we obtained a relational representation for ortholattices by extending the relational representation for De

Morgan lattices to include the condition (O) to deal with the identity (Int). However, since ortholattices may also be considered as extensions of weakly pseudo-complemented lattices by the identity (DeM), one would expect a connection between the two representations.

Let $\boldsymbol{X} = \langle X, \leqslant_1, \leqslant_2, N \rangle$ be a relational structure in $\mathcal{R}_O$, i.e., $\boldsymbol{X}$ satisfies (M1), (M2), (M3) and (O). We shall show that $\boldsymbol{X}$ is equivalent to a relational structure $\langle X, \leqslant_1, \leqslant_2, C \rangle$ in $\mathcal{R}_W$. For this we need to define a relation C in terms of N, $\leqslant_1$ and $\leqslant_2$. To find the connection, consider the canonical frame of an ortholattice $\boldsymbol{W}$. This is $\langle X(\boldsymbol{W}), \leqslant_1, \leqslant_2, N \rangle$, where

$$
\begin{aligned}
x \leqslant_1 y \quad &\text{iff} \quad x_1 \subseteq y_1, \\
x \leqslant_2 y \quad &\text{iff} \quad x_2 \subseteq y_2, \\
N(x) &= (\neg x_2, \neg x_1).
\end{aligned}
$$

Since $\boldsymbol{W}$ is also a weakly pseudo-complemented lattice it also has a canonical frame in $\mathcal{R}_W$, which is $\langle X(\boldsymbol{W}), \leqslant_1, \leqslant_2, C \rangle$ where $\leqslant_1$ and $\leqslant_2$ are as above and

$$
xCy \quad \text{iff} \quad (\forall a)(\neg a \in x_1 \Rightarrow a \in y_2).
$$

In the presence of both (Int) and (DeM) we claim that the following relationship holds between N and C:

$$
xCy \quad \Leftrightarrow \quad N(x) \leqslant_2 y.
$$

To see this, suppose $x, y \in X(\boldsymbol{W})$. If xCy, then

$$
\neg a \in \neg x_1 \Leftrightarrow a \in x_1 \Leftrightarrow \neg\neg a \in x_1 \Rightarrow \neg a \in y_2,
$$

so $\neg x_1 \subseteq y_2$, i.e., $N(x) \leqslant_2 y$. Conversely, suppose $N(x) \leqslant_2 y$ and let $\neg a \in x_1$. Then $a = \neg\neg a \in \neg x_1$, so $a \in y_2$, hence xCy.

Starting with the relational structure $\langle X, \leqslant_1, \leqslant_2, N \rangle \in \mathcal{R}_O$, define a binary relation C on $X(\boldsymbol{W})$ by:

$$
xCy \quad \text{iff} \quad N(x) \leqslant_2 y.
$$

Then one may check that the conditions (FC1–FC5) all hold for this C. In particular, (FC1) and (FC2) are straightforward and (FC3) is just (O). For (FC4), take $z = N(y)$ and, for (FC5), take $z = N(t)$. Thus, $\langle X, \leqslant_1, \leqslant_2, N \rangle$ is equivalent to a relational structure $\langle X, \leqslant_1, \leqslant_2, C \rangle \in \mathcal{R}_W$. Moreover, the complex algebra obtained from either of these relational structures is the same. To see this we need only check that the two definitions of $\neg^C$ coincide:

$$
\begin{aligned}
&\{x \in X : N(x) \in r(A)\} \\
&= \{x \in X : N(x) \leqslant_2 y \Rightarrow y \notin A\} \\
&= \{x \in X : xCy \Rightarrow y \notin A\}.
\end{aligned}
$$

Thus, the definition of $\neg^C$ in the De Morgan case coincides with the definition of $\neg^C$ in the Heyting case.

A natural question arising from the above remark is whether a relational semantics for ortholattices can be obtained in the style of the relational structures in $\mathcal{R}_W$. That is, what conditions should be added to (FC1–FC5) in order to ensure that the complex algebra of such a structure also satisfies (DeM), i.e., so that it's an ortholattice.

7   Pseudo-complemented Lattices

Recall that the $\mathcal{P}$ denotes the class of all pseudo-complemented lattices, which are bounded lattices $\boldsymbol{W} = \langle W, \wedge, \vee, \neg, 0, 1\rangle$ with a unary operation $\neg$ satisfying:

$$a \wedge b = 0 \quad \Leftrightarrow \quad a \leq \neg b.$$

Note that (N2) is derivable by

$$a \leq \neg b \Leftrightarrow a \wedge b = 0 \Leftrightarrow b \wedge a = 0 \Leftrightarrow b \leq \neg a.$$

Thus, (N1), (N2') and $\neg 0 = 1$ are also derivable and, from $a \leq \neg\neg a$, we get $a \wedge \neg a = 0$, i.e., (Int) is derivable. So we also have $\neg 1 = 0$. The class $\mathcal{W}$ of weakly pseudo-complemented lattices is easily seen to satisfy the quasi-identity

$$a \leq \neg b \quad \Rightarrow \quad a \wedge b = 0,$$

hence $\mathcal{P}$ is the subclass of $\mathcal{W}$ defined by the quasi-identity

$$a \wedge b = 0 \quad \Rightarrow \quad a \leq \neg b. \tag{8}$$

$\mathcal{P}$ is a proper subclass of $\mathcal{W}$ (see Example 2.3).

We will denote by $\mathcal{R}_P$ the class of all relational structures of type $\boldsymbol{X} = \langle X, \leq_1, \leq_2, C\rangle$, where $\langle X, \leq_1, \leq_2\rangle$ is a doubly ordered set and C is a binary relation on X such that (FC1–FC5) hold, as well as

(FC6) $(\forall x, y)(xCy \Rightarrow (\exists z)(x \leq_1 z \text{ and } y \leq_1 z)).$

That is, $\mathcal{R}_P$ is the subclass of $\mathcal{R}_W$ defined by (FC6).

If $\boldsymbol{W} \in \mathcal{P}$ then $\boldsymbol{W} \in \mathcal{W}$ as well hence its **canonical frame** is the relational structure $\boldsymbol{X}(\boldsymbol{W}) = \langle X(\boldsymbol{W}), \leq_1, \leq_2, C\rangle$, where $X(\boldsymbol{W})$ is the set of all maximal disjoint filter-ideal pairs of $\boldsymbol{W}$ and, for all $x, y \in X(\boldsymbol{W})$,

$$x \leq_1 y \text{ iff } x_1 \subseteq y_1$$
$$x \leq_2 y \text{ iff } x_2 \subseteq y_2$$
$$xCy \text{ iff } (\forall a)(\neg a \in x_1 \Rightarrow a \in y_2).$$

Lemma 7.1. *If $\boldsymbol{W} \in \mathcal{P}$ then $\boldsymbol{X}(\boldsymbol{W}) \in \mathcal{R}_P$.*

Proof. We need only show that (FC6) holds. So, let $x, y \in X(\boldsymbol{W})$ such that xCy. Consider the filter generated by $x_1 \cup y_1$, denoted $Fi(x_1 \cup y_1)$. We claim that $0 \notin Fi(x_1 \cup y_1)$. If we suppose otherwise, then there exist $a_1, \ldots, a_n \in x_1$ and $b_1, \ldots, b_m \in y_1$ such that

$$(\bigwedge_{i=1}^{n} a_i) \wedge (\bigwedge_{j=1}^{m} b_j) = 0.$$

If we set $a = \bigwedge_{i=1}^{n} a_i$ and $b = \bigwedge_{j=1}^{m} b_j$, then $a \in x_1$ and $b \in y_1$ such that $a \wedge b = 0$. But this implies that $a \leq \neg b$ by (8) hence $\neg b \in x_1$. Finally, since xCy and $\neg b \in x_1$, we have $b \in y_2$. Thus, $b \in y_1 \cap y_2$, a contradiction.

This shows that $0 \notin Fi(x_1 \cup y_1)$ so $\langle Fi(x_1 \cup y_1), \{0\} \rangle$ is a disjoint filter-ideal pair. This can be extended to a maximal disjoint filter-ideal pair, say z. Clearly, $x \leq_1 z$ and $y \leq_1 z$, as required.

Let $\boldsymbol{X} = \langle X, \leq_1, \leq_2, C \rangle \in \mathcal{R}_P$ (so $\boldsymbol{X}$ satisfies (FC1–FC6)). Then $\boldsymbol{X}$ is also in $\mathcal{R}_W$ hence we may consider its complex algebra $\boldsymbol{L} = \langle L(\boldsymbol{X}), \wedge^C, \vee^C, \neg^C, 0^C, 1^C \rangle$, where $L(\boldsymbol{X})$ is the set of ℓ-stable sets, the lattice operations are defined as in (4–7) and

$$\neg^C A = \{x \in X : xCy \Rightarrow y \notin A\}.$$

Lemma 7.2. *If $\boldsymbol{X} \in \mathcal{R}_P$ then $L(\boldsymbol{X}) \in \mathcal{P}$.*

Proof. We need only show that $L(\boldsymbol{X})$ satisfies the quasi-identity (8), i.e., for $A, B \in L(\boldsymbol{X})$,

$$A \cap B = \emptyset \quad \Rightarrow \quad A \subseteq \neg^C B = \{x \in X : xCy \Rightarrow y \notin B\}.$$

Suppose that $A \cap B = \emptyset$ and let $x \in A$. Let $y \in X$ such that xCy. By (FC6), there exists $z \in X$ such that $x \leq_1 z$ and $y \leq_1 z$. Since $x \in A$ and A is $\leq_1$-increasing, we have $z \in A$ as well. If $y \in B$ then, since B is $\leq_1$-increasing, it would follow that $z \in B$ and hence that $z \in A \cap B$, contradicting our assumption that $A \cap B = \emptyset$. Thus, $y \notin B$ hence $x \in \neg^C B$, as required.

Thus, we have shown that if $\boldsymbol{W} \in \mathcal{P}$ then so is $L(\boldsymbol{X}(\boldsymbol{W}))$. Moreover, from the previous section we know that h is an embedding of $\boldsymbol{W}$ into $L(\boldsymbol{X}(\boldsymbol{W}))$, hence we have the following result.

Theorem 7.1. *Each $\boldsymbol{W} \in \mathcal{P}$ is embeddable into $L(\boldsymbol{X}(\boldsymbol{W}))$.*

References

1. Allwein, G., Dunn, J.M.: Kripke models for linear logic. J. Symb. Logic **58** (1993) 514–545.
2. Dunn, J.M.: Star and Perp: Two Treatments of Negation. In J. Tomberlin (ed.), Philosophical Perspectives (Philosophy of Language and Logic) **7** (1993) 331–357.
3. Dunn, J.M., Hardegree, G.M.: Algebraic Methods in Philosophical Logic. Clarendon Press, Oxford (2001).
4. Düntsch, I., Orłowska, E., Radzikowska, A.M.: Lattice–based relation algebras and their representability. In: de Swart, C.C.M. et al (eds), Theory and Applications of Relational Structures as Knowledge Instruments, Lecture Notes in Computer Science **2929** Springer–Verlag (2003) 234–258.

5. Düntsch, I., Orłowska, E., Radzikowska, A.M., Vakarelov, D.: Relational representation theorems for some lattice-based structures. Journal of Relation Methods in Computer Science JoRMiCS vol.1, Special Volume, ISSN 1439-2275 (2004) 132–160.
6. Orłowska, E., Radzikowska, A.M.: Information relations and operators based on double residuated lattices. In de Swart, H.C.M. (ed), Proceedings of the 6th Seminar on Relational Methods in Computer Science RelMiCS'2001 (2001) 185–199.
7. Orłowska, E., Radzikowska, A.M.: Double residuated lattices and their applications. In: de Swart, H.C.M. (ed), Relational Methods in Computer Science, Lecture Notes in Computer Science **2561** Springer–Verlag, Heidelberg (2002) 171–189.
8. Orłowska, E., Vakarelov, D. Lattice-based modal algebras and modal logics. In: Hajek, P., Valdes, L., Westerstahl, D. (eds), Proceedings of the 12th International Congress of Logic, Methodology and Philosophy of Science, Oviedo, August 2003, Elsevier, King's College London Publication (2005) 147–170.
9. Urquhart, A.: A topological representation theorem for lattices. Algebra Universalis **8** (1978) 45–58.

Monotonicity Analysis Can Speed Up Verification

Marcelo F. Frias*, Rodolfo Gamarra, Gabriela Steren, and Lorena Bourg

Department of Computer Science
School of Exact and Natural Sciences
Universidad de Buenos Aires
Argentina
{mfrias, rgamarra, gsteren, lbourg}@dc.uba.ar

Abstract. We introduce a strategy for the verification of relational specifications based on the analysis of monotonicity of variables within formulas. By comparing with the Alloy Analyzer, we show that for a relevant class of problems this technique outperforms analysis of the same problems using SAT-solvers, while consuming a fraction of the memory SAT-solvers require.

1 Introduction

The analysis of relational specifications has gained a lot of acceptance with the growing interest on the Alloy specification language [6]. Alloy's syntax and semantics are based on a first-order relational logic. Due to the automatic analysis capabilities offered by the Alloy tool [8], Alloy has become widely accepted by the community interested in automatic software engineering. The Alloy Analyzer wisely transforms Alloy specifications in which domains are bounded to a fix scope, into propositions that are later fed to SAT-solvers such as Berkmin [3], MChaff [10], or Relsat [1]. A different approach was followed in the definition of the language NP [4], where the supporting automatic analysis tool (Nitpick [4]) searched for relation instances for the variables that would violate a provided assertion.

In this paper we depart from the SAT-solving techniques of Alloy, and go back to generation of instances for the relational variables as in Nitpick. We show in this paper that for a class of problems that frequently arise when writing relational specifications, a strategy based on monotonicity analysis outperforms the analysis performed by Alloy using SAT-solving. This shows that the SAT-solvers employed in Alloy do not profit from monotonicity information after the original model is transformed by Alloy to a SAT problem.

In order to show how well this strategy performs when compared to SAT-solving in Alloy, we will introduce a relational specification language (called REL) and present ReMo, a tool that implements the strategy. Nevertheless, this paper is not about relational specification languages design, and it is quite

---

* And CONICET.

R.A. Schmidt (Ed.): RelMiCS /AKA 2006, LNCS 4136, pp. 177–191, 2006.

possible that a reader might have her own preferences about how to define such language. Similarly, the paper is not about tool design, being ReMo a prototype tool that implements the pruning strategy subject of this paper.

The paper is organized as follows. In Section 2 we give a brief description of the Alloy language, as well as of REL, the relational language we will analyze. In Section 3 we discuss the analysis strategy. In Section 4 we present ReMo, the tool that implements our strategy, and compare ReMo's performance with that of Alloy. Finally, in Section 5 we present our conclusions and proposals for further work.

2   On Alloy and REL

2.1   The Alloy Modeling Language

Alloy is a modeling language designed with (fully automated) analyzability of specifications as a priority. Alloy has its roots in the Z formal specification language, and its few constructs and simple semantics are the result of putting together some valuable features of Z and some constructs that are normally found in informal notations. This is done while avoiding incorporation of other features that would increase Alloy's complexity more than necessary.

Alloy is defined on top of what is called *relational logic*, a logic with a clear semantics based on relations. This logic provides a powerful yet simple formalism for interpreting Alloy's modeling constructs. The simplicity of both the relational logic and the language as a whole makes Alloy suitable for automatic analysis. The main analysis technique associated with Alloy is essentially a counterexample extraction mechanism, based on SAT solving. Basically, given a system specification and a statement about it, a counterexample of this statement (under the assumptions of the system description) is exhaustively searched for. Since first-order logic is not decidable (and the relational logic is a proper extension of first-order logic), SAT solving cannot be used in general to guarantee the consistency of (or, equivalently, the absence of counterexamples for) a theory; then, the exhaustive search for counterexamples has to be performed up to certain bound k in the number of elements in the universe of the interpretations. Thus, this analysis procedure can be regarded as a *validation* mechanism, rather than a *verification* procedure. Its usefulness for validation is justified by the interesting idea that, in practice, if a statement is not true, there often exists a counterexample of it of small size:

> "First-order logic is undecidable, so our analysis cannot be a decision procedure: if no model is found, the formula may still have a model in a larger scope. Nevertheless, the analysis is useful, since many formulas that have models have small ones." (cf. [5, p. 1])

The above described analysis has been implemented by the Alloy Analyzer [8], a tool that incorporates state-of-the-art SAT solvers in order to search for counterexamples of specifications.

In Fig. 1, we describe the grammar and semantics of Alloy's relational logic, which is based on relations of arbitrary rank. Composition of binary relations is well understood; but for relations of different rank, the following definition for the composition of relations has to be considered:

$$R;S = \{\langle a_1, \ldots, a_{i-1}, b_2, \ldots, b_j \rangle :$$
$$\exists b\, (\langle a_1, \ldots, a_{i-1}, b \rangle \in R \;\wedge\; \langle b, b_2, \ldots, b_j \rangle \in S)\} \ .$$

Operations for transitive closure and transposition are only defined for binary relations.

```
form ::=                          expr ::=
expr in expr (inclusion)          iden (identity)
|!form (neg)                      | expr + expr (union)
| form && form (conj)             | expr & expr (intersection)
| form || form (disj)             | expr − expr (difference)
| all v : type | form (univ)      |~ expr (transpose)
| some v : type | form (exist)    | expr.expr (composition)
                                  | +expr (transitive closure)
                                  | {v : t | form} (set former)
                                  | Var
```

Fig. 1. Grammar and semantics of Alloy

The Alloy language provides, on top of the kernel, different idioms that greatly simplify writing models. Let us consider, as a means to introduce notation, a simple example based on memories. In order to specify a data type for memories, data types for data and addresses are especially necessary. We can then start by indicating the existence of disjoint sets (of atoms) for data and addresses, which in Alloy are specified using signatures.

sig *Addr* { }   sig *Data* { }

These are basic signatures. We do not assume any special properties regarding the structure of data and addresses.

With data and addresses already defined, we can now specify what constitutes a memory. A possible way of defining memories is by saying that a memory consists of a partial mapping from addresses to data values:

sig *Memory* {
    map: Addr -> lone *Data* }

The multiplicity marking *lone* in signature Memory establishes that for each address a there must be zero or one data related.

Once data is defined, it is possible to define predicates and operations, or constrain models adding *facts* (axioms), as follows.

pred NotFull (m : Memory) {
    some a : Addr | no (a -> Data) }

In predicate NotFull, expression "no R" means that R is empty. Also, for unary relations (sets) S_1 and S_2, S_1->S_2 = { $\langle a, b \rangle : a \in S_1 \wedge b \in S_2$ }.

```
fun Write (m : Memory, a : Addr, d : Data) : Memory {
    (m.map - (a -> Data)) + (a -> d) }
```

```
fact WriteTwice {
    all m : Memory, a : Addr, d : Data |
        Write(m,a,d) = Write(Write(m,a,d),a,d) }
```

Once the model is specified, assertions about the model can be written down and then be analyzed with the Alloy tool. The following flawed assertion (flawed in the sense that there exist counterexamples) asserts that writing twice in a memory address does not modify the memory.

```
assert ClearVsTwice {
    all m : Memory, a : Addr, d : Data |
        m = Write(Write(m,a,d),a,d)
}
```

More syntactic sugar is available in order to provide idioms ubiquitous in object orientation that greatly simplify writing models [6].

2.2   The REL Modeling Language

In this section we introduce REL, a purely relational specification language. REL's syntax is introduced in Fig. 2. Notice that there are minor differences between REL and Alloy's relational logic. Due to the definition of the semantics of variables, these are in REL binary relations. Also, REL does not allow quantification. Actually, these two shortcomings of REL are overcome with the use of fork [2] (fork is not present in Alloy's relational logic).

$$
\begin{aligned}
&\text{expr} ::= \\
&\quad 1_t \ (\text{universal with type } t) \\
&\quad | \ id_t \ (\text{identity with type } t) \\
\text{form} ::= \qquad\qquad &\quad | \ \text{expr} + \text{expr} \ (\text{union}) \\
\text{expr} <= \text{expr} \ (\text{inclusion}) &\quad | \ \text{expr} \ \& \ \text{expr} \ (\text{intersection}) \\
| \ !\text{form} \ (\text{neg}) &\quad | \ [\text{expr}, \text{expr}] \ (\text{fork}) \\
| \ \text{form} \ \&\& \ \text{form} \ (\text{conj}) &\quad | \ -\text{expr} \ (\text{complement}) \\
| \ \text{form} \ || \ \text{form} \ (\text{disj}) &\quad | \sim \text{expr} (\text{transpose}) \\
&\quad | \ \text{expr} \cdot \text{expr} \ (\text{composition}) \\
&\quad | +\text{expr} \ (\text{transitive closure}) \\
&\quad | \ Var \ (\text{variable})
\end{aligned}
$$

Fig. 2. Grammar and semantics of REL

Notice that REL formulas are Boolean combinations of equations. In the forthcoming sections, we will need the following result proved in [11, p. 26].

Theorem 1. *Every REL formula is equivalent to a REL formula of the form* $T = 1$, *for an appropriate term* T.

In particular, we will use the property $T_1 <= T_2 \ \rightsquigarrow \ (-T_1) + T_2 = 1$.

3   Monotonicity Analysis for Verification of Relational Specifications

Let us consider a relational specification *Spec* in a suitable relational language (as for instance Alloy or REL). In order to validate this specification, we want to automatically analyze whether a given property α follows from *Spec*. Expressive enough relational languages are undecidable. Thus, in order to make automatic analysis feasible we will impose bounds on the size of domains. The analysis procedure reduces then to finding instances (concrete relations among elements from the bounded domains) for the relational variables that satisfy *Spec*, yet falsify α.

Notice that given a family of domains $D_1, \ldots, D_k$ with bounds $b_1, \ldots, b_k$, respectively, and a relational variable R, there are $2^{b_1 \times \cdots \times b_k}$ possible values (relations) for R on $D_1 \times \cdots \times D_k$. Even for small values of $b_1, \ldots, b_k$, exhaustive search of appropriate instances is in general unfeasible (even more if we consider that the previous number scales up exponentially when more relational variables $R_1, \ldots, R_k$ are considered). Therefore, strategies that allow us to prune the state space are mandatory in order to make automatic analysis feasible. Some strategies are *general*, in the sense that are either specification independent, or in general improve analysis performance. Example of such techniques is automorphisms elimination [7], which avoids generating those models obtained by permutations of the underlying domains. Other strategies, as the one we present in this paper, work for certain specifications.

Definition 1. *Given a relational variable R and a term $t(R)$, we say that R is positive (in t) if all the occurrences of R lay under an even number of complement symbols. It is negative (in t) if all the occurrences of R lay under an odd number of complement symbols. If R is neither positive nor negative in t, it is then said to be undefined in t.*

As an example, let us consider the terms

$$(a) - ((-R) \cdot S), \quad (b) - ((\sim R) \cdot R), \quad (c) (R \cdot R) \ \& \ (-R) \ .$$

In term (a), variable R is positive, while variable S is negative. In (b), R is negative. Finally, in (c), R is undefined.

Definition 2. *Given a relational variable R and a term $t(R)$, t is isotonic with respect to R if for all concrete relations $r, s, r \subseteq s \Rightarrow t(r) \subseteq t(s)$. Similarly, t is antitonic with respect to R if for all concrete relations $r, s, r \subseteq s \Rightarrow t(r) \supseteq t(s)$.*

Definition 2 is similar to the notion of syntactic monotonicity from the modal μ–calculus and other logics.

Proposition 1. *Let $t(R)$ be a relational term on the variable R. If R is positive in t, then t is isotonic with respect to R. Similarly, if R is negative in t, then t is antitonic with respect to R.*

Proof. The (easy) proof proceeds by induction on the structure of relational terms.

In order to introduce our strategy, we will assume that the specification *Spec* consists of a sequence of formulas on a single variable R. We will later drop this assumption and generalize to variables $R_1, \ldots, R_n$. Moreover, we will assume that formulas are equations of the form $T = 1$. Notice that from Thm. 1, there is no loss of generality in adopting this assumption. We will also assume that variable R ranges over binary relations. Since fork allows us to simulate relations of arity greater than 2 [2], there is no loss of generality in this assumption either.

The set of all relations on the domain $A \times B$, ordered by inclusion, is a lattice. Since in the worst case it will be necessary (according to our strategy) to explore the whole lattice, it is essential to explore it in a way such that each relation is visited (at most) once.

We will traverse the lattice in a depth-first search (DFS) manner. Actually, we will present two DFS traversals of the lattice. One from the bottom, and one from the top. If A contains elements $a_1, \ldots, a_n$ and B contains elements $b_1, \ldots, b_m$, we can impose on A and B the total orderings $a_1 < \cdots < a_n$ and $b_1 < \cdots < b_m$. Then, the relation on $A \times B$ defined by

$$\langle a_1, b_1 \rangle < \langle a_2, b_2 \rangle \quad \Longleftrightarrow \quad a_1 < a_2 \vee (a_1 = a_2 \wedge b_1 < b_2)$$

is a total ordering, called the *lexicographic* ordering. For the traversal from the bottom, notice that any given relation R has, as immediate *successors*, relations of the form $R \cup \{\langle a, b \rangle\}$ ($a \in A$, $b \in B$), where for every $\langle a', b' \rangle \in R$, $\langle a, b \rangle > \langle a', b' \rangle$.

Once the successors of a given a relation are defined, if we are given two different successors of R, namely $R \cup \{\langle a, b \rangle\}$ and $R \cup \{\langle c, d \rangle\}$, an ordering between them is induced by the ordering between $\langle a, b \rangle$ and $\langle c, d \rangle$. Therefore, in order to traverse the lattice, the successors of R will be visited according to this ordering. Figure 3 shows an example. Each matrix represents a relation contained in the set $\{0, 1\} \times \{0, 1\}$. A dark square in position $\langle i, j \rangle$ means pair $\langle i, j \rangle$ belongs to the relation modeled by the matrix. The number attached to each matrix gives the traversal ordering.

In order to traverse the lattice in a descending order, we define the *predecessors* of a relation R as the set $\{-P : P \text{ is a successor of } -R\}$. Notice that since P is a successor of $-R$, $-R \subseteq P$, and therefore, $-P \subseteq R$. Also, predecessors differ from the parent relation in that the latter has one extra pair. The ordering in which relations are visited in the descending traversal follows from the ordering in the ascending traversal. Figure 3 shows an example of a descending traversal.

Let us consider now an equation of the form $t(R) = 1$ in which variable R is negative in t. As a running example, consider the following equations stating that R is a total (cf. (2)) functional (cf. (1)) relation. Notice that R is negative in (1).

$$- ((\sim R) \cdot R) + Id = 1 \tag{1}$$
$$R \cdot 1 = 1 \tag{2}$$

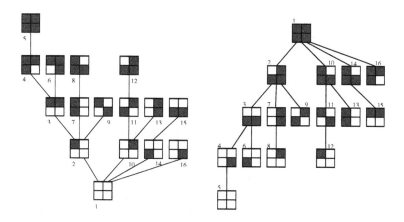

Fig. 3. Ascending and descending traversal

Since we want to satisfy an equation of the form $t(R) = 1$, we want to maximize the value of $t(R)$ (notice that we are strongly using the assumption on the shape of the equation). Since R is negative in t, $t(R)$ reaches a maximum when $R = 0$. Notice that in the example, while (1) is satisfied, (2) is not. Therefore, it is necessary to search for another model. It is clear at this point that values of R near the bottom of the lattice are more likely to satisfy (1).

Proposition 2. *Let $t(R) = 1$ be an equation on the variable R. Assume R is negative in t. Let r be a concrete relation such that $t(r) = 1$, and for a successor r' of r, $t(r') \neq 1$. Then, for every relation $x \supseteq r'$, $t(x) \neq 1$.*

Proof. Since R is negative in t, by Prop. 1 t is antitonic with respect to R. Then, $x \supseteq r' \Rightarrow t(x) \subseteq t(r') \neq 1$.

Proposition 2 provides us with a sufficient criterion for pruning part of the lattice. If in an ascending traversal of the lattice we reach a relation r' for which $t(r') \neq 1$, the branch with origin in r' does not need to be traversed because it cannot produce a model.

Thus, we can conclude that:

1. if we are given an equation of the form $t(R) = 1$,
2. variable R is negative in t,
3. we are performing an ascending traversal of the lattice,
4. we have reached a relation r' in the lattice for which $t(r') \neq 1$,

then the branch of the lattice with origin in r' can be pruned.
Similarly,

1. if we are given an equation of the form $t(R) = 1$,
2. variable R is positive in t,
3. we are performing a descending traversal of the lattice,
4. we have reached a relation r' in the lattice for which $t(r') \neq 1$,

then the branch of the lattice with origin in r' can be pruned.

Let us analyze now how general or restrictive are the hypothesis we are assuming. Notice that so far we have only discussed the situation where a single equation is being analyzed. If we are given equations $E_1, \ldots, E_k$, from Thm. 1 we can assume they are all of the form $T_i(R) = 1$ $(1 \leq i \leq k)$. At this point we are still considering the case in which there is a single relational variable R. In each equation, R may be positive, negative or undefined. Let us assume, without loss of generality, that there are more equations in which R is negative. Then, an ascending traversal of the lattice will allow us to prune a branch when one of the negative equations fails. Notice that the traversal ordering is chosen upon establishing what is the prevailing monotonicity. Therefore, the only real assumption we are making, is that variable R has a defined monotonicity in some of the equations. Thus, this is the context in which our pruning strategy can be applied.

Let us remove now the remaining assumption, namely, the restriction to a single variable R. Let us consider now relational variables $R_1, \ldots, R_n$; and equations $E_1, \ldots, E_k$, which, from Thm. 1, we can assume are all of the form $T_i(R_1, \ldots, R_n) = 1$ $(1 \leq i \leq k)$. We compute for each variable R_i $(1 \leq i \leq n)$ the amount of equations in which it is positive or negative, and call R_i positive (negative) if it appears positive (negative) in more equations. We now define for each variable a traversal ordering of the lattice as follows: if R_i is positive then the lattice is traversed from the top, and if it is negative the lattice is traversed from the bottom. Under these conditions we can prove the following theorem.

Theorem 2. *Let $T_i(R_1, \ldots, R_n) = 1$ $(1 \leq i \leq k)$ be an equation from Spec. Let the sign of each variable in T agree with the sign in the specification (that is, if R_i is positive (negative) in more equations, then it is also positive (negative) in T). If $r_1, \ldots, r_n$ are concrete relations such that $T_i(r_1, \ldots, r_n) \neq 1$, and $r'_1, \ldots, r'_n$ are concrete relations such that $r'_j \supseteq r_j$ $(r'_j \subseteq r_j)$ if R_j is negative (positive), then $T_i(r'_1, \ldots r'_n) \neq 1$.*

Proof. A simple proof by induction on the structure of term T_i allows us to show that $T_i(r'_1, \ldots r'_n) \subseteq T_i(r_1, \ldots, r_n)$. Since $T_i(r_1, \ldots, r_n) \neq 1$, it follows that $T_i(r'_1, \ldots r'_n) \neq 1$.

Theorem 2 allows us to prune the lattice as soon as a configuration as the one described in the hypotheses is reached. In Section 4 we present ReMo, a tool implementing this strategy, and evaluate its performance.

4   ReMo: RElational Verification Through MOnotonicity Analysis

ReMo is an application that implements the analysis strategy described in Section 3. The structure of the relational specifications that ReMo analyzes is shown in Fig. 4.

```
\domains         \identities       \universals       \properties
   D1 [m1:n1]       Id1 D1            Unit1 D1*D1        Formula
   :                :                 :                  :
   Dk [mk:nk]       Ids Dk            Unitu Dk*D3        Formula

\constants       \empties          \axioms
   C1 < D1*D2       Zero1 D1*D2       Formula
   :                :                 :
   Cr < Dk*D1       Zerot Dk*Dk       Formula
```

Fig. 4. Structure of a ReMo Specification

After the \domains keyword, we list the domains in the specification, as well as a range (lower and upper bound) for their size. After the \constants keyword, we list the relational variables in the specification, as well as their type (in Fig. 4 all the relational variables are to be interpreted as binary relations on the corresponding domains). Under the \identities keyword, we list those identity relations that will be required in the specification, together with their type. Similarly, we declare empty and universal relations under the appropriate keywords. Finally, the specification contains the axioms and the assertions to be verified.

ReMo receives a specification as input and transforms, using the translation defined by Thm. 1, each axiom and assertion to an equation of the form $T = 1$. It then computes the monotonicity of each relational variable, and determines a traversal order for each one. Values for the variables are then generated for the variables according to the traversal order, and the pruning strategy is applied whenever possible. ReMo deals with binary relations. These were implemented using Reduced Ordered Binary Decision Diagrams (ROBDDs) [12].

In the remaining part of this section we present several problems for which we provide Alloy and REL specifications. We then analyze running times for different domain sizes. In order to obtain the running times we have used a personal computer with an AMD 3200, 64 bits processor; 2GB, dual channel memory, and Linux Mandriva 10.2, 64 bits. We compared ReMo with the Alloy Analyzer, Version 3.0 Beta, March 5, 2005.

4.1   First Example: Total Injective Functions

Notice that functions play an important role in specification. This is how Alloy, for instance, provides idioms for declaring relations as being functional. Using multiplicity idioms it is possible to define a relation as being a total injective function (cf. (3)).

$$TotInjFun : A \text{ lone } -> \text{ one } B \tag{3}$$

In Fig. 5 we present, side by side, specifications in Alloy and REL for this problem. Despite the fact Alloy can use multiplicity markings in order to specify a relation as being a total function (cf. (3)), in order to have a fair comparison between the SAT-solvers and ReMo we will use the very same axioms. Notice that the assertion in the Alloy specification is always false because iden is the identity on all the domains (untyped). Therefore, the Alloy tool will look for some total injective function. Similarly, the property we use in ReMo is also false. While this looks as a simple problem, it is worthwhile to mention that total injective functions lay somewhere in the middle of the lattice. That is, they are neither close to the top of the lattice (those relations tend not to be functional or injective) nor close to the bottom of the lattice (where relations tend not to be total). Therefore, there is a real challenge in efficiently getting to them.

```
module InjTotFun                        \domains
                                          A[scope:scope]
sig elem { }                            \constants
                                          F < A*A
sig function {                          \identities
  F: elem -> elem                         id A*A
}                                       \empties
{                                         O A*A
  ((~F).F) in elem<:iden                \universals
  (F.(~F)) in elem<:iden                  1 A*A
  (F.(elem<:univ))=(elem<:univ)         \axioms
}                                         ((~F).F)<=id
                                          (F.(~F))<=id
assert funInyTotal {                      F.1 = 1
  all f:function | f.F=iden }           \properties
                                          O = 1
check funInyTotal for 1
but exactly [scope] elem
```

Fig. 5. Specification of total injective functions in Alloy and REL

After translating the axioms according to Thm. 1, we obtain the following equations:

$$-((\sim F).F) + id = 1 \tag{4}$$
$$-(F.(\sim F)) + id = 1 \tag{5}$$
$$F.1 = 1 \tag{6}$$

Since in equations (4) and (5) F is negative, ReMo will traverse the lattice from the bottom for this variable. In Table 1 we present the running times for various scopes for ReMo, and for the same scopes for Alloy using the SAT-solvers it provides.

Table 1. Running times for Alloy and ReMo

scope	Berkmin	MChaff	RelSat	ReMo
30	00:23	11:56	00:45	**00:09**
35	00:42	52:48	01:26	**00:33**
40	01:32	> 60'	02:46	**00:54**
45	02:36	> 60'	04:59	**01:37**
50	02:42	> 60'	08:37	**02:14**
55	06:52	> 60'	14:26	**03:40**
60	08:18	> 60'	23:17	**04:45**
65	Crash	Crash	Crash	**13:02**

Regarding memory consumption, for a scope of 60 in the case of Berkmin and RelSat, 35 for MChaff (the largest for which each SAT-solver found a model within 60'), and 65 for ReMo we obtained the following data:

Berkmin	MChaff	RelSat	ReMo
444.5 MB	163.8 MB	469 MB	**10.24 MB**

4.2   Second Example: Total Orderings

Total orderings are frequently used in specifications. For instance, they come within the library of standard modules for Alloy, and are commonly used when specifying properties of executions of operations [9].

In this section we will deal with two different problems involving total orderings.

Finding a Total Ordering. Total orderings are binary relations O that are reflexive ($id \subseteq O$), antisymmetric ($O\& \sim O = id$), transitive ($O.O \subseteq O$) and total ($O+ \sim O = 1$). This is an interesting problem because, as in the case for functions, these orderings lay somewhere in the middle of the lattice (relations near the top of the lattice are not antisymmetric, and those near the bottom are not total). Given a total ordering, there is a relation $next$ that relates each element to its successor. Given a total order O, $next$ (known as the Hasse–diagram for O) is defined by the equation

$$next = O\& (-id) \& (- ((O\& (-id)) . (O\& (-id)))) \ .$$

Is it true that this relation is functional, injective and total? The answer, confirmed by both Alloy and ReMo, is "No". While $next$ is functional and injective, it is not total. The last element in the ordering does not belong to the domain of $next$.

The specifications in Alloy and ReMo are given in Figs. 6 and 7.

After applying the translation from Thm. 1 to the axioms not involving $next$ (because those involving $next$ will always be undefined), we obtain the following equations:

$$- id + O = 1 \tag{7}$$
$$-(O\&(\sim O)) + id = 1 \tag{8}$$
$$-(O.O) + O = 1 \tag{9}$$
$$O + (\sim O) = 1 \tag{10}$$

Since O is positive in (7) and (10), negative in (8) and undefined in (9), a descending traversal of the lattice is chosen for variable O.

In Table 2 we show that, for this problem, ReMo outperforms the Alloy Analyzer both in running time and memory consumption.

```
module totalOrder
sig elem{ }
sig order {
    O : elem -> elem,
    next : elem -> elem
}{
    elem<:iden in O
    O&(~O) in iden
    O.O in O
    O+(~O) = elem->elem
    next = (O - (elem<:iden)) -
        ((O - (elem<:iden)).(O - (elem<:iden)))
}
assert nextTotInjFun {
    all o:order |
        ~(o.next).(o.next) in elem<:iden &&
        (o.next).~(o.next) in elem<:iden &&
        (o.next).(elem<:univ) = elem<:univ
}

check nextTotInjFun for 1
    but exactly [scope] elem
```

Fig. 6. Specification in Alloy of Total Ordering

```
\domains                \axioms
  elem[scope:scope]       id <= O
                          O&(~O) <= id
\constants                O.O <= O
  O < elem*elem           O+(~O) = 1
  next < elem*elem        next = O & (-id) & (-((O&(-id)).(O&(-id))))
\identities             \properties
  id elem*elem            ~next.next <= id
\universals             next.~next <= id
  1 elem*elem             next.1 = 1
```

Fig. 7. Specification in REL of Total Ordering

Table 2. Running times and memory consumption for total orderings in Alloy and ReMo

scope	Berkmin	MChaff	RelSat	ReMo
10	00:02	00:03	00:03	**00:00**
15	00:03	00:05	00:13	**00:00**
20	00:09	00:11	00:45	**00:01**
25	00:25	00:21	02:07	**00:04**
30	01:16	01:37	05:37	**00:07**
35	01:44	02:54	10:11	**00:19**
40	02:46	05:38	19:20	**00:32**
45	06:31	04:05	34:51	**00:55**
50	10:50	16:02	crashed	**01:23**
Memory	481.3 MB	231.4 MB	307.2 MB	**81.9 MB**

Testing a Valid Property. Since testing a valid property will not produce any counterexamples, all total orderings will be visited. Notice that there are $n!$ different total orderings on a n elements set. Notice also that given a total ordering O on the set *elem*, any other total ordering O' can be obtained from O via a permutation of *elem*. As an instance, for $elem = \{1, 2, 3, 4, 5\}$, let

$$O = 1 < 2 < 3 < 4 < 5, \text{ and } O' = 4 < 3 < 5 < 1 < 2.$$

If p is the permutation: $1 \mapsto 4, 2 \mapsto 3, 3 \mapsto 5, 4 \mapsto 1, 5 \mapsto 2,,$ then $O' = \{ \langle p(x), p(y) \rangle : \langle x, y \rangle \in O \} = p(O)$.

Since all relational operators from REL are *algebraic*[1] [11, p. 57], i.e., invariant under permutations of the atomic domains, testing the equations on a single linear ordering suffices.

Since Alloy eliminates many of the models obtained by permutations, it is not surprising that for these problems Alloy outperforms ReMo considerably.

We will come back to this problem in Section 5 when discussing further work.

4.3 Third Example: A Specification with Two Relational Variables

The examples we have presented so far, all deal with a single relational variable. Even in Example 4.2, where variables O and *next* are declared, *next* is introduced as an abbreviation for a complex term and replaced by its definition in both tools. Thus, this will be the first example where two meaningful variables appear.

The problem we will deal with is the following:

> Let us consider a system in which n processes and m resources are available. Every process has either requested a resource, or a resource has already been allocated to the process. Every resource has already been allocated to a process, or is being requested by one. Resources can be allocated to exactly one process. Finally, the system is deadlock free (no cycles in the relation `reqs.allocs`).
>
> We want to analyze whether is it true that every process is requesting a resource, and every resource has been allocated to a process.

In Figs. 8 we present the Alloy and REL specifications.
If we translate the axioms according to Thm. 1, the axioms become:

$$- idP + (reqs. \sim reqs) + (\sim allocs.allocs) = 1 \tag{11}$$
$$-idR + (\sim reqs.reqs) + (allocs. \sim allocs) = 1 \tag{12}$$
$$-(+(reqs.allocs)) + -idP = 1 \tag{13}$$
$$-(\sim allocs.allocs) + idP = 1 \tag{14}$$

Similarly, since the assertions must be falsified, we translate the negation of the equations. We then obtain:

$$1.(idP \ \& \ -(reqs. \sim reqs)).1 = 1 \tag{15}$$
$$1.(idR \ \& \ -(allocs. \sim allocs)).1 = 1 \tag{16}$$

Variable *reqs* appears positive in (11) and (12), and negative in (13) and (15). Variable *allocs* appears positive in (11) and (12), and negative in (13), (14) and (16). Therefore, variable *allocs* is assigned an ascending traversal. Since the number of times *reqs* is positive (or negative) is greater than zero, it is reasonable

---

[1] Even fork is invariant under permutations of the atomic domains. It is not invariant under arbitrary permutations of the field of a fork algebra.

```
module System                          \domains
sig Process{ }                           Process [scope:scope]
sig Resource{ }                          Resource [scope:scope]
sig System {                           \constants
  reqs: Process->Resource,               reqs < Process*Resource
  allocs: Resource->Process              allocs < Resource*Process
}{                                     \identities
  Process<:iden in                       idP Process*Process
    (reqs.~reqs)+(~allocs.allocs)        idR Resource*Resource
  Resource<:iden in                    \empties
    (~reqs.reqs)+(allocs.~allocs)        zeroP Process*Process
  ~allocs.allocs in Process<:iden       zeroR Resource*Resource
  (Process<:iden).(reqs.allocs) in    \axioms
    (Process->Process)-(Process<:iden)   idP <= (reqs.~reqs) + (~allocs.allocs)
}                                        idR <= (~reqs.reqs) + (allocs.~allocs)
assert allWaiting{                       +(reqs.allocs) & idP = zeroP
all s:System |                           ~allocs.allocs <= idP
  iden & (Process->Process) in        \properties
    (s.reqs).(~(s.reqs)) &&              idP <= reqs.~reqs
  iden & (Resource->Resource) in        idR <= allocs.~allocs
    (s.allocs).(~(s.allocs))
}
check allWaiting for 1 but
  exactly [scope] Process,
  exactly [scope] Resource
```

Fig. 8. Specification in Alloy and ReMo

to use one of the proposed traversals. Since in this case there is a tie between the number of times *reqs* is positive and the number of times it is negative, ReMo chooses (by default) an ascending traversal.

In Table 3 we present the running times and memory consumption for Alloy and ReMo.

Table 3. Running times and memory consumption

scope	Berkmin	MChaff	RelSat	ReMo
10	00.17	01.12	05.36	**00.03**
15	00.23	03.38	48.88	**00.06**
20	01.42	17.51	>60 min	**00.12**
25	03.51	44.75		**00.27**
30	07.73	49.98		**00.52**
Memory scope 15	73.73 MB	49.15 MB	61.5 MB	**10.24 MB**

5   Conclusions and Further Work

In this paper we have presented a pruning strategy that allows us to considerably reduce the time and memory required in order to find a counterexample in a relational specification. We conclude that the strategy is very efficient in those cases where it can be applied. At the same time, this paper leaves open the following problems.

1. It is essential to combine this strategy with another that allows us to handle specifications where our strategy cannot be applied. We are currently combining our strategy with isomorph free model enumeration [7].

2. Since Alloy does not profit so far from monotonicity information as ReMo does, it seems necessary to improve the analysis capabilities of Alloy in this direction.

References

1. Bayardo Jr, R. J. and Schrag R. C., *Using CSP look-back techniques to solve real world SAT instances*. In Proc. of the 14th National Conf. on Artificial Intelligence, pp. 203–208, 1997.
2. Frias M. F., Lopez Pombo C. G., Baum G. A., Aguirre N. and Maibaum T. S. E., *Reasoning About Static and Dynamic Properties in Alloy: A Purely Relational Approach*, to appear in ACM TOSEM, in press.
3. Goldberg E. and Novikov Y., *BerkMin: a Fast and Robust SAT-Solver*, in proceedings of DATE-2002, 2002, pp. 142–149.
4. Jackson, D. *Nitpick: A checkable specification language*. In Proceedings of the Workshop on Formal Methods in Software Practice (San Diego, Calif., Jan. 1996).
5. Jackson D., *Automating First-Order Relational Logic*, in Proceedings of SIGSOFT FSE 2000, pp. 130-139, Proc. ACM SIGSOFT Conf. Foundations of Software Engineering. San Diego, November 2000.
6. Jackson D., *Alloy: A Lightweight Object Modelling Notation*, ACM Transactions on Software Engineering and Methodology (TOSEM), Volume 11, Issue 2 (April 2002), pp. 256-290.
7. Jackson D., Jha S. and Damon C. A., *Isomorph-Free Model Enumeration: A New Method for Checking Relational Specifications*, ACM TOPLAS, Vol. 20, No. 2, 1998, pp. 302–343.
8. Jackson D., Schechter I. and Shlyakhter I., *Alcoa: the Alloy Constraint Analyzer*, Proceedings of the International Conference on Software Engineering, Limerick, Ireland, June 2000.
9. Jackson, D., Shlyakhter, I., and Sridharan, M., *A Micromodularity Mechanism*. Proc. ACM SIGSOFT Conf. Foundations of Software Engineering/European Software Engineering Conference (FSE/ESEC '01), Vienna, September 2001.
10. Moskewicz M., Madigan C., Zhao Y., Zhang L. and Malik S., *Chaff: Engineering an Efficient SAT Solver*, 39th Design Automation Conference (DAC 2001), Las Vegas, June 2001.
11. Tarski, A. and Givant, S.,*A Formalization of Set Theory without Variables*, A.M.S. Coll. Pub., vol. 41, 1987.
12. Wegener I., *Branching Programs and Binary Decision Diagrams*, SIAM Discrete Mathematics and Applications, SIAM, 2000.

Max-Plus Convex Geometry

Stéphane Gaubert[1] and Ricardo Katz[2]

[1] INRIA, Domaine de Voluceau, 78153 Le Chesnay Cédex, France
`Stephane.Gaubert@inria.fr`
[2] CONICET, Instituto de Matemática "Beppo Levi", Universidad Nacional de
Rosario, Av. Pellegrini 250, 2000 Rosario, Argentina
`rkatz@fceia.unr.edu.ar`

Abstract. Max-plus analogues of linear spaces, convex sets, and polyhedra have appeared in several works. We survey their main geometrical
properties, including max-plus versions of the separation theorem, existence of linear and non-linear projectors, max-plus analogues of the
Minkowski-Weyl theorem, and the characterization of the analogues of
"simplicial" cones in terms of distributive lattices.

1 Introduction

The max-plus semiring, $\mathbb{R}_{\max}$, is the set $\mathbb{R} \cup \{-\infty\}$ equipped with the addition
$(a, b) \mapsto \max(a, b)$ and the multiplication $(a, b) \mapsto a + b$. To emphasize the
semiring structure, we write $a \oplus b := \max(a, b)$, $ab := a + b$, $\mathbb{0} := -\infty$ and $\mathbb{1} := 0$.

Many classical notions have interesting max-plus analogues. In particular,
semimodules over the max-plus semiring can be defined essentially like linear
spaces over a field. The most basic examples consist of *subsemimodules of func-
tions* from a set X to $\mathbb{R}_{\max}$, which are subsets $\mathscr{V}$ of $\mathbb{R}_{\max}^X$ that are stable by
max-plus linear combinations, meaning that:

$$\lambda u \oplus \mu v \in \mathscr{V} \tag{1}$$

for all $u, v \in \mathscr{V}$ and for all $\lambda, \mu \in \mathbb{R}_{\max}$. Here, for all scalars λ and functions
u, λu denotes the function sending x to the max-plus product $\lambda u(x)$, and the
max-plus sum of two functions is defined entrywise. Max-plus semimodules have
many common features with convex cones. This analogy leads to define max-plus
convex subsets $\mathscr{V}$ of $\mathbb{R}_{\max}^X$ by the requirement that (1) holds for all $u, v \in \mathscr{V}$ and
for all $\lambda, \mu \in \mathbb{R}_{\max}$ such that $\lambda \oplus \mu = \mathbb{1}$. The finite dimensional case, in which
$X = \{1, \ldots, n\}$, is already interesting.

Semimodules over the max-plus semiring have received much attention [1],
[2], [3], [4], [5]. They are of an intrinsic interest, due to their relation with lattice
and Boolean matrix theory, and also with abstract convex analysis [6]. They
arise in the geometric approach to discrete event systems [7], and in the study
of solutions of Hamilton-Jacobi equations associated with deterministic optimal
control problems [8,4,9,10]. Recently, relations with phylogenetic analysis have
been pointed out [11].

R.A. Schmidt (Ed.): RelMiCS /AKA 2006, LNCS 4136, pp. 192–206, 2006.

In this paper, we survey the basic properties of max-plus linear spaces, convex sets, and polyhedra, emphasizing the analogies with classical convex geometry. We shall present a synopsis of the results of [5,12], including separation theorems, as well as new results, mostly taken from the recent works [13,14]. Some motivations are sketched in the next section. The reader interested specifically in applications to computer science might look at the work on fixed points problems in static analysis of programs by abstract interpretation [28], which is briefly discussed at the end of Section 2.3.

2 Motivations

2.1 Preliminary Definitions

Before pointing out some motivations, we give preliminary definitions. We refer the reader to [5] for background on semirings with an idempotent addition (idempotent semirings) and semimodules over idempotent semirings. In particular, the standard notions concerning modules, like linear maps, are naturally adapted to the setting of semimodules.

Although the results of [5] are developed in a more general setting, we shall here only consider semimodules of functions. A *semimodule of functions* from a set X to a semiring $\mathscr{K}$ is a subset $\mathscr{V} \subset \mathscr{K}^X$ satisfying (1), for all $u, v \in \mathscr{V}$ and $\lambda, \mu \in \mathscr{K}$. When $X = \{1, \ldots, n\}$, we write $\mathscr{K}^n$ instead of $\mathscr{K}^X$, and we denote by u_i the i-th coordinate of a vector $u \in \mathscr{K}^n$.

We shall mostly restrict our attention to the case where $\mathscr{K}$ is the *max-plus semiring*, $\mathbb{R}_{\max}$, already defined in the introduction, or the *completed max-plus semiring*, $\overline{\mathbb{R}}_{\max}$, which is obtained by adjoining to $\mathbb{R}_{\max}$ a $+\infty$ element, with the convention that $(-\infty) + (+\infty) = -\infty$. Some of the results can be stated in a simpler way in the completed max-plus semiring.

The semirings $\mathbb{R}_{\max}$ and $\overline{\mathbb{R}}_{\max}$ are equipped with the usual order relation. Semimodules of functions with values in one of these semirings are equipped with the product order.

We say that a set of functions with values in $\overline{\mathbb{R}}_{\max}$ is *complete* if the supremum of an arbitrary family of elements of this set belongs to it. A *convex* subset $\mathscr{V}$ of $\overline{\mathbb{R}}_{\max}^X$ is defined like a convex subset of $\mathbb{R}_{\max}$, by requiring that (1) holds for all $u, v \in \mathscr{V}$ and $\lambda, \mu \in \mathbb{R}_{\max}$ such that $\lambda \oplus \mu = \mathbf{1}$.

If $\mathscr{X}$ is a set of functions from X to $\mathbb{R}_{\max}$, we define the semimodule that it generates, span $\mathscr{X}$, to be the set of max-plus linear combinations of a finite number of functions of $\mathscr{X}$. In other words, every function f of span $\mathscr{X}$ can be written as

$$f(x) = \max_{i \in I} \lambda_i + g_i(x) \ , \tag{2}$$

where I is a finite set, g_i belongs to $\mathscr{X}$, and λ_i belongs to $\mathbb{R} \cup \{-\infty\}$.

If $\mathscr{X}$ is a set of functions from X to $\overline{\mathbb{R}}_{\max}$, we define the complete semimodule that it generates, $\overline{\text{span}}\, \mathscr{X}$, to be the set of arbitrary max-plus linear combinations of functions of $\mathscr{X}$, or equivalently, the set of arbitrary suprema of elements of span $\mathcal{X}$. Thus, every function of $\overline{\text{span}}\,\mathcal{X}$ can be written in the form (2), if we allow I to be infinite, with $\lambda_i \in \mathbb{R} \cup \{\pm\infty\}$, and if replace the "max" by a "sup". Then, we say that f is an infinite linear combination of the functions g_i.

2.2   Solution Spaces of Max-Plus Linear Equations

An obvious motivation to introduce semimodules over $\mathbb{R}_{\max}$ or $\overline{\mathbb{R}}_{\max}$ is to study the spaces of solutions of max-plus linear equations. Such equations arise naturally in relation with discrete event systems and dynamic programming.

For instance, let $A = (A_{ij})$ denote a $p \times q$ matrix with entries in $\mathbb{R}_{\max}$, and consider the relation

$$y = Ax \ .$$

Here, Ax denotes the max-plus product, so that $y_i = \max_{1 \leq k \leq q} A_{ik} + x_k$. This can be interpreted as follows. Imagine a system with q initial events (arrival of a part in a workshop, entrance of a customer in a network, etc.), and p terminal events (completion of a task, exit of a customer, etc.). Assume that the terminal event i cannot be completed earlier than A_{ij} time units after the initial event j has occurred. Then, the vector $y = Ax$ represents the earliest completion times of the terminal events, as a function of the vector x of occurrence times of the initial events. The image of the max-plus linear operator A, $\mathcal{V} := \{Ax \mid x \in \mathbb{R}_{\max}^q\}$ is a semimodule representing all the possible completion times. More sophisticated examples, relative to the dynamical behavior of discrete event systems, can be found in [7,15].

Other interesting semimodules arise as *eigenspaces*. Consider the eigenproblem

$$Ax = \lambda x \ ,$$

that is, $\max_{1 \leq j \leq q} A_{ij} + x_j = \lambda + x_i$. We assume here that A is square. We look for the *eigenvectors* $x \in \mathbb{R}_{\max}^q$ and the *eigenvalues* $\lambda \in \mathbb{R}_{\max}$. The eigenspace of λ, which is the set of all x such that $Ax = \lambda x$, is obviously a semimodule. In dynamic programming, A_{ij} represents the reward received when moving from state i to state j. If $Ax = \lambda x$ for some vector x with finite entries, it can be checked that the eigenvalue λ gives the maximal mean reward per move, taken over all infinite trajectories. The eigenvector x can be interpreted as a fair relative price vector for the different states. See [16,10,17] for more details on the eigenproblem. The extreme generators of the eigenspace (to be defined in Section 5) correspond to optimal stationary strategies or infinite "geodesics" [10].

The infinite dimensional version of the equation $y = Ax$ and of the spectral equation $Ax = \lambda x$ respectively arise in large deviations theory [18] and in optimal control [10]. When the state space is non compact, the representation of max-plus eigenvectors is intimately related with the compactification of metric spaces in terms of horofunctions [10,19].

2.3   From Classical Convexity to Max-Plus Convexity

The most familiar examples of semimodules over the max-plus semiring arise in classical convex analysis. In this section, unlike in the rest of the paper, the words "convex", "linear", "affine", and "polyhedra", and the notation "·" for the scalar product of $\mathbb{R}^n$, have their usual meaning.

Recall that the *Legendre-Fenchel transform* of a map f from $\mathbb{R}^n$ to $\mathbb{R} \cup \{\pm\infty\}$ is the map $f^\star$ from $\mathbb{R}^n$ to $\mathbb{R} \cup \{\pm\infty\}$ defined by:

$$f^\star(p) = \sup_{x \in \mathbb{R}^n} p \cdot x - f(x) \ . \tag{3}$$

Legendre-Fenchel duality [20, Cor. 12.2.1] tells that $(f^\star)^\star = f$ if f is convex, lower semicontinuous and if $f(x) \in \mathbb{R} \cup \{+\infty\}$ for all $x \in \mathbb{R}^n$. Making explicit the identity $(f^\star)^\star = f$, we get $f(x) = \sup_{p \in \mathbb{R}^n} p \cdot x - f^\star(p)$. This classical result can be restated as follows, in max-plus terms.

Property 1 (Semimodule of convex functions). The set of convex lower semicontinuous convex functions from $\mathbb{R}^n$ to $\mathbb{R} \cup \{+\infty\}$ is precisely the set of infinite max-plus linear combinations of (conventional) linear forms on $\mathbb{R}^n$.

The numbers $-f^\star(p)$, for $p \in \mathbb{R}^n$, may be thought of as the "coefficients", in the max-plus sense, of f with respect to the "basis" of linear forms $x \mapsto p \cdot x$. These coefficients are not unique, since there may be several functions g such that $f = g^\star$. However, the map g giving the "coefficients" is unique if it is required to be lower semicontinuous and if f is *essentially smooth*, see [21, Cor. 6.4]. The semimodule of finite max-plus linear combinations of linear forms is also familiar: it consists of the convex functions from $\mathbb{R}^n$ to $\mathbb{R}$ that are polyhedral [20], together with the identically $-\infty$ map.

By changing the set of generating functions, one obtains other spaces of functions. In particular, an useful space consists of the maps f from $\mathbb{R}^n$ to $\mathbb{R}$ that are order preserving, meaning that $x \leq y \implies f(x) \leq f(y)$, where $\leq$ denotes the standard product ordering of $\mathbb{R}^n$, and commute with the addition with a constant, meaning that $f(\lambda + x_1, \ldots, \lambda + x_n) = \lambda + f(x_1, \ldots, x_n)$. These maps play a fundamental role in the theory of Markov decision processes and games: they arise as the coordinate maps of dynamic programming operators. They are sometimes called *topical maps* [22]. Topical maps include min-plus linear maps sending $\mathbb{R}^n$ to $\mathbb{R}$, which can be written as

$$x \mapsto \min_{1 \leq j \leq n} a_j + x_j \ , \tag{4}$$

where $a_1, \ldots, a_n$ are numbers in $\mathbb{R} \cup \{+\infty\}$ that are not all equal to $+\infty$. Of course, topical maps also include max-plus linear maps sending $\mathbb{R}^n$ to $\mathbb{R}$, which can be represented in a dual way. The following observation was made by Rubinov and Singer [23], and, independently by Gunawardena and Sparrow (personal communication).

Property 2 (Semimodule of topical functions). The set of order preserving maps from $\mathbb{R}^n$ to $\mathbb{R}$ that commute with the addition of a constant coincides, up to the functions identically equal to $-\infty$ or $+\infty$, with the set of infinite max-plus linear combinations of the maps of the form (4).

A map from $\mathbb{R}^n$ to $\mathbb{R}^n$ is called a *min-max function* if each of its coordinates is a *finite* max-plus linear combination of maps of the form (4). Min-max functions arise as dynamic operators of zero-sum deterministic two player games with finite state and action spaces, and also, in the performance analysis of discrete event systems [24,25]. The decomposition of a min-max function as a supremum of min-plus linear maps (or dually, as an infimum of max-plus linear maps) is used in [26,27,25] to design *policy iteration algorithms*, allowing one to solve fixed points problems related to min-max functions. These techniques are applied in [28] to the static analysis of programs by abstract interpretation.

Another application of semimodules of functions, to the discretization of Hamilton-Jacobi equations associated with optimal control problems, can be found in [29].

3   Projection on Complete Semimodules

We now survey some of the main properties of the max-plus analogues of modules (or cones) and convex sets. In the case of Hilbert spaces, a possible approach is to define first the projection on a closed convex set, and then to show the separation theorem. We shall follow here a similar path.

Definition 1 (Projector on a complete semimodule). *If $\mathscr{V}$ is a complete semimodule of functions from X to $\overline{\mathbb{R}}_{\max}$, for all functions u from X to $\overline{\mathbb{R}}_{\max}$, we define:*

$$P_{\mathscr{V}}(u) := \sup\{v \in \mathscr{V} \mid v \leq u\} \ .$$

Since $\mathscr{V}$ is complete, $P_{\mathscr{V}}(u) \in \mathscr{V}$, and obviously, $P_{\mathscr{V}}$ has all elements of $\mathscr{V}$ as fixed points. It follows that

$$P_{\mathscr{V}} \circ P_{\mathscr{V}} = P_{\mathscr{V}} \ .$$

The projector $P_{\mathscr{V}}$ can be computed from a generating family of $\mathscr{V}$. Assume first that $\mathscr{V}$ is generated only by one function $v \in \overline{\mathbb{R}}_{\max}^X$, meaning that $\mathscr{V} = \overline{\mathbb{R}}_{\max} v :=$ $\{\lambda v \mid \lambda \in \overline{\mathbb{R}}_{\max}\}$. Define, for $u \in \overline{\mathbb{R}}_{\max}^X$,

$$u/v := \sup\{\lambda \in \overline{\mathbb{R}}_{\max} \mid u \geq \lambda v\} \ .$$

One can easily check that

$$u/v = \inf\{u(x) - v(x) \mid x \in X\} \ ,$$

with the convention that $(+\infty) - (+\infty) = (-\infty) - (-\infty) = +\infty$. Of course, $P_{\mathscr{V}}(u) = (u/v)v$. More generally, we have the following elementary result.

Proposition 1 ([5]). *If $\mathscr{V}$ is a complete subsemimodule of $\overline{\mathbb{R}}_{\max}^X$ generated by a subset $\mathscr{X} \subset \overline{\mathbb{R}}_{\max}^X$, we have*

$$P_{\mathscr{V}}(u) = \sup_{v \in \mathscr{X}} (u/v)v \ .$$

When $\mathscr{X}$ is finite and $X = \{1, \ldots, n\}$, this provides an algorithm to decide whether a function u belongs to $\mathscr{V}$: it suffices to check whether $P_{\mathscr{V}}(u) = u$.

Example 1. We use here the notation of Section 2.3. When $\mathscr{V}$ is the complete semimodule generated by the set of conventional linear maps $x \mapsto p \cdot x$, $P_{\mathscr{V}}(u)$ can be written as

$$[P_{\mathscr{V}}(u)](x) = \sup_{p \in \mathbb{R}^n} \left(\inf_{y \in \mathbb{R}^n} u(y) - p \cdot y \right) + p \cdot x = (u^\star)^\star(x) \ ,$$

where $u^\star$ is the Legendre-Fenchel transform of u. Hence, $P_{\mathscr{V}}(u)$ is the lower-semicontinuous convex hull of u ([20, Th. 12.2]).

When $\mathscr{V}$ is the complete semimodule generated by the set of functions of the form $x \mapsto -\|x - a\|_\infty$, with $a \in \mathbb{R}^n$, it can be checked that

$$[P_{\mathscr{V}}(u)](x) = \sup_{a \in \mathbb{R}^n} \left(\left(\inf_{y \in \mathbb{R}^n} u(y) + \|y - a\|_\infty \right) - \|x - a\|_\infty \right) = \inf_{a \in \mathbb{R}^n} u(a) - \|x - a\|_\infty \ .$$

This is the "1-Lipschitz regularization" of u. More generally, one may consider semimodules of maps with a prescribed continuity modulus, like Hölder continuous maps, see [21].

The projection of a vector of a Hilbert space on a (conventional) closed convex set minimizes the Euclidean distance of this vector to it. A similar property holds in the max-plus case, but the Euclidean norm must be replaced by the Hilbert seminorm (the additive version of Hilbert's projective metric). For any scalar $\lambda \in \mathbb{R}_{\max}$, define $\lambda^- := -\lambda$. For all vectors $u, v \in \mathbb{R}_{\max}^X$, we define

$$\delta_H(u, v) := ((u/v)(v/u))^- \ ,$$

where the product is understood in the max-plus sense. When $X = \{1, \ldots, n\}$ and u, v take finite values, $\delta_H(u, v)$ can be written as

$$\delta_H(u, v) = \sup_{1 \le i, j \le n} (u_i - v_i + v_j - u_j) \ ,$$

with the usual notation.

Theorem 1 (The projection minimizes Hilbert's seminorm, [5]). *If $\mathscr{V}$ is a complete semimodule of functions from a set X to $\overline{\mathbb{R}}_{\max}$, then, for all functions u from X to $\overline{\mathbb{R}}_{\max}$, and for all $v \in \mathscr{V}$,*

$$\delta_H(u, P_{\mathscr{V}}(u)) \le \delta_H(u, v) \ .$$

This property does not uniquely define $P_{\mathscr{V}}(u)$, even up to an additive constant, because the balls in Hilbert's projective metric are not "strictly convex".

Example 2. Consider

$$A = \begin{bmatrix} 0 & 0 & 0 & -\infty & 0.5 \\ 1 & -2 & 0 & 0 & 1.5 \\ 0 & 3 & 2 & 0 & 3 \end{bmatrix} \ , \qquad u = \begin{bmatrix} 1 \\ 0 \\ 0.5 \end{bmatrix} \ . \tag{5}$$

The semimodule $\mathscr{V}$ generated by the columns of the matrix A is represented in Figure 1 (left). A non-zero vector $v \in \mathbb{R}^3_{\max}$ is represented by the point that is the barycenter with weights $(\exp(\beta v_i))_{1 \leq i \leq 3}$ of the vertices of the simplex, where $\beta > 0$ is a fixed scaling parameter. Observe that vectors that are proportional in the max-plus sense are represented by the same point. Every vertex of the simplex represents one basis vector e_i. The point p_i corresponds to the i-th column of A. The semimodule $\mathscr{V}$ is represented by the closed region in dark grey and by the bold segments joining the points p_1, p_2, p_4 to it.

We deduce from Proposition 1 that

$$
P_{\mathscr{V}}(u) = (-1) \begin{bmatrix} 0 \\ 1 \\ 0 \end{bmatrix} \oplus (-2.5) \begin{bmatrix} 0 \\ -2 \\ 3 \end{bmatrix} \oplus (-1.5) \begin{bmatrix} 0 \\ 0 \\ 2 \end{bmatrix} \oplus (0) \begin{bmatrix} -\infty \\ 0 \\ 0 \end{bmatrix} \oplus (-2.5) \begin{bmatrix} 0.5 \\ 1.5 \\ 3 \end{bmatrix}
$$

$$
= \begin{bmatrix} -1 \\ 0 \\ 0.5 \end{bmatrix} .
$$

Since $P_{\mathscr{V}}(u) < u$, u does not belong to $\mathscr{V}$. The vector u and its projection $P_{\mathscr{V}}(u)$ are represented in Figure 1 (right). The ball in Hilbert's metric centered at point u the boundary of which contains $P_{\mathscr{V}}(u)$ is represented in light grey. The fact that $P_{\mathscr{V}}(u)$ is one of the points of $\mathscr{V}$ that are the closest to u (Theorem 1) is clear from the figure.

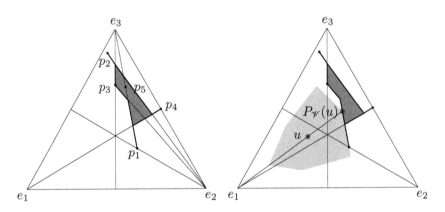

Fig. 1. A max-plus semimodule (left). A point u, its projection $P_{\mathscr{V}}(u)$, and the corresponding ball in Hilbert's projective metric (right).

4   Separation Theorems

We first state separation theorems for complete subsemimodules and complete convex subsets of $\overline{\mathbb{R}}^X_{\max}$, since the results are simpler in this setting. Then, we shall see how the completeness assumptions can be dispensed with.

Several max-plus separation theorems have appeared in the literature: the first one is due to Zimmermann [2]. Other separation theorems appeared in [30], in [5,12], and, in the polyhedral case, in [11,31]. We follow here the approach of [5,12], in which the geometrical interpretation is apparent.

We call *half-space* of $\overline{\mathbb{R}}_{\max}^X$ a set of the form

$$\mathcal{H} = \{v \in \overline{\mathbb{R}}_{\max}^X \mid a \cdot v \leq b \cdot v\} , \tag{6}$$

where $a, b \in \overline{\mathbb{R}}_{\max}^X$ and $\cdot$ denotes here the max-plus scalar product:

$$a \cdot v := \sup_{x \in X} a(x) + v(x) .$$

We extend the notation $\cdot^-$ to functions $v \in \overline{\mathbb{R}}_{\max}^X$, so that v^- denotes the function sending $x \in X$ to $-v(x)$. The following theorem is proved using residuation (or Galois correspondence) techniques.

Theorem 2 (Universal Separation Theorem, [5, Th. 8]). *Let $\mathcal{V} \subset \overline{\mathbb{R}}_{\max}^X$ denote a complete subsemimodule, and let $u \in \overline{\mathbb{R}}_{\max}^X \setminus \mathcal{V}$. Then, the half-space*

$$\mathcal{H} = \{v \in \overline{\mathbb{R}}_{\max}^X \mid (P_{\mathcal{V}}(u))^- \cdot v \leq u^- \cdot v\} \tag{7}$$

contains $\mathcal{V}$ and not u.

Since $P_{\mathcal{V}}(u) \leq u$, the inequality can be replaced by an equality in (7). A way to remember Theorem 2 is to interpret the equality

$$(P_{\mathcal{V}}(u))^- \cdot v = u^- \cdot v$$

as the "orthogonality" of v to the direction $(u, P_{\mathcal{V}}(u))$. This is analogous to the Hilbert space case, where the difference between a vector and its projection gives the direction of a separating hyperplane.

Example 3. Let $\mathcal{V}$, A, and u be as in Example 2. The half-space separating u from $\mathcal{V}$ is readily obtained from the value of u and $P_{\mathcal{V}}(u)$:

$$\mathcal{H} = \{v \in \overline{\mathbb{R}}_{\max}^3 \mid 1v_1 \oplus v_2 \oplus (-0.5)v_3 \leq (-1)v_1 \oplus v_2 \oplus (-0.5)v_3\} .$$

This half-space is represented by the zone in medium gray in Figure 2.

An *affine half-space* of $\overline{\mathbb{R}}_{\max}^X$ is by definition a set of the form

$$\mathcal{H} = \{v \in \overline{\mathbb{R}}_{\max}^X \mid a \cdot v \oplus c \leq b \cdot v \oplus d\} , \tag{8}$$

where $a, b \in \overline{\mathbb{R}}_{\max}^X$ and $c, d \in \overline{\mathbb{R}}_{\max}$. For any complete convex subset $\mathcal{C}$ of $\overline{\mathbb{R}}_{\max}^X$ and $u \in \overline{\mathbb{R}}_{\max}^X$, we define

$$\nu_{\mathcal{C}}(u) := \sup_{v \in \mathcal{C}}(u/v \wedge \mathbb{1}), \qquad Q_{\mathcal{C}}(u) := \sup_{v \in \mathcal{C}}(u/v \wedge \mathbb{1})v ,$$

where $\wedge$ denotes the pointwise minimum of vectors.

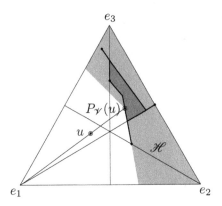

Fig. 2. Separating half-space

Corollary 1 ([5, Cor. 15]). *If $\mathscr{C}$ is a complete convex subset of $\overline{\mathbb{R}}_{\max}^{X}$, and if $u \in \overline{\mathbb{R}}_{\max}^{X} \setminus \mathscr{C}$, then the affine half-space*

$$\mathscr{H} = \{v \in \overline{\mathbb{R}}_{\max}^{X} \mid (Q_{\mathscr{C}}(u))^{-} \cdot v \oplus (\nu_{\mathscr{C}}(u))^{-} \leq u^{-} \cdot v \oplus \mathbb{1}\} \tag{9}$$

contains $\mathscr{C}$ and not u.

This corollary is obtained by projecting the vector $(u, \mathbb{1})$ on the complete sub-semimodule of $\overline{\mathbb{R}}_{\max}^{X} \times \overline{\mathbb{R}}_{\max}$ generated by the vectors $(v\lambda, \lambda)$, where $v \in \mathscr{C}$ and $\lambda \in \overline{\mathbb{R}}_{\max}$. The projection of this vector is precisely $(Q_{\mathscr{C}}(u), \nu_{\mathscr{C}}(u))$. The operator $u \mapsto (\nu_{\mathscr{C}}(u))^{-} Q_{\mathscr{C}}(u)$ defines a projection on the convex set $\mathscr{C}$ [12]. (We note that the scalar $\nu_{\mathscr{C}}(u)$ is invertible, except in the degenerate case where u cannot be bounded from below by a non-zero scalar multiple of an element of $\mathscr{C}$.)

We deduce as an immediate corollary of Theorem 2 and Corollary 1.

Corollary 2. *A complete subsemimodule (resp. complete convex subset) of $\overline{\mathbb{R}}_{\max}^{X}$ is the intersection of the half-spaces (resp. affine half-spaces) of $\overline{\mathbb{R}}_{\max}^{X}$ in which it is contained.*   □

We now consider subsemimodules and convex subsets arising from the max-plus semiring $\mathbb{R}_{\max}$, rather than from the completed max-plus semiring $\overline{\mathbb{R}}_{\max}$. Results of the best generality are perhaps still missing, so we shall restrict our attention to subsemimodules and convex subsets of $\mathbb{R}_{\max}^{n}$. By analogy with convex analysis, we call *cone* a subsemimodule of $\mathbb{R}_{\max}^{n}$.

We equip $\mathbb{R}_{\max}^{n}$ with the usual topology, which can be defined by the metric

$$d(u, v) := \max_{1 \leq i \leq n} |\exp(u_i) - \exp(v_i)|, \qquad \forall u, v \in (\mathbb{R} \cup \{-\infty\})^{n} \ .$$

A *half-space* of $\mathbb{R}_{\max}^{n}$ is a set of the form $\mathscr{H} = \{v \in \mathbb{R}_{\max}^{n} \mid a \cdot v \leq b \cdot v\}$, where $a, b \in \mathbb{R}_{\max}^{n}$. An *affine half-space* of $\mathbb{R}_{\max}^{n}$ is a set of the form $\mathscr{H} = \{v \in \mathbb{R}_{\max}^{n} \mid a \cdot v \oplus c \leq b \cdot v \oplus d\}$, where $a, b \in \mathbb{R}_{\max}^{n}$ and $c, d \in \mathbb{R}_{\max}$. Note that the restriction to $\mathbb{R}_{\max}^{n}$ of an (affine) half-space of $\overline{\mathbb{R}}_{\max}^{n}$ need not be an (affine) half-space of $\mathbb{R}_{\max}^{n}$, because the vectors a, b in (6) and (8) can have entries equal to $+\infty$, and the scalars c, d in (8) can be equal to $+\infty$. However, we have the following refinement of Theorem 2 in the case of closed cones of $\mathbb{R}^n_{\max}$, which is slightly more precise than the result stated in [12], and can be proved along the same lines.

Theorem 3. *Let $\mathscr{V}$ be a closed cone of $\mathbb{R}^n_{\max}$ and let $u \in \mathbb{R}^n_{\max} \setminus \mathscr{V}$. Then, there exist $a \in \mathbb{R}^n_{\max}$ and disjoint subsets I and J of $\{1, \ldots, n\}$ such that the half-space of $\mathbb{R}^n_{\max}$*

$$\mathscr{H} = \{v \in \mathbb{R}^n_{\max} \mid \oplus_{i \in I} a_i v_i \leq \oplus_{j \in J} a_j v_j\} \tag{10}$$

contains $\mathscr{V}$ and not u.

Further information on half-spaces can be found in [31].

Example 4. The restriction to $\mathbb{R}^3_{\max}$ of the separating half-space constructed in Example 3 can be rewritten as:

$$\mathscr{H} = \{v \in \mathbb{R}^3_{\max} \mid 1v_1 \leq v_2 \oplus (-0.5)v_3\} \ ,$$

which is clearly of the form (10). To illustrate the technical difficulty concerning supports, which is solved in [12] and in Theorem 3 above, let us separate the point $u = [-\infty, 1, 0]^T$ from the semimodule $\mathscr{V}$ of Example 2. We have $P_{\mathscr{V}}(u) = [-\infty, 0, 0]^T$, and the half-space of $\overline{\mathbb{R}}^3_{\max}$ defined in Theorem 2 is

$$\{v \in \overline{\mathbb{R}}^3_{\max} \mid (+\infty)v_1 \oplus v_2 \oplus v_3 \leq (+\infty)v_1 \oplus (-1)v_2 \oplus v_3\} \ .$$

Note that due to the presence of the $+\infty$ coefficient, the restriction of this half-space to $\mathbb{R}^3_{\max}$ is not closed. The proof of [12] and of Theorem 3 introduces a finite perturbation of u, for instance, $w = [\epsilon, 1, 0]^T$, where ϵ is a finite number sufficiently close to $-\infty$ (here, any $\epsilon < 0$ will do), and shows that the restriction to $\mathbb{R}^n_{\max}$ of the half-space of $\overline{\mathbb{R}}^n_{\max}$ constructed in the universal separation theorem (Theorem 2), which is a half-space of $\mathbb{R}^n_{\max}$, separates u from $\mathscr{V}$. For instance, when $\epsilon = -1$, we obtain $P_{\mathscr{V}}(w) = [-1, 0, 0]^T$, which gives the half-space of $\mathbb{R}^3_{\max}$

$$\mathscr{H} = \{v \in \mathbb{R}^3_{\max} \mid 1v_1 \oplus v_3 \geq v_2\}$$

containing $\mathscr{V}$ and not u.

Corollary 3. *Let $\mathscr{C} \subset \mathbb{R}^n_{\max}$ be a closed convex set and let $u \in \mathbb{R}^n_{\max} \setminus \mathscr{C}$. Then, there exist $a \in \mathbb{R}^n_{\max}$, disjoint subsets I and J of $\{1, \ldots, n\}$ and $c, d \in \mathbb{R}_{\max}$, with $cd = \mathbb{0}$, such that the affine half-space of $\mathbb{R}^n_{\max}$*

$$\mathscr{H} = \{v \in \mathbb{R}^n_{\max} \mid \oplus_{i \in I} a_i v_i \oplus c \leq \oplus_{j \in J} a_j v_j \oplus d\}$$

contains $\mathscr{C}$ and not u.

This is proved by applying the previous theorem to the point $(u, \mathbb{1}) \in \mathbb{R}^{n+1}_{\max}$ and to the following closed cone:

$$\mathscr{V} := \mathrm{clo}(\{(v\lambda, \lambda) \mid v \in \mathscr{C}, \lambda \in \mathbb{R}_{\max}\}) \subset \mathbb{R}^{n+1}_{\max} \ .$$

We deduce as an immediate corollary of Theorem 3 and Corollary 3.

Corollary 4. *A closed cone of* $\mathbb{R}^n_{\max}$ *is the intersection of the half-spaces of* $\mathbb{R}^n_{\max}$ *in which it is contained. A closed convex subset of* $\mathbb{R}^n_{\max}$ *is the intersection of the affine half-spaces of* $\mathbb{R}^n_{\max}$ *in which it is contained.*

5   Extreme Points of Max-Plus Convex Sets

Definition 2. *Let* $\mathscr{C}$ *be a convex subset of* $\mathbb{R}^n_{\max}$. *An element* $v \in \mathscr{C}$ *is an extreme point of* $\mathscr{C}$, *if for all* $u, w \in \mathscr{V}$ *and* $\lambda, \mu \in \mathbb{R}_{\max}$ *such that* $\lambda \oplus \mu = 1$, *the following property is satisfied*

$$v = \lambda u \oplus \mu w \implies v = u \text{ or } v = w .$$

The set of extreme points of $\mathscr{C}$ *will be denoted by* $\mathrm{ext}(\mathscr{C})$.

Definition 3. *Let* $\mathscr{V} \subset \mathbb{R}^n_{\max}$ *be a cone. An element* $v \in \mathscr{V}$ *is an* extreme generator *of* $\mathscr{V}$ *if the following property is satisfied*

$$v = u \oplus w, \ u, w \in \mathscr{V} \implies v = u \text{ or } v = w .$$

We define an extreme ray *of* $\mathscr{V}$ *to be a set of the form* $\mathbb{R}_{\max} v = \{\lambda v \mid \lambda \in \mathbb{R}_{\max}\}$ *where* v *is an extreme generator of* $\mathscr{V}$. *The set of extreme generators of* $\mathscr{V}$ *will be denoted by* $\mathrm{ext\text{-}g}\,(\mathscr{V})$.

Note that extreme generators correspond to *join irreducible* elements in the lattice theory literature.

We denote by $\mathrm{cone}\,(\mathscr{X})$ the smallest cone containing a subset $\mathscr{X}$ of $\mathbb{R}^n_{\max}$, and by $\mathrm{co}(\mathscr{X})$ the smallest convex set containing it. So $\mathrm{cone}\,(\mathscr{X})$ coincides with span $\mathscr{X}$, if the operator "span" is interpreted over the semiring $\mathbb{R}_{\max}$.

Theorem 4. *Let* $\mathscr{V} \subset \mathbb{R}^n_{\max}$ *be a non-empty closed cone. Then* $\mathscr{V}$ *is the cone generated by the set of its extreme generators, that is,*

$$\mathscr{V} = \mathrm{cone}\,(\mathrm{ext\text{-}g}\,(\mathscr{V})) .$$

The proof of Theorem 4, and of Corollary 5 and Theorem 5 below, can be found in [13]. After the submission of the present paper, a preprint of Buktovič, Schneider, and Sergeev has appeared [32], in which Theorem 4 is established independently. Their approach also yields informations on non-closed cones.

Corollary 5 (Max-Plus Minkowski's Theorem). *Let* $\mathscr{C}$ *be a non-empty compact convex subset of* $\mathbb{R}^n_{\max}$. *Then* $\mathscr{C}$ *is the convex hull of the set of its extreme points, that is,*

$$\mathscr{C} = \mathrm{co}(\mathrm{ext}(\mathscr{C})) .$$

This is more precise than Helbig's max-plus analogue of Krein-Milman's theorem [33], which only shows that a non-empty compact convex subset of $\mathbb{R}^n_{\max}$ is the *closure* of the convex hull of its set of extreme points. Unlike Helbig's proof, our proof of Theorem 4 and Corollary 5 does not use the separation theorem.

If v is a point in a convex set $\mathscr{C}$, we define the *recession cone* of $\mathscr{C}$ at point v to be the set:

$$\mathrm{rec}(\mathscr{C}) = \{u \in \mathbb{R}_{\max}^n \mid v \oplus \lambda u \in \mathscr{C} \text{ for all } \lambda \in \mathbb{R}_{\max}\} \ .$$

If $\mathscr{C}$ is a closed convex subset of $\mathbb{R}_{\max}^n$, it can be checked that the recession cone is independent of the choice of $v \in \mathscr{C}$, and that it is closed.

Theorem 5. *Let $\mathscr{C} \subset \mathbb{R}_{\max}^n$ be a closed convex set. Then,*

$$\mathscr{C} = \mathrm{co}(\mathrm{ext}(\mathscr{C})) \oplus \mathrm{rec}(\mathscr{C}) \ .$$

Corollary 4 suggests the following definition.

Definition 4. *A max-plus polyhedron is an intersection of finitely many affine half-spaces of $\mathbb{R}_{\max}^n$.*

Theorem 6 (Max-Plus Minkowski-Weyl Theorem). *The max-plus polyhedra are precisely the sets of the form*

$$\mathrm{co}(\mathscr{X}) \oplus \mathrm{cone}\,(\mathscr{Y})$$

where $\mathscr{X}, \mathscr{Y}$ are finite subsets of $\mathbb{R}_{\max}^n$.

Note that our notion of max-plus polyhedra is more general than the notion of tropical polyhedra which is considered in [11]: tropical polyhedra can be identified with sets of the form $\mathrm{cone}\,(\mathscr{Y})$ where $\mathscr{Y}$ is a finite set of vectors with only *finite* entries.

Finally, we shall consider the max-plus analogues of simplicial convex cones, which are related to the important notion of regular matrix. We need to work again in the completed max-plus semiring, $\overline{\mathbb{R}}_{\max}$, rather than in $\mathbb{R}_{\max}$. We say that a matrix $A \in \overline{\mathbb{R}}_{\max}^{n \times p}$ is *regular* if it has a generalized inverse, meaning that there exists a matrix $X \in \overline{\mathbb{R}}_{\max}^{p \times n}$ such that $A = AXA$. Regularity is equivalent to the existence of a linear projector onto the cone generated by the columns (or the rows) of A, see [34,35].

A finitely generated subsemimodule $\mathscr{V}$ of $\overline{\mathbb{R}}_{\max}^n$ is a complete lattice, in which the supremum coincides with the supremum in $\overline{\mathbb{R}}_{\max}^n$, and the infimum of any subset of $\mathscr{V}$ is the greatest lower bound of this subset that belongs to $\mathscr{V}$. The following result extends a theorem proved by Zaretski [36] (see [37, Th. 2.1.29] for a proof in English) in the case of the Boolean semiring.

Theorem 7 ([14]). *A matrix $A \in \overline{\mathbb{R}}_{\max}^{n \times p}$ is regular if and only if the subsemimodule of $\overline{\mathbb{R}}_{\max}^n$ generated by its columns is a completely distributive lattice.*

Of course, a dual statement holds for the rows of A. In fact, we know that the semimodule generated by the rows of A is anti-isomorphic to the semimodule generated by its columns [5].

As an illustration of Theorem 5, consider the closed convex set $\mathscr{C} \subset \mathbb{R}_{\max}^2$ depicted in Figure 5. We have $\mathrm{ext}(\mathscr{C}) = \{a, b, c, d, e\}$, where $a = [5,2]^T$, $b =$

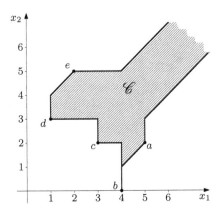

Fig. 3. An unbounded max-plus convex set

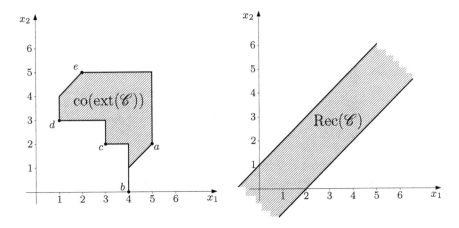

Fig. 4. The sets $\mathrm{co}(\mathrm{ext}(\mathscr{C}))$ and $\mathrm{rec}(\mathscr{C})$ of Theorem 5 for the unbounded convex set depicted in Figure 5

$[4,0]^T$, $c = [3,2]^T$, $d = [1,3]^T$, $e = [2,5]^T$, and $\mathrm{rec}(\mathscr{C}) = \mathrm{cone}\left\{[0,1]^T, [2,0]^T\right\}$. Then,

$$\mathscr{C} = \mathrm{co}\left\{a, b, c, d, e\right\} \oplus \mathrm{cone}\left\{[0,1]^T, [2,0]^T\right\}$$

by Theorem 5. The sets $\mathrm{co}(\mathrm{ext}(\mathscr{C}))$ and $\mathrm{rec}(\mathscr{C})$ are depicted in Figure 4. The cone $\mathrm{rec}(\mathscr{C})$ is a distributive lattice, since the infimum and supremum laws coincide with those of $\overline{\mathbb{R}}^2_{\max}$. Note that any $n \times 2$ or $2 \times n$ matrix is regular, in particular, finitely generated cones which are not distributive lattices cannot be found in dimension smaller than 3, see [34].

Acknowledgment. We thank the referees for their careful reading and for their suggestions.

References

1. Korbut, A.A.: Extremal spaces. Dokl. Akad. Nauk SSSR **164** (1965) 1229–1231
2. Zimmermann, K.: A general separation theorem in extremal algebras. Ekonom.-Mat. Obzor **13**(2) (1977) 179–201
3. Maslov, V.P., Samborskiĭ, S.N.: Idempotent analysis. Volume 13 of Advances in Soviet Mathematics. Amer. Math. Soc., Providence (1992)
4. Litvinov, G., Maslov, V., Shpiz, G.: Idempotent functional analysis: an algebraic approach. Math. Notes **69**(5) (2001) 696–729
5. Cohen, G., Gaubert, S., Quadrat, J.P.: Duality and separation theorems in idempotent semimodules. Linear Algebra and Appl. **379** (2004) 395–422 arXiv:math.FA/0212294.
6. Rubinov, A.M.: Abstract convexity and global optimization. Kluwer (2000)
7. Cohen, G., Gaubert, S., Quadrat, J.: Max-plus algebra and system theory: where we are and where to go now. Annual Reviews in Control **23** (1999) 207–219
8. Kolokoltsov, V.N., Maslov, V.P.: Idempotent analysis and applications. Kluwer Acad. Publisher (1997)
9. Fathi, A.: Weak KAM theorem in Lagrangian dynamics. Lecture notes, preliminary version (Cambridge University Press, to appear.) (2005)
10. Akian, M., Gaubert, S., Walsh, C.: The max-plus Martin boundary. arXiv:math.MG/0412408 (2004)
11. Develin, M., Sturmfels, B.: Tropical convexity. Doc. Math. **9** (2004) 1–27 (Erratum pp. 205–206).
12. Cohen, G., Gaubert, S., Quadrat, J., Singer, I.: Max-plus convex sets and functions. In Litvinov, G.L., Maslov, V.P., eds.: Idempotent Mathematics and Mathematical Physics. Contemporary Mathematics. American Mathematical Society (2005) 105–129. Also ESI Preprint 1341, arXiv:math.FA/0308166.
13. Gaubert, S., Katz, R.: The Minkowski theorem for max-plus convex sets. arXiv:math.GM/0605078 (2006)
14. Cohen, G., Gaubert, S., Quadrat, J.P.: Regular matrices in max-plus algebra. Preprint (2006)
15. Katz, R.D.: Max-plus (A,B)-invariant spaces and control of timed discrete event systems. (2005) E-print arXiv:math.OC/0503448, to appear in IEEE-TAC.
16. Akian, M., Gaubert, S., Walsh, C.: Discrete max-plus spectral theory. In Litvinov, G.L., Maslov, V.P., eds.: Idempotent Mathematics and Mathematical Physics. Contemporary Mathematics. American Mathematical Society (2005) 19–51. Also ESI Preprint 1485, arXiv:math.SP/0405225.
17. Akian, M., Bapat, R., Gaubert, S.: Max-plus algebras. In Hogben, L., Brualdi, R., Greenbaum, A., Mathias, R., eds.: Handbook of Linear Algebra. Chapman & Hall (2006)
18. Akian, M., Gaubert, S., Kolokoltsov, V.: Solutions of max-plus linear equations and large deviations. In: Proceedings of the joint 44th IEEE Conference on Decision and Control and European Control Conference ECC 2005 (CDC-ECC'05), Seville, Espagne (2005) arXiv:math.PR/0509279.
19. Walsh, C.: The horofunction boundary of finite-dimensional normed spaces. To appear in the Math. Proc. of the Cambridge. Phil. Soc., arXiv:math.GT/0510105 (2005)
20. Rockafellar, R.T.: Convex analysis. Princeton University Press Princeton, N.J. (1970)

21. Akian, M., Gaubert, S., Kolokoltsov, V.N.: Set coverings and invertibility of functional Galois connections. In Litvinov, G.L., Maslov, V.P., eds.: Idempotent Mathematics and Mathematical Physics. Contemporary Mathematics. American Mathematical Society (2005) 19–51 Also ESI Preprint 1447, arXiv:math.FA/0403441.

22. Gaubert, S., Gunawardena, J.: The Perron-Frobenius theorem for homogeneous, monotone functions. Trans. of AMS **356**(12) (2004) 4931–4950

23. Rubinov, A.M., Singer, I.: Topical and sub-topical functions, downward sets and abstract convexity. Optimization **50**(5-6) (2001) 307–351

24. Gunawardena, J.: From max-plus algebra to nonexpansive maps: a nonlinear theory for discrete event systems. Theoretical Computer Science **293** (2003) 141–167

25. Dhingra, V., Gaubert, S., Gunawardena, J.: Policy iteration algorithm for large scale deterministic games with mean payoff. Preprint (2006)

26. Cochet-Terrasson, J., Gaubert, S., Gunawardena, J.: A constructive fixed point theorem for min-max functions. Dynamics and Stability of Systems **14**(4) (1999) 407–433

27. Gaubert, S., Gunawardena, J.: The duality theorem for min-max functions. C. R. Acad. Sci. Paris. **326, Série I** (1998) 43–48

28. Costan, A., Gaubert, S., Goubault, E., Martel, M., Putot, S.: A policy iteration algorithm for computing fixed points in static analysis of programs. In: Proceedings of the 17th International Conference on Computer Aided Verification (CAV'05). Number 3576 in LNCS, Edinburgh, Springer (2005) 462–475

29. Akian, M., Gaubert, S., Lakhoua, A.: The max-plus finite element method for solving deterministic optimal control problems: basic properties and convergence analysis. arXiv:math.OC/0603619 (2006)

30. Samborskiĭ, S.N., Shpiz, G.B.: Convex sets in the semimodule of bounded functions. In: Idempotent analysis. Amer. Math. Soc., Providence, RI (1992) 135–137

31. Joswig, M.: Tropical halfspaces. In: Combinatorial and computational geometry. Volume 52 of Math. Sci. Res. Inst. Publ. Cambridge Univ. Press, Cambridge (2005) 409–431

32. Butkovic, P., Schneider, H., Sergeev, S.: Generators, extremals and bases of max cones. arXiv:math.RA/0604454 (2006)

33. Helbig, S.: On Caratheodory's and Kreĭn-Milman's theorems in fully ordered groups. Comment. Math. Univ. Carolin. **29**(1) (1988) 157–167

34. Cohen, G., Gaubert, S., Quadrat, J.: Linear projectors in the max-plus algebra. In: Proceedings of the IEEE Mediterranean Conference, Cyprus, IEEE (1997)

35. Cohen, G., Gaubert, S., Quadrat, J.P.: Projection and aggregation in maxplus algebra. In Menini, L., Zaccarian, L., Abdallah, C.T., eds.: Current Trends in Nonlinear Systems and Control, in Honor of Petar Kokotovic and Turi Nicosia. Systems & Control: Foundations & Applications. Birkhauser (2006)

36. Zaretski, K.: Regular elements in the semigroup of binary relations. Uspeki Mat. Nauk **17**(3) (1962) 105–108

37. Kim, K.: Boolean Matrix Theory and Applications. Marcel Dekker, New York (1982)

Lazy Semiring Neighbours and Some Applications

Peter Höfner* and Bernhard Möller

Institut für Informatik, Universität Augsburg
D-86135 Augsburg, Germany
{hoefner, moeller}@informatik.uni-augsburg.de

Abstract. We extend an earlier algebraic approach to Neighbourhood Logic (NL) from domain semirings to lazy semirings yielding lazy semiring neighbours. Furthermore we show three important applications for these. The first one extends NL to intervals with infinite length. The second one applies lazy semiring neighbours in an algebraic semantics of the branching time temporal logic CTL*. The third one sets up a connection between hybrid systems and lazy semiring neighbours.

1 Introduction

Chop-based interval temporal logics like ITL [5] and IL [3] are useful for specification and verification of safety properties of real-time systems. However, as it is shown in [15], these logics cannot express all desired properties, like (un-bounded) liveness properties. Hence Zhou and Hansen proposed *Neighbourhood Logic* (NL) [14], a first-order interval logic with extra atomic formulas. In [7] NL has been embedded and extended into the algebraic framework of semirings. But neither NL nor the algebraic version handle intervals with infinite length. Therefore we transfer the neighbour concept to lazy semirings [10]. This provides a combination of NL and interval logic with infinite intervals on a uniform algebraic basis. Surprisingly, lazy semiring neighbours are not only useful for the extension of NL; they occur in different situations and structures.

The paper is structured into two main parts. The first one presents the algebraic theory. Therefore we recapitulate the basic notions, like lazy semirings, in Section 2. In Section 3 we define domain and codomain and give some important properties. In the next section we introduce and discuss lazy semiring neighbours and boundaries. That section contains the main contribution from a theoretical point of view. The second part presents three different applications for the theory. It starts by extending Neighbourhood Logic to intervals with infinite length in Section 5. Afterwards, in Section 6, we show that in the algebraic characterisation of the branching time temporal logic CTL* of [11], the existential and universal path quantifiers E and A correspond to lazy semiring neighbours. The last application is presented in Section 7 and shows how to transfer lazy semiring neighbours to the algebraic model of hybrid systems presented in [8]; some of them guarantee liveness, others non-reachability, i.e., a form of safety.

* This research was supported by DFG (German Research Foundation).

R.A. Schmidt (Ed.): RelMiCS /AKA 2006, LNCS 4136, pp. 207–221, 2006.

2   Algebraic Foundations

A *lazy semiring* (L-*semiring* or *left semiring*) is a quintuple $(S, +, \cdot, 0, 1)$ where $(S, +, 0)$ is a commutative monoid and $(S, \cdot, 1)$ is a monoid such that $\cdot$ is left-distributive over $+$ and *left-strict*, i.e., $0 \cdot a = 0$. A lazy semiring structure is also at the core of process algebra frameworks. The lazy semiring is *idempotent* if $+$ is idempotent and $\cdot$ is right-isotone, i.e., $b \leq c \Rightarrow a \cdot b \leq a \cdot c$, where the *natural order* $\leq$ on S is given by $a \leq b \Leftrightarrow_{df} a + b = b$. Left-isotony of $\cdot$ follows from its left-distributivity. Moreover, 0 is the $\leq$-least element and $a + b$ is the join of a and b. Hence every idempotent L-semiring is a join semilattice. A *semiring* (for clarity sometimes also called *full semiring*) is a lazy semiring in which $\cdot$ is also right-distributive and right-strict. An L-semiring is *Boolean* if it is idempotent and its underlying semilattice is a Boolean algebra. Every Boolean L-semiring has a greatest element $\top$.

A *lazy quantale* is an idempotent L-semiring that is also a complete lattice under the natural order with $\cdot$ being universally disjunctive in its left argument. A *quantale* is a lazy quantale in which $\cdot$ is universally disjunctive also in its right argument. Following [1], one might also call a quantale a *standard Kleene algebra*. A lazy quantale is *Boolean* if it is right-distributive and a Boolean L-semiring.

An important lazy semiring (that is even a Boolean quantale) is REL, the algebra of binary relations over a set under relational composition.

To model assertions in semirings we use the idea of tests as introduced into Kleene algebras by Kozen [9]. In REL a set of elements can be modelled as a subset of the identity relation; meet and join of such partial identities coincide with their composition and union. Generalising this, one defines a *test* in a (left) quantale to be an element $p \leq 1$ that has a complement q relative to 1, i.e., $p + q = 1$ and $p \cdot q = 0 = q \cdot p$. The set of all tests of a quantale S is denoted by $\mathsf{test}(S)$. It is not hard to show that $\mathsf{test}(S)$ is closed under $+$ and $\cdot$ and has 0 and 1 as its least and greatest elements. Moreover, the complement $\neg p$ of a test p is uniquely determined by the definition. Hence $\mathsf{test}(S)$ forms a Boolean algebra. If S itself is Boolean then $\mathsf{test}(S)$ coincides with the set of all elements below 1. We will consistently write $a, b, c \ldots$ for arbitrary semiring elements and $p, q, r, \ldots$ for tests.

With the above definition of tests we deviate slightly from [9], in that we do not allow an arbitrary Boolean algebra of sub identities as $\mathsf{test}(S)$ but only the maximal complemented one. The reason is that the axiomatisation of domain to be presented below forces this maximality anyway (see [2]).

In the remainder we give another important example of an L-semiring (especially with regard to temporal logics like CTL* and hybrid systems). It is based on trajectories (cf. e.g. [12]) that reflect the values of the variables over time and was introduced in [8].

Let V be a set of *values* and D a set of *durations* (e.g. $\mathbb{N}$, $\mathbb{Q}$, $\mathbb{R}$, ...). We assume a cancellative addition $+$ on D and an element $0 \in D$ such that $(D, +, 0)$ is a commutative monoid and the relation $x \leq y \Leftrightarrow_{df} \exists z \,.\, x + z = y$ is a linear order on D. Then 0 is the least element and $+$ is isotone w.r.t. $\leq$. Moreover, 0 is indivisible, i.e., $x + y = 0 \Leftrightarrow x = y = 0$. D may include the special value ∞.

It is required to be an annihilator w.r.t. $+$ and hence the greatest element of D (and cancellativity of $+$ is restricted to elements in $D - \{\infty\}$). For $d \in D$ we define the interval $\mathsf{intv}\, d$ of admissible times as

$$\mathsf{intv}\, d =_{df} \begin{cases} [0, d] & \text{if } d \neq \infty \\ [0, d[& \text{otherwise .} \end{cases}$$

A *trajectory* t is a pair (d, g), where $d \in D$ and $g : \mathsf{intv}\, d \to V$. Then d is the *duration* of the trajectory. This view models *oblivious* systems in which the evolution of a trajectory is independent of the history before the starting time.

The set of all trajectories is denoted by TRA. Composition of trajectories (d_1, g_1) and (d_2, g_2) is defined by

$$(d_1, g_1) \cdot (d_2, g_2) =_{df} \begin{cases} (d_1 + d_2, g) & \text{if } d_1 \neq \infty \wedge g_1(d_1) = g_2(0) \\ (d_1, g_1) & \text{if } d_1 = \infty \\ \text{undefined} & \text{otherwise} \end{cases}$$

with $g(x) = g_1(x)$ for all $x \in [0, d_1]$ and $g(x + d_1) = g_2(x)$ for all $x \in \mathsf{intv}\, d_2$.

For a value $v \in V$, let $\underline{v} =_{df} (0, g)$ with $g(0) = v$ be the corresponding zero-length trajectory. Moreover, set $I =_{df} \{\underline{v} \mid v \in V\}$.

A *process* is a set of trajectories. The *infinite and finite parts* of a process A are the processes $\inf A =_{df} \{(d, g) \in A \mid d = \infty\}$ and $\mathsf{fin}\, A =_{df} A - \inf A$. Composition is lifted to processes as follows:

$$A \cdot B =_{df} \inf A \cup \{a \cdot b \mid a \in \mathsf{fin}\, A, b \in B\} .$$

Then we obtain the lazy Boolean quantale

$$\mathrm{PRO} =_{df} (\mathcal{P}(\mathrm{TRA}), \cup, \cdot, \emptyset, I) ,$$

which can be extended to a test quantale by setting $\mathsf{test}(\mathrm{PRO}) =_{df} \mathcal{P}(I)$.

For a discrete infinite set D, e.g. $D = \mathbb{N}$, trajectories are isomorphic to nonempty finite or infinite words over the value set V. If V consists of states of computations, then the elements of PRO can be viewed as sets of computation streams; therefore we also write $\mathrm{STR}(V)$ instead of PRO in this case.

Note that $A \in \mathrm{PRO}$ consists of infinite trajectories only, i.e., $A = \inf A$, iff $A \cdot B = A$ for all $B \in \mathrm{PRO}$. We call such a process *infinite*, too. Contrarily, A consists of finite trajectories only, i.e., $A = \mathsf{fin}\, A$, iff $A \cdot \emptyset = \emptyset$. We call such a process *finite*, too.

We now generalise these notions from PRO to an arbitrary L-semiring S. An element $a \in S$ is called *infinite* if it is a left zero, i.e., $a \cdot b = a$ for all $b \in S$, which is equivalent to $a \cdot 0 = a$. By this property, $a \cdot 0$ may be considered as the *infinite part* of a, i.e., the part consisting just of infinite computations (if any). We assume that there exists a largest infinite element N, i.e.,

$$a \leq \mathsf{N} \Leftrightarrow_{df} a \cdot 0 = a .$$

Dually, we call an element a *finite* if its infinite part is trivial, i.e., if $a \cdot 0 = 0$. We also assume that there is a largest finite element F, i.e.,

$$a \leq \mathsf{F} \Leftrightarrow_{df} a \cdot 0 = 0 .$$

In Boolean quantales N and F always exist[1] and satisfy $N = \top \cdot 0$ and $F = \overline{N}$, where $^{-}$ denotes complementation. Moreover, every element can be split into its finite and infinite parts: $a = \text{fin } a + \text{inf } a$, where $\text{fin } a =_{df} a \sqcap F$ and $\text{inf } a =_{df} a \sqcap N$. In particular, $\top = N + F$.

3   Domain and Codomain in L-Semirings

Domain and codomain abstractly characterise, in the form of tests, the sets of initial and final states of a set of computations. In contrast to the domain and codomain operators of full semirings and Kleene algebras [2] the operators for L-semirings are not symmetric. Therefore we recapitulate their definitions [10] and establish some properties which we need afterwards.

Definition 3.1. A *lazy semiring with domain* ($^{\ulcorner}$*-L-semiring*) is a structure $(S, ^{\ulcorner})$, where S is an idempotent lazy test semiring and the *domain operation* $^{\ulcorner}: S \rightarrow \text{test}(S)$ satisfies for all $a, b \in S$ and $p \in \text{test}(S)$

$$a \leq {^{\ulcorner}a} \cdot a \quad \text{(d1)}, \qquad {^{\ulcorner}(p \cdot a)} \leq p \quad \text{(d2)}, \qquad {^{\ulcorner}(a \cdot {^{\ulcorner}b})} \leq {^{\ulcorner}(a \cdot b)} \quad \text{(d3)}.$$

The axioms are the same as in [2]. Since the domain describes all possible starting states of an element, it is easy to see that "laziness" of the underlying semiring doesn't matter. Most properties of [2,10] can still be proved in L-semirings with domain. We only give some properties which we need in the following sections. First, the conjunction of (d1) and (d2) is equivalent to each of

$${^{\ulcorner}a} \leq p \Leftrightarrow a \leq p \cdot a \quad \text{(llp)}, \qquad {^{\ulcorner}a} \leq p \Leftrightarrow \neg p \cdot a \leq 0 \quad \text{(gla)}.$$

(llp) says that $^{\ulcorner}a$ is the least left preserver of a; (gla) that $\neg^{\ulcorner}a$ is the greatest left annihilator of a. By Boolean algebra, (gla) is equivalent to

$$p \cdot {^{\ulcorner}a} \leq 0 \Leftrightarrow p \cdot a \leq 0 . \tag{1}$$

Lemma 3.2. *[10] Let S be a $^{\ulcorner}$-L-semiring.*

(a) $^{\ulcorner}$ *is isotone.*
(b) $^{\ulcorner}$ *is universally disjunctive;*
   in particular $^{\ulcorner}0 = 0$ and $^{\ulcorner}(a + b) = {^{\ulcorner}a} + {^{\ulcorner}b}$.
(c) $^{\ulcorner}a \leq 0 \Leftrightarrow a \leq 0$. $\qquad\qquad\qquad\qquad\qquad\qquad$ (Full Strictness)
(d) $^{\ulcorner}p = p$. $\qquad\qquad\qquad\qquad\qquad\qquad\qquad\qquad\qquad$ (Stability)
(e) $^{\ulcorner}(p \cdot a) = p \cdot {^{\ulcorner}a}$. $\qquad\qquad\qquad\qquad\qquad\qquad$ (Import/Export)
(f) $^{\ulcorner}(a \cdot b) \leq {^{\ulcorner}a}$.

We now turn to the dual case of the domain operation. In the case where we have (as in full semirings) right-distributivity and right-strictness, a codomain operation $^{\urcorner}$ is easily defined as a domain operation in the opposite L-semiring (i.e., the one that swaps the order of composition). But due to the absence of right-distributivity and right-strictness we need an additional axiom.

---

[1] In general N and F need not exist. In [10] lazy semirings where these elements exist are called *separated*.

Definition 3.3. A *lazy semiring with codomain* ($\ulcorner$-*L-semiring*) is a structure $(S, \ulcorner)$, where S is an idempotent lazy test semiring and the *codomain operation* $\ulcorner : S \rightarrow \mathsf{test}(S)$ satisfies for all $a, b \in S$ and $p \in \mathsf{test}(S)$

$$a \leq a \cdot a^{\urcorner} \quad (\text{cd1}), \qquad\qquad (a \cdot p)^{\urcorner} \leq p \quad (\text{cd2}),$$
$$(a^{\urcorner} \cdot b)^{\urcorner} \leq (a \cdot b)^{\urcorner} \quad (\text{cd3}), \qquad\qquad (a + b)^{\urcorner} \geq a^{\urcorner} + b^{\urcorner} \quad (\text{cd4}).$$

(cd4) guarantees isotony of the codomain operator. As for domain, the conjunction of (cd1) and (cd2) is equivalent to

$$a^{\urcorner} \leq p \Leftrightarrow a \leq a \cdot p , \tag{lrp}$$

i.e., $a^{\urcorner}$ is the least right preserver of a. However, due to lack of right-strictness $\neg a^{\urcorner}$ need not be the greatest right annihilator; we only have the weaker equivalence

$$a^{\urcorner} \leq p \Leftrightarrow a \cdot \neg p \leq a \cdot 0 . \tag{wgra}$$

Lemma 3.4. *Let S be a $\ulcorner$-L-semiring.*
(a) $\ulcorner$ *is isotone.*
(b) $\ulcorner$ *is universally disjunctive;*
    in particular $0^{\urcorner} = 0$ and $(a + b)^{\urcorner} = a^{\urcorner} + b^{\urcorner}$.
(c) $a^{\urcorner} \leq 0 \Leftrightarrow a \leq \mathsf{N}$.
(d) $p^{\urcorner} = p$.   (Stability)
(e) $(a \cdot p)^{\urcorner} = a^{\urcorner} \cdot p$.   (Import/Export)
(f) $(a \cdot b)^{\urcorner} \leq b^{\urcorner}$.

Lemma 3.2(c) and Lemma 3.4(c) show the asymmetry of domain and codomain.
    As in [10], a *modal lazy semiring* (ML-semiring) is an L-semiring with domain and codomain. The following lemma has some important consequences for the next sections, and illustrates again the asymmetry of L-semirings.

Lemma 3.5. *In an ML-semiring with a greatest element $\top$, we have*
(a) $\neg p \cdot a \leq 0 \Leftrightarrow \ulcorner a \leq p \Leftrightarrow a \leq p \cdot a \Leftrightarrow a \leq p \cdot \top$.
(b) $a \cdot \neg p \leq a \cdot 0 \Leftrightarrow a^{\urcorner} \leq p \Leftrightarrow a \leq a \cdot p \Leftrightarrow a \leq \top \cdot p$.
(c) $a \leq \mathsf{F} \Leftrightarrow (a \leq a \cdot p \Leftrightarrow a \cdot \neg p \leq 0) \Leftrightarrow (a \leq \top \cdot p \Leftrightarrow a \cdot \neg p \leq 0)$.
    Therefore, in general, $a \leq a \cdot p \not\Leftrightarrow a \cdot \neg p \leq 0$ and $a \leq \top \cdot p \not\Leftrightarrow a \cdot \neg p \leq 0$.

Proof.
(a) The first equivalence is (gla), the second (llp). $a \leq p \cdot a \Rightarrow a \leq p \cdot \top$ holds by isotony of $\cdot$ and $a \leq p \cdot \top \Rightarrow \ulcorner a \leq p$ by isotony of domain and $\ulcorner(p \cdot \top) \overset{3.2(e)}{=} p \cdot \ulcorner \top = p$, since $\ulcorner \top \geq \ulcorner 1 = 1$ by Lemma 3.2(d).
(b) Symmetrically to (a).
(c) $a \leq \mathsf{F} \Rightarrow (a \leq a \cdot p \Leftrightarrow a \cdot \neg p \leq 0)$ holds by (b) and $a \cdot 0 \leq 0 \Leftrightarrow a \leq \mathsf{F}$. The converse implication is shown by setting $p = 1$, Boolean algebra and definition of F: $a \leq a \Rightarrow a \cdot \neg 1 \leq 0 \Leftrightarrow a \cdot 0 \leq 0 \Leftrightarrow a \leq \mathsf{F}$.
    The second equivalence follows from $a \leq a \cdot p \Leftrightarrow a \leq \top \cdot p$ (see (b)).   □

(c) says that we do not have a law for codomain that is symmetric to (a).
    Further properties of (co)domain and ML-semirings can be found in [2,10].

4 Neighbours — Definitions and Basic Properties

In [7] semiring neighbours and semiring boundaries are motivated by Neighbour-hood Logic [14,15]. The definitions there require full semirings as the underlying algebraic structure. In this section we use the same axiomatisation as in [7] to define neighbours and boundaries in L-semirings. Since the domain and codomain operators are not symmetric we also discuss some properties and consequences of the lack of right-distributivity and right-strictness. Note that in [7] the semiring neighbours and boundaries work on predomain and precodomain, i.e., assumed only (d1)–(d2) and (cd1)–(cd2), resp. Here we assume (d3)/(cd3) as well.

In the remainder some proofs are done only for one of a series of similar cases.

Definition 4.1. Let S be an ML-semiring and $a, b \in S$. Then
(a) a is a *left neighbour* of b (or $a \leq \otimes_l b$ for short) iff $\ulcorner a \urcorner \leq \ulcorner b \urcorner$,
(b) a is a *right neighbour* of b (or $a \leq \otimes_r b$ for short) iff $\ulcorner a \urcorner \leq \ulcorner b \urcorner$,
(c) a is a *left boundary* of b (or $a \leq \oslash_l b$ for short) iff $\ulcorner a \urcorner \leq \ulcorner b \urcorner$,
(d) a is a *right boundary* of b (or $a \leq \oslash_r b$ for short) iff $\ulcorner a \urcorner \leq \ulcorner b \urcorner$.

We will see below that the notation using $\leq$ is justified. By *lazy semiring neighbours* we mean both, left/right neighbours and boundaries. Most of the properties given in [7] use Lemma 3.5(a) in their proofs and a symmetric version of it for codomain which holds in full semirings. Unfortunately, by Lemma 3.5(b) and 3.5(c), we do not have this symmetry. Hence we have to check all properties in the setting of L-semirings again. Definition 4.1 works for all ML-semirings. However, most of the interesting properties postulate a greatest element $\top$. Therefore we assume the existence of such an element in the remainder.

Lemma 4.2. *Neighbours and boundaries can be expressed explicitly as*

$$\otimes_l b = \top \cdot \ulcorner b \urcorner , \quad \otimes_r b = \ulcorner b \urcorner \cdot \top , \quad \oslash_l b = \ulcorner b \urcorner \cdot \top , \quad \oslash_r b = \top \cdot \ulcorner b \urcorner .$$

Proof. We use the principle of indirect (in)equality.
By definition and Lemma 3.5(b) we get

$$a \leq \otimes_l b \Leftrightarrow \ulcorner a \urcorner \leq \ulcorner b \urcorner \Leftrightarrow a \leq \top \cdot \ulcorner b \urcorner . \qquad \Box$$

For nested neighbours we have the following cancellation properties.

Lemma 4.3.
(a) $\otimes_l \otimes_r b = \otimes_r b$ and $\otimes_r \otimes_l b = \otimes_l b,$
(b) $\oslash_l \otimes_r b = \otimes_r b$ and $\oslash_r \otimes_l b = \otimes_l b,$
(c) $\otimes_l \oslash_l b = \oslash_l b$ and $\otimes_r \oslash_r b = \oslash_r b,$
(d) $\oslash_l \oslash_l b = \oslash_l b$ and $\oslash_r \oslash_r b = \oslash_r b.$

Proof. The proof of [7] can immediately be adopted, since it only uses the explicit representations of neighbours and boundaries, which are identical for L-semirings and full semirings. E.g., by definition (twice), $\ulcorner p \urcorner \cdot \top = p$ and definition again,

$$\otimes_l \otimes_r b = \otimes_l (\ulcorner b \urcorner \cdot \top) = \top \cdot \ulcorner (\ulcorner b \urcorner \cdot \top) \urcorner = \top \cdot \ulcorner b \urcorner = \otimes_r b . \qquad \Box$$

Now we draw some conclusions when S is Boolean.

Lemma 4.4. *For a Boolean* ML-*semiring S, we have*
(a) $\neg\ulcorner a \leq \ulcorner\bar{a}$ *and* $\neg\bar{a}^{\urcorner} \leq \bar{a}^{\urcorner}$.
(b) $\overline{p \cdot \top} = \neg p \cdot \top$
(c) *If S is right-distributive,* $\overline{\top \cdot p} = \mathsf{F} \cdot \neg p$

Proof.
(a) By Boolean algebra and additivity of domain, $1 = \ulcorner\top = \ulcorner(a + \bar{a}) = \ulcorner a + \ulcorner\bar{a}$, and the first claim follows by shunting. The second inequality can be shown symmetrically.
(b) By Boolean algebra we only have to show that $\neg p \cdot \top + p \cdot \top = \top$ and $\neg p \cdot \top \sqcap p \cdot \top = 0$. The first equation follows by left-distributivity, the second one by Boolean algebra and the law [10]

$$p \cdot a \sqcap q \cdot a = p \cdot q \cdot a . \tag{2}$$

(c) By left and right distributivity, Boolean algebra and N being a left zero,

$$\mathsf{F} \cdot \neg p + \top \cdot p = \mathsf{F} \cdot \neg p + (\mathsf{F} + \mathsf{N}) \cdot p = \mathsf{F} \cdot \neg p + \mathsf{F} \cdot p + \mathsf{N} \cdot p$$
$$= \mathsf{F} \cdot (\neg p + p) + \mathsf{N} = \mathsf{F} + \mathsf{N} = \top .$$

Next, again by distributivity,

$$\mathsf{F} \cdot \neg p \sqcap \top \cdot p = \mathsf{F} \cdot \neg p \sqcap (\mathsf{F} + \mathsf{N}) \cdot p = \mathsf{F} \cdot \neg p \sqcap (\mathsf{F} \cdot p + \mathsf{N} \cdot p)$$
$$= (\mathsf{F} \cdot \neg p \sqcap \mathsf{F} \cdot p) + (\mathsf{F} \cdot \neg p \sqcap \mathsf{N} \cdot p) .$$

The first summand is 0, since the law symmetric to (2) holds for finite a and hence for F. The second summand is, by $p, \neg p \leq 1$ and isotony, below $\mathsf{F} \sqcap \mathsf{N} = 0$ and thus 0, too.   □

Similarly to [7], we now define perfect neighbours and boundaries.

Definition 4.5. Let S be a Boolean ML-semiring and $a, b \in S$.
(a) a is a *perfect left neighbour* of b (or $a \leq \boxed{n}_l b$ for short) iff $\bar{a}^{\urcorner} \cdot \ulcorner b \leq 0$,
(b) a is a *perfect right neighbour* of b (or $a \leq \boxed{n}_r b$ for short) iff $\bar{b}^{\urcorner} \cdot \ulcorner a \leq 0$,
(c) a is a *perfect left boundary* of b (or $a \leq \boxed{b}_l b$ for short) iff $\ulcorner a \cdot \ulcorner b \leq 0$,
(d) a is a *perfect right boundary* of b (or $a \leq \boxed{b}_r b$ for short) iff $\bar{a}^{\urcorner} \cdot \bar{b}^{\urcorner} \leq 0$.

From this definition, we get the following exchange rule for perfect neighbours.

$$a \leq \boxed{n}_l b \Leftrightarrow \bar{b} \leq \boxed{n}_r \bar{a} . \tag{3}$$

Lemma 4.6. *Perfect neighbours and perfect boundaries have the following explicit forms:*

$$\boxed{n}_l b = \top \cdot \neg\ulcorner b , \quad \boxed{n}_r b = \neg\bar{b}^{\urcorner} \cdot \top , \quad \boxed{b}_l b = \neg\bar{b}^{\urcorner} \cdot \top , \quad \boxed{b}_r b = \top \cdot \neg\ulcorner b .$$

Proof. By definition, shunting and Lemma 3.5(b)
$$a \leq \boxed{n}_l b \Leftrightarrow \bar{a}^{\urcorner} \cdot \ulcorner b \leq 0 \Leftrightarrow \bar{a}^{\urcorner} \leq \neg\ulcorner b \Leftrightarrow a \leq \top \cdot \neg\ulcorner b .$$ □

Lemma 4.7. *Each perfect neighbour (boundary) is a neighbour (boundary):*

$$\boxed{n}_l b \le \Diamond_l b\,, \qquad \boxed{n}_r b \le \Diamond_r b\,, \qquad \boxed{b}_l b \le \Diamond_l b\,, \qquad \boxed{b}_r b \le \Diamond_r b\,.$$

Proof. The claim follows by definition, shunting, Lemma 4.4(a), Boolean algebra and definition again:

$$a \le \boxed{n}_l b \Leftrightarrow a^{\urcorner} \cdot {}^{\urcorner}\!b \le 0 \Leftrightarrow a^{\urcorner} \le \neg{}^{\urcorner}\!b \Rightarrow a^{\urcorner} \le {}^{\urcorner}\!b \Leftrightarrow a \le \Diamond_l b\,. \qquad \square$$

Similarly to Lemma 4.3, we have cancellative laws for all box-operators. By $\Box\Box a = \Diamond\Diamond \overline{a}$ for all kinds of perfect lazy semiring neighbours, we have

Corollary 4.8.

(a) $\boxed{n}_l \boxed{n}_r b = \boxed{b}_r b$ *and* $\boxed{n}_r \boxed{n}_l b = \boxed{b}_l b,$

(b) $\boxed{b}_l \boxed{n}_r b = \boxed{n}_r b$ *and* $\boxed{b}_r \boxed{n}_l b = \boxed{n}_l b,$

(c) $\boxed{b}_l \boxed{b}_l b = \boxed{b}_l b$ *and* $\boxed{b}_r \boxed{b}_r b = \boxed{b}_r b,$

(d) $\boxed{n}_l \boxed{b}_l b = \boxed{n}_l b$ *and* $\boxed{n}_r \boxed{b}_r b = \boxed{n}_r b.$

There are also cancellation rules for mixed diamond/box expressions, e.g.,

$$\Diamond_l \boxed{b}_l b = \boxed{b}_l b \quad and \quad \boxed{b}_l \Diamond_l b = \Diamond_l b\,. \tag{4}$$

By straightforward calculations we get the de Morgan duals of right neighbours and left boundaries, respectively.

$$\overline{\Diamond_r \overline{b}} = \boxed{n}_r b \quad and \quad \overline{\boxed{n}_r \overline{b}} = \Diamond_r b\,, \tag{5}$$
$$\overline{\Diamond_l \overline{b}} = \boxed{b}_l b \quad and \quad \overline{\boxed{b}_l \overline{b}} = \Diamond_l b\,.$$

Furthermore, we have the following Galois connections.

Lemma 4.9. *We have* $\Diamond_r a \le b \Leftrightarrow a \le \boxed{n}_l b$ *and* $\Diamond_l a \le b \Leftrightarrow a \le \boxed{b}_r b$.

Proof. By de Morgan duality, Boolean algebra and the exchange rule (3)
$$\Diamond_r a \le b \Leftrightarrow \overline{\boxed{n}_r \overline{a}} \le b \Leftrightarrow \overline{b} \le \boxed{n}_r \overline{a} \Leftrightarrow a \le \boxed{n}_l b\,. \qquad \square$$

Since Galois connections are useful as theorem generators and dualities as theorem transformers we get many properties of (perfect) neighbours and (perfect) boundaries for free. For example we have

Corollary 4.10.

(a) $\Diamond_r$, $\Diamond_l$ *and* $\boxed{n}_l$, $\boxed{b}_r$ *are isotone.*

(b) $\Diamond_r$, $\Diamond_l$ *are disjunctive and* $\boxed{n}_l$, $\boxed{b}_r$ *are conjunctive.*

(c) *We also have cancellative laws:*
$$\Diamond_r \boxed{n}_l a \le a \le \boxed{n}_l \Diamond_r a \quad and \quad \Diamond_l \boxed{b}_r a \le a \le \boxed{b}_r \Diamond_l a.$$

But, because of Lemma 4.4(c), we do not have the full semiring de Morgan dualities of left neighbours and right boundaries, respectively. We only obtain

Lemma 4.11. *Let S be right-distributive.*

(a) $\overline{\Diamond_l \overline{b}} \le \boxed{n}_l b$ *and* $\overline{\boxed{n}_l \overline{b}} \le \Diamond_l b\,,$

(b) $\overline{\Diamond_r \overline{b}} \le \boxed{b}_r b$ *and* $\overline{\boxed{b}_r \overline{b}} \le \Diamond_r y\,.$

Proof. (a) By Lemma 4.2, 4.4(c), isotony and Lemma 4.6,

$$\overline{\Diamond_l \overline{b}} = \overline{\top \cdot \ulcorner \overline{b}} = \mathsf{F} \cdot \neg \ulcorner \overline{b} \leq \top \cdot \neg \ulcorner \overline{b} = \boxed{n}_l b.$$

The equation $\overline{\boxed{n}_l \overline{b}} \leq \Diamond_l b$ then follows by shunting.   □

The converse inequations do not hold. For example, setting $b = \top$ implies $\overline{\Diamond_l \top} = \overline{\top \cdot \ulcorner 0} = \overline{\top \cdot 0} = \overline{\mathsf{N}} = \mathsf{F}$ and $\boxed{n}_l \top = \top \cdot \neg \ulcorner 0 = \top$. But in general, $\top \leq \mathsf{F}$ is false (if there is at least one infinite element $a \neq 0$). Also, the Galois connections of [7] are not valid for left neighbours and right boundaries, but one implication can still be proved.

Lemma 4.12. *Let S be right-distributive, then*

$$\Diamond_l a \leq b \Rightarrow a \leq \boxed{n}_r b \,, \qquad \Diamond_r a \leq b \Rightarrow a \leq \boxed{b}_r b \,.$$

Proof. By Lemma 4.11(a), Boolean algebra and the exchange rule (3)

$$\Diamond_l a \leq b \Rightarrow \overline{\boxed{n}_l \overline{a}} \leq b \Leftrightarrow \overline{b} \leq \boxed{n}_l \overline{a} \Leftrightarrow a \leq \boxed{n}_r b \,.$$ □

By lack of Galois connections, we do not have a full analogue to Corollary 4.10.

Lemma 4.13.

(a) $\Diamond_l$, $\Diamond_r$, $\boxed{n}_r$ and $\boxed{b}_l$ *are isotone.*
(b) *If S is right-distributive, then $\Diamond_l$, $\Diamond_r$ are disjunctive and $\boxed{n}_r$, $\boxed{b}_l$ are conjunctive.*

Proof.

(a) The claim follows directly by the explicit representation of (perfect) neighbours and boundaries (Lemma 4.2 and Lemma 4.6).
(b) By Lemma 4.2, additivity of domain and right-distributivity we get
$$\Diamond_l (a + b) = \top \cdot \ulcorner (a + b) = \top \cdot (\ulcorner a + \ulcorner b) = \top \cdot \ulcorner a + \top \cdot \ulcorner b = \Diamond_l a + \Diamond_l b \,.$$ □

Until now, we have shown that most of the properties of [7] hold in L-semirings, too. At some points, we need additional assumptions like right-distributivity. Many more properties, like $\overline{b} \leq \Diamond_r b$, can be shown. Most proofs use the explicit forms for lazy semiring neighbours or the Galois connections (Lemma 4.9) and Lemma 4.12. However, since L-semirings reflect some aspects of infinity, we get some useful properties, which are different from all properties given in [7]. Some are summarised in the following lemma.

Lemma 4.14.

(a) $\Diamond_l \mathsf{F} = \Diamond_r \mathsf{F} = \dot{\Diamond}_l \mathsf{F} = \dot{\Diamond}_r \mathsf{F} = \top$.
(b) $b \leq \mathsf{N} \Leftrightarrow \Diamond_r b \leq 0 \Leftrightarrow \Diamond_r b \leq \mathsf{N}$.
(c) $\boxed{n}_l \mathsf{N} = \boxed{b}_r \mathsf{N} = \mathsf{N}$ *and* $\boxed{n}_r \mathsf{N} = \boxed{b}_l \mathsf{N} = 0$.
(d) $\overline{b} \leq \mathsf{N} \Leftrightarrow \mathsf{F} \leq b \Leftrightarrow \boxed{n}_r b = \top \Leftrightarrow \boxed{b}_r b = \top$.

Proof. First we note that by straightforward calculations using Lemma 3.2 and 3.4, we get

$$\top \cdot p \leq \top \cdot q \Leftrightarrow p \leq q \Leftrightarrow p \cdot \top \leq q \cdot \top \,. \tag{6}$$

(a) Directly by Lemma 4.2 and $\ulcorner F = \bar{F}\urcorner = 1$, since $1 \leq F$:
$$\diamondsuit_l F = \top \cdot \ulcorner F = \top \cdot 1 = \top .$$

(b) By Lemma 3.4, (6), left-strictness and definition of $\diamondsuit_l$
$$b \leq N \Leftrightarrow \bar{b} \leq 0 \Leftrightarrow \bar{b} \cdot \top \leq 0 \cdot \top \Leftrightarrow \diamondsuit_r b \leq 0 .$$

(c) By Lemma 4.6 and $\ulcorner F = 1$ we get
$$\boxed{m}_l N = \top \cdot \neg \ulcorner \overline{N} = \top \cdot \neg \ulcorner F = \top \cdot 0 = N.$$

(d)Similar to (b).   □

Note that (a) implies $\diamondsuit_l \top = \diamondsuit_r \top = \diamondsuit_l \top = \diamondsuit_r \top = \top$ using isotony. (c) shows again that the inequations of Lemma 4.11 cannot be strengthened to equations.

Since the above theory concerning lazy semiring neighbours is based on lazy semirings, it is obvious that one can use it also in the framework of lazy Kleene algebra and lazy omega algebra [10]. The former one provides, next to the L-semiring operators, an operator for finite iteration. The latter one has an additional operator for infinite iteration.

5 Neighbourhood Logic with Infinite Durations

Using the theory of the previous section, we can now formulate a generalisation of NL, which includes infinite elements (intervals with infinite duration). Those intervals are not included in the original Neighbourhood Logic of [14,15], i.e., if we compose two intervals $[a, b]$ and $[b, c]$ (where intervals are defined, as usual, as $[a, b] =_{df} \{x \,|\, a \leq x \leq b, a \leq b\}$), it is assumed that the points of $[b, c]$ are reached after finite duration $b - a$. However, for many applications, e.g. for hybrid systems, as we will see in Section 7, a time point ∞ of infinity is reasonable. But then the composition of the intervals $[a, \infty[$ and $[b, c]$ never reaches the second interval. This gives rise to an L-semiring.

Neighbourhood Logic and Its Embedding. In this paragraph the Neighbourhood Logic [14,15] and its embedding [7] are briefly recapitulated.

Chop-based interval temporal logics, such as ITL [5] and IL [3] are useful for the specification and verification of safety properties of real-time systems. In these logics, one can easily express a lot of properties such as "if ϕ holds for an interval, then there is a subinterval where ψ holds". As shown in [15], these logics cannot express all desired properties. E.g., (unbounded) liveness properties such as "eventually there is an interval where ϕ holds" are not expressible in these logics. As it is shown in [15] the reason is that the modality *chop* $\frown$ is a *contracting* modality, in the sense that the truth value of $\phi \frown \psi$ on $[a, b]$ only depends on subintervals of $[a, b]$:

$$\phi \frown \psi \text{ holds on } [a, b] \text{ iff}$$
there exists $c \in [a, b]$ such that ϕ holds on $[a, c]$ and ψ holds on $[c, b]$.

Hence Zhou and Hansen proposed a first-order interval logic called *Neighbourhood Logic* (NL) in 1996 [14]. In this logic they introduce *left* and *right* *neighbourhoods* as new primitive intervals to define other unary and binary modalities of intervals in a first-order logic. The two proposed simple expanding modalities $\Diamond_l \phi$ and $\Diamond_r \phi$ are defined as follows:

$$\Diamond_l \phi \text{ holds on } [a, b] \text{ iff there exists } \delta \geq 0 \text{ such that } \phi \text{ holds on } [a - \delta, a], \quad (7)$$
$$\Diamond_r \phi \text{ holds on } [a, b] \text{ iff there exists } \delta \geq 0 \text{ such that } \phi \text{ holds on } [b, b + \delta], \quad (8)$$

where ϕ is a *formula*[2] of NL. These modalities can be illustrated by

where $c = a - \delta$ where $d = b + \delta$

With $\Diamond_r (\Diamond_l)$ one can reach the *left* (*right*) *neighbourhood* of the beginning (ending) point of an interval. In contrast to the chop operator, the neighbourhood modalities are *expanding* modalities, i.e., $\Diamond_l$ and $\Diamond_r$ depend not only on subintervals of an interval $[a, b]$, but also on intervals "outside". In [14] it is shown that the modalities of [6] and [13] as well as the chop operator can be expressed by the neighbourhood modalities.

In [7] we present an embedding and extension of NL into the framework of full semirings. There, (perfect) neighbours and boundaries are defined on full semirings in the same way as we have done this for L-semirings in Section 4. Consider the structure

$$\text{INT} =_{df} (\mathcal{P}(\text{Int}), \cup, ;, \emptyset, \mathbb{1}) ,$$

where $\mathbb{1} =_{df} \{[a, a]\}$ denotes the set of all intervals consisting of one single point and Int is the set of all intervals $[a, b]$ with $a, b \in$ Time and Time is a totally ordered poset, e.g. $\mathbb{R}$. Further we assume that there is an operation $-$ on Time, which gives us the duration of an interval $[a, b]$ by $b - a$. By this operation $\mathbb{1}$ consists of all 0-length intervals.

For the moment we exclude intervals with infinite duration. The symbol $;$ denotes the pointwise lifted composition of intervals which is defined by

$$[a, b] ; [c, d] =_{df} \begin{cases} [a, d] & \text{if } b = c \\ \text{undefined} & \text{otherwise} . \end{cases}$$

It can easily be checked that INT forms a full semiring. In [7] we have shown

$$\Diamond_l \phi \text{ holds on } [a, b] \iff \{[a, b]\} \leq \circledast_r \mathbb{I}_\phi ,$$
$$\Diamond_r \phi \text{ holds on } [a, b] \iff \{[a, b]\} \leq \circledast_l \mathbb{I}_\phi ,$$

where $\mathbb{I}_\phi =_{df} \{i \mid i \in \text{Int}, \phi \text{ holds on } i\}$. This embedding gives us the possibility to use the structure of a semiring to describe NL. Many simplifications of NL and properties concerning the algebraic structure are given in [7].

---

[2] The exact definition of the syntax of formulas can be found e.g. in [14].

Adding Infinite Durations. Now, we assume a point of infinity $\infty \in$ Time, e.g. Time $= \mathbb{R} \cup \{\infty\}$. If there is such an element, it has to be the greatest element. Consider the slightly changed structure

$$\text{INT}^i =_{df} (\mathcal{P}(\text{Int}), \cup, ;, \emptyset, \mathbb{1}) ,$$

where ; is now the pointwise lifted composition defined as

$$[a,b] ; [c,d] =_{df} \begin{cases} [a,d] & \text{if } b = c, b \neq \infty \\ [a,\infty[& \text{if } b = \infty \\ \text{undefined otherwise .} \end{cases}$$

Again, it is easy to check that INT^i forms an L-semiring, which even becomes an ML-semiring by setting, for $A \in \mathcal{P}(\text{Int})$,

$$\ulcorner A =_{df} \{[a,a] \mid [a,b] \in A\} \quad \text{and} \quad \overline{A}^\urcorner =_{df} \{[b,b] \mid [a,b] \in A, b \neq \infty\} .$$

Note that INT^i is right-distributive, so that all Lemmas and Corollaries of Section 4 hold in this model.

Thereby we have defined a new version NL^i of NL which handles intervals with infinite durations. NL^i also subsumes the theory presented in [16]. In particular, it builds a bridge between NL and a duration calculus for infinite intervals.

6   Lazy Semiring Neighbours and CTL*

The branching time temporal logic CTL* (see e.g. [4]) is a well-known tool for analysing and describing parallel as well as reactive and hybrid systems. In CTL* one distinguishes state formulas and path formulas, the former ones denoting sets of states, the latter ones sets of computation traces.

The language Ψ of CTL^* *formulas* over a set Φ of atomic propositions is defined by the grammar

$$\Psi ::= \bot \mid \Phi \mid \Psi \to \Psi \mid \mathsf{X}\Psi \mid \Psi \mathsf{U} \Psi \mid \mathsf{E}\Psi ,$$

where X and U are the next-time and until operators and E is the existential quantifier on paths. As usual,

$$\neg\varphi =_{df} \varphi \to \bot , \qquad \varphi \wedge \psi =_{df} \neg(\varphi \to \neg\psi) ,$$
$$\varphi \vee \psi =_{df} \neg\varphi \to \psi , \qquad \mathsf{A}\varphi =_{df} \neg\mathsf{E}\neg\varphi .$$

In [11] a connection between CTL* and Boolean modal quantales is presented. Since these are right-distributive, again all the lemmas of the previous sections are available. If A is a set of states one could, e.g., use the algebra STR(A) (cf. Section 2) of finite and infinite streams of A-states as a basis. For an arbitrary Boolean modal quantale S, the concrete standard semantics for CTL* is generalised to a function $[\![\_]\!] : \Psi \to S$ as follows, where $[\![\varphi]\!]$ abstractly represents the set of paths satisfying formula φ. One fixes an element n (n standing for "next") as representing the transition system underlying the logic and sets

$$[\![\perp]\!] = 0 \,,$$
$$[\![p]\!] = p \cdot \top \,,$$
$$[\![\varphi \rightarrow \psi]\!] = \overline{[\![\varphi]\!]} + [\![\psi]\!] \,,$$
$$[\![\mathsf{X}\,\varphi]\!] = \mathsf{n} \cdot [\![\varphi]\!] \,,$$
$$[\![\varphi \,\mathsf{U}\, \psi]\!] = \bigsqcup_{j \geq 0} (\mathsf{n}^j \cdot [\![\psi]\!] \sqcap \bigsqcap_{k < j} \mathsf{n}^k \cdot [\![\varphi]\!]) \,,$$
$$[\![\mathsf{E}\varphi]\!] = \lceil[\![\varphi]\!]\rceil \cdot \top \,.$$

Using these definitions, it is straightforward to check that $[\![\varphi \vee \psi]\!] = [\![\varphi]\!] + [\![\psi]\!]$, $[\![\varphi \wedge \psi]\!] = [\![\varphi]\!] \sqcap [\![\psi]\!]$ and $[\![\neg\varphi]\!] = \overline{[\![\varphi]\!]}$.

By simple calculations we get the following result.

Lemma 6.1. *[11] Let φ be a state formula of* CTL*. *Then*

$$[\![\mathsf{A}\varphi]\!] = \neg\lceil(\overline{[\![\varphi]\!]})\rceil \cdot \top \,.$$

Hence we see that $[\![\mathsf{E}\varphi]\!]$ corresponds to a left boundary and $[\![\mathsf{A}\varphi]\!]$ to a perfect left boundary, i.e.,

$$[\![\mathsf{E}\varphi]\!] = \diamondsuit_l[\![\varphi]\!] \quad \text{and} \quad [\![\mathsf{A}\varphi]\!] = \boxdot_l[\![\varphi]\!] \,.$$

With these equations we have connected lazy neighbours with CTL*. From Lemma 4.3, Corollary 4.8 and equations (4) we obtain immediately

$$[\![\mathsf{EE}\varphi]\!] = [\![\mathsf{E}\varphi]\!] \,, \qquad\qquad [\![\mathsf{AA}\varphi]\!] = [\![\mathsf{A}\varphi]\!] \,,$$
$$[\![\mathsf{EA}\varphi]\!] = [\![\mathsf{A}\varphi]\!] \,, \qquad\qquad [\![\mathsf{AE}\varphi]\!] = [\![\mathsf{E}\varphi]\!] \,.$$

The other two boundaries as well as all variants of (perfect) neighbours do not occur in CTL* itself.

A connection to hybrid systems will be set up in the next section.

7   Lazy Semiring Neighbours and Hybrid Systems

Hybrid systems are dynamical heterogeneous systems characterised by the interaction of discrete and continuous dynamics. In [8] we use the L-semiring PRO of processes from Section 2 for the description of hybrid systems.

Hybrid systems and NL. In PRO the left/right neighbours describe a kind of composability, i.e., for processes A, B,

$$A \leq \diamondsuit_l B \quad \text{iff} \quad \forall a \in A : \exists b \in B : a \cdot b \text{ is defined,} \qquad (9)$$
$$A \leq \diamondsuit_r B \quad \text{iff} \quad \forall a \in A : \exists b \in \mathsf{fin}\,(B) : b \cdot a \text{ is defined.} \qquad (10)$$

These equivalences are closely related to (7) and (8), respectively. $\diamondsuit_r$ and $\diamondsuit_l$ each guarantee existence of a composable element. Especially, $\diamondsuit_r \neq 0$ guarantees that there exists a process, and therefore a trajectory, that can continue the current process (trajectory). Therefore it is a form of liveness assertion. In particular, the process $\Diamond_r B$ contains all trajectories that are composable with the "running" one. If $\Diamond_r B = \emptyset$, we know that the system will terminate if all trajectories of the running process have finite durations. Note that in the above characterisation of $\Diamond_l$ the composition $a \cdot b$ is defined if either $f(d_1) = g(0)$ (assuming $a = (d_1, f)$ and $b = (d_2, g)$) or a has infinite duration, i.e., $d = \infty$. The next paragraph will show that left and right boundaries of lazy semirings are closely connected to temporal logics for hybrid systems. But, by Lemma 4.3, they are also useful as operators that simplify nestings of semiring neighbours.

The situation for right/left perfect neighbours is more complicated. As shown in [7], $\boxed{n}_r B$ is the set of those trajectories which can be reached only from B, not from $\overline{B}$. Hence it describes a situation of guaranteed non-reachability from $\overline{B}$. The situation with $\boxed{n}_l$ is similar for finite processes, because of the symmetry between left and right perfect neighbours.

Hybrid Systems and CTL*. Above we have shown how lazy semiring neighbours are characterised in PRO. Although a next-time operator is not meaningful in continuous time models, the other operators of CTL* still make sense. Since PRO is a Boolean modal quantale, we simply re-use the above semantic equations (except those for X and U) and obtain a semantics of a fragment of CTL* for hybrid systems. In particular, the existential quantifier E is a left boundary also in hybrid systems. The operators F, G and U can be realised as

$$[\![F\varphi]\!] =_{df} \mathsf{F} \cdot [\![\varphi]\!]^3 \,, \quad \mathsf{G}\varphi =_{df} \neg \mathsf{F} \neg \varphi \,, \quad [\![\varphi \,\mathsf{U}\, \psi]\!] =_{df} (\mathsf{fin}\, [\![\mathsf{G}\varphi]\!]) \cdot [\![\psi]\!] \,.$$

Of course all other kinds of left and right (perfect) neighbours and boundaries have their own interpretation in PRO and in (the extended) CTL*, respectively. A detailed discussion of all these interpretations is part of our future work (cf. Section 8).

8   Conclusion and Outlook

In the paper we have presented a second extension of Neighbourhood Logic. Now this logic is able to handle intervals which either have finite or infinite length. For this we have established semiring neighbours over lazy semirings. During the development of lazy semiring neighbours it turned out that they are not only useful and necessary for NL but also in other areas of computer science; we have sketched connections to temporal logics and to hybrid systems.

We have only given a short overview over the connections between lazy semiring neighbours, CTL* and hybrid systems. One of our aims for further work is a more elaborate treatment of this. Further, it will be interesting to see if there are even more applications for semiring neighbours.

Acknowledgement. We are grateful to Kim Solin and the anonymous referees for helpful discussions and remarks.

---

[3] On the right hand side F is the largest finite element.

References

1. J. H. Conway: *Regular Algebra and Finite State Machines*. Chapman & Hall, 1971
2. J. Desharnais, B. Möller, G. Struth: Kleene Algebra with Domain. ACM Trans. Computational Logic (to appear 2006). Preliminary version: Universität Augsburg, Institut für Informatik, Report No. 2003-07, June 2003
3. B. Dutertre: Complete Proof Systems for First-Order Interval Temporal Logic. In IEEE Press, editor, *Tenth Annual IEEE Symp. on Logic in Computer Science*, 1995, 36–43
4. E.A. Emerson: Temporal and Modal Logic. In J. van Leeuwen (ed.): *Handbook of Theoretical Computer Science. Vol. B: Formal Models and Semantics*. Elsevier 1991, 995–1072
5. J.Y. Halpern, B. Moszkowski, Z. Manna: A Hardware Semantics Based on Temporal Intervals. In J. Diaz (ed.) *Proc. ICALP'83*. LNCS 154. Springer 1983, 278–291
6. J.Y. Halpern, Y. Shoham: A Propositional Modal Logic of Time Intervals. Proceedings of the First IEEE Symposium on Logic in Computer Science. IEEE Press, Piscataway, NJ, 279–292.
7. P. Höfner: Semiring Neighbours — An Algebraic Embedding and Extension of Neighbourhood Logic. In J. van de Pol, J. Romijn, G. Smith (eds.): IFM 2005 Doctoral Symposium on Integrated Formal Methods, 6–13, 2005. Extended version: P. Höfner: Semiring Neighbours. Technical Report 2005-19, Universität Augsburg, 2005
8. P. Höfner, B. Möller: Towards an Algebra of Hybrid Systems. In W. MacCaull, M. Winter and I. Düntsch (eds.): Relational Methods in Computer Science. LNCS 3929. Springer 2006, 121–133
9. D. Kozen: Kleene Algebra with Tests. ACM Trans. Programming Languages and Systems 19(3), 427–443 (1997)
10. B. Möller: Kleene Getting Lazy. Science of Computer Programming, Special issue on MPC 2004 (to appear). Previous version: B. Möller: Lazy Kleene algebra. In D. Kozen (ed.): Mathematics of program construction. LNCS 3125. Springer 2004, 252–273
11. B. Möller, P. Höfner, G. Struth: Quantales and Temporal Logics. In M. Johnson, V. Vene (eds.): AMAST 2006. LNCS 4019. Springer 2006, 263–277
12. M. Sintzoff: Iterative Synthesis of Control Guards Ensuring Invariance and Inevitability in Discrete-Decision Games. In O. Owe, S. Krogdahl, T. Lyche (eds.): From Object-Orientation to Formal Methods — Essays in Memory of Ole-Johan Dahl. LNCS 2635. Springer 2004, 272–301
13. Y. Venema: A Modal Logic for Chopping Intervals. J. of Logic and Computation 1(4):453–476, 1990
14. C. Zhou, M.R. Hansen: An Adequate First Order Interval Logic. In W.-P. de Roever, H. Langmaack, A. Pnueli (eds.): Compositionality: The Significant Difference: International Symposium, COMPOS'97. LNCS 1536. Springer 1998, 584–608
15. C. Zhou, M.R. Hansen: Duration Calculus – A Formal Approach to Real-Time Systems. Monographs in Theoretical Computer Science. Springer 2004
16. C. Zhou, D. Van Hung, L. Xiaoshan: Duration Calculus with Infinite Intervals. In H. Reichel (ed.): Fundamentals of Computation Theory. LNCS 965. Springer 1995, 16–41

Omega Algebra, Demonic Refinement Algebra and Commands

Peter Höfner[1,*], Bernhard Möller[1], and Kim Solin[1,2]

[1] Institut für Informatik, Universität Augsburg, D-86135 Augsburg, Germany
{hoefner, moeller}@informatik.uni-augsburg.de
[2] Turku Centre for Computer Science
Lemminkäinengatan 14 A, FIN-20520 Åbo, Finland
kim.solin@utu.fi

Abstract. Weak omega algebra and demonic refinement algebra are two ways of describing systems with finite and infinite iteration. We show that these independently introduced kinds of algebras can actually be defined in terms of each other. By defining modal operators on the underlying weak semiring, that result directly gives a demonic refinement algebra of commands. This yields models in which extensionality does not hold. Since in predicate-transformer models extensionality always holds, this means that the axioms of demonic refinement algebra do not characterise predicate-transformer models uniquely. The omega and the demonic refinement algebra of commands both utilise the convergence operator that is analogous to the halting predicate of modal μ-calculus. We show that the convergence operator can be defined explicitly in terms of infinite iteration and domain if and only if domain coinduction for infinite iteration holds.

1 Introduction

An omega algebra [2] is an extension of Kleene algebra [10] adding an infinite iteration operator to the signature. Demonic refinement algebra is an extension of a relaxed version of Kleene algebra (right-strictness, $a \cdot 0 = 0$, does not hold in general) adding a strong iteration operator to the signature. Demonic refinement algebra was devised in [20] for reasoning about total-correctness preserving program transformations. A structure satisfying all the axioms of omega algebra except right strictness (called a weak omega algebra [14]) always has a greatest element $\top$. As one of the main contributions of this paper, we show that weak omega algebra with the extra axiom $\top x = \top$ is equivalent to demonic refinement algebra in the sense that they can be defined in terms of each other.

We then consider commands, that is, pairs (a, p) such that a describes the state transition behaviour and p characterises the states with guaranteed termination. Möller and Struth have already shown how the addition of modal operators on the underlying semiring facilitates definitions of operators on commands such

---

* This research was supported by DFG (German Research Foundation).

that they form a weak Kleene and a weak omega algebra, respectively [14]. The definitions of these operators use modal operators, defined from the domain operator of Kleene algebra with domain [4]. To define a demonic refinement algebra of commands, we need a strong iteration operator on commands [19]. We define this operator with the aid of the above-mentioned result. The demonic refinement algebra of commands gives rise to a model that is not extensional, thus showing that the axioms of demonic refinement algebra do not characterise predicate-transformer models uniquely.

The definition of infinite iteration and strong iteration on commands both utilise the convergence operator of [13], that is, the underlying structure is actually assumed to be a convergence algebra. The convergence operator is analogous to the halting predicate of modal μ-calculus [8]. As the third result in this paper, we show that the convergence operator can be explicitly defined in terms of infinite iteration and domain if and only if domain coinduction for the infinite iteration operator is assumed to hold in general.

The historic development of this paper has it starting point in Kozen's axiomatisation of Kleene algebra and his injection of tests into the algebra [11], rendering reasoning about control structures possible. As mentioned earlier, Cohen [2] conservatively extends Kleene algebra with an infinite iteration operator. Von Wright's demonic refinement algebra, introducing the strong iteration operator, was the first algebra that was genuinely an algebra intended for total-correctness reasoning about programs. Desharnais, Möller and Struth's domain-operator extension [4] was the seminal work for modal operators in Kleene algebra. The domain operator was investigated in the context of refinement algebra in [18]. Möller later weakened the axiomatisation to form left semirings and left Kleene algebras [12]. The former is one of the most foundational structures found in this paper.

The paper is organised as follows. We begin in Sect. 2 by the result concerning the equivalence of top-left-strict weak omega algebra and demonic refinement algebra, upon which in Sect. 3 we construct the demonic refinement algebra of commands and relate it to the demonic algebras with domain of de Carufel and Desharnais [3]. In Sect. 4 we give some remarks on refinement algebra in the light of Sect. 3. Before concluding, we consider the explicit definition of the convergence operator in Sect. 5.

2 Omega and Demonic Refinement Algebra

We begin by recapitulating some basic definitions. By a *left semiring* we shall understand a structure $(+, 0, \cdot, 1)$ such that the reduct $(+, 0)$ is a commutative and idempotent monoid, and the reduct structure $(\cdot, 1)$ is a monoid such that $\cdot$ distributes over $+$ in its left argument and is left-strict, i.e., $0 \cdot a = 0$. A *weak semiring* is a left semiring that is also right-distributive. A weak semiring with right-strictness is called a *full semiring* or simply *semiring*. When no risk for confusion arises $\cdot$ is left implicit. We define the *natural order* $\leq$ on a left semiring by $a \leq b \Leftrightarrow_{df} a + b = b$ for all a and b in the carrier set. With respect to that order, 0 is the least element and multiplication as well as addition are isotone. Moreover, $a + b$ is the join of a and b.

A *(weak) Kleene algebra* is a structure $(+, 0, \cdot, 1, ^*)$ such that the reduct $(+, 0, \cdot, 1)$ is a (weak) semiring and the star * satisfies the axioms

$$1 + aa^* \leq a^* \,, \qquad\qquad 1 + a^*a \leq a^* \,, \qquad\qquad (* \text{ unfold})$$
$$b + ac \leq c \Rightarrow a^*b \leq c \,, \qquad b + ca \leq c \Rightarrow ba^* \leq c \,, \qquad (* \text{ induction})$$

for a, b and c in the carrier set of the structure. A *(weak) omega algebra* [14] is a structure $(+, 0, \cdot, 1, ^*, {}^\omega)$ such that the reduct $(+, 0, \cdot, 1, ^*)$ is a (weak) Kleene algebra and the infinite iteration ${}^\omega$ satisfies the axioms

$$a^\omega = aa^\omega \,, \qquad\qquad\qquad (\omega \text{ unfold})$$
$$c \leq b + ac \Rightarrow c \leq a^\omega + a^*b \,, \qquad (\omega \text{ coinduction})$$

for a, b and c in the carrier set of the structure. In particular, a^ω is the greatest fixpoint of the function $f(x) = ax$. The element 1^ω is the greatest element and we denote it by $\top$. Since, by the ω unfold law, $a^\omega \top$ is a fixpoint of f, we have $a^\omega = a^\omega \top$ for all a. We call a weak omega algebra *top-left-strict* iff the equation $\top a = \top$ holds for all a. In that case we get

$$a^\omega b = a^\omega \top b = a^\omega \top = a^\omega \,. \qquad\qquad (1)$$

In general omega algebra only the inequation $a^\omega b \leq a^\omega$ holds. The above derivation (1) strengthens it to an equation. In fact we have the following result.

Proposition 2.1. *Top-left-strictness is equivalent to left ω annihilation, i.e.,*

$$\top b = \top \Leftrightarrow (\forall a \bullet a^\omega \leq a^\omega b) \,.$$

Proof. The implication ($\Rightarrow$) follows from (1), whereas ($\Leftarrow$) can be calculated by

$$(\forall a \bullet a^\omega \leq a^\omega b)$$
$$\Rightarrow \quad \{\!\{ \text{ set } a = 1 \}\!\}$$
$$\quad 1^\omega \leq 1^\omega b$$
$$\Leftrightarrow \quad \{\!\{ 1^\omega = \top \}\!\}$$
$$\quad \top \leq \top b \,.$$

The other inequation ($\top b \leq \top$) holds since $\top$ is the greatest element. $\qquad\square$

A *demonic refinement algebra* [19] is a structure $(+, 0, \cdot, 1, ^*, {}^{\overline{\omega}})$ such that the reduct $(+, 0, \cdot, 1, ^*)$ is a weak Kleene algebra and the strong iteration operator ${}^{\overline{\omega}}$ satisfies the axioms

$$a^{\overline{\omega}} = aa^{\overline{\omega}} + 1 \,, \qquad\qquad (\overline{\omega} \text{ unfold})$$
$$a^{\overline{\omega}} = a^* + x^{\overline{\omega}}0 \,, \qquad\qquad (\overline{\omega} \text{ isolation})$$
$$c \leq ac + b \Rightarrow c \leq a^{\overline{\omega}}b \,, \qquad (\overline{\omega} \text{ coinduction})$$

for a, b and c in the carrier set of the structure. It is easily shown that $1^{\overline{\omega}}$ is the greatest element and satisfies $1^{\overline{\omega}}a = 1^{\overline{\omega}}$ for all a in the carrier set [20]. This element is again denoted by $\top$.

In the remainder of this section we present one of our main contributions, namely that top-left-strict weak omega algebra is equivalent to demonic refinement algebra in the sense that they can be defined in terms of each other. This is done in two steps: First we show that weak omega algebra subsumes demonic refinement algebra, then we show the converse subsumption.

Lemma 2.2. *Top-left-strict weak omega algebra subsumes demonic refinement algebra.*

Proof. Given a top-left-strict weak omega algebra, the strong iteration is defined by $a^{\overline{\omega}} =_{df} a^* + a^{\omega}$. It is sufficient to show that this definition satisfies the axioms of strong iteration; the other axioms of demonic refinement algebra are immediate from the axioms of top-left-strict weak omega algebra.

1. $\overline{\omega}$ unfold:

$$a^{\overline{\omega}}$$
$$= \quad \{\!\!\{ \text{ definition } \}\!\!\}$$
$$a^* + a^{\omega}$$
$$= \quad \{\!\!\{ * \text{ and } \omega \text{ unfold } \}\!\!\}$$
$$aa^* + 1 + aa^{\omega}$$
$$= \quad \{\!\!\{ \text{ commutativity } \}\!\!\}$$
$$aa^* + aa^{\omega} + 1$$
$$= \quad \{\!\!\{ \text{ distributivity } \}\!\!\}$$
$$a(a^* + a^{\omega}) + 1$$
$$= \quad \{\!\!\{ \text{ definition } \}\!\!\}$$
$$aa^{\overline{\omega}} + 1$$

2. isolation:

$$a^{\overline{\omega}}$$
$$= \quad \{\!\!\{ \text{ definition } \}\!\!\}$$
$$a^* + a^{\omega}$$
$$= \quad \{\!\!\{ \text{ neutrality of } 0 \text{ and } (1) \}\!\!\}$$
$$a^*(1 + 0) + a^{\omega}0$$
$$= \quad \{\!\!\{ \text{ right-distributivity } \}\!\!\}$$
$$a^* + a^*0 + a^{\omega}0$$
$$= \quad \{\!\!\{ \text{ left-distributivity } \}\!\!\}$$
$$a^* + (a^* + a^{\omega})0$$
$$= \quad \{\!\!\{ \text{ definition } \}\!\!\}$$
$$a^* + a^{\overline{\omega}}0$$

3. $\overline{\omega}$ coinduction:

$$c \leq a^{\overline{\omega}}b$$
$\Leftrightarrow$ { definition }
$$c \leq (a^* + a^\omega)b$$
$\Leftrightarrow$ { left-distributivity }
$$c \leq a^*b + a^\omega b$$
$\Leftrightarrow$ { (1) }
$$c \leq a^*b + a^\omega$$
$\Leftarrow$ { ω coinduction }
$$c \leq ac + b$$

$\square$

In a concrete predicate-transformer algebra, the same definition of $^{\overline{\omega}}$ is made by Back and von Wright [1]. In the present paper the definition is given in an abstract setting for which (conjunctive) predicate transformers constitute a model.

Lemma 2.3. *Demonic refinement algebra subsumes top-left-strict weak omega algebra.*

Proof. Given a demonic refinement algebra, infinite iteration is defined as $a^\omega =_{df} a^{\overline{\omega}}0$. It is sufficient to show that this definition satisfies the axioms for infinite iteration; the other axioms of the top-left-strict weak omega algebra are immediate from demonic refinement algebra.

1. ω unfold:

$$a^\omega$$
$=$ { definition }
$$a^{\overline{\omega}}0$$
$=$ { $\overline{\omega}$ unfold }
$$(aa^{\overline{\omega}} + 1)0$$
$=$ { left-distributivity and neutrality of 1 }
$$aa^{\overline{\omega}}0 + 0$$
$=$ { neutrality of 0 }
$$aa^{\overline{\omega}}0$$
$=$ { definition }
$$aa^\omega$$

2. top-left-strictness:

$$\top \leq \top a$$
$\Leftrightarrow$ { $\top = 1^{\overline{\omega}}$ }
$$\top \leq 1^{\overline{\omega}}a$$

$$\Leftarrow \quad \{\!\lceil \bar{\omega} \text{ coinduction } \}\!\}$$
$$\top \leq \top + a$$
$$\Leftrightarrow \quad \{\!\lceil \text{ join } \}\!\}$$
$$\text{true}$$

$\top a \leq \top$ holds since $\top$ is the greatest element.

3. ω coinduction:

$$c \leq a^*b + a^\omega$$
$$\Leftrightarrow \quad \{\!\lceil \text{ definition } \}\!\}$$
$$c \leq a^*b + a^{\bar{\omega}}0$$
$$\Leftrightarrow \quad \{\!\lceil \text{ annihilation } \}\!\}$$
$$c \leq a^*b + a^{\bar{\omega}}0b$$
$$\Leftrightarrow \quad \{\!\lceil \text{ distributivity } \}\!\}$$
$$c \leq (a^* + a^{\bar{\omega}}0)b$$
$$\Leftrightarrow \quad \{\!\lceil \text{ isolation } \}\!\}$$
$$c \leq a^{\bar{\omega}}b$$
$$\Leftarrow \quad \{\!\lceil \bar{\omega} \text{ coinduction } \}\!\}$$
$$c \leq ac + b$$

$\square$

The above lemmas directly yield the following theorem.

Theorem 2.4. *Top-left-strict weak omega algebra and demonic refinement algebra are equivalent in the sense that they can be defined in terms of each other.*

3   The Demonic Refinement Algebra of Commands

So far, our semiring elements could be viewed as abstract representations of state transition systems. We now want to introduce a way of dealing with sets of states in an abstract algebraic way. This is done using tests. A *test semiring* is a structure $(\mathcal{S}, \text{test}(S))$, where $\mathcal{S} = (S, +, 0, \cdot, 1)$ is a semiring and $\text{test}(S)$ is a Boolean subalgebra of the interval $[0, 1] \subseteq S$ with $0, 1 \in \text{test}(S)$. Join and meet in $\text{test}(S)$ coincide with $+$ and $\cdot$, the complement is denoted by $\neg$, 0 is the least and 1 is the greatest element. Furthermore, this definition of test semiring coincides with the definition on Kleene algebras given in [11]. We use $a, b, \ldots$ for general semiring elements and $p, q, \ldots$ for tests.

On a test semiring we axiomatise a domain operator $^\lceil : S \to \text{test}(S)$ by

$$a \leq {}^\lceil a \cdot a , \tag{d1}$$
$${}^\lceil(pa) \leq p , \tag{d2}$$
$${}^\lceil(a^\lceil b) \leq {}^\lceil(ab) , \tag{d3}$$

for all $a \in S$ and $p \in \text{test}(S)$. Inequations (d1) and (d3) can be strengthened to equations. Many properties of domain can be found in [4]. For example, we have stability of tests and additivity of domain, i.e.,

$$\ulcorner p = p \, , \tag{2}$$
$$\ulcorner(a + b) = \ulcorner a + \ulcorner b \, . \tag{3}$$

With the aid of this operator, we can define modal operators by

$$|a\rangle p =_{df} \ulcorner(ap) \quad \text{and} \quad |a]p =_{df} \neg|a\rangle\neg p \, .$$

This is the reason why we shall call a test semiring with a domain operator *modal*. All the structures above extending a weak semiring are called *modal* when the underlying weak semiring is modal.

Given a modal semiring $\mathcal{S} = (S, +, 0, \cdot, 1)$ we define the set of commands (over S) as $\mathrm{COM}(S) =_{df} S \times \mathrm{test}(S)$. Three basic non-iterative commands and two basic operators on commands are defined by

$$\begin{aligned}
\mathsf{fail} &=_{df} (0, 1) \\
\mathsf{skip} &=_{df} (1, 1) \\
\mathsf{loop} &=_{df} (0, 0) \\
(a, p) [\!] (b, q) &=_{df} (a + b, pq) \\
(a, p) \, ; (b, q) &=_{df} (ab, p \cdot [a]q)
\end{aligned}$$

As noted by Möller and Struth in [14] the structure $(\mathrm{COM}(S), [\!], \mathsf{fail}, ; , \mathsf{skip})$ forms a weak semiring. The natural order on the command weak semiring is given by $(a, p) \leq (b, q) \Leftrightarrow a \leq b \wedge q \leq p$. We will discuss below how it connects to the usual refinement relation.

If $\mathcal{S}$ is even a weak Kleene algebra, a star operator can be defined by

$$(a, p)^* =_{df} (a^*, |a^*]p)$$

and then $(\mathrm{COM}(S), [\!], \mathsf{fail}, ; , \mathsf{skip},^*)$ forms a weak Kleene algebra [14].

Defining an omega operator over the set of commands does not work as simply as for star. To do this, we also need to assume that the underlying modal omega algebra $(S, +, 0, \cdot, 1,^*,^\omega)$ comes equipped with a convergence operator [14] $\triangle : S \rightarrow \mathrm{test}(S)$ satisfying

$$|a](\triangle a) \leq \triangle a \, , \tag{$\triangle$ unfold}$$
$$q \cdot |a]p \leq p \Rightarrow \triangle a \cdot |a^*]q \leq p \, . \tag{$\triangle$ induction}$$

In [14] it is shown that $\triangle a$ is the least (pre-)fixed point of $|a]$. The test $\triangle a$ characterises the states from which no infinite transition paths emanate. It corresponds to the halting predicate of the modal μ-calculus [8].

The infinite iteration operator on commands can then be defined by

$$(a, p)^\omega =_{df} (a^\omega, \triangle a \cdot [a^*]p) \, .$$

The greatest command is $\mathsf{chaos} =_{df} \mathsf{skip}^\omega = (\top, 0)$.

The semiring of commands reflects the view of general correctness as introduced in [17]. Therefore it is not to be expected that it forms a demonic refinement algebra which was designed for reasoning about total correctness. Indeed, top-left-strictness fails unless it is already satisfied in the underlying semiring $\mathcal{S}$, since $\mathsf{chaos}\,;(a,p) = (\top a, 0) = \mathsf{chaos}$ iff $\top a = \top$.

There is, however, another possibility. One can define a refinement preorder on commands by

$$(a,p) \sqsubseteq (b,q) \Leftrightarrow_{df} q \leq p \wedge qa \leq b .$$

This is the converse of the usual refinement relation: $k \sqsubseteq l$ for any two commands k, l means that k refines l. We have chosen this direction, since by straightforward calculation we get the implication $k \leq l \Rightarrow k \sqsubseteq l$. The associated equivalence relation $\equiv$ is defined by

$$k \equiv l \Leftrightarrow_{df} k \sqsubseteq l \wedge l \sqsubseteq k .$$

Componentwise, it works out to $(a,p) \equiv (b,q) \Leftrightarrow p = q \wedge pa = pb$. The equivalence classes correspond to the designs of the Unifying Theories of Programming of [9] and hence represent a total correctness view.

It has been shown in [7] (in the setting of condition semirings that is isomorphic to that of test semirings) that the set of these classes forms again a left semiring and can be made into a weak Kleene and omega algebra by using exactly the same definitions as above (as class representatives).

Now top-left-strictness holds, since $\mathsf{chaos} \equiv \mathsf{loop}$ and loop is a left zero by the definition of command composition. Therefore the set of $\equiv$-classes of commands can be made into a demonic refinement algebra. Let $\mathrm{CCOM}(S)$ be the set of all these classes.

By Lemma 2.2 the strong iteration of commands is

$$(a,p)^{\overline{\omega}} = (a,p)^* \,[\!]\, (a,p)^{\omega} ,$$

and thus $(\mathrm{CCOM}(S), [\!], \mathsf{fail}, ; , \mathsf{skip},^*,^{\overline{\omega}})$ constitutes a demonic refinement algebra of commands. The above expression can be simplified by

$$\begin{aligned}
&(a,p)^* \,[\!]\, (a,p)^{\omega} \\
=\quad &\{\!\!\{ \text{ definition of } ^* \text{ and } ^{\omega} \text{ on commands } \}\!\!\} \\
&(a^*, [a^*]p) \,[\!]\, (a^{\omega}, \triangle a \cdot [a^*]p) \\
=\quad &\{\!\!\{ \text{ definition of } [\!]\, \}\!\!\} \\
&(a^* + a^{\omega}, [a^*]p \cdot \triangle a \cdot [a^*]p) \\
=\quad &\{\!\!\{ \text{ definition of } ^{\overline{\omega}}, \text{ commutativity and idempotence of tests } \}\!\!\} \\
&(a^{\overline{\omega}}, \triangle a \cdot [a^*]p) .
\end{aligned}$$

Thus strong iteration of commands can also be expressed as

$$(a,p)^{\overline{\omega}} = (a^{\overline{\omega}}, \triangle a \cdot [a^*]p) .$$

We conclude this section by relating the command algebra to the demonic algebras (DA) of [3]. These are intended to capture the notion of total correctness in an algebraic fashion. Since their axiomatisation is extensive, we do not want to repeat it here. We only want to point out that a subalgebra of the command algebra yields a model of DA. This is formed by the $\equiv$-classes of *feasible* commands which are pairs (a, p) with $p \leq \ulcorner a$. So these model programs where no miraculous termination can occur; they correspond to the feasible designs of [9]. In [7] it is shown that the set $F(S)$ classes of feasible commands can isomorphically be represented by simple semiring elements. The mediating functions are

$$E : F(S) \rightarrow S , \qquad\qquad D : S \rightarrow F(S) ,$$
$$E((a, p)) =_{df} pa , \qquad\qquad D(a) =_{df} (a, \ulcorner a) .$$

Then one has $E(D(a)) = a$ and $D(E(a, p)) \equiv (a, p)$. Moreover, the demonic refinement ordering of [3] is induced on S by

$$a \sqsubseteq b \Leftrightarrow_{df} D(a) \sqsubseteq D(b) \Leftrightarrow \ulcorner b \leq \ulcorner a \wedge \ulcorner b \cdot a \leq b$$

and demonic join and composition by

$$a \sqcup b =_{df} E(D(a) \ [\!]\ D(b)) = \ulcorner a \cdot \ulcorner b \cdot (a + b) ,$$
$$a \circ b =_{df} E(D(a) ; D(b)) = |a|\ulcorner b \cdot a \cdot b .$$

Using pairs (p, p) as demonic tests in $F(S)$ one even obtains a DA with domain. Further details are left to a future publication.

4   Two Remarks on Refinement Algebra

In this section we remark that demonic refinement algebra does not characterise predicate transformer models uniquely. We also remark that an equivalence similar to that of Theorem 2.4 cannot be established between general refinement algebra [20] and a top-left-strict strong left omega algebra.

Characterisation of the predicate transformer models. To connect the algebra of commands to predicate transformer models we first define

$$\mathsf{wp}.(a, p).q =_{df} p \cdot [a]q$$

and get

$$\mathsf{wp.fail}.q = 1 \qquad \text{and} \qquad \mathsf{wp.chaos}.q = 0 .$$

Hence fail can be interpreted as magic in the refinement calculus tradition and chaos as abort. Indeed, chaos is refined by every command and every command is refined by fail. Furthermore, we have the implications, for commands k, l,

$$k \leq l \Rightarrow k \sqsubseteq l \Rightarrow (\forall p \in \mathsf{test}(S) \bullet \mathsf{wp}.k.p \geq \mathsf{wp}.l.p) .$$

However, the command model of demonic refinement algebra is, unlike predicate transformer models as presented in [19,20], in general not extensional in that we do not necessarily have the converse implications. In particular,

$$(\forall p \in \mathsf{test}(S) \bullet \mathsf{wp}.k.p = \mathsf{wp}.l.p) \Rightarrow k = l$$

holds iff already the underlying semiring S is *extensional*, i.e., satisfies, for $a, b \in S$,

$$[a] = [b] \Rightarrow a = b .$$

Contrarily, in concrete predicate transformer models the elements are mappings $T, U : \wp(\Sigma) \to \wp(\Sigma)$, where Σ is any set. They can be seen as semantic values that arise by applying the wp operator to concrete programming constructs. Their equality is defined by

$$T = U \Leftrightarrow_{df} (\forall p \in \wp(\Sigma) \bullet T.p = U.p) .$$

Hence in concrete predicate transformer models extensionality always holds.

Since the command model of DRA is non-extensional, this observation shows that the DRA axioms do not restrict their models to algebras isomorphic to predicate transformer algebras and hence do not uniquely capture this type of algebras.

A similar move for general refinement algebra? A *left Kleene algebra* is a left semiring extended with two axioms for *

$$1 + aa^* \leq a^* \qquad \text{and} \qquad b + ac \leq c \Rightarrow a^* b \leq c ,$$

laid down in Sect. 2. A *left omega algebra* is a left Kleene algebra extended with an infinite iteration operator ω axiomatised as in Sect. 2. Clearly, every left omega algebra has a greatest element $\top$, and along the lines above we call a left omega algebra *top-left-strict* when $\top$ satisfies $\top a = \top$. A *general refinement algebra* [20] is a left Kleene algebra extended with the axioms for $^{\overline{\omega}}$ found in Sect. 2, except the isolation axiom, i.e., $a^{\overline{\omega}} = a^* + a^{\overline{\omega}}0$ does not hold in general. A general refinement algebra becomes a demonic refinement algebra by adding the other two axioms for * of Sect. 2, right-distributivity and isolation.

It is tempting to try to show that top-left-strict left omega algebra corresponds to general refinement algebra in a similar way as top-left-strict weak omega algebra corresponds to demonic refinement algebra (Theorem 2.4). However, this is not possible as the following argument shows.

Let Σ be any set and let $T : \wp(\Sigma) \to (\Sigma)$ be any predicate transformer. If $p, q \in \wp(\Sigma)$ and T satisfies $p \subseteq q \Rightarrow T.p \subseteq T.q$ then T is *isotone*[1]. If T satisfies $T.(\bigcap_{i \in I} p_i) = \bigcap_{i \in I}(T.p_i)$, for any index set I, it is *conjunctive*. The isotone predicate transformers constitute a model for general refinement algebra [20]. The reason why isolation is dropped is that it does not hold for isotone predicate transformers in general [1,20]. Since isolation is an essential property needed for proving ω coinduction under the interpretation $a^{\overline{\omega}} =_{df} a^{\overline{\omega}}0$, it is not possible to prove that demonic refinement algebra subsumes top-left-strict strong left omega algebra. For the same reason, one cannot define strong iteration as $a^{\overline{\omega}} =_{df} a^* + a^{\omega}$ since this is valid only for conjunctive predicate transformers [1]. I.e., one cannot prove that top-left-strict strong left omega algebra subsumes general refinement algebra in an analogous way to the proof of Lemma 2.3.

---

[1] In the literature these predicate transformers are usually called monotone [1]. However, in other contexts the term monotone can mean isotone *or* antitone.

5 Making Convergence Explicit

In this section, we prove a result concerning the convergence operator of Sect. 3: having a convergence operator such that $\triangle a = \neg \ulcorner a^\omega$ is equivalent to having ω coinduction for the domain operator. Since $\triangle a = \neg \ulcorner a^\omega$ does not hold in all models of weak omega algebra [5], we also know that ω coinduction for domain does not follow from the axioms of omega algebra.

Proposition 5.1. *Omega coinduction for the domain operator, i.e.,*

$$p \leq \ulcorner(q + ap) \Rightarrow p \leq \ulcorner(a^\omega + a^*q) \ ,$$

holds if and only if $\triangle a = \neg \ulcorner a^\omega$ does.

Proof. The convergence operator is given by the implicit axiomatisation of Sect. 2. It is unique by the fact that it is a least fixpoint. We show that $\neg \ulcorner a^\omega$ always satisfies the $\triangle$ unfold axiom and that it satisfies the $\triangle$ induction axiom if and only if ω coinduction for the domain operator holds:

1. $\qquad |a]\neg\ulcorner a^\omega \leq \neg\ulcorner a^\omega$
   $\Leftrightarrow \quad$ { definition of $|\_]$ and Boolean algebra }
   $\qquad \neg|a\rangle\ulcorner a^\omega \leq \neg\ulcorner a^\omega$
   $\Leftrightarrow \quad$ { shunting }
   $\qquad \ulcorner a^\omega \leq \langle a\rangle\ulcorner a^\omega$
   $\Leftrightarrow \quad$ { definition of $|\_\rangle$ }
   $\qquad \ulcorner a^\omega \leq \ulcorner(a\ulcorner a^\omega)$
   $\Leftrightarrow \quad$ { (d3) }
   $\qquad \ulcorner a^\omega \leq \ulcorner(aa^\omega)$
   $\Leftrightarrow \quad$ { ω unfold }
   $\qquad \ulcorner a^\omega \leq \ulcorner a^\omega$
   $\Leftrightarrow \quad$ { reflexivity }
   $\qquad$ true

2. $\qquad q \cdot |a]p \leq p \Rightarrow \neg\ulcorner a^\omega \cdot [a^*]q \leq p$
   $\Leftrightarrow \quad$ { Boolean algebra }
   $\qquad \neg p \leq \neg|a]p + \neg q \Rightarrow \neg p \leq \ulcorner a^\omega + \neg|a^*]q$
   $\Leftrightarrow \quad$ { definition of $|\_]$ and Boolean algebra }
   $\qquad \neg p \leq |a\rangle\neg p + \neg q \Rightarrow \neg p \leq \ulcorner a^\omega + |a^*\rangle\neg q$
   $\Leftrightarrow \quad$ { definition of $|\_\rangle$ }
   $\qquad \neg p \leq \ulcorner(a\neg p) + \neg q \Rightarrow \neg p \leq \ulcorner a^\omega + \ulcorner(a^*\neg q)$
   $\Leftrightarrow \quad$ { set $\neg p = r$ and $\neg q = s$ }
   $\qquad r \leq \ulcorner(ar) + s \Rightarrow r \leq \ulcorner a^\omega + \ulcorner(a^*s)$
   $\Leftrightarrow \quad$ { (2) and (3) }
   $\qquad r \leq \ulcorner(ar + s) \Rightarrow r \leq \ulcorner(a^\omega + a^*s)$

Assume now that ω coinduction for the domain operator holds. By the above calculations $\neg\ulcorner a^\omega$ then satisfies both $\triangle$ unfold and $\triangle$ induction. Since these axioms impose uniqueness, we have that $\triangle a = \neg\ulcorner a^\omega$. If, conversely, $\triangle a = \neg\ulcorner a^\omega$ is assumed then the implication in the first line of the above calculation for 2. is true by $\triangle$ induction and hence ω coinduction for domain holds.   □

This means that in a command omega or demonic refinement algebra based on an omega algebra where ω coinduction for the domain operator holds, infinite and strong iteration can be defined as

$$(a,p)^\omega \ =_{df} \ (a^\omega, \neg\ulcorner a^\omega \cdot [a^*]p) \qquad \text{and} \qquad (a,p)^{\overline{\omega}} \ =_{df} \ (a^{\overline{\omega}}, \neg\ulcorner a^\omega \cdot [a^*]p) \ ,$$

respectively.

We finally note that the special case $q = 0$ of the ω coinduction rule for domain (Prop. 5.1) has been termed *cycle rule* and used as an additional postulate in the computation calculus of R. Dijkstra [6].

6   Conclusion

Top-left-strict omega algebra and demonic refinement algebra are equivalent in the sense that they can be defined in terms of each other. In particular, results from one of these frameworks can now be reused in the other. The equivalence also facilitates the definition of a demonic refinement algebra of commands, yielding a model in which extensionality does not hold. Since extensionality always holds in predicate-transformer models, it can be concluded that demonic refinement algebra does not characterise predicate transformers uniquely. A similar equality between general refinement algebra and top-left-strict left omega algebra as between demonic refinement algebra and top-left-strict weak omega algebra cannot be shown. The demonic refinement algebra and the omega algebra of commands are based on the convergence operator. In a modal demonic refinement or omega algebra that satisfies domain coinduction for infinite iteration, the convergence operator can be defined explicitly in terms of infinite iteration and domain.

Having set up the connections between various algebraic structures allows mutual re-use of the large existing body of results about Kleene/ω algebra with tests and modal Kleene/ω algebra as well as demonic refinement algebra and action systems. Having embedded the command algebras we can also apply the general algebraic results to UTP and related systems.

References

1. R.J. Back, J. von Wright: Refinement calculus: a systematic introduction. Springer 1998
2. E. Cohen: Separation and reduction. In R. Backhouse, J. Oliveira (eds.): Mathematics of Program Construction. LNCS 1837. Springer 2000, 45–59

3. J.-L. de Carufel, J. Desharnais: Demonic algebra with domain. In: R. Schmidt, G. Struth (eds.): Relations and Kleene Algebra in Computer Science. LNCS (this volume). Springer 2006 (to appear)

4. J. Desharnais, B. Möller, G. Struth: Kleene algebra with domain. Technical Report 2003-7, Universität Augsburg, Institut für Informatik, 2003. Revised version to appear in ACM TOCL

5. J. Desharnais, B. Möller, G. Struth: Termination in modal Kleene algebra. In J.-J. Lévy, E. Mayr, J. Mitchell (eds.): Exploring new frontiers of theoretical informatics. IFIP International Federation for Information Processing Series 155. Kluwer 2004, 653–666

6. R.M. Dijkstra: Computation calculus bridging a formalisation gap. Science of Computer Programming 37, 3-36 (2000)

7. W. Guttmann, B. Möller: Modal design algebra. In S. Dunne, B. Stoddart (eds.): Proc. First International Symposium on Unifying Theories of Programming. LNCS 4010. Springer 2006, 236–256

8. D. Harel, D. Kozen, J. Tiuryn: Dynamic Logic. MIT Press 2000

9. C.A.R. Hoare, J. He: Unifying theories of programming. Prentice Hall 1998

10. D. Kozen: A completeness theorem for Kleene algebras and the algebra of regular events. Inf. Comput. 110, 366–390 (1994)

11. D. Kozen: Kleene algebra with tests. ACM Transactions on Programming Languages and Systems 19, 427–443 (1997)

12. B. Möller: Lazy Kleene algebra. In D. Kozen (ed.): Mathematics of Program Construction. LNCS 3125. Springer 2004, 252–273. Revised version: B. Möller: Kleene getting lazy. Sci. Comput. Prog. (to appear)

13. B. Möller, G. Struth: Modal Kleene algebra and partial correctness. In C. Rattray, S. Maharaj, C. Shankland (eds.): Algebraic methodology and software technology. LNCS 3116. Springer 2004, 379–393. Revised and extended version: B. Möller, G. Struth: Algebras of modal operators and partial correctness. Theoretical Computer Science 351, 221–239 (2006)

14. B. Möller, G. Struth: wp is wlp. In W. MacCaull, M. Winter, I. Düntsch (eds.): Relational methods in computer Science. LNCS 3929. Springer 2006, 200-211

15. C. Morgan: Data Refinement by Miracles. Inf. Process. Lett. 26, 243-246 (1988)

16. J.M. Morris, Laws of data refinement, Acta Informatica (26), 287-308 (1989)

17. G. Nelson: A generalization of Dijkstra's calculus. ACM TOPLAS 11, 517–561 (1989)

18. K. Solin and J. von Wright: Refinement algebra with operators for enabledness and termination. In T. Uustalu (ed.): Mathematics of Program Construction. LNCS 4014. Springer 2006, 397–415

19. J. von Wright: From Kleene algebra to refinement algebra. In E. Boiten, B. Möiller (eds.): Mathematics of Program Construction. LNCS 2386. Springer 2002, 233–262

20. J. von Wright: Towards a refinement algebra. Sci. Comput. Prog. 51, 23–45 (2004)

Semigroupoid Interfaces for
Relation-Algebraic Programming in Haskell

Wolfram Kahl

McMaster University, Hamilton, Ontario, Canada
kahl@cas.mcmaster.ca

Abstract. We present a Haskell interface for manipulating finite binary relations as data in a point-free relation-algebraic programming style that integrates naturally with the current Haskell collection types. This approach enables seamless integration of relation-algebraic formulations to provide elegant solutions of problems that, with different data organisation, are awkward to tackle.

Perhaps surprisingly, the mathematical foundations for dealing with *finite* relations in such a context are not well-established, so we provide an appropriate generalisation of relational categories to semigroupoids to serve as specification for our interface.

After having established an appropriate interface for relation-algebraic programming, we also need an efficient implementation; we find this in BDD-based kernel library KURE of recent versions of the Kiel RelView system. We show how this combination enables high-level declarative and efficient relational programming in Haskell.

1   Introduction

After a small demonstration of relation-algebraic programming in Haskell, we give a quick overview over the programmer's interface to our relation library in Sect. 1.2. We then show in Sect. 1.3 how standard relation-algebraic theories are not an appropriate specification for this kind of relation library; as a solution to these problems, we define in Sect. 2 an appropriate generalisation of the relation-algebraic framework. In Sect. 3 we explain the options we offer concerning support from the Haskell type system for relational programming; we summarise our current implementation in Sect. 4.

1.1   Motivating Example: Dependency Graphs

One particularly useful visualisation aid when exploring software products is a tool that produces a module dependency graph. This is, as such, not very complex — most compilers can produce dependency output in a format that can be used by the make utility; it is easy to parse this and then produce the input format of a graph layout tool such as dot [9].

For extracting the dependency graph of the theorem prover Agda [5], which is implemented in the pure functional programming language Haskell [23], my tool HsDep [14] (also implemented in Haskell) takes 0.850 seconds (including the dependency-generating call to the Glasgow Haskell compiler); dot then takes an additional 45 seconds to produce a layout for this graph:

R.A. Schmidt (Ed.): RelMiCS /AKA 2006, LNCS 4136, pp. 235–250, 2006.
© Springer-Verlag Berlin Heidelberg 2006

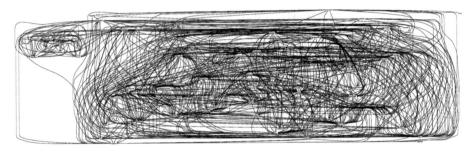

Even at considerable magnification, the usefulness of this graph drawing is rather limited. One naturally would like to see only the transitively irreducible kernel of this graph — the latest version of HsDep supports an additional flag, with which it produces that kernel in 0.952 seconds; dot then produces the following layout in less than half a second:

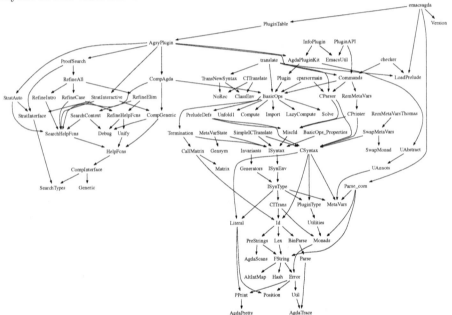

This new drawing is immediately useful (at appropriate magnification); obviously, reduction to the transitively irreducible kernel makes a real qualitative difference, and altogether even has negative run-time cost.

Most experienced Haskell programmers would find producing a tool like HsDep without kernel calculation to be a fairly straight-forward exercise, but would probably consider adding the kernel calculation to be quite a challenge.

In this paper, we show how to add an appropriate relation datatype to the libraries so that this challenge completely disappears.

For an example of the resulting programming style, let us look at the source of HsDep, which is a two-page literate Haskell document. It uses lists of pairs of strings to represent the dependency graph as parsed from the compiler-generated dependency file — this is completely natural in Haskell. Although calculating the transitively irreducible kernel of a finite relation is a simple relation-algebraic operation, it is rather non-trivial to implement this directly on the association list representation of the relation. Using conversion to an abstract datatype of finite relations, we achieve full realisation of the relation-algebraic simplicity in the Haskell source — the following is the central part of the HsDep source code:

> Then we parse the file contents into dependency pairs:
>
>   deps ← fmap parseDepFile $ readFile depfile
>
> If "-k" was given, we have to calculate for the dependency relation D the intransitive kernel $D \cap \overline{D_i D^+}$.
>
>   let deps' = if null kernel then deps
>   else let d :: Rel String String
>   d = fromList deps in toList (d —=—(d *** transClos d))
>
> The dependencies are then output as dot graph.
>
>   writeFile dotfile ∘ show ∘ dotOfDeps name $ deps'

Only the second code block here is concerned with the calculation of the intransitive part of the dependency graph; after seeing the mathematical formula for this, it should be easy to follow the corresponding Haskell expressions.

1.2   Overview of the Haskell Relation Library Data.Rel

The type *Rel a b* implements finite relations between *a* and *b*, where a relation can be understood as a set of pairs, and pair components are restricted to those elements on which the equality and ordering functions provided by the *Ord* interface terminate. This means roughly that *Rel a b* is intended to be used to establish relationships only between finite, fully defined elements of *a* and *b*, but no relationships with the (partially) undefined (and infinite) elements that also inhabit most Haskell datatypes.

The relation-algebraic part of HsDep in Sect. 1.1 is connected with the list part via the following conversion functions (a *constraint* "Ord a" expresses that a linear ordering needs to be available on type *a*):

$$fromList :: (Ord\ a,\ Ord\ b) \Rightarrow [(a,\ b)] \rightarrow Rel\ a\ b$$
$$toList\ \ :: (Ord\ a,\ Ord\ b) \Rightarrow Rel\ a\ b \rightarrow [(a,\ b)]$$

Besides these, the data type constructor *Rel* exposes a relation-algebraic interface, part of which is listed here (omitting the *Ord* constraints for readability).

(<==)	:: Rel a b → Rel a b → Bool	inclusion ⊆
(&&&)	:: Rel a b → Rel a b → Rel a b	meet (intersection)
(‖‖)	:: Rel a b → Rel a b → Rel a b	join (union)
(—=—)	:: Rel a b → Rel a b → Rel a b	difference
(***)	:: Rel a b → Rel b c → Rel a c	composition ⨾
(/*/)	:: Rel a c → Rel b c → Rel a b	restricted left residual
(*\)	:: Rel a b → Rel a c → Rel b c	restricted right residual

($\backslash\!\!\!\ast\!/$)	:: $Rel\ a\ b \rightarrow Rel\ a\ c \rightarrow Rel\ b\ c$	restricted symmetric quotient
converse	::	$Rel\ a\ b \rightarrow Rel\ b\ a$	converse $\breve{\_}$
symClos	::	$Rel\ a\ a \rightarrow Rel\ a\ a$	symmetric closure $R \cup R^{\breve{}}$
transClos	::	$Rel\ a\ a \rightarrow Rel\ a\ a$	transitive closure $\_^{+}$
dom	::	$Rel\ a\ b \rightarrow Rel\ a\ a$	domain dom
ran	::	$Rel\ a\ b \rightarrow Rel\ b\ b$	range ran
isUnivalent	::	$Rel\ a\ b \rightarrow Bool$	univalence predicate
isInjective	::	$Rel\ a\ b \rightarrow Bool$	injectivity predicate
isTransitive	::	$Rel\ a\ a \rightarrow Bool$	transitivity predicate
isSymmetric	::	$Rel\ a\ a \rightarrow Bool$	symmetry predicate
isAntisymmetric	::	$Rel\ a\ a \rightarrow Bool$	antisymmetry predicate

Besides these, we also provide some point-level functions following the naming and argument order conventions of *Data.Set* and *Data.Map*:

empty   :: $Rel\ a\ b$
member :: $a \rightarrow b \rightarrow Rel\ a\ b \rightarrow Bool$
insert   :: $a \rightarrow b \rightarrow Rel\ a\ b \rightarrow Rel\ a\ b$

1.3   Finite Relations Between Infinite Types

From the HsDep fragment in the previous sections, one naturally gains the impression that the formal foundations for this style of programming should be sought in relation algebra, with its tradition of concise, point-free formulations based on abstractly axiomatised operations on binary relations. We are obviously in the heterogeneous setting, which has partial composition in the same way as category theory. However, the finite relations accommodated by the type constructor *Rel* do *not* form a heterogeneous relation algebra, not even any kind of allegory [8] or category.

The problem lies deeper, and already occurs with finite sets and finite maps. All these finite collections are, in most programming languages, set in the context of infinite types. For example, even in strict object-oriented or functional languages, list types have infinitely many fully defined elements. In Haskell, also types like *Integer* are infinite. If the types *a* and *b* are infinite, then the types[1] *Set a*, *Map a b*, and *Rel a b* together with their algebraic interfaces do not satisfy the "standard" specifications of their unrestricted counterparts. In order to provide a background for the choices made in the rather complex case of relations, we first consider the simpler cases of sets and partial functions.

Finite subsets of an infinite set do *not* form a Boolean algebra: there is no largest finite set, and therefore the concept of complement does not make sense. Only relative complements exist — the function *Data.Set.difference*, which calculates $A - B = A \cap \overline{B}$, can be defined in several ways without recurring to

---

[1] *Set a* implements finite sets of elements of type *a*, and *Map a b* implements finite partial functions from type *a* to type *b*. Both require the key type *a* to have a linear ordering, specified via *Ord a* constraints. Both are provided, in the modules *Data.Set* respectively *Data.Map*, by the "hierarchical libraries" currently included in the distributions of most Haskell implementations.

complement, most typically as *pseudo-complement* of B in the sub-lattice of finite sets contained in A, i.e., the largest finite set $X \subseteq A$ for which $X \cap B = \varnothing$ hold. The problem of missing universal and complement sets of course immediately applies to sets of pairs, i.e., relations, too.

Turning our attention to finite maps, i.e., finite partial functions, we notice that, when infinite types are considered, these do not form a category, because identity functions on infinite types are not finite. Composition is still associative, so the appropriate algebraic structure is that of **semi-groupoids** (see for example [26, 27], and Def. 2.1.1 below), which are related to categories in the same way as semigroups are related to monoids, that is, identities are omitted.[2] It is interesting to note that most libraries of finite maps, including *Data.Map*, do not even provide map composition in their interface.

A (heterogeneous) relation algebra, which is a category where every homset is a Boolean algebra and some additional laws hold [24], can be obtained from any collection of sets by taking these sets as objects and arbitrary relations between them as morphisms. If we move to *finite* relations, then we again loose complements, largest elements, and identities, so where infinite sets are involved, we have no category of finite relations, only a semigroupoid.

Finiteness is also *not* preserved by the residuals of composition; this is obvious from the complement expressions that calculate these residuals in full relation algebras (shown in the right-most column):

$$X \subseteq (Q \backslash S) \quad \text{iff} \quad Q \mathbin{;} X \subseteq S \quad \text{right-residual} \quad S/R = \overline{\overline{S} \mathbin{;} R^{\smile}}$$
$$X \subseteq (S/R) \quad \text{iff} \quad X \mathbin{;} R \subseteq S \quad \text{left-residual} \quad Q \backslash S = \overline{Q^{\smile} \mathbin{;} \overline{S}}$$

Residuals are important since they provide the standard means to translate predicate logic formulae involving universal quantification into complement-free relational formalisms:

$$(x, y) \in S/R \quad \text{iff} \quad \forall z.(y, z) \in R \Rightarrow (x, z) \in S$$
$$(y, z) \in Q \backslash S \quad \text{iff} \quad \forall x.(x, y) \in Q \Rightarrow (x, z) \in S$$

Although to the newcomer to relational algebra, residuals may appear to be a rather strange construction, the fact that they are *the* tool to translate universal quantifications into relation-algebraic formulae frequently makes them indispensable for point-free formulations.

In the next section, we show how most relational formalisations can be adapted into a generalised framework that avoids these problems.

---

[2] I would have much preferred to use the name "semicategory", but this is already used for "categories with identities, but with only partial composition" [25], and the name "semi-groupoid" seems to be reasonably well-established in the mathematical literature.

2   Relational Semigroupoids

Above, we showed how set difference can be defined in a point-free way, without reference to the complement operation available only in the superstructure of arbitrary sets, only in terms of the theory of finite sets. This way of directly defining concepts that are better-known as derived in more general settings has the advantage that it guarantees a certain "inherent conceptual coherence".

In order to achieve this coherence also for the interface to our finite relation library, we now step back from concrete finite relations and consider instead a hierarchy of semigroupoid theories geared towards relational concepts in a similar way as Freyd and Scedrov's hierarchy of allegories [8] does this for category theory. Our exposition will, however, be structured more as a generalisation of the theory organisation of [13] from categories to semigroupoids.

This section serves simultaneously as formal specification for the point-free aspects of our interface to the relation datatypes in Haskell, and as (rather concise) introduction to reasoning using these tools.

2.1   Semigroupoids, Identities, and Categories

Definition 2.1.1. A **semi-groupoid** (Obj, Mor, src, trg, $\,\mathrm{\scriptstyle\circ}\,$) is a graph with a set Obj of objects as vertices, a set Mor of morphisms as edges, with src, trg : Mor $\to$ Obj assigning source and target object to each morphism (we write "$f : \mathcal{A} \to \mathcal{B}$" instead of "$f \in Mor$ and src $f = \mathcal{A}$ and trg $f = \mathcal{B}$), and an additional partial operation of composition such that the following hold:

- For $f : \mathcal{A} \to \mathcal{B}$ and $g : \mathcal{B}' \to \mathcal{C}$, the composition $f \,\mathrm{\scriptstyle\circ}\, g$ is defined iff $\mathcal{B} = \mathcal{B}'$, and if it is defined, then $(f \,\mathrm{\scriptstyle\circ}\, g) : \mathcal{A} \to \mathcal{C}$.
- Composition is associative, i.e., if one of $(f \,\mathrm{\scriptstyle\circ}\, g) \,\mathrm{\scriptstyle\circ}\, h$ and $f \,\mathrm{\scriptstyle\circ}\, (g \,\mathrm{\scriptstyle\circ}\, h)$ is defined, then so is the other and they are equal.

For two objects $\mathcal{A}$ and $\mathcal{B}$, the collections of morphisms $f : \mathcal{A} \to \mathcal{B}$ is also called the *homset* from $\mathcal{A}$ to $\mathcal{B}$, and written Hom$(\mathcal{A}, \mathcal{B})$. $\square$

A morphism is called an *endomorphism* iff its source and target objects coincide. In typed relation algebras, such morphisms are often called *homogeneous*. An endomorphism $R : \mathcal{A} \to \mathcal{A}$ is called *idempotent* if $R \,\mathrm{\scriptstyle\circ}\, R = R$.

Definition 2.1.2. In a semigroupoid, a morphism $I : \mathcal{A} \to \mathcal{A}$ is called an *identity on* $\mathcal{A}$ (or just an *identity*) iff it is a local left- and right-identity for composition, i.e., iff for all objects $\mathcal{B}$ and for all morphisms $F : \mathcal{A} \to \mathcal{B}$ and $G : \mathcal{B} \to \mathcal{A}$ we have $I \,\mathrm{\scriptstyle\circ}\, F = F$ and $G \,\mathrm{\scriptstyle\circ}\, I = G$.

If there is an identity on $\mathcal{A}$, we write it $\mathbb{I}_{\mathcal{A}}$.

A *category* is a semigroupoid where each object has an identity. $\square$

As in semigroups, identities are unique where they exist, and whenever we write $\mathbb{I}_{\mathcal{A}}$ without further comment, we imply the assumption that it exists.

2.2   Ordered Semigroupoids, Subidentities

A hallmark of the relational flavour of reasoning is the use of inclusion; the abstract variant of this directly corresponds to the categorical version:

Definition 2.2.1. A *locally ordered semigroupoid*, or just *ordered semigroupoid*, is a groupoid in which on each homset $\mathsf{Hom}(\mathcal{A}, \mathcal{B})$, there is an ordering $\sqsubseteq_{\mathcal{A},\mathcal{B}}$, and composition is monotonic in both arguments.   □

Some familiar concepts are available unchanged: In an ordered semigroupoid, a morphism $R : \mathcal{A} \to \mathcal{A}$ is called *transitive* iff $R \!;\! R \sqsubseteq R$, and *co-transitive* iff $R \sqsubseteq R \!;\! R$. The usual definitions of reflexivity and co-reflexivity, however, involve an identity; we work around this by essentially defining the concept "included in an identity" without actually referring to identity morphisms:

Definition 2.2.2. In an ordered semigroupoid, if for a morphism $p : \mathcal{A} \to \mathcal{A}$, and for all objects $\mathcal{B}$ and all morphisms $R : \mathcal{A} \to \mathcal{B}$ and $S : \mathcal{B} \to \mathcal{A}$, we have

- $p \!;\! R \sqsubseteq R$ and $S \!;\! p \sqsubseteq S$, then p is called a *subidentity*, and if we have
- $R \sqsubseteq p \!;\! R$ and $S \sqsubseteq S \!;\! p$, then p is called a *superidentity*.   □

In ordered categories (or monoids), subidentities are normally defined as elements included in identities, see e.g. [6]. If an object $\mathcal{A}$ in an ordered semigroupoid does have an identity $\mathbb{I}_{\mathcal{A}}$, then each subidentity $p : \mathcal{A} \to \mathcal{A}$ is indeed contained in the identity, since $p = p \!;\! \mathbb{I}_{\mathcal{A}} \sqsubseteq \mathbb{I}_{\mathcal{A}}$. Therefore, we also call subidentities *co-reflexive*, and, dually, we call superidentities *reflexive*.

If the homset $\mathsf{Hom}(\mathcal{A}, \mathcal{B})$ has a least element, then this will be denoted $\perp\!\!\!\perp_{\mathcal{A},\mathcal{B}}$. Existence of least morphisms is usually assumed together with the *zero law*:

Definition 2.2.3. An *ordered semigroupoid with zero morphisms* is an ordered semigroupoid such that each homset $\mathsf{Hom}(\mathcal{A}, \mathcal{B})$ has a least element $\perp\!\!\!\perp_{\mathcal{A},\mathcal{B}}$, and each least element $\perp\!\!\!\perp_{\mathcal{A},\mathcal{B}}$ is a left- and right-zero for composition.   □

In contexts where the inclusion ordering $\sqsubseteq$ is not primitive, but defined using meet, the meet-subdistributivity of composition is usually listed as an axiom; here it follows from monotonicity of composition:

Definition 2.2.4. A *lower semilattice semigroupoid* is an ordered semigroupoid such that each homset is a lower semilattice with binary meet $\sqcap$.   □

By demanding *strict* distributivity of composition over join, upper semilattice semigroupoids are *not* completely dual to lower semilattice semigroupoids:

Definition 2.2.5. An *upper semilattice semigroupoid* is an ordered semigroupoid such that each homset is an upper semilattice with binary join $\sqcup$, and composition distributes over joins from both sides.   □

If we consider *upper* or *lower semilattice semigroupoids with converse*, then the involution law for join respectively meet follows from isotony of converse.

2.3 Ordered Semigroupoids with Converse (OSGCs)

We now introduce converse into ordered semigroupoids *before* we consider join or meet, and we shall see that, even with only the basic axiomatisation of converse as an involution, the resulting theory is quite expressive.

Definition 2.3.1. An *ordered semigroupoid with converse (OSGC)* is an ordered semigroupoids such that each morphism $R : \mathcal{A} \to \mathcal{B}$ has a *converse* $R^{\smile} : \mathcal{B} \to \mathcal{A}$; conversion is monotonic with respect to $\sqsubseteq$, and for all $R : \mathcal{A} \to \mathcal{B}$ and $S : \mathcal{B} \to \mathcal{C}$, the *involution equations* $(R^{\smile})^{\smile} = R$ and $(R\,\mathbin{;}\,S)^{\smile} = S^{\smile}\,\mathbin{;}\,R^{\smile}$ hold. $\square$

Many standard properties of relations can be characterised in the context of OSGCs — not significantly hindered by the absence of identities. Those relying on superidentities are, of course, only of limited use in semigroupoids of finite relations between potentially infinite sets.

Definition 2.3.2. For a morphism $R : \mathcal{A} \to \mathcal{B}$ in an OSGC we define:

- R is *univalent* iff $R^{\smile}\,\mathbin{;}\,R$ is a subidentity.
- R is *injective* iff $R\,\mathbin{;}\,R^{\smile}$ is a subidentity.
- R is *difunctional* iff $R\,\mathbin{;}\,R^{\smile}\,\mathbin{;}\,R \sqsubseteq R$,
- R is *co-difunctional* iff $R \sqsubseteq R\,\mathbin{;}\,R^{\smile}\,\mathbin{;}\,R$.
- R is *total* iff $R\,\mathbin{;}\,R^{\smile}$ is a superidentity,
- R is *surjective* iff $R^{\smile}\,\mathbin{;}\,R$ is a superidentity,
- R is a *mapping* iff R is univalent and total,
- R is *bijective* iff R is injective and surjective. $\square$

All concrete relations, including all finite relations, are co-difunctional.

For endomorphisms, there are a few additional properties of interest:

Definition 2.3.3. For a morphism $R : \mathcal{A} \to \mathcal{A}$ in an OSGC we define:

- R is *symmetric* iff $R^{\smile} \sqsubseteq R$,
- R is a *partial equivalence* iff R is symmetric and idempotent.
- R is an *equivalence* iff R is reflexive, transitive, and symmetric. $\square$

In the categorical context, a number of connections between the properties introduced above has been shown in [13, Sect. 3.4]; is easy to see these all carry over directly to the semigroup-based definitions presented here.

2.4 Domain

Related to the introduction of "Kleene algebras with tests" [16], which allow the study of pre- and postconditions in a Kleene algebra setting, domain (and range) operators have been studied in the Kleene algebra setting [21, 6].[3] Much of the material there can be transferred into our much weaker setting of ordered semigroupoids by replacing preservation of joins with monotonicity and using our subidentity concept of Def. 2.2.2. The definition of "predomain" is given as a special residual of composition with respect to the ordering $\sqsupseteq$:

---

[3] It is important not to confuse these *domain* and *range* operations, which only make sense in *ordered* semigroupoids, with the semigroupoid (or categorical) concepts of source and target of a morphism!

Definition 2.4.1. An *ordered semigroupoid with predomain* is an ordered semi-groupoid where for every $R : \mathcal{A} \to \mathcal{B}$ there is a subidentity $\mathsf{dom}\,R : \mathcal{A} \to \mathcal{A}$ such that for every $X : \mathcal{A} \to \mathcal{A}$, we have $X \mathbin{;} R \sqsupseteq R$ iff $X \sqsupseteq \mathsf{dom}\,R$.

In an *ordered semigroupoid with domain*, additionally the "locality" condition $\mathsf{dom}\,(R \mathbin{;} S) = \mathsf{dom}\,(R \mathbin{;} \mathsf{dom}\,S)$ has to hold. □

Already in ordered semigroupoids with predomain, we have $(\mathsf{dom}\,R) \mathbin{;} R = R$. Range can be defined analogously; an OSGC with domain also has range, and range is then related with domain via converse: $\mathsf{ran}\,R = (\mathsf{dom}\,(R^{\smile}))^{\smile}$. (Without co-difunctionality, subidentities need not be symmetric.)

2.5   Kleene Semigroupoids

Kleene algebras are a generalisation of the algebra of regular languages; the typed version [17] is an extension of upper semilattice categories with zero morphisms. Since the reflexive aspect of the Kleene star is undesirable in semigroupoids, we adapt the axiomatisation by Kozen [15] to only the transitive aspect:

Definition 2.5.1. A *Kleene semigroupoid* is an upper semilattice semigroupoid with zero morphisms such that on homsets of endomorphisms there is an additional unary operation $\_^{+}$ which satisfies the following axioms for all $R : \mathcal{A} \to \mathcal{A}$, $Q : \mathcal{B} \to \mathcal{A}$, and $S : \mathcal{A} \to \mathcal{C}$:

$$R \sqcup R^{+} \mathbin{;} R^{+} = R^{+} \qquad \text{recursive definition}$$
$$Q \mathbin{;} R \sqsubseteq Q \;\Rightarrow\; Q \mathbin{;} R^{+} \sqsubseteq Q \qquad \text{right induction}$$
$$R \mathbin{;} S \sqsubseteq S \;\Rightarrow\; R^{+} \mathbin{;} S \sqsubseteq S \qquad \text{left induction} \qquad □$$

It is interesting to note that the only change from Kozen's definition is the omission of the join with the identity from the left-hand side of the recursive definition; Kozen also states the induction laws with inclusions in the conclusion, although for reflexive transitive closure, equality immediately ensues. This is not the case here, so this definition of transitive closure is in some sense more "satisfactory" than the the reflexive transitive variant.

While transitive closure of concrete relations does preserve finiteness, this is not the case for language-based models, so the usefulness of Kleene semi-groupoids as such may be limited.

Definition 2.5.2. A *Kleene semigroupoid with converse* is a Kleene semi-groupoid that is at the same time an OSGC, and the involution law for transitive closure holds: $(R^{+})^{\smile} = (R^{\smile})^{+}$. □

In Kleene semigroupoids with converse, difunctional closures always exist, and can be calculated as $R^{\boxplus} := R \sqcup (R \mathbin{;} R^{\smile})^{+} \mathbin{;} R$.

2.6   Semi-allegories

In direct analogy with allegories and distributive allegories, we define:

Definition 2.6.1. A *semi-allegory* is a lower semilattice semigroupoid with converse and domain such that for all $Q : \mathcal{A} \to \mathcal{B}$, $R : \mathcal{B} \to \mathcal{C}$, and $S : \mathcal{A} \to \mathcal{C}$, the *Dedekind rule* $Q \,\mathbin{\text{\textcent}}\, R \sqcap S \sqsubseteq (Q \sqcap S \,\mathbin{\text{\textcent}}\, R^{\smile}) \,\mathbin{\text{\textcent}}\, (R \sqcap Q^{\smile} \,\mathbin{\text{\textcent}}\, S)$ holds.

A *distributive semi-allegory* is a semi-allegory that is also an upper semilattice semigroupoid with zero morphisms. $\square$

The inclusion of domain is inspired by Gutiérrez' graphical calculus for allegories [7]; a contributing factor is that besides the Dedekind formula, in the absence of identities we also need domain to be able to show that, in semi-allegories, all morphisms are co-difunctional:

$$R = \operatorname{dom} R \,\mathbin{\text{\textcent}}\, R \sqcap R \sqsubseteq (\operatorname{dom} R \sqcap R \,\mathbin{\text{\textcent}}\, R^{\smile}) \,\mathbin{\text{\textcent}}\, R \sqsubseteq R \,\mathbin{\text{\textcent}}\, R^{\smile} \,\mathbin{\text{\textcent}}\, R \ .$$

In a semi-allegory we can define $R : \mathcal{A} \to \mathcal{A}$ to be *antisymmetric* iff $R^{\smile} \sqcap R$ is a subidentity.

2.7 Division Semi-allegories

We have seen in Sect. 1.3 that, in the interesting semigroupoid of finite relations between arbitrary sets, residuals do not in general exist. But we can define a set of *restricted* residuals that do exist for finite relations:

Definition 2.7.1. For morphisms $S : \mathcal{A} \to \mathcal{C}$ and $Q : \mathcal{A} \to \mathcal{B}$ and $R : \mathcal{B} \to \mathcal{C}$ in an OSGC with domain, we define:

- the *restricted right-residual* $Q \backslash\!\!\backslash S$ and the *restricted left-residual* $S /\!\!*\!/ R$:

$$Y \sqsubseteq (Q \backslash\!\!\backslash S) \quad \text{iff} \quad Q \,\mathbin{\text{\textcent}}\, Y \sqsubseteq S \quad \text{and} \quad \operatorname{dom} Y \sqsubseteq \operatorname{ran} Q \ ,$$
$$X \sqsubseteq (S /\!\!*\!/ R) \quad \text{iff} \quad X \,\mathbin{\text{\textcent}}\, R \sqsubseteq S \quad \text{and} \quad \operatorname{ran} X \sqsubseteq \operatorname{dom} R \ ,$$

- and the *restricted symmetric quotient* $Q \backslash\!\!*\!/ S := (Q \backslash\!\!\backslash S) \sqcap (Q^{\smile} /\!\!*\!/ S^{\smile})$ $\square$

For concrete relations, we then have (using infix notation for relations):

$$y(Q \backslash\!\!\backslash S)x \quad \text{iff} \quad \forall x . \, xQy \Rightarrow xSz \quad \text{and} \quad \exists x . \, xQy$$
$$x(S /\!\!*\!/ R)y \quad \text{iff} \quad \forall z . \, yRz \Rightarrow xSz \quad \text{and} \quad \exists z . \, yRz$$

For finite relations between (potentially) infinite types, this definition chooses the largest domain, respectively range, on which each residual is still guaranteed to be finite if its arguments are both finite. Where residuals exist, the restricted residuals can be defined using the unrestricted residuals:

$$Q \backslash\!\!\backslash S = \operatorname{ran} Q \,\mathbin{\text{\textcent}}\, (Q \backslash S) \ , \qquad S /\!\!*\!/ R = (S / R) \,\mathbin{\text{\textcent}}\, \operatorname{dom} R \ .$$

This "definedness restriction" essentially takes away from the standard residuals only the "uninteresting part", where the corresponding universally quantified formula is trivially true, and therefore is still useful in relational formalisations in essentially the same way as the "full" residuals. Therefore, we use these restricted residuals now for division semi-allegories:

Definition 2.7.2. A *division semi-allegory* is a distributive semi-allegory in which all restricted residuals exist. $\square$

We then complete the hierarchy of semigroupoids by joining the semi-allegory branch with the Kleene branch, adding pseudo-complements, and a "finite replacement" for largest elements in homsets:

Definition 2.7.3. A *Dedekind semigroupoid* is a division semi-allegory that is also a Kleene semigroupoid and has pseudo-complements, and where for any two subidentities $p : \mathcal{A} \to \mathcal{A}$ and $q : \mathcal{B} \to \mathcal{B}$ there is a largest morphism $_p\overline{\mathbb{T}}_q$ such that $p = \mathsf{dom}\left(_p\overline{\mathbb{T}}_q\right)$ and $q = \mathsf{ran}\left(_p\overline{\mathbb{T}}_q\right)$ □

2.8 Direct Products, Direct Sums, Direct Powers

For product and sum types we have the problem that the natural access relations, namely the projection respectively injection mappings, will be infinite for infinite types. Therefore, a formalisation that is useful in the context of finite relations between infinite types has to work without projections and injections. The natural starting point for such a formalisation are monoidal categories [19], which easily generalise to monoidal semigroupoids.

Since the details are beyond the scope of this paper, we only shortly indicate how we deal with product types. Since duplication and termination can again be infinite, we axiomatise finiteness-preserving "usage patterns" of the potentially infinite projection functions $\pi : \mathcal{A} \times \mathcal{B} \to \mathcal{A}$ and $\rho : \mathcal{A} \times \mathcal{B} \to \mathcal{B}$

- The fork operation as introduced in the context of relation algebras by Haeberer *et al.* [10] can be defined by $R \nabla S := R_{\,\flat}\pi^{\smallsmile} \sqcap S_{\,\flat}\rho^{\smallsmile}$.
- The "target projection" operations $P^\pi := P_{\,\flat}\pi$ and $P^\rho := P_{\,\flat}\rho$ also preserve finiteness.
- These can be axiomatised without projections by $(R \nabla S)^\pi = (\mathsf{dom}\, S)_{\,\flat} R$ and $(R \nabla S)^\rho = (\mathsf{dom}\, R)_{\,\flat} S$ and $(P \nabla Q)_{\,\flat} (R \nabla S)^{\smallsmile} = P_{\,\flat}R \sqcap Q_{\,\flat}S$.

In Dedekind semigroupoid with the monoidal product bifunctor and these operations, most product-related relational programming can be adequately expressed.

Direct sums with injections and direct powers with element relations are dealt with similarly; for the latter, the use of restricted residuals implies that set comprehension is typically restricted to non-empty sets.

3 Programming in Different Semigroupoids

There are three ways to situate the objects of the relation semigroupoid underlying a relation datatype with respect to the host language (Haskell) type system:

"Types as objects" guarantees full type safety.

"Sets as objects" offers finer granularity at the expense of dynamic compatibility checks for relations on possibly different subsets of the same types.

"Elements as objects" uses elements of a single datatype as objects, with no support from the type system for relation compatibility.

The last approach has been taken by the relation-algebraic experimentation toolkit RATH [11], and is motivated by a point of view that considers whole relation algebras as data items.

Here we are concerned with concrete relation algebraic operations on finite relations as a programming tool in a polymorphically typed programming language. In this context, both of the first two views have natural applications, so we support both, and we support a uniform programming style across the two views by organising all relational operations, including those listed in Sect. 1.2, into a wide range of Haskell type classes exported by *Data.Rel*, with even finer granularity than the hierarchy of definitions in Sect. 2, and supplemented by classes for the corresponding structures with identities, including categories, allegories, and relation algebras.

Exporting *all* relational operations as class members first of all makes relational programming implementation independent: Applications written only against these class interfaces can be used without change on any new implementation.

Providing a class hierarchy with very fine granularity in addition extends the scope of possible models that can be implemented; currently we only have implementations of concrete relations, but the machinery can easily be extended to, for example, relational graph homomorphisms [12], or fuzzy relations.

3.1   Types as Objects

It is quite obvious from the presentation of the *Data.Rel* interface that the choice of relation semigroupoid here is essentially the same as in the specification notations Z and B, where only certain sets are types: If *different* subsets $A_1, A_2 : \mathbb{P}\, A$ and $B_1, B_2 : \mathbb{P}\, B$ of two types A and B are given, the relations in $A_1 \leftrightarrow B_1$ are still considered as having the *same type* as the relations in $A_2 \leftrightarrow B_2$, namely the type $A \leftrightarrow B$. Therefore, if $R : A_1 \leftrightarrow B_1$ and $S : A_2 \leftrightarrow B_2$, writing for example $R \cap S$ is perfectly legal and well-defined.

This means that in this view we are operating in a relation semigroupoid that has only types as objects — we realise this in the *Rel* relation type constructor. This has the advantage that Haskell type checking implements semigroupoid morphism compatibility checking, so relation-algebraic *Rel* expressions are completely semigroupoid-type-safe. Since some Haskell types are infinite, *Rel* can implement only semigroupoid interfaces, up to Dedekind semigroupoids, but no category interfaces. Also, *Rel* can only provide pseudo-complements (difference), not complements, just like the de-facto-standard library module *Data.Set*.

3.2   Sets as Objects

The situation described above is different from the point of view taken by the category *Rel* which has *all sets* as objects. If an implementation wants to realise this point of view, then the empty relation $\varnothing : \{0,1\} \leftrightarrow \{0,1,2\}$, for example, must be different from the empty relation $\varnothing : \{0,1,2,3\} \leftrightarrow \{0\}$, since in a category or semigroupoid, source and target objects need to be accessible for every morphism, and operations on incompatible morphisms should not be defined.

Realising static morphism compatibility checking for this view would normally involve dependent types. One could also use Haskell type system extensions as implemented in GHC, the most popular compiler, to achieve most of this type safety, but the interface would definitely become less intuitive.

Realising this "arbitrary sets as objects" view in Haskell naturally uses *finite* subsets of types as objects; we provide this in the *SetRel* type constructor. This still has to resort to dynamic relation compatibility checking. This forces programmers either to move all relational computations into an appropriate monad, or to employ the common semigroupoid interface, where the operations provided by the *SetRel* implementation become partial, with possible run-time failures in the case of morphism incompatibility errors.

The "sets as objects" view has the advantage that the full relation algebra interface becomes available, and, in the BDD-based implementations, an implementation with partial operations can be realised with much lower overhead than the total operations of the "types as objects" view.

4   Implementation

The main reason why previously no significant relation library existed for Haskell is, in my opinion, that all "obvious" implementation choices inside the language are unsatisfactory.

More space- and time-efficient representations that also can make use of certain regularities in the structure to achieve more compact representations are based on binary decision diagrams (BDDs) [4]. Several BDD packages are freely available, but the only known Haskell implementations are still rather inefficient and incomplete. Even with a Haskell BDD library, or with a complete Haskell binding to an external BDD library, there still would be considerable way to go to implement relation algebraic operations; we are aware of two BDD-based implementations of relational operations: gbdd [22] is a C++ library providing relational operations using a choice of underlying BDD C libraries, and KURE [20] is the BDD-based kernel library of the RelView system [1]; KURE is written in C, and provides many special-purpose functions such as producing element relations between sets and their powersets [18]. Since C++ is notoriously hard to interface with Haskell, KURE remains as the natural choice for implementing *Data.Rel* with reasonable effort.

However, it turned out that producing a Haskell binding *KureRel* to KURE still was a non-trivial task, mainly because of heavily imperative APIs motivated by the graphical user interaction with RelView. In addition, RelView and KURE do not support relations where at least one dimension is zero; we take care of this entirely on the Haskell side.

On top of *KureRel*, we have implemented instances for (the appropriate parts of) the semigroupoid class hierarchy for three datatype constructors.

CRel is used for the relation algebra of finite relations between finite sets. For this, all provided interfaces have been implemented. A *CRel* is implemented as a triple consisting of two *Carriers* representing the source and target sets (together with eventual sum, product, or powerset structure), and one *KureRel* with the dimensions of the two carriers. *SetRel* is a special case of this, where both source and target are plain set carriers.

TypeRel is used for the Dedekind semigroupoid of finite relations between Haskell types. *Carriers* provide support for choices of sum and product, and *Rel* is the special case of *TypeRel* for unstructured carriers. The implementation of *TypeRel* is just a wrapper around *CRel*, and the implementations of the relational operations automatically generate adaptation injections as necessary.

NRel is used for finite relations between the sets $\underline{n}$ for $n \in \mathbb{N}$, where $\underline{0} = \varnothing$ and $\underline{n+1} = \{0, \ldots, n\}$. This gives rise to a relation algebra, but since no choice of products or sums is injective, the product- and sum-related classes cannot be implemented. *NRel* is a simple wrapper around *KureRel* that is necessary only for typing reasons.

FinMap, finally, is a first example of an implementation that is not based on *KureRel*; it is used for finite partial functions between Haskell types, which form a lower semilattice semigroupoid with domain, range, zero-morphisms, pseudo-complements, and a large part of the product and sum interface. *FinMap* uses *Data.Map.Map* for its implementation; since it uses $(,)$ and *Either* as choice for sum and product, some of the product and sum interfaces currently apparently cannot be implemented in Haskell for constraint propagation reasons.

Our library can be used interactively from the Haskell interpreter *GHCi*, which provides a very flexible environment for experimentation. For example, using a small utility function *classGraph* written using the Haskell syntax datatypes and parsing functions included with the GHC distribution, we can extract the subclass relation for the semigroupid classes of our library by passing its the relevant source file location, and then find out about the type of the produced relation, its numbers of nodes and edges, and, just as an example, display those edges that are the only incoming edges at their target *and* the only outgoing edges at their source, once producing a RelView-style bit matrix drawing, and once using dot to layout the produced subgraph; finally use a 3D graph layout algorithm to present relation *g1* in an OpenGL viewing window:

```
> cg <- classGraph ["Data/Rel/Classes.lhs"]
> :t cg
cg :: SetRel HsId HsId
> Carrier.size $ source cg
81
> length $ Data.Rel.toList cg
144
> gv $ tighten $ injectivePart cg &&& univalentPart cg
> dot $ tightenEndo $ injectivePart cg &&& univalentPart cg
> gl' g1
```

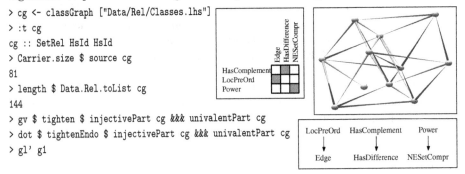

5   Conclusion

Starting from the insight that, with relations as data, the usual model is one of finite relations between both finite and infinite types, we showed that a hierarchy of relational theories based on semigroupoids instead of categories still captures essentially all expressivity of relation algebraic formalisations at only minimal cost of working around the absence of identities. We believe that, in this context, our axiomatisations of subidentities, restricted residuals, restricted top elements, and transitive closure are interesting contributions.

We used this hierarchy of theories to guide the design of a collection of Haskell type class interfaces, and provided implementations both for the rather intuitive "types as objects" view where we have to live with the absence of identities, and for a "finite sets as objects" view, where we have the full theory and interface of relation algebras at our disposal. These implementations of concrete relations are based on the efficient BDD routines of the RelView kernel library KURE.

To Haskell programmers, this offers a standard data type for finite relations that had been sorely missing, with an implementation that is so efficient that for many uses it will now be perfectly feasible to just write down a point-free relation-algebraic formulation, without spending any effort on selecting or developing a non-point-free algorithm which usually would be much less perspicuous. Even for hard problems this can be a viable method; Berghammer and Milanese describe how to implement a direct SAT solver in RelView, and report that this performs quite competitively for satisfiable problems [2]. It is straightforward to translate such RelView algorithms into Haskell using our library; this essentially preserves performance, and in many cases also adds type safety.

To those interested in programming with relations, we offer an interface to *the* state-of-the-art BDD-based relation-algebraic toolkit KURE in *the* state-of-the-art pure functional programming language Haskell. In comparison with for example the imperative special-purpose programming language of RelView, this has obvious advantages in flexibility, interoperability, and accessibility. Especially those who are mathematically inclined will feel more at home in Haskell than in the RelView programming language or in C or Java, which are the other alternatives for access to KURE.

Acknowledgements. I am grateful to Scott West for his collaboration in the implementation of the KURE Haskell binding, supported by an NSERC USRA and a McMaster Workstudy position, and to Shiqi Cao for his work on the OpenGL visualisation and 3D graph layout, supported by McMaster Engineering UROP and Workstudy positions, and to NSERC for additional funding.

References

[1] R. BERGHAMMER, T. HOFFMANN, B. LEONIUK, U. MILANESE. *Prototyping and Programming with Relations.* ENTCS **44**(3) 3.1–3.24, 2003.

[2] R. BERGHAMMER, U. MILANESE. *Relational Approach to Boolean Logic Problems.* In I. DÜNTSCH, W. MCCAULL, M. WINTER, eds., 8th Intl. Conf. Relational Methods in Computer Science, RelMiCS 8, LNCS **3929**. Springer, 2006.

[3] C. BRINK, W. KAHL, G. SCHMIDT, eds. *Relational Methods in Computer Science.* Advances in Computing Science. Springer, Wien, New York, 1997.

[4] R. E. BRYANT. *Graph-Based Algorithms for Boolean Function Manipulation.* IEEE Transactions on Computers **C-35**(8) 677–691, 1986.

[5] C. COQUAND. *Agda*, 2000. http://www.cs.chalmers.se/~catarina/agda/.

[6] J. DESHARNAIS, B. MÖLLER, G. STRUTH. *Kleene Algebra with Domain*. ACM Transactions on Computational Logic , 2006.

[7] D. DOUGHERTY, C. GUTIÉRREZ. *Normal Forms and Reduction for Theories of Binary Relations*. In L. BACHMAIR, ed., Rewriting Techniques and Applications, Proc. RTA 2000, LNCS **1833**, pp. 95–109. Springer, 2000.

[8] P. J. FREYD, A. SCEDROV. *Categories, Allegories*, North-Holland Mathematical Library **39**. North-Holland, Amsterdam, 1990.

[9] E. R. GANSNER, E. KOUTSOFIOS, S. C. NORTH, K.-P. VO. *A Technique for Drawing Directed Graphs*. IEEE-TSE **19** 214–230, 1993.

[10] A. HAEBERER *et al. Fork Algebras*. In [3], Chapt. 4, pp. 54–69.

[11] W. KAHL, G. SCHMIDT. *Exploring (Finite) Relation Algebras Using Tools Written in Haskell*. Technical Report 2000-02, Fakultät für Informatik, Universität der Bundeswehr München, 2000. http://ist.unibw-muenchen.de/relmics/tools/RATH/.

[12] W. KAHL. *A Relation-Algebraic Approach to Graph Structure Transformation*, 2001. Habil. Thesis, Fakultät für Informatik, Univ. der Bundeswehr München, Techn. Bericht 2002-03.

[13] W. KAHL. *Refactoring Heterogeneous Relation Algebras around Ordered Categories and Converse*. J. Relational Methods in Comp. Sci. **1** 277–313, 2004.

[14] W. KAHL. *HsDep: Dependency Graph Generator for Haskell*. Available at http://www.cas.mcmaster.ca/~kahl/Haskell/, 2004.

[15] D. KOZEN. *A Completeness Theorem for Kleene Algebras and the Algebra of Regular Events*. Inform. and Comput. **110**(2) 366–390, 1991.

[16] D. KOZEN. *Kleene Algebra with Tests*. ACM Transactions on Programming Languages and Systems pp. 427–443, 1997.

[17] D. KOZEN. *Typed Kleene Algebra*. Technical Report 98-1669, Computer Science Department, Cornell University, 1998.

[18] B. LEONIUK. *ROBDD-basierte Implementierung von Relationen und relationalen Operationen mit Anwendungen*. PhD thesis, Institut für Informatik und Praktische Mathematik, Christian-Albrechts-Universität Kiel, 2001.

[19] S. MAC LANE. *Categories for the Working Mathematician*. Springer-Verlag, 1971.

[20] U. MILANESE. *KURE: Kiel University Relation Package, Release 1.0*. http://www.informatik.uni-kiel.de/~progsys/relview/kure, 2004.

[21] B. MÖLLER. *Typed Kleene algebras*. Technical Report 1999-8, Institut für Informatik, Universität Augsburg, 1999.

[22] M. NILSSON. *GBDD — A package for representing relations with BDDs*. Available from http://www.regularmodelchecking.com/, 2004.

[23] S. PEYTON JONES et al. *The Revised Haskell 98 Report*. Cambridge Univ. Press, 2003. Also on http://haskell.org/.

[24] G. SCHMIDT, C. HATTENSPERGER, M. WINTER. *Heterogeneous Relation Algebra*. In [3], Chapt. 3, pp. 39–53.

[25] L. SCHRÖDER. *Isomorphisms and splitting of idempotents in semicategories*. Cahiers de Topologie et Géométrie Différentielle catégoriques **41** 143–153, 2000.

[26] B. TILSON. *Categories as algebra: an essential ingredient in the theory of monoids*. Journal of Pure and Applied Algebra **48** 83–198, 1987.

[27] P. WEIL. *Profinite methods in semigroup theory*. Intl. J. Algebra Comput. **12** 137–178, 2002.

On the Cardinality of Relations

Yasuo Kawahara

Department of Informatics, Kyushu University 33, Fukuoka 812-8581, Japan
kawahara@i.kyushu-u.ac.jp

Abstract. This paper will discuss and characterise the cardinality of boolean (crisp) and fuzzy relations. The main result is a Dedekind inequality for the cardinality, which enables us to manipulate the cardinality of the composites of relations. As applications a few relational proofs for the basic theorems on graph matchings, and fundamentals about network flows will be given.

1 Introduction

The obvious relationship between relations and graphs has been recoginsed by many researchers. Given a graph, the numbers of its nodes and edges, that is, the cardinalities of the sets of nodes and edges respectively, are fundamental data to analyse and characterise it. Since graphs are one of importnant data structure in computer science, their formal or computational study from a relational point of view is interesting for computer science. The book "Relations and Graphs" [4] by Schmidt and Ströhlein is an excellent exposition for the subjects in computer science and applied mathematics. Unfortunately, the cardinality of relations is treated rather implicitly or intuitively treated in the book. The aim of this paper is to find a law on the cardinality of relations, which enables us to solve problems on graphs and algorithms by relational methods. To achieve the subject the author recalls Dedekind formula (or the law of modularity [2]), namely the most significant law of relations, and has found out that an inequality, which we will call Dedekind inequality, effectively dominates the behaviors of cardinalities of fuzzy relations as well as boolean (crisp) relations. The soundness of formulation for cardinalities in the paper will be seen as characterisation Theorems 2 and 6.

The paper will be organised as follows. In Section 2 we prove the Dedekind inequality and some basic properties of the cardinality as consequences of Dedekind inequality. Also a characterisation of the cardinality of relations between finite sets will be given. In Section 3, Dedekind inequality will be applied to basic graph theory. That is, Hall's theorem and König's theorem will be demonstrated using the relational calculus. In Section 4 we recall some fundamantals on fuzzy relations. In Section 5 the cardinality of fuzzy relations between finite sets will be defined and a Dedekind inequality will be showed for fuzzy relations. Also we will show a characterisation of the cardinality of fuzzy relations between finite sets. In Section 6 we will try to give a relational framework for network flows. Finally another proof of Hall's thoerem is given as an application of relational theory of network flows.

R.A. Schmidt (Ed.): RelMiCS /AKA 2006, LNCS 4136, pp. 251–265, 2006.

2   Cardinality of Boolean Relations

In this section we intend to define and discuss the cardinality of relations. First we recall the elements on the cardinality of sets. For two sets A and B the notation $|A| = |B|$ means that there is a bijection from A onto B, and $|A| \leq |B|$ means that there is an injection from A into B. It is a basic fact that (1) $|A| \leq |A|$, (2) if $|A| \leq |B|$ and $|B| \leq |C|$ then $|A| \leq |C|$, (3) $|A| = |B|$ iff $|A| \leq |B|$ and $|B| \leq |A|$ (Bernstein and Schröder's Theorem). On the other hand the cardinality of a finite set may be regarded as a natural number.

We now review some relational notations and terminology used in this paper. A (boolean) relation $\alpha : X \rightharpoonup Y$ of a set X into a set Y is usually defined to be a subset of the cartesian product $X \times Y$. Let $\alpha, \alpha' : X \rightharpoonup Y$ and $\beta : Y \rightharpoonup Z$ be relations. The composite of α followed by β will be written as $\alpha\beta$. The join and the meet of α and α' are denoted by $\alpha \sqcup \alpha'$ and $\alpha \sqcap \alpha'$, respectively. The converse and the complement of α are presented by $\alpha^\sharp$ and α^-, respectively. Also $\mathrm{id}_X : X \rightharpoonup X, 0_{XY} : X \rightharpoonup Y$ and $\nabla_{XY} : X \rightharpoonup Y$ denote the identity relation, the least (or zero) relation and the greatest (or universal) relation, respectively. α is *univalent* if $\alpha^\sharp\alpha \sqsubseteq \mathrm{id}_Y$, and it is *total* if $\mathrm{id}_X \sqsubseteq \alpha\alpha^\sharp$. A *function* is a univalent and total relation. It will be introduced as $f : X \rightarrow Y$. (Univalent relations are often called *partial functions*. In this view a partial function is not always a function, unless it is total.) A function $f : X \rightarrow Y$ is called a *surjection* if $f^\sharp f = \mathrm{id}_Y$, and an *injection* if $ff^\sharp = \mathrm{id}_X$. In what follows, the word *boolean* relation is a synonym for relation on sets.

As a relation $\alpha : X \rightharpoonup Y$ is a subset of $X \times Y$, the cardinality $|\alpha|$ of α is defined to be its cadinality as a set. For relations $\alpha, \alpha' : X \rightharpoonup Y$ it trivially holds that (i) $|\alpha^\sharp| = |\alpha|$ and (ii) if $\alpha \sqsubseteq \alpha'$ then $|\alpha| \leq |\alpha'|$.

It is the most fundamental question to find out how the cardinality of relations behaves under the composition of relations. The next theorem answers the question by establishing an inequality for the cardinality of relations that we will call *Dedekind inequality*. The proof of Theorem 1 is the only non-calculational argument in the paper.

Theorem 1. *Let $\alpha : X \rightharpoonup Y$, $\beta : Y \rightharpoonup Z$ and $\gamma : X \rightharpoonup Z$ be relations. If α is univalent, i.e. if $\alpha^\sharp\alpha \sqsubseteq \mathrm{id}_Y$, then the following inequalities hold:*

$$|\beta \sqcap \alpha^\sharp\gamma| \leq |\alpha\beta \sqcap \gamma| \quad and \quad |\alpha \sqcap \gamma\beta^\sharp| \leq |\alpha\beta \sqcap \gamma|.$$

Proof. Let $(x, z) \in \alpha\beta \sqcap \gamma$. Then by the definition of composition of relations and the univalency of α, there exists a unique element $y \in Y$ such that $(x, y) \in \alpha \wedge (y, z) \in \beta$. Hence we have the following two mapping

$$\phi : \alpha\beta \sqcap \gamma \rightarrow \alpha \sqcap \gamma\beta^\sharp \quad and \quad \psi : \alpha\beta \sqcap \gamma \rightarrow \alpha^\sharp\gamma \sqcap \beta$$

by $\phi(x, z) = (x, y)$ and $\psi(x, z) = (y, z)$, respectively. It is clear that both of ϕ and ψ are surjective, which shows the desired inequalites.   □

A relation $\alpha : X \rightharpoonup Y$ is called a *matching* (or partial bijection) if $\alpha^\sharp\alpha \sqsubseteq \mathrm{id}_Y$ and $\alpha\alpha^\sharp \sqsubseteq \mathrm{id}_X$. Matchings are closed under composition and converse. Let I denote the singleton set $\{*\}$. It is readily seen that $\mathrm{id}_I = \nabla_{II}$ and $\nabla_{XI}\nabla_{IX} = \nabla_{XX}$ for all sets X.

The Dedekind inequality is very fruitful as we will see below:

Corollary 1. *Let $\alpha : X \rightharpoonup Y$, $\beta : Y \rightharpoonup Z$ and $\gamma : X \rightharpoonup Z$ be relations. Then the following holds:*

(a) *If α and β are univalent, then $|\alpha\beta \sqcap \gamma| = |\alpha \sqcap \gamma\beta^{\sharp}|$.*
(b) *If α is a matching, then $|\alpha\beta \sqcap \gamma| = |\beta \sqcap \alpha^{\sharp}\gamma|$.*
(c) *If α is univalent and β is a function, then $|\alpha\beta| = |\alpha|$.*
(d) *If α is a matching, then $|\alpha^{\sharp}\alpha\beta| = |\alpha\beta|$.*
(e) *If α is a matching, then $|\nabla_{IX}\alpha| = |\alpha|$.*
(f) *If $u \sqsubseteq \mathrm{id}_X$, then $|\nabla_{IX}u| = |u|$. In particular $|\nabla_{IX}| = |\mathrm{id}_X|\,(= |X|)$.*
(g) *If β is an injection, then $|\alpha| = |\alpha\beta|$.*
(h) *If α is an injection, then $|\nabla_{IX}| \leq |\nabla_{IY}|$.*

Proof. (a) Let α and β be univalent. Then we have

$$
\begin{aligned}
|\alpha \sqcap \gamma\beta^{\sharp}| &\leq |\alpha\beta \sqcap \gamma| && \{ \text{ Theorem 1, } \alpha : \text{univalent } \} \\
&= |\gamma^{\sharp} \sqcap \beta^{\sharp}\alpha^{\sharp}| && \{ \text{ (i), } (\alpha\beta \sqcap \gamma)^{\sharp} = \gamma^{\sharp} \sqcap \beta^{\sharp}\alpha^{\sharp} \} \\
&\leq |\beta\gamma^{\sharp} \sqcap \alpha^{\sharp}| && \{ \text{ Theorem 1, } \beta : \text{univalent } \} \\
&= |\alpha \sqcap \gamma\beta^{\sharp}|. && \{ \text{ (i), } (\beta\gamma^{\sharp} \sqcap \alpha^{\sharp})^{\sharp} = \alpha \sqcap \gamma\beta^{\sharp} \}
\end{aligned}
$$

(b) Let α be a matching. Then it follows that

$$
\begin{aligned}
|\beta \sqcap \alpha^{\sharp}\gamma| &\leq |\alpha\beta \sqcap \gamma| && \{ \text{ Theorem 1, } \alpha : \text{univalent } \} \\
&\leq |\alpha^{\sharp}\gamma \sqcap \beta|. && \{ \text{ Theorem 1, } \alpha^{\sharp} : \text{univalent } \}
\end{aligned}
$$

(c) Let α be univalent and β a function. It follows from the totality of β that $\nabla_{XZ}\beta^{\sharp} = \nabla_{XY}$. Hence we have

$$
\begin{aligned}
|\alpha\beta| &= |\alpha \sqcap \nabla_{XZ}\beta^{\sharp}| && \{ \text{ (a) } \} \\
&= |\alpha|. && \{ \nabla_{XZ}\beta^{\sharp} = \nabla_{XY} \}
\end{aligned}
$$

(d) Assume α is a matching. Then $\alpha^{\sharp}$ is a matching too and so

$$
\begin{aligned}
|\alpha^{\sharp}\alpha\beta| &= |\alpha\beta \sqcap \alpha\nabla_{YZ}| && \{ \text{ (b), } \alpha^{\sharp} : \text{matching } \} \\
&= |\alpha\beta|. && \{ \alpha\beta \sqsubseteq \alpha\nabla_{YZ} \}
\end{aligned}
$$

(e) Assume α is a matching. Then we have

$$
\begin{aligned}
|\nabla_{IX}\alpha| &= |\alpha^{\sharp}\nabla_{XI}| && \{ \text{ (i) } \} \\
&= |\alpha^{\sharp}| && \{ \text{ (c), } \alpha^{\sharp} : \text{univalent, } \nabla_{XI} : \text{function } \} \\
&= |\alpha|. && \{ \text{ (i) } \}
\end{aligned}
$$

(f) Every subidentity $u \sqsubseteq \mathrm{id}_X$ is a matching. Hence it holds by (e).
(g) Assume β is an injection. Then it follows that

$$
\begin{aligned}
|\alpha| &= |\alpha\beta\beta^{\sharp}| && \{ \mathrm{id}_Y = \beta\beta^{\sharp} \} \\
&= |\alpha\beta|. && \{ \text{ (d), } \beta : \text{matching } \}
\end{aligned}
$$

(h) Assume α is an injection. Noticing that ∇_{XI} is a function, we have

$$|\nabla_{IX}| = |\nabla_{IX}\alpha| \ \{ \ (c) \ \}$$
$$\leq |\nabla_{IY}|. \ \{ \ (ii), \nabla_{IX}\alpha \sqsubseteq \nabla_{IY} \ \}$$

$\square$

Now recall an interesting example given by Tarski [5]. Let $\alpha, \beta : X \rightharpoonup X$ be a pair of univalent relations with $\alpha^{\sharp}\beta = \nabla_{XX}$. Then it follows that

$$|\nabla_{XX}| = |\alpha^{\sharp}\beta \sqcap \nabla_{XX}| \qquad \{ \ \alpha^{\sharp}\beta = \nabla_{XX} \ \}$$
$$\leq |\beta \sqcap \alpha\nabla_{XX}| \qquad \{ \ \text{Dedekind inequality} \ \}$$
$$= |\beta^{\sharp}\text{id}_X \sqcap \nabla_{XX}\alpha^{\sharp}|$$
$$\leq |\text{id}_X \sqcap \beta\nabla_{XX}\alpha^{\sharp}| \ \{ \ \text{Dedekind inequality} \ \}$$
$$\leq |\text{id}_X|,$$

which means $|X \times X| \leq |X|$. But it is obvious that the last inequality holds only if X is the empty set, a singleton set or an infinite set. The example suggests that the validity of a relational formula

$$\alpha^{\sharp}\beta \sqcap [(\alpha^{\sharp}\alpha)^{-} \sqcup \text{id}] \sqcap [(\beta^{\sharp}\beta)^{-} \sqcup \text{id}] = \nabla$$

depends on the interpretation set X of relations.

Throughout the rest of the paper we assume that symbols X, Y, Z and V denote finite sets. Thus all relations $\alpha : X \rightharpoonup Y$ are finite sets and so the cardinality $|\alpha|$ can be represented by a natural number. For finite sets X and Y the set of all relations from X into Y will be denoted by $Rel(X, Y)$. The cardinalities of relations on finite sets are characterised as follows.

Theorem 2. *A family of mappings* $| \cdot | : Rel(X, Y) \rightarrow \mathbb{N}$ *defined for all pairs* (X, Y) *of finite sets coincides with the cardinality of relations iff it satisfies the following four conditions:*

(a) $|\alpha| = 0$ *iff* $\alpha = 0_{XY}$.
(b) $|\text{id}_I| = 1$ *and* $|\alpha^{\sharp}| = |\alpha|$.
(c) $|\alpha \sqcup \alpha'| = |\alpha| + |\alpha'| - |\alpha \sqcap \alpha'|$. *In particular* $\alpha \sqsubseteq \alpha'$ *implies* $|\alpha| \leq |\alpha'|$.
(d) *(Dedekind inequality) If* α *is univalent, then* $|\beta \sqcap \alpha^{\sharp}\gamma| \leq |\alpha\beta \sqcap \gamma|$ *and* $|\alpha \sqcap \gamma\beta^{\sharp}| \leq |\alpha\beta \sqcap \gamma|$ *hold.* $\square$

For the proof of the last theorem we note that every element of X can be identified with a point relation $\hat{x} : I \rightharpoonup X$ defined by $\hat{x} = \{(*, x)\}$. Note that all point relations $x = \hat{x} : I \rightharpoonup X$ are injections. And the so-called point axioms hold: (PA1) $x \sqcap x' = 0_{IX}$ iff $x \neq x'$, and (PA2) for all relations $\rho : I \rightharpoonup X$ an identity $\rho = \sqcup_{x \sqsubseteq \rho}x$ holds. In particular $\sqcup_{x \in X}x = \nabla_{IX}$. (Remark that the subscript $x \sqsubseteq \rho$ in (PA2) is an abbreviation of $x \in X$ and $x \sqsubseteq \rho$.) Note that by the point axiom a relation $\rho : I \rightharpoonup X$ bijectively corresponds to a subset S of X such that $S = \{x \in X \mid x \sqsubseteq \rho\}$.

The following proposition seems to be an expansion formula for the cardinality of relations and gives a proof of Theorem 2. (Cf. Proof of Theorem 6 in Section 4.)

Proposition 1. *Assume a family of mappings* $|\cdot| : Rel(X, Y) \to \mathbb{N}$ *defined for all pairs* (X, Y) *of finite sets satisfies all the conditions in Theorem 2. Then for relations* $\alpha : X \to Y$, $\rho, \rho_0, \rho_1 : I \to X$ *and* $\mu : I \to Y$ *the following holds:*

(a) $0 \le |\alpha| \le |\nabla_{XY}| = |X||Y|$.

(b) $|\alpha \sqcap x^{\sharp}\mu| = |x\alpha \sqcap \mu|$.

(c) $|\alpha \sqcap (\rho_0 \sqcup \rho_1)^{\sharp}\mu| = |\alpha \sqcap \rho_0^{\sharp}\mu| + |\alpha \sqcap \rho_1^{\sharp}\mu| - |\alpha \sqcap (\rho_0 \sqcap \rho_1)^{\sharp}\mu|$.

(d) $|\alpha \sqcap \rho^{\sharp}\mu| = \sum_{x \sqsubseteq \rho} |x\alpha \sqcap \mu| = \sum_{x \sqsubseteq \rho} \sum_{y \sqsubseteq \mu} |x\alpha y^{\sharp}|$.

(e) $|\alpha| = \displaystyle\sum_{x \in X} \sum_{y \in Y} |x\alpha y^{\sharp}|$.   □

Remark. (1) Let $\alpha : X \to Y$ be a relation. Then $(x, y) \in \alpha$ iff $x\alpha y^{\sharp} = \mathrm{id}_I$.

(2) There are exactly two relations on I, namely 0_{II} and $\mathrm{id}_I \,(= \nabla_{II})$.

3   Application to Graphs

Let $\alpha : X \to Y$ be a relation. A matching $f : X \to Y$ is called a *matching of* α if $f \sqsubseteq \alpha$. The following proposition shows a simple inequality deduced from the condition (c) in Theorem 2 and a fact that a matching is a partial bijection.

Proposition 2. *Let* $\alpha : X \to Y$ *be a relation and* $f : X \to Y$ *a matching of* α. *Then an inequality* $|f| \le |\nabla_{IX}| - (|\rho| - |\rho\alpha|)$ *holds for all relations* $\rho : I \to X$.

Proof. It follows from Dedekind formula that $\rho \sqcap \nabla_{IY} f^{\sharp} = \rho f f^{\sharp}$. Hence we have

$$|\rho \sqcap \nabla_{IY} f^{\sharp}| = |\rho f f^{\sharp}| \ \{ \ \rho \sqcap \nabla_{IY} f^{\sharp} = \rho f f^{\sharp} \ \}$$
$$= |\rho f|. \ \{ \text{ Corollary 1(d) } \}$$

and

$$|\nabla_{IY} f^{\sharp}| = |f^{\sharp}| \ \{ \text{ Corollary 1(e) } \}$$
$$= |f|. \ \{ \text{ (i) } \}$$

Therefore we have an inequality

$$|\nabla_{IX}| \ge |\rho \sqcup \nabla_{IY} f^{\sharp}| \qquad\qquad \{ \ \rho \sqcup \nabla_{IY} f^{\sharp} \sqsubseteq \nabla_{IX} \ \}$$
$$= |\rho| + |\nabla_{IY} f^{\sharp}| - |\rho \sqcap \nabla_{IY} f^{\sharp}|$$
$$= |\rho| + |f| - |\rho f|$$
$$\ge |\rho| + |f| - |\rho\alpha|. \qquad\qquad \{ \ f \sqsubseteq \alpha \ \}$$

□

Corollary 2. *Let* $\alpha : X \to Y$ *be a relation. For all matchings* $f : X \to Y$ *of* α *an inequality*

$$|f| \le |\nabla_{IX}| - \delta(\alpha)$$

holds, where $\delta(\alpha) = \max\{|\rho| - |\rho\alpha| \mid \rho : I \to X\}$.   □

The number $\delta(\alpha)$ is an integer with $0 \le \delta(\alpha) \le |\nabla_{IX}|$ (because, $|\rho| - |\rho\alpha| = 0$ when $\rho = 0_{IX}$). For example, $\delta(0_{XY}) = |\nabla_{IX}|$ and if α is a matching then $\delta(\alpha) = 0$.

Theorem 3 (Hall's Theorem). *Let $\alpha : X \to Y$ be a relation and X a non-empty set. Then there exists an injection (or total matching) $f : X \to Y$ with $f \sqsubseteq \alpha$ iff for all relations $\rho : I \to X$ an inequality $|\rho| \le |\rho\alpha|$ holds.*

Proof. (Necessity) Let f be an injection with $f \sqsubseteq \alpha$. Then it simply holds that

$$|\rho| = |\rho f| \quad \{ \text{ Corollary 1(g) } \}$$
$$\le |\rho\alpha|. \quad \{ f \sqsubseteq \alpha \}$$

(Sufficiency) Few traditional graph theoretic proofs are known. However in the paper we will give another proof using the existence of maximal flows at the end of Section 6.   □

The following theorem shows a relational version of the so-called König's theorem.

Theorem 4. *Let $\alpha : X \to Y$ be a relation. If $\delta(\alpha) > 0$, then there exists a maximal matching $f : X \to Y$ of α such that $|f| = |\nabla_{XI}| - \delta(\alpha)$.*

Proof. Choose a set Z with $|\nabla_{IZ}| = \delta(\alpha)$ and construct a coproduct $Y + Z$ of Y and Z together with injections $i : Y \to Y + Z$ and $j : Z \to Y + Z$. (Recall that $i^\sharp i \sqcup j^\sharp j = \mathrm{id}_{Y+Z}$, $ii^\sharp = \mathrm{id}_Y$, $jj^\sharp = \mathrm{id}_Z$ and $ij^\sharp = 0_{YZ}$.) Define a relation $\hat{\alpha} : X \to Y + Z$ by $\hat{\alpha} = \alpha i \sqcup \nabla_{XZ} j$. Then for all relations $\rho : I \to X$ we have

$$
\begin{aligned}
|\rho\hat{\alpha}| &= |\rho(\alpha i \sqcup \nabla_{XZ} j)| \quad && \{ \hat{\alpha} = \alpha i \sqcup \nabla_{XZ} j \} \\
&= |\rho\alpha| + |\rho\nabla_{XZ}| \quad && \{ |\rho\alpha i| = |\rho\alpha i i^\sharp| = |\rho\alpha| \} \\
&= |\rho\alpha| + |\nabla_{IZ}| \quad && \{ \rho \ne 0_{IX} \text{ implies } \rho\nabla_{XZ} = \nabla_{IZ} \} \\
&= |\rho\alpha| + \delta(\alpha) \quad && \{ |\nabla_{IZ}| = \delta(\alpha) \} \\
&\ge |\rho\alpha| + (|\rho| - |\rho\alpha|) \\
&= |\rho|.
\end{aligned}
$$

By Hall's theorem there is an injection $g : X \to Y + Z$ with $g \sqsubseteq \hat{\alpha}$. Now we set $f = gi^\sharp : X \to Y$. Then f is a matching, because

$$f^\sharp f = (gi^\sharp)^\sharp (gi^\sharp) = ig^\sharp gi^\sharp \sqsubseteq ii^\sharp = \mathrm{id}_Y,$$

$$ff^\sharp = (gi^\sharp)(gi^\sharp)^\sharp = gi^\sharp ig^\sharp \sqsubseteq gg^\sharp = \mathrm{id}_X.$$

Also we have

$$|\nabla_{IX}| = |g| = |g(i^\sharp i \sqcup j^\sharp j)| = |gi^\sharp| + |gj^\sharp|$$

and so

$$|f| = |gi^\sharp| = |\nabla_{IX}| - |gj^\sharp| \ge |\nabla_{IX}| - |\nabla_{IZ}| = |\nabla_{IX}| - \delta(\alpha).$$

On the other hand $|f| \le |\nabla_{IX}| - \delta(\alpha)$ holds by Corollary 2. Therefore $|f| = |\nabla_{IX}| - \delta(\alpha)$ and f is maximal.   □

4   Fuzzy Relations

First we recall some binary operations on the unit interval that are useful for defining opearations on fuzzy relations. The unit interval $[0, 1]$ is a set whose elements are all reals r with $0 \leq r \leq 1$. We use four binary operations "$\vee$", "$\wedge$", "$\rhd$" and "$\oplus$" on $[0, 1]$ defined by $a^{\bullet} = 0$ if $a = 0$ and $a^{\bullet} = 1$ otherwise, $a \vee b = \max\{a, b\}$, $a \wedge b = \min\{a, b\}$, $a \rhd b = \max\{0, a-b\}$ and $a \oplus b = \min\{1, a+b\}$ for all $a, b \in [0, 1]$. A real $a \in [0, 1]$ is called *boolean* if $a = 0$ or $a = 1$. It is easy to see the following basic facts:

Proposition 3. *For reals $a, b, c \in [0, 1]$ the following holds:*

(a) $a \leq a^{\bullet} = a^{\bullet\bullet}$ *and* $(a \rhd b) \wedge (b \rhd a) = 0$.
(b) $a \rhd b = (a \vee b) - b$. *In particular,* $a \rhd b = a - b$ *if* $b \leq a$.
(c) $a = (a \rhd b) \oplus (a \wedge b)$.
(d) *If* $a \leq c$ *and* $b \leq c \rhd a$, *then* $a \oplus b = a + b \leq c$.
(e) *If* a *is boolean, then* $a \wedge (b \rhd c) = (a \wedge b) \rhd (a \wedge c)$.
(f) *If* a *is boolean, then* $a \wedge (b \oplus c) = (a \wedge b) \oplus (a \wedge c)$.   □

Note that $1 \rhd a = 1 - a$, $a \oplus b = 1 \rhd ((1 \rhd a) \rhd (1 \rhd b))$, $a \wedge b = a \rhd (a \rhd b)$ and $a \vee b = 1 \rhd ((1 \rhd a) \rhd ((1 \rhd a) \rhd (1 \rhd b)))$ hold.

A fuzzy relation from X into Y, denoted by $\alpha : X \to Y$, is a function $\alpha : X \times Y \to [0, 1]$. Algebraic properties of fuzzy relations have been discussed in [3] and we prefer to use the same notations as in [3]. Now we define operations which have not been defined in [3] yet. For fuzzy relations $\alpha, \beta : X \to Y$ the following three relations $\alpha \rhd \beta, \alpha \oplus \beta, \alpha^{\bullet} : X \to Y$ are obtained by point-wise lifting as a standard construction:

(a) $\forall x \in X \forall y \in Y. (\alpha \rhd \beta)(x, y) = \alpha(x, y) \rhd \beta(x, y)$,
(b) $\forall x \in X \forall y \in Y. (\alpha \oplus \beta)(x, y) = \alpha(x, y) \oplus \beta(x, y)$,
(c) $\forall x \in X \forall y \in Y. \alpha^{\bullet}(x, y) = \alpha(x, y)^{\bullet}$.

A fuzzy relation $\alpha : X \to Y$ is called *boolean* (or *crisp*) if $\alpha(x, y)$ is boolean for all $x \in X$ and $y \in Y$. Remark that boolean relations are just relations as discussed in Section 2, because a function $\alpha : X \times Y \to \{0, 1\}$ can be identified with a subset of $X \times Y$.

Proposition 4. *Let $\alpha, \beta, \gamma : X \to Y$ and $\mu : V \to X$ be fuzzy relations. Then the following holds:*

(a) $\alpha \rhd \beta \sqsubseteq \alpha$, $\alpha \rhd 0_{XY} = \alpha$ *and* $(\alpha \rhd \beta) \sqcap (\beta \rhd \alpha) = 0_{XY}$.
(b) *If* $\alpha \sqsubseteq \gamma$ *and* $\beta \sqsubseteq \gamma \rhd \alpha$, *then* $\alpha \oplus \beta \sqsubseteq \gamma$.
(c) *If* γ *is boolean, then* $(\alpha \rhd \beta) \sqcap \gamma = (\alpha \sqcap \gamma) \rhd (\beta \sqcap \gamma)$ *and* $(\alpha \oplus \beta) \sqcap \gamma = (\alpha \sqcap \gamma) \oplus (\beta \sqcap \gamma)$.
(d) *If* μ *is boolean and univalent, then* $\mu(\alpha \rhd \beta) = (\mu\alpha) \rhd (\mu\beta)$ *and* $\mu(\alpha \oplus \beta) = (\mu\alpha) \oplus (\mu\beta)$.
(e) $\alpha \sqsubseteq \alpha^{\bullet} = \alpha^{\bullet\bullet}$, $0_{XY}{}^{\bullet} = 0_{XY}$, $\nabla_{XY}{}^{\bullet} = \nabla_{XY}$ *and* $\mathrm{id}_X{}^{\bullet} = \mathrm{id}_X$.
(f) $\alpha^{\sharp\bullet} = \alpha^{\bullet\sharp}$, $(\alpha \sqcap \beta)^{\bullet} = \alpha^{\bullet} \sqcap \beta^{\bullet}$, $(\alpha \sqcup \beta)^{\bullet} = \alpha^{\bullet} \sqcup \beta^{\bullet}$ *and* $(\alpha\beta)^{\bullet} = \alpha^{\bullet}\beta^{\bullet}$.   □

Let M be a natural number with $M \geq 2$. The set $\{\frac{j}{M-1} \mid j = 0, 1, \cdots, M-1\}$ with M elements will be denoted by B_M. A fuzzy relation $\alpha : X \to Y$ is called M-valued if $\alpha(x, y) \in B_M$ for all $x \in X$ and $y \in Y$. It is trivial that a fuzzy relation is 2-valued iff it is boolean. The following proposition is immediate from the definition of M-valued fuzzy relations.

Proposition 5. (a) M-valued fuzzy relations are closed under the join $\sqcup$, the meet $\sqcap$, the composition and the converse of fuzzy relations.
(b) For all reals $k \in B_M$ and boolean relations $\xi : X \to Y$ the semi-scalar multiplication $k \cdot \xi$ is an M-valued fuzzy relation.   □

5   The Cardinality of Fuzzy Relations

In this section the cardinality of fuzzy relations is defined and its fundamental properties are discussed.

Definition 1. The cardinality $|\alpha|$ of a fuzzy relation $\alpha : X \to Y$ is defined by

$$|\alpha| = \sum_{x \in X} \sum_{y \in Y} \alpha(x, y). \quad \text{(finite sum)}$$

Of course the cardinality $|\alpha|$ of a fuzzy relation α is given by a nonnegative real.   □

The following are the basic properties of the cardinality of fuzzy relations.

Proposition 6. Let $\alpha, \beta, \gamma : X \to Y$ be fuzzy relations and $x \in X$ and $y \in Y$. Then the following holds:

(a) $|x\alpha y^\sharp| = \alpha(x, y)$ and $x\alpha y^\sharp = \alpha(x, y) \cdot \mathrm{id}_I$.
(b) $|\alpha \rhd \beta| = |\alpha \sqcup \beta| - |\beta|$. In particular, if $\beta \sqsubseteq \alpha$ then $|\alpha \rhd \beta| = |\alpha| - |\beta|$.
(c) If $\alpha \sqsubseteq \gamma$ and $\beta \sqsubseteq \gamma \rhd \alpha$, then $|\alpha \oplus \beta| = |\alpha| + |\beta|$.

As the Dedekind inequality also holds for fuzzy relations, Proposition 1 is also valid for fuzzy relations.

Theorem 5. Let $\alpha : X \to Y$, $\beta : Y \to Z$ and $\gamma : X \to Z$ be fuzzy relations. If α is univalent, i.e. if $\alpha^\sharp \alpha \sqsubseteq \mathrm{id}_Y$, then the Dedekind inequality holds:

$$|\beta \sqcap \alpha^\sharp \gamma| \leq |\alpha\beta \sqcap \gamma| \quad \text{and} \quad |\alpha \sqcap \gamma\beta^\sharp| \leq |\alpha\beta \sqcap \gamma|.$$

Proof. Let α be univalent. Then it follows from the univalency $\alpha^\sharp \alpha \sqsubseteq \mathrm{id}_Y$ that

$$(\alpha^\sharp \alpha)(y, y') = \vee_{x \in X}[\alpha(x, y) \wedge \alpha(x, y')] \leq \mathrm{id}_Y(y, y').$$

Hence for each $x \in X$ there exists at most one $y \in Y$ such that $\alpha(x, y) > 0$, and so we have

$$|\alpha\beta \sqcap \gamma| = \sum_{x \in X, z \in Z} \vee_{y \in Y}[\alpha(x, y) \wedge \beta(y, z) \wedge \gamma(x, z)]$$
$$= \sum_{x \in X, y \in Y, z \in Z}[\alpha(x, y) \wedge \beta(y, z) \wedge \gamma(x, z)].$$

Remark that the inequality $|\alpha\beta \sqcap \gamma| \leq \sum_{x\in X, y\in Y, z\in Z}[\alpha(x,y) \wedge \beta(y,z) \wedge \gamma(x,z)]$ always holds. Therefore we have

$$|\alpha \sqcap \gamma\beta^\sharp| \leq \sum_{x\in X, y\in Y, z\in Z}[\alpha(x,y) \wedge \beta(y,z) \wedge \gamma(x,z)]$$
$$= |\alpha\beta \sqcap \gamma|,$$

and

$$|\beta \sqcap \alpha^\sharp\gamma| \leq \sum_{x\in X, y\in Y, z\in Z}[\alpha(x,y) \wedge \beta(y,z) \wedge \gamma(x,z)]$$
$$= |\alpha\beta \sqcap \gamma|.$$

□

It is immediate that Theorem 1 and Corollary 1 also hold for fuzzy relations. Similarly we characterise the cardinality of fuzzy relations as follows.

Theorem 6. *Let $\mathbb{R}^+$ be the set of all nonnegative reals. A family of mappings $|\cdot| : Rel(X,Y) \to \mathbb{R}^+$ defined for all pairs (X,Y) of finite sets gives the cardinality iff it satisfies the following five conditions:*

(a) $|\alpha| = 0$ *iff* $\alpha = 0_{XY}$.
(b) $|id_I| = 1$ *and* $|\alpha^\sharp| = |\alpha|$.
(c) $|\alpha \sqcup \alpha'| = |\alpha| + |\alpha'| - |\alpha \sqcap \alpha'|$.
(d) *If α is univalent, then the Dedekind inequalities $|\beta \sqcap \alpha^\sharp\gamma| \leq |\alpha\beta \sqcap \gamma|$ and $|\alpha \sqcap \gamma\beta^\sharp| \leq |\alpha\beta \sqcap \gamma|$ hold.*
(e) $|k \cdot \alpha| = k|\alpha|$ *for all* $k \in [0,1]$.

Proof. It is trivial that the cardinality satisfies the above five conditions. Conversely assume that a family of mappings $|\cdot|$ satisfies the five conditions. Then one easily understands that the family of mapppings $|\cdot|$ has all the properties proved in Proposition 1 as well as Theorem 1 and Corollary 1. Hence for all fuzzy relations $\alpha : X \to Y$ we have

$$|\alpha| = \sum_{x\in X}\sum_{y\in Y}|x\alpha y^\sharp| \qquad \{ \text{ Proposition 1(e) } \}$$
$$= \sum_{x\in X}\sum_{y\in Y}|\alpha(x,y) \cdot id_I| \{ \text{ Proposition 6(a) } \}$$
$$= \sum_{x\in X}\sum_{y\in Y}\alpha(x,y). \qquad \{ \text{ (e) and (b) } \}$$

This completes the proof. □

For a fuzzy relation $\alpha : X \to X$ the reflexive transitive closure $\alpha^* : X \to X$ of α is defined by

$$\alpha^* = \sqcup_{n\geq 0}\alpha^n,$$

where $\alpha^0 = id_X$ and $\alpha^{n+1} = \alpha^n\alpha$ for all natural numbers n.
    The next lemma shows a basic relationship between the reflexive transitive closure and paths (or flows) of relations.

Lemma 1. *Let $\alpha : X \to X$ be a fuzzy relation, and $s,t \in X$ two distinct elements of X. Then there exists a fuzzy relation $\xi : X \to X$ satisfying the following conditions:*

(a) $\xi \sqsubseteq \alpha$, $s\xi^{\sharp} = t\xi = 0_{IX}$ and $|s\xi| = |s\alpha^{*}t^{\sharp}|$.
(b) $|\xi \sqcap \rho_{0}^{\sharp}\nabla_{IX}| = |\xi^{\sharp} \sqcap \rho_{0}^{\sharp}\nabla_{IX}|$ for all boolean relations $\rho_{0} : I \to X$ such that
$\rho_{0} \sqsubseteq (s \sqcup t)^{-}$.
(c) $\xi \sqcap \xi^{\sharp} = 0_{XX}$.

Proof. Assume X has n elements. Then the fundamental argument shows that
$\alpha^{n} \sqsubseteq \bigsqcup_{0 \le j \le n-1} \alpha^{j}$ and hence $\alpha^{*} = \bigsqcup_{0 \le j \le n-1} \alpha^{j}$. Set $k = |s\alpha^{*}t^{\sharp}| = \alpha^{*}(s,t)$ (Cf.
Proposition 6(a)). Then there is a sequence $s = v_{0}, v_{1}, \cdots, v_{p} = t$ of distinct
elements of X such that $k = \bigwedge_{j=1}^{p} \alpha(v_{j-1}, v_{j})$. We now define a fuzzy relation
$\xi : X \to X$ by $\xi = k \cdot (\bigsqcup_{j=1}^{p} v_{j-1}^{\sharp} v_{j})$. Then (b) is clear and (a) follows from a
computation:

$$k \cdot (v_{j-1}^{\sharp} v_{j}) \sqsubseteq \alpha(v_{j-1}, v_{j}) \cdot (v_{j-1}^{\sharp} \, \mathrm{id}_{I} \, v_{j}) \; \{ \; k \le \alpha(v_{j-1}, v_{j}) \; \}$$
$$= v_{j-1}^{\sharp}[\alpha(v_{j-1}, v_{j}) \cdot \mathrm{id}_{I}] v_{j}$$
$$= v_{j-1}^{\sharp} v_{j-1} \, \alpha \, v_{j}^{\sharp} v_{j}$$
$$\sqsubseteq \alpha.$$

Condition (c) can be similarly obtained. □

6   Network Flows

Network flows are usually defined as directed graphs with edges labelled by reals
[1]. But their labels can be restricted to [0,1] without loss of generality. This
idea enables us to regard networks and flows as fuzzy relations, and to develop a
relational method for theory of network flows. It should be noticed that networks
treated here are just simple graphs.

Definition 2. A *network* N is a triple $(\alpha : X \to X, s, t)$ consisting of a fuzzy
relation $\alpha : X \to X$ and two distinct elements s and t of X such that

$$s\alpha^{\sharp} = 0_{XI}, \quad t\alpha = 0_{IX} \quad \text{and} \quad \alpha \sqcap \alpha^{\sharp} = 0_{XX}.$$

The relation α is the capacity relation of N, s is the source of N, and t the exit
of N. □

In the above definition the conditions $s\alpha^{\sharp} = 0_{XI}$ and $t\alpha = 0_{IX}$ intuitively mean
that the network has no capacity into a source and from a target. On the other
hand the last condition $\alpha \sqcap \alpha^{\sharp} = 0_{XX}$ may look like too strong. However, for an
arbitrary relation $\alpha : X \to X$ we can construct a relation $\hat{\alpha} = \alpha \rhd \alpha^{\sharp}$ satisfying
$\hat{\alpha} \sqcap \hat{\alpha}^{\sharp} = 0_{XX}$ and $\alpha = \hat{\alpha} \oplus (\alpha \sqcap \alpha^{\sharp})$. (Cf. Proposition 3 (a) and (c).) Flows of
the network are defined to be an assignment of amount of flow satisfying global
conservation within the capacity.

Definition 3. A *flow* φ of a network $N = (\alpha : X \to X, s, t)$ is a fuzzy relation
$\varphi : X \to X$ such that $\varphi \sqsubseteq \alpha$ and $|\varphi \sqcap \rho_{0}^{\sharp}\nabla_{IX}| = |\varphi^{\sharp} \sqcap \rho_{0}^{\sharp}\nabla_{IX}|$ for all boolean
relations $\rho_{0} : I \to X$ such that $\rho_{0} \sqsubseteq (s \sqcup t)^{-}$. □

The following proposition shows the basic property of network flows.

Proposition 7. *Let $N = (\alpha : X \multimap X, s, t)$ be a network. For each flow $\varphi : X \multimap X$ of N the identity $|s\varphi| = |t\varphi^\sharp|$ holds.*

Proof. Set $\rho_0 = (s \sqcup t)^-$. Then we have $s \sqcup t \sqcup \rho_0 = \nabla_{IX}$ and

$$
\begin{aligned}
|\varphi| &= |\varphi \sqcap (s \sqcup t \sqcup \rho_0)^\sharp \nabla_{IX}| && \{\ \nabla_{XX} = \nabla_{XI} \nabla_{IX}\ \} \\
&= |s\varphi| + |t\varphi| + |\varphi \sqcap \rho_0^\sharp \nabla_{IX}| && \{\ \text{Proposition 1(c)}\ \} \\
&= |s\varphi| + |\varphi \sqcap \rho_0^\sharp \nabla_{IX}|. && \{\ t\varphi = 0_{IX}\ \}
\end{aligned}
$$

Dually $|\varphi^\sharp| = |t\varphi^\sharp| + |\varphi^\sharp \sqcap \rho_0^\sharp \nabla_{IX}|$ is valid. Hence $|s\varphi| = |t\varphi^\sharp|$ holds, because $|\varphi| = |\varphi^\sharp|$ and $|\varphi \sqcap \rho_0^\sharp \nabla_{IX}| = |\varphi^\sharp \sqcap \rho_0^\sharp \nabla_{IX}|$. $\square$

Next we review several fundamental notions on network flows.

Definition 4. Let $N = (\alpha : X \multimap X, s, t)$ be a network.

(a) The value $val(\varphi)$ of a flow φ of N is defined by $val(\varphi) = |s\varphi| = |t\varphi^\sharp|$.
(b) A flow φ of N is *maximal* if $val(\psi) \le val(\varphi)$ holds for all flows ψ of N.
(c) A *cut* ρ of N is a boolean relation $\rho : I \multimap X$ such that $s \sqsubseteq \rho \sqsubseteq t^-$.
(d) A cut ρ of N is *minimal* if $|\alpha \sqcap \rho^\sharp \rho^-| \le |\alpha \sqcap \mu^\sharp \mu^-|$ for all cuts μ of N.
(e) A fuzzy relation $\varphi_\alpha : X \multimap X$ is defined by $\varphi_\alpha = (\alpha \rhd \varphi) \sqcup \varphi^\sharp$. $\square$

Proposition 8. *Let $N = (\alpha : X \multimap X, s, t)$ be a network. For all flows $\varphi : X \multimap X$ of N and all cuts $\rho : I \multimap X$ of N the identity*

$$
val(\varphi) = |\alpha \sqcap \rho^\sharp \rho^-| - |\varphi_\alpha \sqcap \rho^\sharp \rho^-|
$$

holds.

Proof. First note that $\varphi \sqsubseteq \alpha$ and $(\alpha \rhd \varphi) \sqcap \varphi^\sharp \sqsubseteq \alpha \sqcap \alpha^\sharp = 0_{XX}$. Since ρ is boolean it follows from Proposition 4(c) that

$$
\varphi_\alpha \sqcap \rho^\sharp \rho^- = [(\alpha \sqcap \rho^\sharp \rho^-) \rhd (\varphi \sqcap \rho^\sharp \rho^-)] \sqcup (\varphi^\sharp \sqcap \rho^\sharp \rho^-),
$$

and so from Proposition 6(b) that

$$
|\varphi_\alpha \sqcap \rho^\sharp \rho^-| = |\alpha \sqcap \rho^\sharp \rho^-| - |\varphi \sqcap \rho^\sharp \rho^-| + |\varphi^\sharp \sqcap \rho^\sharp \rho^-|.
$$

Set $\rho_0 = s^- \sqcap \rho$. Then we have

$$
\begin{aligned}
val(\varphi) &= |s\varphi| - |s\varphi^\sharp| \\
&\qquad \{\ s\varphi^\sharp \sqsubseteq s\alpha^\sharp = 0_{IX}\ \} \\
&= |\varphi \sqcap (s \sqcup \rho_0)^\sharp \nabla_{IX}| - |\varphi^\sharp \sqcap (s \sqcup \rho_0)^\sharp \nabla_{IX}| \\
&\qquad \{\ \text{Proposition 1(c)},\ |\varphi \sqcap \rho_0^\sharp \nabla_{IX}| = |\varphi^\sharp \sqcap \rho_0^\sharp \nabla_{IX}|\ \} \\
&= |\varphi \sqcap \rho^\sharp \nabla_{IX}| - |\varphi^\sharp \sqcap \rho^\sharp \nabla_{IX}| \\
&\qquad \{\ s \sqcup \rho_0 = \rho\ \} \\
&= |\varphi \sqcap \rho^\sharp \rho| + |\varphi \sqcap \rho^\sharp \rho^-| - |\varphi^\sharp \sqcap \rho^\sharp \rho| - |\varphi^\sharp \sqcap \rho^\sharp \rho^-| \\
&= |\varphi \sqcap \rho^\sharp \rho^-| - |\varphi^\sharp \sqcap \rho^\sharp \rho^-| \\
&\qquad \{\ |\varphi \sqcap \rho^\sharp \rho| = |\varphi^\sharp \sqcap \rho^\sharp \rho|\ \} \\
&= |\alpha \sqcap \rho^\sharp \rho^-| - (|\alpha \sqcap \rho^\sharp \rho^-| - |\varphi \sqcap \rho^\sharp \rho^-| + |\varphi^\sharp \sqcap \rho^\sharp \rho^-|) \\
&= |\alpha \sqcap \rho^\sharp \rho^-| - |\varphi_\alpha \sqcap \rho^\sharp \rho^-|.
\end{aligned}
$$
                    $\square$

Remark. $\varphi_\alpha \sqcap \rho^\sharp \rho^- = 0_{XX}$ iff $\rho\varphi_\alpha \sqcap \rho^- = 0_{IX}$ iff $\rho\varphi_\alpha \sqsubseteq \rho$.

For a network N there exist finitely many cuts $\rho : I \rightharpoonup X$, because X is finite, and consequently a minimal cut exists. The following lemma indicates a construction of a new greater flow when φ_α contains a flow.

Lemma 2. *Let $N = (\alpha : X \rightharpoonup X, s, t)$ be a network and φ a flow of N. If $\xi : X \rightharpoonup X$ is a fuzzy relation satisfying $\xi \sqsubseteq \varphi_\alpha$, $s\xi^\sharp = 0_{IX}$ and $|\xi \sqcap \rho_0^\sharp \nabla_{IX}| = |\xi^\sharp \sqcap \rho_0^\sharp \nabla_{IX}|$ for all boolean relations $\rho_0 : I \rightharpoonup X$ such that $\rho_0 \sqsubseteq (s \sqcup t)^-$, then a fuzzy relation*

$$\psi = [\varphi \rhd (\alpha \sqcap \xi^\sharp)] \oplus (\alpha \sqcap \xi)$$

is a flow of N such that $val(\psi) = val(\varphi) + |s\xi|$.

Proof. First note that

$$\xi^\sharp \sqcap \alpha \sqsubseteq (\alpha^\sharp \sqcup \varphi) \sqcap \alpha \quad \{ \, \xi \sqsubseteq \varphi_\alpha \sqsubseteq \alpha \sqcup \varphi^\sharp \, \}$$
$$= \varphi, \qquad\qquad\quad \{ \, \alpha \sqcap \alpha^\sharp = 0_{XX} \, \}$$

and

$$\xi \sqcap \alpha \sqsubseteq [(\alpha \rhd \varphi) \sqcup \varphi^\sharp] \sqcap \alpha \, \{ \, \xi \sqsubseteq \varphi_\alpha \, \}$$
$$= \alpha \rhd \varphi. \qquad\qquad \{ \, \varphi^\sharp \sqcap \alpha \sqsubseteq \alpha^\sharp \sqcap \alpha = 0_{XX} \, \}$$

Thus, since $\varphi \sqsubseteq \alpha$ and $\xi \sqcap \alpha \sqsubseteq \alpha \rhd \varphi$, it follows from Proposition 4(b) that

$$\psi \sqsubseteq \varphi \oplus (\xi \sqcap \alpha) \sqsubseteq \alpha.$$

Let $\rho_0 : I \rightharpoonup X$ be a boolean relation such that $\rho_0 \sqsubseteq (s \sqcup t)^-$ and set $\hat\rho_0 = \rho_0^\sharp \nabla_{IX}$. Next we will see that $|\psi \sqcap \hat\rho_0| = |\psi^\sharp \sqcap \hat\rho_0|$. As $\hat\rho_0$ is boolean, an equation

$$\psi \sqcap \hat\rho_0 = [(\varphi \sqcap \hat\rho_0) \rhd (\alpha \sqcap \xi^\sharp \sqcap \hat\rho_0)] \oplus (\alpha \sqcap \xi \sqcap \hat\rho_0)$$

holds by Proposition 4(c), and so

$$|\psi \sqcap \hat\rho_0| = |\varphi \sqcap \hat\rho_0| - |\alpha \sqcap \xi \sqcap \hat\rho_0| + |\alpha \sqcap \xi^\sharp \sqcap \hat\rho_0|$$

applying Proposition 4(e), since $\varphi \rhd (\alpha \sqcap \xi^\sharp) \sqsubseteq \alpha$, $\alpha \sqcap \xi \sqsubseteq \alpha \rhd [\varphi \rhd (\alpha \sqcap \xi^\sharp)]$ and $\alpha \sqcap \xi^\sharp \sqsubseteq \varphi$. Hence by using $\xi \sqsubseteq \alpha \sqcup \alpha^\sharp$ we have

$$|\psi \sqcap \hat\rho_0| - |\xi \sqcap \hat\rho_0| = |\varphi \sqcap \hat\rho_0| - |\alpha \sqcap \xi^\sharp \sqcap \hat\rho_0| - |\alpha^\sharp \sqcap \xi \sqcap \hat\rho_0|.$$

Dually $|\psi^\sharp \sqcap \hat\rho_0| - |\xi^\sharp \sqcap \hat\rho_0| = |\varphi^\sharp \sqcap \hat\rho_0| - |\alpha^\sharp \sqcap \xi \sqcap \hat\rho_0| - |\alpha \sqcap \xi^\sharp \sqcap \hat\rho_0|$ holds. Therefore $|\psi \sqcap \hat\rho_0| = |\psi^\sharp \sqcap \hat\rho_0|$ follows from $|\varphi \sqcap \hat\rho_0| = |\varphi^\sharp \sqcap \hat\rho_0|$ and $|\xi \sqcap \hat\rho_0| = |\xi^\sharp \sqcap \hat\rho_0|$. Finally we obtain

$$val(\psi) = |s\varphi| - |s(\alpha \sqcap \xi^\sharp)| + |s(\alpha \sqcap \xi)| \, \{ \, val(\psi) = |s\psi| \, \}$$
$$= |s\varphi| + |s(\alpha \sqcap \xi)| \qquad\qquad \{ \, s\xi^\sharp = 0_{IX} \, \}$$
$$= |s\varphi| + |s[(\alpha \sqcup \alpha^\sharp) \sqcap \xi]| \qquad \{ \, s\alpha^\sharp = 0_{IX} \, \}$$
$$= val(\varphi) + |s\xi|. \qquad\qquad\quad \{ \, \xi \sqsubseteq \alpha \sqcup \alpha^\sharp \, \}$$

$\square$

The next theorem [1] essentially due to Ford and Fulkerson (1956) characterises the maximality of network flows.

Theorem 7. *Let $N = (\alpha : X \rightarrow X, s, t)$ be a network and φ a flow of N. Then the following three statements are equivalent:*

(a) *φ is maximal,*
(b) *$t \sqcap s\varphi_\alpha^* = 0_{IX}$ (or equivalently $|s\varphi_\alpha^* t^\sharp| = 0$),*
(c) *There exists a cut ρ such that $val(\varphi) = |\alpha \sqcap \rho^\sharp \rho^-|$.*

Proof. (a)$\Rightarrow$(b) Assume φ is maximal and set $k = |s\varphi_\alpha^* t^\sharp|$ ($0 \le k \le 1$). From Lemma 1 there is a fuzzy relation $\xi : X \rightarrow X$ such that $\xi \sqsubseteq \varphi_\alpha$, $s\xi^\sharp = 0_{IX}$, $|s\xi| = k$ and $|\xi \sqcap \rho_0^\sharp \nabla_{IX}| = |\xi^\sharp \sqcap \rho_0^\sharp \nabla_{IX}|$ for all boolean relations $\rho_0 : I \rightarrow X$ such that $\rho_0 \sqsubseteq (s \sqcup t)^-$. By the last Lemma 2 a fuzzy relation

$$\psi = [\varphi \triangleright (\xi^\sharp \sqcap \alpha)] \oplus (\xi \sqcap \alpha)$$

is a flow of N with $val(\psi) = val(\varphi) + k$. As φ is maximal we have $k = 0$.

(b)$\Rightarrow$(c) Assume $t \sqcap s\varphi_\alpha^* = 0_{IX}$. Then a boolean relation $\rho = (s\varphi_\alpha^*)^\bullet$ is a cut of N, since $s \sqsubseteq s\varphi_\alpha^* \sqsubseteq \rho$ and $t \sqcap \rho = (t \sqcap s\varphi_\alpha^*)^\bullet = 0_{IX}$. Also it is easy to verify $\varphi_\alpha \sqcap \rho^\sharp \rho^- = 0_{IX}$ from

$$\rho\varphi_\alpha \sqsubseteq (s\varphi_\alpha^*)^\bullet \varphi_\alpha^\bullet = (s\varphi_\alpha^* \varphi_\alpha)^\bullet \sqsubseteq (s\varphi_\alpha^*)^\bullet = \rho$$

Therefore $val(\varphi) = |\alpha \sqcap \rho^\sharp \rho^-|$ holds by Proposition 8. (In fact ρ is minimal.)

(c)$\Rightarrow$(a) Assume a cut ρ satisfies $val(\varphi) = |\alpha \sqcap \rho^\sharp \rho^-|$ and let ψ be a flow. Then we have

$$\begin{aligned}
val(\psi) &= |\alpha \sqcap \rho^\sharp \rho^-| - |\psi_\alpha \sqcap \rho^\sharp \rho^-| \quad \{ \text{ Proposition 8 } \} \\
&\le |\alpha \sqcap \rho^\sharp \rho^-| \qquad\qquad\qquad \{ |\psi_\alpha \sqcap \rho^\sharp \rho^-| \ge 0 \} \\
&= val(\varphi),
\end{aligned}$$

which means that φ is maximal. $\square$

The following theorem is the so-called integral flow theorem.

Theorem 8. *A network $N = (\alpha : X \rightarrow X, s, t)$ over M-valued relations has an M-valued maximal flow φ.*

Proof. Construct a sequence of M-valued flows of N by the following algorithm: (I) Set $\varphi_0 = 0_{XX}$. It is trivial that φ_0 is an M-flow of N. (II) In the case that φ_i has already been defined. Set $k_i = |s[(\alpha \triangleright \varphi_i) \sqcup \varphi_i^\sharp]^* t^\sharp|$. If $k_i = 0$ then φ_i is maximal by Theorem 7. If $k_i > 0$ then by the same construction as in the proof (a)$\Rightarrow$(b) in Theorem 7 we have the next flow φ_{i+1} such that $val(\varphi_{i+1}) = val(\varphi_i) + k_i$. As $k_i \ge 1/M$ this algorithm terminates within $M|X|^2$ steps. $\square$

As promised in Section 3 we now give a sufficiency proof for Hall's Theorem.

[Proof of Hall's theorem] Let $\alpha : X \rightarrow Y$ be a boolean relation such that $|\rho_0| \le |\rho_0 \alpha|$ for all boolean relations $\rho_0 : I \rightarrow X$. First construct the coproduct $\hat{X} = I + X + Y + I$ together with injections $s : I \rightarrow \hat{X}$, $i : X \rightarrow \hat{X}$, $j :$

$Y \to \hat{X}$ and $t : I \to \hat{X}$ such that $ss^\sharp = tt^\sharp = \text{id}_I$, $ii^\sharp = \text{id}_X$, $jj^\sharp = \text{id}_Y$, $s^\sharp s \sqcup i^\sharp i \sqcup j^\sharp j \sqcup t^\sharp t = \text{id}_{\hat{X}}$, $si^\sharp = 0_{IX}$, $sj^\sharp = 0_{IY}$, $st^\sharp = 0_{II}$, $ij^\sharp = 0_{XY}$, $it^\sharp = 0_{XI}$, $jt^\sharp = 0_{YI}$. Set $\hat{\alpha} = s^\sharp \nabla_{IX} i \sqcup i^\sharp \alpha j \sqcup j^\sharp \nabla_{YI} t$. Then it is trivial that $s\hat{\alpha}^\sharp = 0_{I\hat{X}}$, $t\hat{\alpha} = 0_{I\hat{X}}$ and $\hat{\alpha} \sqcap \hat{\alpha}^\sharp = 0_{\hat{X}\hat{X}}$. Hence we have a network $N = (\hat{\alpha} : \hat{X} \to \hat{X}, s, t)$. By Theorem 8 there is a maximal flow φ of N which is boolean. On the other $s : I \to \hat{X}$ is a minimal cut of N with $|\hat{\alpha} \sqcap s^\sharp s^-| = |\nabla_{IX}|$. For an arbitrary cut $\rho = s \sqcup \rho_0 i \sqcup \rho_1 j : I \to \hat{X}$ of N with boolean relations $\rho_0 : I \to X$ and $\rho_1 : I \to Y$, we have $\rho^- = \rho_0^- i \sqcup \rho_1^- j \sqcup t$ and

$$\hat{\alpha} \sqcap \rho^\sharp \rho^- = s^\sharp \rho_1^- i \sqcup i^\sharp (\alpha \sqcap \rho_0^\sharp \rho_1^-) j \sqcup j^\sharp \rho_1^\sharp t,$$

and so

$$
\begin{aligned}
|\hat{\alpha} \sqcap \rho^\sharp \rho^-| &= |s^\sharp \rho_0^- i| + |i^\sharp (\alpha \sqcap \rho_0^\sharp \rho_1^-) j| + |j^\sharp \rho_1^\sharp t| && \\
&= |\rho_0^-| + |\alpha \sqcap \rho_0^\sharp \rho_1^-| + |\rho_1^\sharp| && \{\, s, i, j, t : \text{injections} \,\} \\
&\geq |\rho_0^-| + |\rho_0 \alpha \sqcap \rho_1^-| + |\rho_1| && \{\, \text{Dedekind inequality} \,\} \\
&= |\rho_0^-| + |(\rho_0 \alpha \sqcap \rho_1^-) \sqcup \rho_1| && \{\, \text{Theorem 6(c)} \,\} \\
&\geq |\rho_0^-| + |\rho_0 \alpha| && \{\, (\rho_0 \alpha \sqcap \rho_1^-) \sqcup \rho_1 \sqsupseteq \rho_0 \alpha \,\} \\
&\geq |\rho_0^-| + |\rho_0| && \{\, |\rho_0| \leq |\rho_0 \alpha| \,\} \\
&= |\nabla_{IX}| && \\
&= |\nabla_{IX} i| && \{\, i : \text{injection} \,\} \\
&= |s\hat{\alpha} \sqcap s^-| && \{\, \nabla_{IX} i = s\hat{\alpha} \sqcap s^- \,\} \\
&= |\hat{\alpha} \sqcap s^\sharp s^-|. && \{\, s : \text{injection} \,\}
\end{aligned}
$$

Thus s is a minimal cut and $val(\varphi) = |\hat{\alpha} \sqcap s^\sharp s^-| = |\nabla_{IX}|$ by Theorem 7. Set $f = i\varphi j^\sharp : X \to Y$. It is clear that $f \sqsubseteq \alpha$. Also one can easily see that $|xf| = 1$ for all $x \in X$ and $|yf^\sharp| \leq 1$ for all $y \in Y$, as follows:

(1) $\varphi(s, xi) = 1$;
$$
\begin{aligned}
|\nabla_{IX}| &= val(\varphi) && \\
&= |s\varphi| && \{\, val(\varphi) = |s\varphi| \,\} \\
&= |s\varphi \sqcap \nabla_{IX} i| && \{\, s\varphi \sqsubseteq s\hat{\alpha} \sqsubseteq \nabla_{IX} i \,\} \\
&= |s\varphi i^\sharp| && \{\, i : \text{matching} \,\} \\
&= \textstyle\sum_{x \in X} \varphi(s, xi). && \{\, \varphi(s, xi) = 0 \text{ or } 1 \,\}
\end{aligned}
$$

(2) $fj = i\varphi$; $i\varphi = i\varphi \sqcap \alpha j \sqsubseteq (i\varphi j^\sharp \sqcap \alpha) j \sqsubseteq i\varphi j^\sharp j \sqsubseteq i\varphi$.

(3) $|xf| = 1$;
$$
\begin{aligned}
|xf| &= |xfj| && \{\, j : \text{injection} \,\} \\
&= |xi\varphi| && \{\, fj = i\varphi \,\} \\
&= |\varphi(xi)^\sharp| && \{\, \varphi : \text{flow} \,\} \\
&= |(s^\sharp s \sqcup i^\sharp i \sqcup j^\sharp j \sqcup t^\sharp t)\varphi(xi)^\sharp| && \{\, \text{id}_{\hat{X}} = s^\sharp s \sqcup i^\sharp i \sqcup j^\sharp j \sqcup t^\sharp t \,\} \\
&= |s\varphi(xi)^\sharp| && \{\, \varphi i^\sharp \sqsubseteq s^\sharp \nabla_{IX} \,\} \\
&= \varphi(s, xi) && \\
&= 1. &&
\end{aligned}
$$

(4) $|yf^{\sharp}| \leq 1;$ $|fy^{\sharp}| = |fjj^{\sharp}y^{\sharp}|$ { j : injection }
$\phantom{(4) |yf^{\sharp}| \leq 1; |fy^{\sharp}|} = |i\varphi j^{\sharp}y^{\sharp}|$ { (2) }
$\phantom{(4) |yf^{\sharp}| \leq 1; |fy^{\sharp}|} = |i^{\sharp}i\varphi(yj)^{\sharp}|$ { i : injection }
$\phantom{(4) |yf^{\sharp}| \leq 1; |fy^{\sharp}|} \leq |\varphi(yj)^{\sharp}|$ { $i^{\sharp}i \sqsubseteq \mathrm{id}_{\hat{X}}$ }
$\phantom{(4) |yf^{\sharp}| \leq 1; |fy^{\sharp}|} = |yj\varphi|$ { φ : flow }
$\phantom{(4) |yf^{\sharp}| \leq 1; |fy^{\sharp}|} \leq |t|$ { $yj\varphi \sqsubseteq yj\hat{a} = y\nabla_{YI}t \sqsubseteq t$ }
$\phantom{(4) |yf^{\sharp}| \leq 1; |fy^{\sharp}|} = 1.$

This completes the proof.   □

7   Conclusion

This paper proposed Dedekind inequalities for the cardinality of boolean and fuzzy relations, and illustrated applications to graphs and network flows. Also we reviewed Tarski's example for decision problem on relational formulas.

Future work is to study on proof mechanisms for the cardinality of relations and to look for more applications in mathematics and computer science, for example to greedoids and electrical circuits.

Acknowledgements. The author is grateful to Georg Struth and anonymous referees for helpful discussions and comments.

References

1. R. Diestel, Graph theory, Graduate texts in mathematics **173**, Third Edition, Springer, Berlin, 2005.
2. P. Freyd and A. Scedrov, Categories, allegories, North-Holland, Amsterdam, 1990.
3. Y. Kawahara and H. Furusawa, An algebraic formalization of fuzzy relations. Fuzzy Sets and Systems **101** (1999), 125 - 135.
4. G. Schmidt and T. Ströhlein, Relations and graphs – Discrete Mathematics for Computer Science – (Springer-Verlag, Berlin, 1993).
5. A. Tarski, Some metalogical results concerning the calculus of relations, Journal of Symbolic Logic, **18** (2) (1953) 188–189.

Evaluating Sets of Search Points Using Relational Algebra

Britta Kehden

Christian-Albrechts-Univ. of Kiel, 24098 Kiel, Germany
bk@informatik.uni-kiel.de

Abstract. We model a set of search points as a relation and use relational algebra to evaluate all elements of the set in one step in order to select search points with certain properties. Therefore we transform relations into vectors and prove a formula to translate properties of relations into properties of the corresponding vectors. This approach is applied to timetable problems.

1 Introduction

Randomized search heuristics have found many applications in solving combinatorial optimization problems in recent years. This class of heuristics contains well-known approaches such as Randomized Local Search, the Metropolis Algorithm, Simulated Annealing and Evolutionary Algorithms. Such heuristics are often applied to problems whose structure is not known or if there are not enough resources such as time and money to obtain good specific algorithms. Particularly evolutionary computation has become quite popular in the last years.

Especially in the case of discrete structures, relational algebra has been successfully applied for algorithm development. Relational algebra has a small, but efficiently to implement, set of operations. On the other hand it allows a formal development of algorithms and expressions starting usually with a predicate logic description of the problem, like it is demonstrated for example in [1] and [2].

We think it's worth combining relational methods with evolutionary computation. First steps into this direction have been made in [3] and [4]. By having a relational view on problems of discrete structures, search points are represented as relations, so that parts of the search process can be carried out using relation algebraic expressions. In [4] that approach is applied to well-known graph problems and it is demonstrated, that relational methods can reduce the computation time of evolutionary algorithms.

Evolutionary algorithms work with sets of search points, called populations, that have to be evaluated. This means, as a first step, one has to determine the subset of individuals in the population satisfying the desired properties. By modeling single search points as vectors and populations as relations we can use relational expressions to carry out this evaluation. The aim of this paper is to extend this approach and have a more general view on it. On the one hand, we want to take a closer look at the relational expressions we can use in such

R.A. Schmidt (Ed.): RelMiCS /AKA 2006, LNCS 4136, pp. 266–280, 2006.
© Springer-Verlag Berlin Heidelberg 2006

an evaluation process. Therefore we define in Section 3 a set of predicates for vectors. These are mappings modeling properties of vectors and are represented by certain relational expressions. Each vector predicate can be transformed into a test mapping that can be applied to a population in order to determine the individuals fulfilling the property modeled by the predicate.

On the other hand we want to extend the combination of evolutionary algorithms and relational methods to a larger problem domain. Up to now, this approach has been applied to problems from combinatorial optimization, where the search points are sets of vertices or edges of a given graph or hypergraph. In our relational model, sets are represented as vectors in a natural way. In this work, we extend our considerations to problems, where the search points are binary relations. Typical examples are coloring problems for graphs or hypergraphs, where edges or vertices have to be assigned to colors, and timetable problems, where meetings have to be assigned to time slots. To be able to apply our methods to the search for such relations, it is necessary to model relations as vectors. Hence, the desired properties of the relations have to be transformed into properties of the corresponding vectors. In Section 4 we prove a formula that is helpful in doing this transformation and apply it to several relational properties.

In the last part of the paper, the results are used to discuss timetable problems. In our model, a solution of a timetable problem is a relation assigning meetings to time slots and fulfilling certain properties. By using the results of Section 4, these conditions are transformed into vector properties. So we get a simple criterium in the form of a predicate to test whether a vector represents a solution of a given timetable. With the results of Section 3 we can derive a test mapping, which is applied to a set of vectors to select the solutions for the timetable problem.

2 Relation-Algebraic Preliminaries

In the sequel we introduce the basics of abstract and concrete relation algebra. Starting with a definition of an abstract relation algebra, we state a selection of relational properties and specify the classes of relations that are used in the remainder of the paper. In the second part of this section we give a brief introduction to the algebra of set-theoretic relations.

2.1 Abstract Relational Algebra

A relational algebra is a structure $(\mathcal{R}, \cap, \cup, \overline{}, \subseteq, \cdot)$ over a non-empty set $\mathcal{R}$ of elements called *relations*. Every $R \in \mathcal{R}$ belongs to a subset $\mathcal{R}_R \subseteq \mathcal{R}$, so that the following conditions are fulfilled.

- $(\mathcal{R}_R, \cap, \cup, \overline{}, \subseteq)$ is a complete atomistic Boolean algebra. The null element and the universal element of $\mathcal{R}_R$ are denoted by O and L.
- For every $R \in \mathcal{R}$ there exist a transposed relation $R^\top$ and the products $R^\top R$ and $RR^\top$.

- The multiplication is associative and the existence of RS implies the existence of QS for every $Q \in \mathcal{R}_R$.
- For every $\mathcal{R}_R$, there exist left and right identities, which are denoted by I.
- The Tarski-rule holds, which means

$$R \neq \mathsf{O} \Longleftrightarrow \mathsf{L}R\mathsf{L} = \mathsf{L}.$$

- The Schröder-rule holds, which means

$$RS \subseteq Q \Longleftrightarrow R^\top \overline{Q} \subseteq \overline{S} \Longleftrightarrow \overline{Q}S^\top \subseteq \overline{R},$$

assuming the existence of RS.

A relation R is called *unique* if $R^\top R \subseteq \mathsf{I}$. If R fulfills one of the equivalent conditions $R\mathsf{L} = \mathsf{L}$ or $\mathsf{I} \subseteq RR^\top$, it is called *total*. A *function* is a total and unique relation. We call R *injective* if $R^\top$ is unique, and *surjective* if $R^\top$ is total.

In [5] several properties of abstract relation algebra are shown. The following selection of it is used in Sections 3 and 4:

1. $(Q \cup R)^\top = Q^\top \cup R^\top$ and $(Q \cap R)^\top = Q^\top \cap R^\top$
2. $(SR)^\top = R^\top S^\top$
3. Dedekind-rule: $QR \cap S \subseteq (Q \cap SR^\top)(R \cap Q^\top S)$
4. $(Q \cup R)S = QS \cup RS$ and $S(Q \cup R) = SQ \cup SR$
5. $(Q \cap R)S \subseteq QS \cap RS$ and $S(Q \cap R) \subseteq SQ \cap SR$
6. If S is unique, even $S(Q \cap R) = SQ \cap SR$ holds.
   If S is injective, even $(Q \cap R)S = QS \cap RS$ holds.
7. If S is injective and surjective, $\overline{RS} = \overline{R}S$ holds.

In Section 4 we also use direct products. A pair (π, ρ) of relations is called a *direct product* if it fulfills the following conditions:

1. $\pi^\top \pi = \mathsf{I}$ and $\rho^\top \rho = \mathsf{I}$
2. $\pi\pi^\top \cap \rho\rho^\top = \mathsf{I}$
3. $\pi^\top \rho = \mathsf{L}$ and $\rho^\top \pi = \mathsf{L}$

Obviously π and ρ are surjective functions.

2.2 Concrete Relational Algebra

We write $R : X \leftrightarrow Y$ if R is a (concrete) relation with domain X and range Y, i.e. a subset of $X \times Y$. The set of all relations of the *type* $X \leftrightarrow Y$ is denoted by $[X \leftrightarrow Y]$. In the case of finite carrier sets, we may consider a relation as a Boolean matrix. Since this Boolean matrix interpretation is well suited for many purposes, we often use matrix terminology and matrix notation in the following. Especially, we speak of the rows, columns and entries of R and write R_{ij} instead of $(i, j) \in R$. We assume the reader to be familiar with the basic operations on relations, viz. $R^\top$ (transposition), $\overline{R}$ (negation), $R \cup S$ (union), $R \cap S$ (intersection), RS (composition), $R \subseteq S$ (inclusion), and the special relations O (empty relation), L (universal relation), and I (identity relation).

A relation R is called *vector*, if $RL = R$ holds. As for a vector therefore the range is irrelevant, we consider in the following vectors $v : X \leftrightarrow 1$ with a specific singleton set $1 = \{\perp\}$ as range and omit in such cases the second subscript, i.e. write v_i instead of $v_{i\perp}$. Such a vector can be considered as a Boolean matrix with exactly one column, i.e., as a Boolean column vector, and describes the subset $\{x \in X : v_x\}$ of X. A vector v is called a *point* if it is injective and surjective. For $v : X \leftrightarrow 1$ these properties mean that it describes a singleton set, i.e., an element of X. In the matrix model, a point is a Boolean column vector in which exactly one component is true. The set $[1 \leftrightarrow 1]$ only contains the elements L and O, which can be regarded as boolean values *true* and *false*.
A pair (π, ρ) of *natural projections* of $X \times Y$, i.e.,

$$\pi : X \times Y \leftrightarrow X \text{ and } \rho : X \times Y \leftrightarrow Y$$

with

$$\pi_{<x,y>x'} \iff x = x' \text{ and } \rho_{<x,y>y'} \iff y = y'$$

is a direct product in the concrete relation algebra.

3   Predicates of Vectors

We define a set of *vector predicates* as a special set of mappings which can be applied to vectors and return a boolean value represented by the relations L and O of the type $1 \leftrightarrow 1$. As a first step we introduce a more general set Φ of mappings, that is a generalization of the set of relational functions that model column-wise, described in [6]. The set of vector predicates we are interested in is then a special subset of Φ. In detail, given a set X and a set of sets $\mathcal{U}$, we define a set of mappings

$$\Phi \subseteq \bigcup_{Y \in \mathcal{U}} [[X \leftrightarrow 1] \to [Y \leftrightarrow 1]]$$

so that for each element $\varphi \in \Phi$ of the type $[X \leftrightarrow 1] \to [Y \leftrightarrow 1]$ and every set Z there exists a mapping $\varphi' : [X \leftrightarrow Z] \to [Y \leftrightarrow Z]$ with the property, that for every point $p : Z \leftrightarrow 1$ and every relation $M : X \leftrightarrow Z$ the equation

$$\varphi(Mp) = \varphi'(M)p$$

holds. Assuming Z as the set of the first n numbers $[1..n]$ we can regard the relation M as a set of n vectors $v^{(1)}, \ldots, v^{(n)}$ (e.g. a population of an evolutionary algorithm) of type $X \leftrightarrow 1$ so that $v_x^{(i)}$ if and only if M_{xi} for $i \in [1..n]$ and $x \in X$. Then the relation $\varphi'(M)$ is of type $Y \leftrightarrow Z$ and consists of the vectors $\varphi(v^{(1)}), \ldots, \varphi(v^{(n)}) \in [Y \leftrightarrow 1]$.

First, we define the set Φ inductively and clarify how to construct a suitable φ' from a given φ. After that we prove the equation given above for every $\varphi \in \Phi$. For the definition of the set Φ we need the following two notations concerning certain mappings. Given four sets X, Y, Z, W and a relation $C : Z \leftrightarrow W$ we denote by χ_C the constant mapping of type $[X \leftrightarrow Y] \to [Z \leftrightarrow W]$ with $\chi_C(R) = C$ for every $R : X \leftrightarrow Y$. Furthermore, let $id_{[X \leftrightarrow Y]}$ be the identity mapping of type $[X \leftrightarrow Y] \to [X \leftrightarrow Y]$.

Definition 1. *Let* $\mathcal{U}$ *be a set of sets and* $X \in \mathcal{U}$. *We define the set* Φ *of mappings inductively:*

1. *Identity:*
$$id_{[X \leftrightarrow 1]} \in \Phi$$

2. *Constant mappings: For each set* $Y \in \mathcal{U}$ *and every* $c : Y \leftrightarrow 1$ *it is*
$$\chi_c \in \Phi$$
with $\chi_c : [X \leftrightarrow 1] \to [Y \leftrightarrow 1]$.

3. *Cut, union, complement: For every two mappings* $\varphi_1, \varphi_2 \in \Phi$ *with the same type* $[X \leftrightarrow 1] \to [Y \leftrightarrow 1]$ *there are also*
$$\varphi_1 \cap \varphi_2 \in \Phi,$$
$$\varphi_1 \cup \varphi_2 \in \Phi,$$
$$\overline{\varphi_1} \in \Phi,$$
whereas $\varphi_1 \cap \varphi_2$, $\varphi_1 \cup \varphi_2$ *and* $\overline{\varphi_1}$ *have the type* $[X \leftrightarrow 1] \to [Y \leftrightarrow 1]$ *and are defined by* $(\varphi_1 \cap \varphi_2)(v) = \varphi_1(v) \cap \varphi_2(v)$, $(\varphi_1 \cup \varphi_2)(v) = \varphi_1(v) \cup \varphi_2(v)$ *and* $\overline{\varphi_1}(v) = \overline{\varphi_1(v)}$ *for each vector* $v : [X \leftrightarrow 1]$.

4. *Left-muliplication: Given sets* $W, Y \in \mathcal{U}$, *for every* $R : W \leftrightarrow Y$ *and every* $\varphi \in \Phi$ *with the type* $[X \leftrightarrow 1] \to [Y \leftrightarrow 1]$ *it is*
$$R\varphi \in \Phi,$$
whereas $R\varphi : [X \leftrightarrow 1] \to [W \leftrightarrow 1]$ *is defined by* $(R\varphi)(v) = R(\varphi(v))$.

In the following we use this inductive definition to assign to each mapping $\varphi \in \Phi$ of the type $[X \leftrightarrow 1] \to [Y \leftrightarrow 1]$ a related mapping φ' that can be applied to a relation of the type $[X \leftrightarrow Z]$, representing a population of size $|Z|$.

Definition 2. *Let* Z *be a set. For every* $\varphi \in \Phi$ *we define a mapping* $\varphi' \in \bigcup_{Y \in \mathcal{U}} [[X \leftrightarrow Z] \to [Y \leftrightarrow Z]]$ *inductively.*

1. *Identity:*
$$id'_{[X \leftrightarrow 1]} := id_{[X \leftrightarrow Z]}$$

2. *Constant mappings: For each* $Y \in \mathcal{U}$ *and every vector* $c : Y \leftrightarrow 1$ *we define*
$$\chi'_c := \chi_{cL},$$
where the universal relation L *has the type* $1 \leftrightarrow Z$.

3. *Cut, union, complement: For* $Y \in \mathcal{U}$ *and* $\varphi_1, \varphi_2 \in \Phi$ *with the same type* $[X \leftrightarrow 1] \to [Y \leftrightarrow 1]$ *we define*
$$(\varphi_1 \cap \varphi_2)' := \varphi'_1 \cap \varphi'_2,$$
$$(\varphi_1 \cup \varphi_2)' := \varphi'_1 \cup \varphi'_2,$$
$$\overline{\varphi}' := \overline{\varphi'}.$$

4. *Left-muliplication: For sets W, $Y \in \mathcal{U}$, relations $R : W \leftrightarrow Y$ and mappings $\varphi \in \Phi$ with the type $[X \leftrightarrow 1] \to [Y \leftrightarrow 1]$ we define*

$$(R\varphi)' := R\varphi'.$$

For every $Y \in \mathcal{U}$ and $\varphi \in \Phi$ with $\varphi : [X \leftrightarrow 1] \to [Y \leftrightarrow 1]$ the type of the corresponding mapping φ' is $[X \leftrightarrow Z] \to [Y \leftrightarrow Z]$. There is a close correlation between the two mappings φ and φ', stated in the following theorem. The proof has been moved to the appendix.

Theorem 1. *For every mapping $\varphi \in \Phi$, every point $p : Z \leftrightarrow 1$ and every relation $M : X \leftrightarrow Z$ the equation*

$$\varphi(Mp) = \varphi'(M)p$$

holds.

For every set Y and every $\varphi \in \Phi$ we get the following commutative diagram, where the mapping μ_p denotes the right-multiplication with a point $p : Z \leftrightarrow 1$.

$$
\begin{array}{ccc}
[X \leftrightarrow Z] & \xrightarrow{\;\varphi'\;} & [Y \leftrightarrow Z] \\
\Big\downarrow{\mu_p} & & \Big\downarrow{\mu_p} \\
[X \leftrightarrow 1] & \xrightarrow{\;\varphi\;} & [Y \leftrightarrow 1]
\end{array}
$$

Denoting the composition of mappings as $\circ$, the statement of Theorem 1 can also be expressed by the equation

$$\varphi \circ \mu_p = \mu_p \circ \varphi'.$$

In this work, we are especially interested in the subset of Φ that consists of the mappings with range $[1 \leftrightarrow 1]$. We call

$$\Psi := \Phi \cap [[X \leftrightarrow 1] \to [1 \leftrightarrow 1]]$$

the set of *vector predicates*. Considering the elements L and O of $[1 \leftrightarrow 1]$ as the boolean values 'true' and 'false', each $\psi \in \Psi$ represents a property of vectors in the following way. A vector $v : X \leftrightarrow 1$ has a certain property - modeled by ψ - if and only if $\psi(v) = L$ holds.

Given a $\psi \in \Psi$, the related mapping ψ' has the type $[X \leftrightarrow Z] \to [1 \leftrightarrow Z]$. Modeling a population of $|Z|$ vectors as a relation $M : X \leftrightarrow Z$, the mapping ψ' determines the subset of individuals in the population fulfilling ψ, which means that for every point $p \subseteq \psi'(M)^\top$ the vector Mp satisfies the property modeled by ψ. In other words, with $Z = [1..n]$ and $k \in Z$ we have the following connection between ψ and ψ'. The kth column of M fulfills the predicate ψ if and only if $\psi'(M)_k^\top$ holds. Hence, ψ' is a kind of test mapping, testing the columns of a relation M (representing the individuals in a population) whether they satisfy the predicate ψ. As demonstrated in [3] and [4], the stated approach can be applied to graph theoretical problems where the search points are sets of vertices or edges, for example covering problems. In the next section we will focus on the issue of relations as search points.

4 Vectors and Relations

Some problems in combinatorial optimization (like graph coloring and timetable problems) deal with binary relations that have certain properties. In the case that the search points are relations of the type $X \leftrightarrow Y$, it is useful to encode them as vectors of the type $X \times Y \leftrightarrow \mathbf{1}$, so that sets of n search points can be represented as relations of type $X \times Y \leftrightarrow [1..n]$ to be evaluated as described in Section 3. Hence, the desired properties of a relation have to be translated into properties of the corresponding vector, so that can be decided if the relation R has a certain property only knowing the vector-representation r (and without computing R). In the case that the vector-property can be expressed as a vector predicate in the sense of Section 3, we achieve a test-mapping, that can be applied to a population in order to select the suitable individuals. Theorem 3 states a formula, that is helpful for the transformation of relation-properties into vector-properties. We carry out all proofs in abstract relation algebra so that we don't have to argue with components of concrete relations und achieve more readable and elegant proofs. First, we use direct products to define two constructions on relations, called *parallel composition* and *tupling*.

Definition 3. *For every two relations A and B there exist two direct products (π_1, ρ_1) and (π_2, ρ_2), so that $\pi_1 A \pi_2^\top \cap \rho_1 B \rho_2^\top$ is defined. In the following, let*

$$A \| B := \pi_1 A \pi_2^\top \cap \rho_1 B \rho_2^\top$$

be the parallel composition of A and B. If $AB^\top$ exists, there is a direct product (π, ρ), so that $\pi A \cap \rho B$ is defined. In the following, let

$$[A, B] := \pi A \cap \rho B$$

be the tupling of A and B.

Obviously, it holds $A \| B = [A\pi_2^\top, B\rho_2^\top]$. In the concrete relation algebra the two constructions have the following meaning. Given concrete relations $A : X \leftrightarrow Y$ and $B : Z \leftrightarrow W$, we use the natural projections (π_1, ρ_1) and (π_2, ρ_2) of $X \times Z$ and $Y \times W$ respectively. Then the relation $A \| B$ has the type $X \times Z \leftrightarrow Y \times W$ and it holds

$$(A \| B)_{<x,z><y,w>} \iff A_{xy} \text{ and } B_{zw}.$$

For the tupling, we need the existence of $AB^\top$. In the case of concrete relations that means A and B have the same range, i.e. $A : X \leftrightarrow Y$ and $B : Z \leftrightarrow Y$. With (π, ρ) as the natural projections of $X \times Z$, the relation $[A, B]$ has the type $X \times Z \leftrightarrow Y$ and

$$[A, B]_{<x,z>y} \iff A_{xy} \text{ and } B_{zy}$$

holds.

In the following, we prove several properties concerning parallel composition and tupling. Some similar - but not as general as ours - results can be found in [7]. The first lemma follows immediately from the Dedekind rule.

Lemma 1. *Let A, B, C and D be relations so that $AB \cap C$ is defined. Then from $A^\top C \subseteq D$ it follows $AB \cap C \subseteq A(B \cap D)$.*

The next lemma states a few properties of the parallel composition.

Lemma 2. *Let Q and R be relations and (π, ρ) and (τ, σ) direct products, so that $\pi Q \tau^\top \cap \rho R \sigma^\top$ exists. Then the following statements hold.*

1. $(Q\|R)^\top = Q^\top \| R^\top$
2. $(Q\|R)\tau \subseteq \pi Q$
   If R is total, even $(Q\|R)\tau = \pi Q$ holds.
3. $(Q\|R)\sigma \subseteq \rho R$
   If Q is total, even $(Q\|R)\sigma = \rho R$ holds.

Proof. The first equation follows immediately from the definition of the parallel composition. To show the second statement, we use the properties of direct products stated in Section 2.1. With $\tau^\top \tau = \mathsf{I}$ we achieve

$$(Q\|R)\tau = (\pi Q \tau^\top \cap \rho R \sigma^\top)\tau \subseteq \pi Q \tau^\top \tau = \pi Q$$

for arbitrary relations Q and R. Now let R be total. Because ρ is also total and $\sigma^\top \tau = \mathsf{L}$, it follows

$$\rho R \sigma^\top \tau = \rho R \mathsf{L} = \rho \mathsf{L} = \mathsf{L},$$

and therefore

$$
\begin{aligned}
\pi Q &= \mathsf{L} \cap \pi Q \\
&= \rho R \sigma^\top \tau \cap \pi Q \\
&\subseteq (\rho R \sigma^\top \cap \pi Q \tau^\top)(\tau \cap (\rho R \sigma^\top)^\top \pi Q) \qquad \text{(Dedekind rule)}\\
&\subseteq (Q\|R)\tau.
\end{aligned}
$$

The remaining statement can be proven in the same way.   □

In particular, we obtain the special cases

$$(Q\|\mathsf{I})\tau = \pi Q \quad \text{and} \quad (\mathsf{I}\|R)\sigma = \rho R,$$

since the identity relations are total. Using these equations and Lemma 1, we are able to prove multiplication formulas for parallel composition and tupling for the special cases, where one of the relations in the parallel composition is the identity relation. Later we will prove a formula for arbitrary relations.

Lemma 3. *Let Q, R, (π, ρ) and (τ, σ) be as in Lemma 2 and S, T relations so that $\tau S \cap \sigma T$ exists.*

1. *If $\rho \sigma^\top$ exists, $(Q\|\mathsf{I})[S, T] = [QS, T]$ holds.*
2. *if $\pi \tau^\top$ exists, $(\mathsf{I}\|R)[S, T] = [S, RT]$ holds.*

Proof. We only prove the first equation, because the second statement can be shown in a similar way. The proof of the first inclusion basically uses Lemma 2:

$$\begin{aligned}
(Q\|\mathsf{I})[S,T] &= (Q\|\mathsf{I})(\tau S \cap \sigma T) \\
&\subseteq (Q\|\mathsf{I})\tau S \cap (Q\|\mathsf{I})\sigma T \\
&\subseteq \pi QS \cap \rho \mathsf{I} T \qquad\qquad\qquad\qquad\qquad\text{(Lemma 2)} \\
&= [QS,T].
\end{aligned}$$

To show the second inclusion we use the fact, that

$$(*) \quad (Q\|\mathsf{I})^{\top}\rho T \subseteq \sigma T$$

holds, which follows immediately from Lemma 2. With the inclusion $(*)$, we can apply Lemma 1.

$$\begin{aligned}
[QS,T] &= \pi QS \cap \rho T \\
&= (Q\|\mathsf{I})\tau S \cap \rho T \qquad\qquad\qquad\qquad\text{(Lemma 2)} \\
&\subseteq (Q\|\mathsf{I})(\tau S \cap \sigma T) \qquad\quad \text{(with $(*)$, Lemma 1 can be used)} \\
&= (Q\|\mathsf{I})[S,T].
\end{aligned}$$

$\square$

Lemma 3 enables us to prove a formula concerning a special case of the multiplication of parallel compositions.

Lemma 4. *For arbitrary relations Q and R the following equation holds.*

$$(Q\|\mathsf{I})(\mathsf{I}\|R) = (\mathsf{I}\|R)(Q\|\mathsf{I}) = (Q\|R)$$

Proof. Since $(\mathsf{I}\|R) = [\tau^{\top}, R\sigma^{\top}]$ holds, we can apply Lemma 3:

$$(Q\|\mathsf{I})(\mathsf{I}\|R) = (Q\|\mathsf{I})[\tau^{\top}, R\sigma^{\top}] = [Q\tau^{\top}, R\sigma^{\top}] = (Q\|R).$$

The second equation can be shown in the same way. $\square$

In the following, Lemma 3 and Lemma 4 are used to prove a multiplication formula for parallel composition and tupling. The theorem is a generalization of a statement given in [7], where the formula is shown for injective S and R.

Theorem 2. *Let Q, R, S and T be relations such that $[S,T]$, QS and RT exist. Then the following equation holds:*

$$(Q\|R)[S,T] = [QS, RT].$$

Proof. The proof is a simple application of Lemma 3 and Lemma 4:

$$(Q\|R)[S,T] = (Q\|\mathsf{I})(\mathsf{I}\|R)[S,T] = (Q\|\mathsf{I})[S,RT] = [QS,RT].$$

$\square$

In the following, we use the tupling to define the *vector-representation* of relations. We examine the mapping *vec*, that transforms relations into vectors. After stating several basic properties of *vec*, we use Theorem 2 to prove a formula concerning the vector-representation of the composition of relations, stated in Theorem 3.

Definition 4. *For every relation A there exists an identity relation I and an universal relation L so that $[A, \mathsf{I}]\mathsf{L}$ is defined. In the following, let*

$$vec(A) := [A, \mathsf{I}]\mathsf{L}$$

be the vector-representation of A.

Given a concrete relation $A : X \leftrightarrow Y$ and the universal vector $\mathsf{L} : Y \leftrightarrow 1$ the calculation of $vec(A)$ transforms A into a vector $a : X \times Y \leftrightarrow 1$, so that $a_{<x,y>} \Longleftrightarrow A_{xy}$. In [5] it is shown that the mapping

$$rel : [X \times Y \leftrightarrow 1] \to [X \leftrightarrow Y]$$

defined by

$$rel(a) = \pi^{\top}(\rho \cap s\mathsf{L})$$

is the inverse mapping of

$$vec : [X \leftrightarrow Y] \to [X \times Y \leftrightarrow 1].$$

The properties of vec stated in the following lemma obviously hold in the concrete relation algebra. Their proofs for the abstract relation algebra can be found in [5].

Lemma 5. *For relations A and B with $B \in \mathcal{R}_A$ the following properties hold.*

1. $vec(\overline{A}) = \overline{vec(A)}$
2. $A \subseteq B \Longleftrightarrow vec(A) \subseteq vec(B)$
3. $vec(A \cap B) = vec(A) \cap vec(B)$
4. $vec(A \cup B) = vec(A) \cup vec(B)$
5. $vec(\mathsf{O}) = \mathsf{O}$
6. $vec(\mathsf{L}) = \mathsf{L}$,

We finally need the following lemma, that enables us to prove Theorem 3. It is proven in [5] as well.

Lemma 6. *The equation*

$$[AB, \mathsf{I}]\mathsf{L} = [A, B^{\top}]\mathsf{L}$$

holds, if AB is defined.

The next theorem, which is the main result of this section, solves the following problem. Given a vector s that is a vector-representation of a relation S, we want to compute the vector-representation of the composition of S with other relations, for example $vec(QS)$. Obviously, we can calculate $vec(QS) = vec(Qrel(s))$ but in this case, we have to do the transformations between vector- and relation-representation. The formula stated in the following theorem enables us to compute such expressions without calculating $rel(s)$. The proof basically uses Theorem 2.

Theorem 3. *Let Q, S, R be relations so that QSR exists. Then it holds*

$$vec(QSR) = (Q\|R^\top)vec(S).$$

Proof. The equation follows immediately from Theorem 2:

$$\begin{aligned}
vec(QSR) &= [QSR, \mathsf{I}]\mathsf{L} \\
&= [QS, R^\top]\mathsf{L} && \text{(Lemma 6)} \\
&= (Q\|R^\top)[S, \mathsf{I}]\mathsf{L} && \text{(Theorem 2)} \\
&= (Q\|R^\top)vec(S).
\end{aligned}$$

$\square$

In the case of concrete relations, we can visualize the formula stated above with the following diagram. Therefore, let $Q : Z \leftrightarrow X$, $R : Y \leftrightarrow W$ and ν_Q the left-multiplication with Q.

$$
\begin{array}{ccccc}
[X \leftrightarrow Y] & \xrightarrow{\ \nu_Q\ } & [Z \leftrightarrow Y] & \xrightarrow{\ \mu_R\ } & [Z \leftrightarrow W] \\
\big\downarrow{\scriptstyle vec} & & & & \big\downarrow{\scriptstyle vec} \\
[X \times Y \leftrightarrow 1] & \xrightarrow{\quad \nu_{Q\|R^\top}\quad} & & & [Z \times W \leftrightarrow 1]
\end{array}
$$

The statement of Theorem 3 can also be expressed as

$$vec \circ \mu_R \circ \nu_Q = \nu_{Q\|R^\top} \circ vec.$$

Theorem 3 can be used to describe fundamental properties of relations, like uniqueness, totality, injectivity and surjectivity as properties of their vector-representation. The proof of the following corollary has been moved to the appendix.

Corollary 1. *Let S be a relation, (π, ρ) a direct product so that $\pi S \rho^\top$ is defined and $s = vec(S)$ the vector-representation of S. Then the following equivalences hold.*

1. *S is unique $\Longleftrightarrow (\mathsf{I}\|\bar{\mathsf{I}})s \subseteq \bar{s}$*
2. *S is injective $\Longleftrightarrow (\bar{\mathsf{I}}\|\mathsf{I})s \subseteq \bar{s}$*
3. *S is total $\Longleftrightarrow \bar{s} \subseteq (\mathsf{I}\|\bar{\mathsf{I}})s \Longleftrightarrow \pi^\top s = \mathsf{L}$*
4. *S is surjective $\Longleftrightarrow \bar{s} \subseteq (\bar{\mathsf{I}}\|\mathsf{I})s \Longleftrightarrow \rho^\top s = \mathsf{L}$*

5 Application in Concrete Relational Algebra

In this section we use the formula of Theorem 3 to discuss a simple timetable problem. In our relational model of timetables, which is similar to the model in [8] and [9], meetings have to be assigned to time slots, so that specific conditions are satisfied. This means that we search for relations with certain properties. Timetable problems are typical applications for evolutionary algorithms. In each step a population, which is a set of possible solutions, is created randomly and then evaluated w.r.t. the desired properties. Modeling the possible solutions as vectors enables us to represent a population as a relation, that can be evaluated by applying a test mapping in the sense of Section 3. We apply Theorem 3 to formulate the desired properties of a timetable as a vector predicate. Hence we can derive a test mapping that can be applied to populations in order to determine, which of the individuals are suitable solutions for the given timetable problem.

Definition 5. *A timetable problem is a tuple*

$$\mathcal{T} = (\mathcal{M}, \mathcal{P}, \mathcal{H}, A, P)$$

where

- $\mathcal{M}$ *is a finite set of meetings,*
- $\mathcal{P}$ *is a finite set of participants,*
- $\mathcal{H}$ *is a finite set of hours,*
- $A : \mathcal{M} \leftrightarrow \mathcal{H}$ *and*
- $P : \mathcal{M} \leftrightarrow \mathcal{P}$ *are relations.*

The relation A describes the availabilities of the meetings, i.e. A_{mh} holds if the meeting m can take place in time slot h. The relation P assigns participants to meetings. The participant p takes part in meeting m, if P_{mp} holds. We say that two different meetings m and m' are *in conflict* if they have a common participant, i.e. there is an participant p so that P_{mp} and $P_{m'p}$ holds, which means, that p attends both meetings m and m'. Defining the *conflict relation*

$$C : \mathcal{M} \leftrightarrow \mathcal{M} \qquad \text{by} \qquad C = PP^\mathsf{T} \cap \bar{\mathsf{I}},$$

m and m' are in conflict if and only if $C_{mm'}$ holds. Solving a timetable problem means to assign a time slot to every meeting, so that the meeting is available and two meetings that are in conflict don't take place at the same time.

Definition 6. *A timetable (a solution for the timetable problem $\mathcal{T}$) is a relation*

$$S : \mathcal{M} \leftrightarrow \mathcal{H}$$

that satisfies the following four conditions.

1. $\forall m, h : S_{mh} \rightarrow A_{mh}$
2. $\forall m, m', h : (C_{m'm} \wedge S_{mh}) \rightarrow \neg S_{m'h}$
3. *S is unique*
4. *S is total*

The first property describes that each meeting is available in the time slot it is assigned to, the second property ensures that no meetings in conflict are assigned to the same time slot. The univalence and totality of S means that each meeting takes place in exactly one time slot. Translated into relational expressions, S is a timetable if and only if

1. $S \subseteq A$
2. $CS \subseteq \overline{S}$
3. $S\overline{I} \subseteq \overline{S}$
4. $SL = L$

A relation S fulfilling only the first three conditions is called a state or a partial solution.

By assuming A to be the universal relation $L : \mathcal{M} \leftrightarrow \mathcal{H}$, we can ignore the first condition. Then the problem to find a solution for $\mathcal{T}$ corresponds to the problem of graph coloring in the following sense. With C being irreflexive and symmetric, it can be interpreted as the adjacency relation of an undirected graph with the vertex set $\mathcal{M}$. Viewing $\mathcal{H}$ as a set of $k = |\mathcal{H}|$ colors, the task is to find a coloring of the vertices with k colors, so that two vertices that are connected by an edge don't have the same color. This means we have to find a relation $S : \mathcal{M} \leftrightarrow \mathcal{H}$ with the properties (2) - (4) as given above. In [10] it is shown that the problem of coloring a graph with k colors is NP-complete for $k \geq 3$, therefore the timetable problem is also NP-complete.

To simplify matters we call a vector $s : \mathcal{M} \times \mathcal{H} \leftrightarrow \mathbf{1}$ a solution or a state of a timetable problem $\mathcal{T}$ if and only if $rel(s)$ is a solution or a state of $\mathcal{T}$. In the following, Theorem 3 will be used to translate the conditions 1 - 4 for a relation S into conditions for the corresponding vector $s = vec(S)$. Theorem 4 enables us decide whether a vector s is a solution or state without computing $rel(s)$.

Theorem 4. *Let $a := vec(A)$ and $s := vec(S)$. Then s is a timetable if and only if the following 4 conditions hold.*

1. $\overline{aa^\top}s \subseteq \overline{s}$
2. $(C\|I)s \subseteq \overline{s}$
3. $(I\|\overline{I})s \subseteq \overline{s}$
4. $\pi^\top s = L$

If s fulfills the first three conditions, it is a state.

Proof. It is easy to show that the four conditions correspond to the four properties of Definition 6. The equivalence $S \subseteq A \Longleftrightarrow \overline{aa^\top}s \subseteq \overline{s}$ follows immediately from Lemma 5 and the Schröder-rule. As a consequence of Theorem 3 we achieve

$$CS \subseteq \overline{S} \Longleftrightarrow vec(CS) \subseteq vec(\overline{S}) \Longleftrightarrow (C\|I)s \subseteq \overline{s}.$$

The remaining conditions are already stated in Corollary 1.   □

By combining the first three conditions we achieve a simple criteria to test whether a vector is a state. Defining the verification relation

$$V := \overline{aa^\top} \cup (C\|I) \cup (I\|\overline{I})$$

of the type $V : \mathcal{M} \times \mathcal{H} \leftrightarrow \mathcal{M} \times \mathcal{H}$, the following equivalence holds:

$$s \text{ is a state} \Longleftrightarrow Vs \subseteq \overline{s}.$$

Note, that a similar result, but with a much more complicated proof, can be found in [8].

The verification relation V enables us to model the timetable problem $\mathcal{T} = (\mathcal{M}, \mathcal{P}, \mathcal{H}, A, P)$ as a 3-tuple $(\mathcal{M}, \mathcal{H}, V)$. For V being symmetric and irreflexive, it can be regarded as a adjacency relation of an undirected graph G. A vector s with $Vs \subseteq \overline{s}$ then represents an independent set of G. Hence, the problem to find a state of $\mathcal{T}$ with a maximum number of assigned meetings can be transformed into the problem to find an independent set of a graph with a maximum number of vertices.

It is quite simple to derive a vector predicate in the sense of Section 3 of the inclusion stated above. It holds

$$Vs \subseteq \overline{s} \Longleftrightarrow \overline{L(Vs \cap s)} = \mathsf{L},$$

where the first universal relation is of the type $\mathbf{1} \leftrightarrow \mathcal{M} \times \mathcal{H}$ and the second one of the type $\mathbf{1} \leftrightarrow \mathbf{1}$. We obtain the vector predicate $\psi_{st} : [\mathcal{M} \times \mathcal{H} \leftrightarrow \mathbf{1}] \to [\mathbf{1} \leftrightarrow \mathbf{1}]$ defined by

$$\psi_{st}(s) = \overline{\mathsf{L}(Vs \cap s)}$$

to test, whether a vector s is a state of the timetable problem $\mathcal{T}$. Following Section 3, we derive for an arbitrary $n \in \mathbb{N}$ the corresponding test mapping $\psi'_{st} : [\mathcal{M} \times \mathcal{H} \leftrightarrow [1..n]] \to [\mathbf{1} \leftrightarrow [1..n]]$ by

$$\psi'_{st}(M) = \overline{\mathsf{L}(VM \cap M)},$$

that enables us to filter all states of $\mathcal{T}$ out of a set of n vectors, modeled as a relation $M : \mathcal{M} \times \mathcal{H} \leftrightarrow [1..n]$.

To select the complete solutions we determine a second test mapping to decide whether a vector represents a total relation. From Corollary 1 we know that S is total if and only if $\pi^\top s = \mathsf{L}$ holds for $s = vec(S)$. This condition can easily be transformed into the predicate $\psi_{to} : [\mathcal{M} \times \mathcal{H} \leftrightarrow \mathbf{1}] \to [\mathbf{1} \leftrightarrow \mathbf{1}]$ defined by

$$\psi_{to}(s) = \overline{\mathsf{L}\overline{\pi^\top s}}$$

to test whether the a vector corresponds to a total relation. We can immediately derive the test mapping $\psi'_{to} : [\mathcal{M} \times \mathcal{H} \leftrightarrow [1..n]] \to [\mathbf{1} \leftrightarrow [1..n]]$ defined by

$$\psi'_{to}(M) = \overline{\mathsf{L}\overline{\pi^\top M}},$$

that select all columns of M representing total relations. Hence we achieve the predicate $\psi_{sol} = \psi_{st} \cap \psi_{to}$ to decide, whether a vector is a solution of the timetable problem $\mathcal{T}$. According to Section 3, the test mapping $\psi'_{sol} = \psi'_{st} \cap \psi'_{to}$ can be used to select timetables of a set of possible solutions.

6   Conclusions

The combination of relational algebra and evolutionary algorithms is a promising research direction. Sets of search points can be modeled and evaluated with relational algebra. We interpret relations columnwise as sets of vectors and use relational expressions to evaluate all included vectors in one step. Therefore we have defined a set of vector predicates that can be easily transformed into test mappings for sets of search points, modeled as relations. By applying these mappings to relations we can select the vectors fulfilling certain properties.

In this work we have extended this approach to a greater class of problems, where the search points are binary relations. We have discussed the transformation of relations to vectors and have proven a formula to translate certain properties of relations into properties of the corresponding vectors.

By applying our approach to a simple timetable problem we have given an example of its usefulness. Our results lead to a simple criteria to decide whether search points are solutions or partial solutions of a timetable problem.

References

1. Berghammer R.: A generic program for minimal subsets with applications. Leuschel M., editor, "Logic-based Program Development and Transformation" (LOPSTR '02) (proceedings), LNCS 2664, Springer, 144-157, 2003.
2. Berghammer R., Milanese U.: Relational approach to Boolean logic problems. W.MacCaull, M.Winter and I.Duentsch: Relational Methods in Computer Science, LNCS 3929, Springer, 2006
3. Kehden B., Neumann F., Berghammer R.: Relational implementation of simple parallel evolutionary algorithms. W.MacCaull, M.Winter and I.Duentsch: Relational Methods in Computer Science, LNCS 3929, Springer, 2006
4. Kehden, B., Neumann F.: A Relation-Algebraic View on Evolutionary Algorithms for Some Graph Problems. Gottlieb and Raidl (Eds.): 6th European Conference on Evolutionary Computation in Combinatorial Optimization, LNCS 3906, Springer, 147 - 158, 2006.
5. Schmidt G., Ströhlein T.: Relations and graphs. Springer,1993.
6. Berghammer R.: Relational-algebraic computation of fixed points with applications. The Journal of Logic and Algebraic Programming 66, 112 - 126, 2006.
7. Berghammer R., Zierer H.: Relational algebraic semantics of deterministic and nondeterministic programs. Theoretical Computer Science 43, 1986.
8. Schmidt, G. Ströhlein, T: Some aspects in the construction of timetables. Information processing 74 , Proc. IFIP Congress, Stockholm, 516 - 520, 1074 North-Holland, Amsterdam, 1974.
9. Schmidt G., Ströhlein, T.: A Boolean matrix iteration in timetable construction. Linear Algebra and Appl. 15, no. 1, 27 - 51, 1976.
10. Wegener I.: Complexity Theory. Springer, 2005.

Algebraization of Hybrid Logic with Binders

Tadeusz Litak

School of Information Science, JAIST
Asahidai 1–1, Nomi-shi, Ishikawa-ken
923-1292 Japan
litak@jaist.ac.jp

Abstract. This paper introduces an algebraic semantics for hybrid logic with binders $\mathcal{H}(\downarrow, @)$. It is known that this formalism is a modal counterpart of the bounded fragment of the first-order logic, studied by Feferman in the 1960's. The algebraization process leads to an interesting class of boolean algebras with operators, called *substitution-satisfaction algebras*. We provide a representation theorem for these algebras and thus provide an algebraic proof of completeness of hybrid logic.

1 Introduction

1.1 Motivation

The aim of this paper is to provide an algebraic semantics for *hybrid logic with binders* $\mathcal{H}(\downarrow, @)$. This formalism is, as was proven in the 1990's [1], the modal counterpart of *the bounded fragment of first-order logic*. Hence, an algebraization of $\mathcal{H}(\downarrow, @)$ provides also an algebraic insight into the nature of bounded quantification, i.e., quantification of the form $\forall x(tRx \rightarrow \phi)$ and $\exists x(tRx \wedge \phi)$, where t is a term not containing x. The fragment of first-order logic obtained by allowing only such quantifiers was investigated in the 1960's by Feferman and Kreisel [2], [3]. A discovery they made is that formulas in this fragment are exactly those which are preserved by formation of *generated submodels*, as modal logicians would say, or — to use Feferman's term — *outer extensions*.

The aim of this paper is to present a class of algebras which are hybrid (or bounded) equivalent of cylindric algebras for first-order logic. These algebras are *substitution-satisfaction algebras (SSA's)*, boolean algebras equipped with three kinds of operators: $\downarrow^k$ corresponding to *binding* of variable i_k to the present state, $@_k$ saying that a formula is *satisfied* in the state named by i_k and standard modal operator $\Diamond$, corresponding to restricted quantification itself. The theory of cylindric algebras proves to be an important source of insights and methods, but not all techniques can be applied directly to SSA's. For example, cylindric algebras often happen to be simple. For locally finite dimensional ones, subdirect irreducibility is equivalent to simplicity and in the finitely dimensional case, we even have a discriminator term. SSA's are not so well-behaved. Another example: in cylindric algebras, the operation of substitution of one variable for another is always definable in terms of quantifier operators. SSA's do not allow such a

R.A. Schmidt (Ed.): RelMiCS /AKA 2006, LNCS 4136, pp. 281–295, 2006.

feat. And yet, it turns out that their representation theory is not much more complicated than in the cylindric case.

Algebraic operators formalizing substitutions in first-order logic have been studied since Halmos started working on polyadic algebras [4]. In particular, they play a prominent role in formalisms developed by Pinter in the 1970's, cf., e.g., [5]. Nevertheless, algebras studied in the present paper do not have full substitution algebras as reducts — certain substitution operators are missing. Besides, as Halmos himself observed, the most interesting thing about satisfaction operators is their interplay with quantifiers — and bounded quantifiers do not interact with substitution operators in the same way as standard quantifiers do.

The structure of the present paper is as follows. In Section 1.2, we introduce the bounded fragment and $\mathcal{H}(\downarrow, @)$ as well as the truth preserving translation that show they are expressively equivalent. In Section 2, we introduce concrete, set-theoretical instantiation of SSA's — our counterpart of cylindric set algebras. In Section 3 we characterize SSA's axiomatically. Also, we prove some useful arithmetic facts and characterize basic algebraic notions, such as congruence filters or subdirect irreducibility. Section 4 contains main results of the paper. First, we identify Lindenbaum-Tarski algebras of hybrid theories as those which are *properly generated*. It is a more restrictive notion than the notion of *local finiteness* in the case of cylindric algebras. Then we show that every properly generated algebra of infinite (countable) dimension can be represented as a subdirect product of set algebras. In other words, we provide a representation theorem for SSA's and thus an algebraic proof of completeness of $\mathcal{H}(\downarrow, @)$. The proof was inspired by a concise proof of representation theorem for cylindric algebras by Andréka and Németi [6].

The author wishes to thank heartfully Ian Hodkinson for inspiration to begin the present research and for invaluable suggestions how to tackle the issue. The author can only hope that this advice was not entirely wasted. Thanks are also due to Patrick Blackburn for his ability to seduce people into doing hybrid logic and to the anonymous referee for suggestions and comments on the first version of this paper.

1.2 $\mathcal{H}(\downarrow, @)$ and the Bounded Fragment

This subsection briefly recalls some results of Areces et al. [1]; cf. also ten Cate [7]. For any ordinal α, define $\alpha^+ = \alpha - \{0\}$. It will become clear soon that zero is going to play the role of *a distinguished variable*. Fix a countable supply of *propositional variables* $\{p_a\}_{a \in PROP}$ (the restriction on cardinality is of no importance here) and *nominal variables* $\{i_k\}_{k \in \alpha^+}$; most of the time, we assume $\alpha = \omega$. Formulas of hybrid language are given by

$$\phi ::= p_a \mid i_k \mid \neg\phi \mid \phi \wedge \psi \mid \Diamond\phi \mid @_{i_k}\phi \mid \downarrow^{i_k}.\phi$$

$\Box$, $\vee$ and $\rightarrow$ are introduced as usual. Some papers introduced one more kind of syntactic objects: nominal constants, which cannot be bound by $\downarrow$. They do not increase the expressive power of the language and for our present goal the disadvantages of introducing such objects would outweigh the merits. They can be replaced by free unquantified variables.

Hybrid formulas are interpreted in *models*. A model $\mathfrak{M} := \langle W, R, V \rangle$ consists of an arbitrary non–empty set W, a binary *accessibility relation* $R \subseteq W \times W$ and a *(propositional) valuation* $V : p_a \mapsto A \in \mathbb{P}(W)$ mapping propositional variables to subsets of W. A *(nominal) assignment* in a model is any mapping $v : i_k \mapsto w \in W$ of nominal variables to elements of W. For an assignment v, $k \in \alpha^+$ and $w \in W$, define v_w^k to be the same assignment as v except for $v(i_k) = w$. The notion of satisfaction of formula at a point is defined inductively:

$$w \vDash_{\mathfrak{M},v} i_k \quad \text{if } w = v(i_k) \qquad\qquad w \vDash_{\mathfrak{M},v} p_a \quad \text{if } w \in V(p_a)$$
$$w \vDash_{\mathfrak{M},v} \psi \wedge \phi \ \text{if } w \vDash_{\mathfrak{M},v} \psi \text{ and } w \vDash_{\mathfrak{M},v} \phi \quad w \vDash_{\mathfrak{M},v} \neg\psi \quad \text{if not } w \vDash_{\mathfrak{M},v} \psi$$
$$w \vDash_{\mathfrak{M},v} \Diamond\psi \quad \text{if } \exists y.(wRy \text{ and } w \vDash_{\mathfrak{M},v} \psi)$$
$$w \vDash_{\mathfrak{M},v} @_{i_k}\psi \ \text{if } v(i_k) \vDash_{\mathfrak{M},v} \psi \qquad\qquad w \vDash_{\mathfrak{M},v} \downarrow^{i_k}.\psi \text{ if } w \vDash_{\mathfrak{M},v_w^k} \psi.$$

Fix any first order-language with a fixed binary relation constant R, unary predicate constants $\{P_a\}_{a \in PROP}$ and variables in $VAR := \{x_k\}_{k \in \alpha^+} \cup \{x, y\}$. *The bounded fragment* is generated by the following grammar:

$$\phi ::= P_a(v) \mid vRv' \mid v \approx v' \mid \neg\phi \mid \phi \wedge \psi \mid \exists v.(tRv\&\psi),$$

where $v, v' \in VAR$ and t is a term which do not contain v. The last requirement is crucial. Define the following mapping from hybrid language to first-order language by mutual recursion between two functions ST_x and ST_y:

	ST_x	ST_y
i_k	$x \approx x_k$	$y \approx x_k$
p_a	$P_a(x)$	$P_a(y)$
$\psi \wedge \phi$	$ST_x(\psi) \wedge ST_x(\phi)$	$ST_y(\psi) \wedge ST_y(\phi)$
$\neg\psi$	$\neg ST_x(\psi)$	$\neg ST_y(\psi)$
$\Diamond\phi$	$\exists y.(xRy \wedge ST_y(\phi))$	$\exists x.(yRx \wedge ST_x(\phi))$
$@_{i_k}\phi$	$\exists x.(x \approx x_k \wedge ST_x(\phi))$	$\exists y.(y \approx x_k \wedge ST_y(\phi))$
$\downarrow^{i_k}.\phi$	$\exists x_k.(x \approx x_k \wedge ST_x(\phi))$	$\exists x_k.(y \approx x_k \wedge ST_y(\phi))$

This mapping is known as *the standard translation*.

Theorem 1. *Let $\mathfrak{M} := \langle W, R, V \rangle$ be a hybrid model, v a nominal assignment. Let also ν be a valuation of first-order individual variables satisfying $\nu(x_k) = v(i_k)$, $\nu(x)$ and $\nu(y)$ being arbitrary; recall that unary predicate constants correspond to propositional variables. For every $w \in W$ and every hybrid formula ϕ, $w \vDash_{\mathfrak{M},v} \phi$ iff $\nu_w^x, w \vDash ST_x(\phi)$.*

Proof: See, e.g., Section 3.1 of Areces et al. [1] or Section 9.1 of ten Cate [7].   ⊣

The special role, then, is played by x: we sometimes call it *the distinguished variable* and identify it with x_0. The role of y is purely auxiliary. It is never used as a non-bound variable.

The apparatus of binders and satisfaction operators makes also the reverse translation possible. Let the supply of individual variables be $\{x_k\}_{k \in \alpha^+}$; no distinguished variables this time. Define

$$HT(P_a(x_k)) := @_{i_k} p_a, \; HT(x_k \approx x_l) := @_{i_k} i_l, \; HT(x_k R x_l) := @_{i_k} \Diamond i_l,$$
$$HT(\neg \phi) := \neg H(\phi), \quad HT(\phi \wedge \psi) := HT(\phi) \wedge HT(\psi),$$
$$HT(\exists x_k . x_l R x_k \wedge \psi) := @_{i_l} \Diamond \downarrow^{i_k} . HT(\psi),$$

Theorem 2. *Let $\mathfrak{M}$ be a first-order model in the signature $\{P_a\} \cup \{R\}$ and ν be a valuation of first-order individual variables. Define a nominal assignment $v(i_k) := \nu(x_k)$. For every formula ψ in the bounded fragment and every point $x \in \mathfrak{M}, \; \nu, x \vDash \phi$ iff $x \vDash_{\mathfrak{M}, v} HT(\psi)$.*

Proof: See Section 3.1 of Areces et al. [1] or Section 9.1 of ten Cate [7].   ⊣

In short, $\mathcal{H}(\downarrow, @)$ and the bounded fragment of first-order logic have the same expressive power. There is a beautiful semantic characterization of first-order formulas equivalent to those in the bounded fragment: these are exactly formulas *invariant for generated submodels*. Unfortunately, we cannot enter into details here: cf. Feferman [3] or Areces et al. [1].

2   Concrete Algebras

A set substitution-satisfaction algebra or *a concrete substitution-satisfaction algebra of dimension α (CSSA$_\alpha$) with base $\langle W, R \rangle$,* where $R \subseteq W^2$, is defined as a structure $\mathfrak{A} := \langle A, \vee, \neg, \emptyset, A, @_i, \mathbf{s}_0^i, \mathbf{d}_i, \Diamond \rangle_{i \in \alpha^+}$, where A is a field of subsets of W^α closed under all operations defined below and for every $i \in \alpha^+$ and $X \in \mathbb{P}(W^\alpha)$:

- $\mathbf{d}_i := \{x \mid x_0 = x_i\}$,
- $@_i X := \{y \mid \exists x \in X . x_0 = x_i \& \forall j \neq 0 . x_j = y_j\}$,
- $\downarrow^i X := \{y \mid \exists x \in X . x_0 = x_i \& \forall j \neq i . x_j = y_j\}$,
- $\Diamond X := \{y \mid \exists x \in X . y_0 R x_0\}$.

$\mathbf{d}_i$ corresponds to $\mathbf{d}_{0i}$ in cylindric algebras, hence our present notation. $\emptyset$ is often denoted as $\perp$ and $\mathbb{P}(W^\alpha)$ as $\top$. The zero coordinate of an element from W^α is called *the distinguished axis*. Geometrically, $@_k X$ corresponds to the effect of intersecting X with the hyperplane $\mathbf{d}_k$ and moving the set thus obtained parallel to the distinguished axis. Analogously, $\downarrow^k X$ corresponds to the effect of intersecting X with the hyperplane $\mathbf{d}_k$ and moving the set thus obtained parallel to the k-axis. Logical counterparts of these operations will be made explicit in the next section, but the notation should suggest the proper interpretation. Those CSSA$_\alpha$'s whose universe consists of the whole W^α form the class of *full-set substitution-satisfaction algebras of dimension α*, denoted by FSSA$_\alpha$. Thus, CSSA$_\alpha = S(\text{FSSA}_\alpha)$. The class of *representable substitution-satisfaction algebras of dimension α* is defined as RSSA$_\alpha = ISP(\text{FSSA}_\alpha)$.

2.1   Connection with Logic

With every model $\mathfrak{M} = \langle W, R, V \rangle$ we can associate the structure $\mathbf{Ss}\mathfrak{M}$. Namely, with every formula ϕ whose nominal variables are in $\{i_k\}_{k \in \alpha^+}$ we can associate the set $\phi^{\mathfrak{M}} = \{v \in W^{\alpha} \mid v_0 \vDash_{\mathfrak{M}, v|_{\alpha^+}} \phi\}$; $v|_{\alpha^+}$ is identified with the corresponding assignment of nominal variables. Such sets form a field of sets closed under $\downarrow^k$, $@_k$, $\diamondsuit_R$ and all diagonals. This is exactly the algebra $\mathbf{Ss}\mathfrak{M} \in \mathrm{CSSA}_{\alpha}$. Let us record the following basic

Fact 1. *For all hybrid formulas ϕ, ψ, every $k \in \alpha^+$, every hybrid model $\mathfrak{M}$,* $i_k^{\mathfrak{M}} = \mathbf{d}_k$, $(\psi \wedge \phi)^{\mathfrak{M}} = \psi^{\mathfrak{M}} \wedge \phi^{\mathfrak{M}}$, $(\neg \psi)^{\mathfrak{M}} = \neg \psi^{\mathfrak{M}}$, $(\diamondsuit \psi)^{\mathfrak{M}} = \diamondsuit \psi^{\mathfrak{M}}$, $(@_{i_k} \psi)^{\mathfrak{M}} = @_k \psi^{\mathfrak{M}}$, $(\downarrow^{i_k} . \psi)^{\mathfrak{M}} = \downarrow^k \psi^{\mathfrak{M}}$.

In order to characterize those CSSA_{α}'s which are of the form $\mathbf{Ss}\mathfrak{M}$ for some $\mathfrak{M}$, let us introduce the notion of *a dimension set* $\Delta a := \{i \in \alpha^+ \mid \downarrow^k a \neq a\}$. An element a is *zero-dimensional* if $\Delta a = 0$. The family of all zero-dimensional elements of $\mathfrak{A}$ is denoted by $A^{[0]}$. The algebra generated (as CSSA_{α}; of course, all constant elements are also treated as generators) from $A^{[0]}$ is denoted as $[A^0]$. $\mathfrak{A}$ is called *properly generated*[1] if $\mathfrak{A} = [A^0]$. $\mathfrak{A}$ is called *locally finitely dimensional* if $\#\Delta a < \omega$ for every a. Finally, $\mathfrak{A} \in \mathrm{CSSA}_{\alpha}$ with base $\langle W, R \rangle$ is called *0-regular* if for every $a \in A$, every $v, w \in W^{\alpha}$, $v \in a$ and $v_0 = w_0$ implies $w \in a$.

Lemma 2. *Every algebra of the form $\mathbf{Ss}\mathfrak{M}$, where $\mathfrak{M} = \langle W, R, V \rangle$ is some hybrid model, is properly generated and 0-regular.*

Proof: The proof of proper generation consists of three straightforward claims. First, every hybrid formula is by definition built from $\{p_a\}_{a \in PROP}$ and $\{i_k\}_{k \in \alpha^+}$ by finitely many applications of $\neg$, $\wedge$, $\diamondsuit$, $@_{i_k}$ and $\downarrow^{i_k}$. Second, by Fact 1, the connectives are interpreted by corresponding operations in algebra. Third, for every propositional variable p, $p^{\mathfrak{M}}$ is zero-dimensional, as $v \in p^{\mathfrak{M}}$ iff $v_0 \in V(p)$ iff $(v_{v_0}^i)_0 \in V(p)$ iff $v \in \downarrow^k p^{\mathfrak{M}}$.

For 0-regularity, for any 0-dimensional $\phi^{\mathfrak{M}}$ let us assume that $var(\phi) = \{k \in \alpha^+ \mid i_k \text{ occurs in } \phi\}$. If $f \in \phi^{\mathfrak{M}}$, $g|_{var(\phi)} = f|_{var(\phi)}$, then $g \in \phi^{\mathfrak{M}}$, as it is irrelevant what values g assigns to variables which do not occur in ϕ. Assume now $g_0 = f_0$. In case $var(\phi)$ is non-empty, let $i_{v(0)}, \ldots, i_{v(n)}$ be an enumeration of it. Let f' (g') be a valuation obtained from f (g) by substituting $f_0(= g_0)$ for every $i_{v(k)}$, where $k \in \{0, \ldots, n\}$. $f \in \phi^{\mathfrak{M}} = \downarrow^{v(0)} \ldots \downarrow^{v(k)} \phi^{\mathfrak{M}}$, hence $f' \in \phi^{\mathfrak{M}}$, by the above observation $g' \in \phi^{\mathfrak{M}}$, thus $g \in \downarrow^{v(0)} \ldots \downarrow^{v(k)} \phi^{\mathfrak{M}} = \phi^{\mathfrak{M}}$.   ⊣

The above observation can be strengthened to an equivalence.

Theorem 3. *$\mathfrak{A} \in \mathrm{CSSA}_{\alpha}$ based on $\langle W, R \rangle$ is of the form $\mathbf{Ss}\mathfrak{M}$ for some hybrid model $\mathfrak{M}$ with the same base iff it is properly generated and regular.*

---

[1] We avoid the notion *zero-generated* as it could be misleading: algebraists usually call this way the smallest subalgebra, i.e., the algebra generated from constants.

Proof: The left-to-right direction has already been proven. For the converse, let $\mathfrak{A} \in \text{FSSA}_\alpha$ be a properly generated and regular algebra based on $\mathfrak{F} = \langle W, R \rangle$. For any $a \in A^{[0]}$, let $V(p_a) = \{w \in W \mid \exists v \in a.w = v_0\}$. Let $\mathfrak{M} = \langle \mathfrak{F}, V \rangle$. We want to show $\mathfrak{A} = \mathbf{Ss}\mathfrak{M}$. The bases of both algebras and hence the fundamental operations on the intersection of both universes coincide. Thus, in order to show the $\subseteq$-direction, it is enough to show that for every $a \in A^{[0]}$, $a = p_a^{\mathfrak{M}}$. For every $v \in W^\alpha$, $v \in p_a^{\mathfrak{M}}$ iff $v_0 \in V(p_a)$ iff $v_0 = w_0$ for some $w \in a$ iff (by 0-regularity) $v \in a$. For the reverse inclusion, observe that the atomic formulas in the language of $\mathfrak{M}$ are always of the form p_a or i_k. The proof proceeds then by standard induction on the complexity of formulas. $\dashv$

3   Abstract Approach

3.1   Axioms and Basic Arithmetics

Let $i, j, k \ldots$ be arbitrary ordinals in α^+. The *class of substitution-satisfaction algebras of dimension* α SSA_α is defined as the class of algebras satisfying the following axioms

Ax1. Axioms for boolean algebras

Ax2. Axioms for the modal operator
  - (a) $\Diamond \bot = \bot$
  - (b) $\Diamond (p \vee q) = \Diamond p \vee \Diamond q$

Ax3. Axioms governing $@_k$
  - (a) $\neg @_k p = @_k \neg p$
  - (b) $@_k (p \vee q) = @_k p \vee @_k q$
  - (c) $@_k \mathbf{d}_k = \top$
  - (d) $@_k @_j p = @_j p$
  - (e) $\mathbf{d}_k \leq p \leftrightarrow @_k p$

Ax4. Interaction of $\Diamond$ and $@_k$: $\Diamond @_k p \leq @_k p$

Ax5. Axioms governing $\downarrow^k$
  - (a) $\neg \downarrow^k p = \downarrow^k \neg p$
  - (b) $\downarrow^k (p \vee q) = \downarrow^k p \vee \downarrow^k q$
  - (c) $\downarrow^k \downarrow^j p = \downarrow^j \downarrow^k p$
  - (d) $\downarrow^k \downarrow^k p = \downarrow^k p$
  - (e) $\downarrow^j \mathbf{d}_k = \mathbf{d}_k$ for $j \neq k$ and $\downarrow^k \mathbf{d}_k = \top$

Ax6. Interaction of $\downarrow^k$ and $@_j$
  - (a) $\downarrow^k @_j \downarrow^k p = @_j \downarrow^k p$
  - (b) $\downarrow^k @_k p = \downarrow^k p$
  - (c) $@_k \downarrow^k p = @_k p$

Ax7. Interaction of $\downarrow^k$ and $\Diamond$: $\downarrow^k \Diamond \downarrow^k p = \Diamond \downarrow^k p$

Ax8. The Blackburn-ten Cate axiom BG: $@_k \Box \downarrow^j @_k \Diamond \mathbf{d}_j = \top$

Fact 3. $\text{FSSA}_\alpha \subseteq \text{SSA}_\alpha$ *and thus* $\text{RSSA}_\alpha \subseteq \text{SSA}_\alpha$.

Lemma 4. *The following are derivable:*

Ar1. $\Box(p \to q) \le \Diamond p \to \Diamond q$
Ar2. $@_k(p \to q) = @_k p \to @_k q$
Ar3. $\downarrow^k (p \to q) = \downarrow^k p \to \downarrow^k q$
Ar4. $@_k \mathbf{d}_j \le @_k p \leftrightarrow @_j x$
Ar5. $@_k \mathbf{d}_j \le @_j \mathbf{d}_k$
Ar6. $\mathbf{d}_j \wedge p \le @_j p$
Ar7. $@_k \Diamond \mathbf{d}_j \wedge @_j p \le @_k \Diamond p.$

Proof: The only one which requires some calculation is Ar7 and we will need this one later. From Ax3e we get that $@_j p \le \mathbf{d}_j \to p$, this by Ax4 gives us $@_j p \le \Box(\mathbf{d}_j \to p)$. Using Ax3d, we get $@_j p \le @_k \Box(\mathbf{d}_j \to p)$. By Ar1, this gives us $@_j p \le @_k(\Diamond \mathbf{d}_j \to \Diamond p)$. By Ar2, we get the desired conclusion.   ⊣

3.2 Proper Generation and Finite Dimensionality

The notions of *dimension* of an element, *locally finitely dimensional* and *properly generated* algebra are introduced in exactly the same way as in the concrete case. The class of locally finite algebras of dimension α is denoted as $\mathbf{Lf}_\alpha$, the properly generated ones — as $\mathbf{Prop}_\alpha$

Fact 5. $\Delta \mathbf{d}_k = \{k\}$, $\Delta \Diamond a \subseteq \Delta a$, $\Delta \neg a = \Delta a$, $\Delta(a \wedge b) \subseteq \Delta a \cup \Delta b$, $\Delta @_k a \subseteq \Delta a \cup \{k\}$, $\Delta \downarrow^k a \subseteq \Delta a - \{k\}$

Corollary 1. $\mathbf{Prop}_\alpha \subseteq \mathbf{Lf}_\alpha$

From now on, we use the fact that $\downarrow^k$ and $@_k$ distribute over all boolean connectives without explicit reference to Ax3a, Ax3b, Ax5a, Ax5b and Ar2. A straightforward consequence of Ax6b is

Fact 6. *For every* $k \in \Delta p$, $p \ne \bot$ *implies* $@_k p \ne \bot$. *Consequently for any* $a \in \mathfrak{A} \in \mathbf{Lf}_\alpha$, $a \ne \bot$ *iff there is* k *s.t.* $@_k a \ne \bot$.

The following result is an algebraic counterpart of an observation of ten Cate and Blackburn. [8], [7]

Lemma 7. *Assume* $\alpha \ge \omega$, $\mathfrak{A} \in \mathbf{Lf}_\alpha$, $p \in \mathfrak{A}$. *Then*

$$@_j \Diamond p = \bigvee_{l \in \Delta p} (@_j \Diamond \mathbf{d}_l \wedge @_l p) \tag{1}$$

Proof: $@_j \Diamond \mathbf{d}_l \wedge @_l p \le @_j \Diamond p$ by Ar7. Thus, in order to show 1, it is enough to prove that for any z, if $@_j \Diamond \mathbf{d}_l \wedge @_l p \le z$ for every $l \in \Delta p$, then $@_j \Diamond p \le z$. Choose some $l, k \notin \Delta p \cup \Delta z$ (here is where we use assumptions on $\mathfrak{A}$). By assumption and by Ax3d, $@_j \Diamond \mathbf{d}_l \wedge @_l p \le @_k z$. This in turn, by $l \notin \Delta z \cup \Delta p$ and Lemma 5 implies $\downarrow^l @_j \Diamond \mathbf{d}_l \wedge p \to @_k z = \top$, from which we get $@_j \Box(\downarrow^l @_j \Diamond \mathbf{d}_l \to (p \to @_k z)) = \top$. Here is where we use Ax8 to obtain $@_j \Box(p \to @_k z) = \top$. This implies $@_j \Diamond p = @_j \Diamond p \wedge @_j \Box(p \to @_k z)$. By Ar1, we get thus $@_j \Diamond p \le @_j \Diamond @_k z$ and by Ax4 and Ax3d we obtain $@_j \Diamond p \le @_k z$. By $k \notin \Delta z \cup \Delta p$ and Lemma 5, the conclusion follows.   ⊣

3.3   Ideals, Homomorphisms, The Rasiowa-Sikorski Lemma

Let us introduce several standard algebraic notions concerning the structure of SSA_α's. *An open ideal* is a lattice-theoretical ideal closed under $\Diamond$, all $@_i$ and $\downarrow^i$. It is a standard observation that congruence ideals correspond to homomorphisms. *An ideal generated by* p is the smallest open ideal containing p; it is denoted by $\mathbf{Gen}(p)$. Let $\mathbf{Mod}_\alpha$ be the set of words in the alphabet $\{\Diamond, \downarrow^i, @_i \mid i \in \alpha^+\}$.

Fact 8. $\mathbf{Gen}(p) = \{q \mid q \leq \blacklozenge_1 p \vee \cdots \vee \blacklozenge_n p, \blacklozenge_1, \dots, \blacklozenge_n \in \mathbf{Mod}_\alpha\}$

A subdirectly irreducible algebra is one which contains smallest nontrivial open ideal. By the above observation, we can reformulate it as follows.

Corollary 2. $\mathfrak{A} \in \text{SSA}_\alpha$ *is subdirectly irreducible iff there exists* $\mathsf{o} \neq \bot$ *s.t. for every* $p \neq \bot$ *there are* $\blacklozenge_1, \dots, \blacklozenge_n \in \mathbf{Mod}_\alpha$ *s.t.* $\mathsf{o} \leq \blacklozenge_1 p \vee \cdots \vee \blacklozenge_n p$. o *is called a (dual) opremum element.*

Of course, we don't really have to consider all members of $\mathbf{Mod}_\alpha$; it is possible to restrict the set significantly. In particular, for zero-dimensional p we can restrict attention to $\blacklozenge_1, \dots, \blacklozenge_n \in \{\Diamond^n, @_i \Diamond^n \mid i \in \alpha^+, n \in \omega\}$.

Combining Corollary 2 and Lemma 6 we arrive at the following:

Corollary 3. $\mathfrak{A} \in \mathbf{Lf}_\alpha$ *is subdirectly irreducible iff there is* $\mathsf{o} \in \mathfrak{A}$ *and* $k \in \alpha^+$ *s.t.* $@_k \mathsf{o} \neq \bot$ *and for every* $p \neq \bot$ *there are* $\blacklozenge_1, \dots, \blacklozenge_n \in \mathbf{Mod}_\alpha$ *s.t.* $@_k \mathsf{o} \leq @_k \blacklozenge_1 p \vee \cdots \vee @_k \blacklozenge_n p$. *Of course, we can take* $@_k \mathsf{o}$ *to be opremum itself.*

Definition 1. *Let* $\mathfrak{A} \in \text{SSA}_\alpha$. *An ultrafilter* H *of* $\mathfrak{A}$ *is* elegant *if for every* $i \in \alpha^+$ *and every* $p \in \mathfrak{A}$, $@_i \Diamond p \in H$ *iff there is* $j \in \alpha^+$ *s.t.* $@_i \Diamond \mathbf{d}_j \wedge @_j p \in H$.

Lemma 9. *Let* α *be countably infinite,* $\mathfrak{A} \in \mathbf{Lf}_\alpha$, $\#\mathfrak{A} \leq \omega$. *For every* $a \neq \bot$, *there exists an elegant ultrafilter containing* a.

Proof: Follows from Lemma 7 and The Rasiowa-Sikorski Lemma: cf. Koppelberg [9, Theorem 2.21] . We briefly sketch the proof here to make the paper more self-contained. Let $b_0, b_1, b_2 \dots$ be an enumeration of all elements of the form $@_j \Diamond p$ for some $j \in \alpha^+$ and $p \in \mathfrak{A}$: here is where we use the fact that universe of $\mathfrak{A}$ is countable. Define $a_0 := a$. If a_n is defined, let $a_{n+1} := a_n$ if $a_n \wedge b_n = \bot$. Otherwise, assume $b_n = @_j \Diamond p$. Lemma 7 implies there is $k \in \alpha^+$ s.t. $a_{n+1} := a_n \wedge @_j \Diamond \mathbf{d}_k \wedge @_k p \neq \bot$. In this way we obtain an infinite descending chain of nonzero points. Any ultrafilter containing $\{a_n\}_{n \in \omega}$ is elegant.   ⊣

4   The Representation Theorem

This section proves the main result of the paper. We identify those SSA's which correspond to Lindenbaum-Tarski algebras of $\mathcal{H}(\downarrow, @)$-theories and prove a representation theorem for them.

4.1   Transformations, Retractions, Replacements

Halmos [4] developed general theory of transformations and used it as a foundation for theory of polyadic algebras. Let us recall some basic results. A *transformation* of α^+ is any mapping of α^+ into itself. We call a transformation τ *finite* if $\tau(i) = i$ for almost all i (i.e., cofinitely many). We will be interested only in finite transformations. The intuitive reason is that transformations will correspond to substitutions of variables — and, for a given formula, only finitely many variables are relevant. From now on, finiteness of transformations is assumed tacitly.

A transformation τ is called *a transposition* if for some k and l, $\tau(k) = l$, $\tau(l) = k$ and for all other arguments, τ is equal to identity. Such mappings are denoted as (k, l). A product of transpositions is called *a permutation*. τ is called *a replacement* if τ is different from identity for exactly one argument, say $\tau(l) = k$. τ is then written as (l/k). A transformation τ is called *a retraction* if $\tau^2 = \tau$. It is a well-known mathematical fact that every bijection of α^+ onto itself is a permutation. Halmos generalized this fact as follows:

Lemma 10. *Every retraction is a product of replacements and every transformation is a product of a permutation and a retraction.*

In case of locally finite algebras of infinite dimension, we can restrict our attention only to retractions, i.e., products of replacements: this will be justified further on. A similar observation for locally finite polyadic algebras was made by Halmos [4]. Finally, a bit of notation. For τ a transformation, τ_k^l be the substitution defined as $\tau_k^l(j) = \tau(j)$ for $j \neq l$ and $\tau_k^l(l) = k$. Also, let $\tau - l$ be the transformation which is the same as τ except that it leaves l unchanged. Thus, τ_k^l is the composition of $\tau - l$ and (l/k).

4.2   Axioms for $\mathcal{H}(\downarrow, @)$

We present an axiom system for $\mathcal{H}(\downarrow, @)$ taken from Blackburn and ten Cate [8], [7]. A nominal variable i_k is called *bound* in a formula ϕ if it occurs within the scope of some $\downarrow^{i_k}$ and *free* otherwise.

Definition 2. *Let $\tau : \alpha^+ \mapsto \alpha^+$ be a transformation. The nominal substitution associated with τ of formulas of hybrid language is a function $\psi \mapsto \psi^\tau$ which replaces all free occurrences of i_k and $@_{i_k}$ with $i_{\tau(k)}$ and $@_{i_{\tau(k)}}$, respectively, in those places which are not in the scope of some $\downarrow^{i_{\tau(k)}}$.*

A $\mathcal{H}(\downarrow, @)$-*theory* is any set of hybrid formulas T containing all instances of

H1. classical tautologies,
H2. $\Box(\psi \to \phi) \to (\Box\psi \to \Box\phi)$,
H3. $@_{i_k}(\psi \to \phi) \to (@_{i_k}\psi \to @_{i_k}\phi)$,
H4. $@_{i_k}\psi \leftrightarrow \neg@_{i_k}\neg\psi$,
H5. $@_{i_k}i_k$,

H6. $@_{i_j}@_{i_k}\psi \leftrightarrow @_{i_k}\psi$,
H7. $i_k \rightarrow (\psi \leftrightarrow @_{i_k}\psi)$,
H8. $\Diamond @_{i_k}\psi \rightarrow @_{i_k}\psi$,
H9. $@_{i_k}(\downarrow^{i_l}.\psi \leftrightarrow \psi^{(l/k)})$,
H10. $\downarrow^{i_k}.(i_k \rightarrow \psi) \rightarrow \psi$, if i_k does not occur free in ψ,
H11. $@_{i_j}\Box \downarrow^{i_k}.@_{i_j}\Diamond i_k$

and closed under Modus Ponens, Substitution (i.e., if $\psi \in T$, then $\psi^\tau \in T$ for every substitution τ), and Generalization for all operators (i.e., if $\psi \in T$, then $@_{i_k}\psi \in T$, $\downarrow^{i_k}\psi \in T$ and $\Box\psi \in T$). It is easy to see that our Ax1–Ax4 and Ax8 are direct translations of corresponding $\mathcal{H}(\downarrow, @)$ axioms. Those governing $\downarrow$ could not be translated straightforwardly into equations. See the concluding section for further comments on the relationship between these two axiomatizations.

Blackburn and ten Cate [8], [7] prove the following

Theorem 4 (Hybrid Completeness). *For every consistent $\mathcal{H}(\downarrow, @)$-theory T, there is a hybrid model $\mathfrak{M}_T$ such that T is exactly the set of all formulas whose value under all assignments is equal to the universe of $\mathfrak{M}_T$.*

In this work, we provide an algebraic counterpart of their result. First, let us characterize Lindenbaum-Tarski algebras of $\mathcal{H}(\downarrow, @)$-theories.

4.3 Lindenbaum-Tarski Algebras

From now on, we always assume we work with a countably infinite α. Fix a supply a propositional variables and denote the set of all hybrid formulas as *Form*. With every $\mathcal{H}(\downarrow, @)$-theory T we can associate an equivalence relation on the set of hybrid formulas: $[\psi]_T = \{\phi \mid \psi \leftrightarrow \phi \in T\}$. With every connective, we can associate a corresponding operator on equivalence classes, i.e., $\Diamond[\phi] = [\Diamond\phi]$, $@_k[\phi] = [@_{i_k}\phi]$, $\downarrow^k[\phi] = [\downarrow^{i_k}\phi]$ etc. It is a matter of routine verification that this definition is correct, i.e., independent of the choice of representatives. We have to show that *Form*/T with operators corresponding to logical connectives and constants is an element of **Prop**$_\alpha$. Such a structure is called *the Lindenbaum-Tarski algebra* of T. Verification that Ax1–Ax4 and Ax8 hold does not pose any problems by the remark above. Verification of axioms governing $\downarrow^k$ can be done in uniform manner: first, use H9 and axioms governing $@_{i_j}$ to prove an instance of the axiom preceded by arbitrary $@_{i_j}$, e.g. $@_{i_j}(\neg \downarrow^{i_k}\phi \leftrightarrow \downarrow^{i_k}\neg\phi) \in T$. As i_j can be chosen such that i_j does not appear in ϕ, we can use H7, generalization rule for $\downarrow^{i_j}$ and H10 to get rid of initial $@_{i_j}$. The same strategy can be used to show that equivalence classes of propositional variables are zero-dimensional and thus the algebra is properly generated.

In the reverse direction, we use the same strategy as in the proof of Theorem 3. With every element $a \in A^{[0]}$, associate a distinct propositional variable p_a. Hybrid formulas ϕ in the language whose propositional variables are p_a's and nominal variables are in α^+ are in $1 - 1$ correspondence with constant terms in the language of SSA$_\alpha$'s extended with a name for every $a \in [A^0]$. And so, for every such formula ϕ, let Φ be the corresponding term. The substitution associated with τ for terms is defined in the same way as for hybrid formulas with $\mathbf{d}_k$, $\downarrow^k$ and $@_k$ replacing, respectively, i_k, $\downarrow^{i_k}$ and $@_{i_k}$. Define $T_{\mathfrak{A}} := \{\phi \in Form \mid \Phi = \top\}$. First, we show this is a $\mathcal{H}(\downarrow, @)$-theory. The only part which is not immediate is showing that all instances of H9 belong to T.

Lemma 11. $\Psi^{(l/k)} = \downarrow^l \Psi^{(l/k)}$ for $l \neq k$.

Proof: The only relevant information for the basic inductive step is that p_a's correspond to $a \in A^{[0]}$. The inductive steps are trivial for booleans and use Ax7, Ax6a and Ax5c for modal, satisfaction and substitution operators. $\dashv$

Lemma 12. For every retraction τ, $@_k \Phi^{\tau_k^l} = @_k \downarrow^l \Phi^{\tau - l}$.

Proof: To prove the lemma, it is enough to show that

$$@_l \mathbf{d}_k \leq \Psi \leftrightarrow \Psi^{(l/k)}. \tag{2}$$

For then we get that $\mathbf{d}_k \leq \downarrow^l \Psi \leftrightarrow \downarrow^l \Psi^{(l/k)}$. By Lemma 11, it is equivalent to $\mathbf{d}_k \leq \downarrow^l \Psi \leftrightarrow \Psi^{(l/k)}$. By laws of boolean algebras, it is equivalent to $\mathbf{d}_k \wedge \downarrow^l \Psi = \mathbf{d}_k \wedge \Psi^{(l/k)}$. But then

$$@_k \downarrow^l \Psi = @_k (\mathbf{d}_k \wedge \downarrow^l \Psi) = @_k (\mathbf{d}_k \wedge \Psi^{(l/k)}) = @_k \Psi^{(l/k)}.$$

Thus, let us prove 2 by induction on the complexity of Ψ. For nominals, it's a consequence of Ax3c and Ar5. For propositional variables, $p_a^{(l/k)} = p_a$. For booleans, the inductive step is trivial. For $\Diamond$, it follows from Ax4 and Ar1. For $@_j$, it follows from Ax3d. Finally, for $\downarrow^j$, we use the fact that either $j \in \Delta(@_l \mathbf{d}_k)$ or $j \in \{k, l\}$. If the latter is the case, then $\downarrow^j \Psi^{(l/k)} = \downarrow^j \Psi$, by definition of nominal substitution. $\dashv$

We have proven that T is indeed a $\mathcal{H}(\downarrow, @)$-theory, but before proceeding with the proof that $\mathfrak{A}$ is isomorphic to $Form/T_{\mathfrak{A}}$ let us record two useful consequences of the Lemma just proven.

Corollary 4. If $\Psi = \top$, then for arbitrary retraction τ and arbitrary k in the range of τ, $@_k \Psi^\tau = \top$.

We can also justify the observation made before: that in locally finite algebras of infinite dimension, the only kind of transformations which are relevant are retractions, i.e., products of replacements. In view of Lemma 10, it is enough to show the following.

Corollary 5. For arbitrary transposition (k, l) and for every Φ, there exists a retraction τ s.t. $\Phi^{(k,l)} = \Phi^\tau$.

Proof: Choose any $m \notin \Delta\Phi \cup \{k, l\}$ (here is where we use the fact that $\mathfrak{A}$ is locally finite and of infinite dimension). Define $\tau = (m/k)(k/l)(l/m)$. The only argument where τ can possibly differ from (k, l) is m, for $\tau(m) = k$ and $(k, l)(m) = m$. But then $\Psi^\tau = \downarrow^m \Psi^\tau = \downarrow^m @_m \downarrow^m \Psi^\tau = \downarrow^m @_m \Psi^{(k,l)} = \downarrow^m \Psi^{(k,l)} = \Psi^{(k,l)}$. $\dashv$

Now, arbitrary $a \in \mathfrak{A}$ is named by a certain term Ψ. Thus, for arbitrary a we can arbitrarily choose one Ψ_a and define a mapping $f(a) = [\Psi_a]$. It is straightforward to observe this mapping is correctly defined, $1 - 1$ and onto. Hence, we have shown

Theorem 5. *For every* $\mathcal{H}(\downarrow, @)$-*theory* T, $Form/T$ *with operators corresponding to logical connectives and constants is an element of* **Prop**$_\alpha$. *Conversely, for every* $\mathfrak{A} \in$ **Prop**$_\alpha$ *there is a* $\mathcal{H}(\downarrow, @)$-*theory* $T_\mathfrak{A}$ *s.t.* $Form/T_\mathfrak{A}$ *is isomorphic to* $\mathfrak{A}$.

4.4   The Main Result

In this section, we finally prove the main result of the paper: a representation theorem for SSA's after the manner of Andréka and Németi [6].

Definition and Lemma 13. *Let* α *be countably infinite,* $\mathfrak{A} \in$ SSA$_\alpha$ *and* F *be any filter. Define* $\sim_F$ *on* α^+ *as* $k \sim_F l$ *if* $@_k \mathbf{d}_l \in F$. *This is a congruence relation. Define also* R_F *on* $\alpha^+/\sim_F$ *as* $[k]_F R_F[l]_F$ *if* $@_k \diamond \mathbf{d}_l \in F$. *This is a correct definition.*

Proof: That $\sim_F$ is a congruence relation follows from Ax3c, Ar4 and Ar5. Correctness of the definition of R_F follows from Ar4 and Ar7.   ⊣

Theorem 6 (Countable Representation). *Let* α *be countably infinite,* $\mathfrak{A} \in$ **Prop**$_\alpha$ *be a subdirectly irreducible algebra,* $\#\mathfrak{A} \leq \omega$. $\mathfrak{A}$ *is embeddable in a* FSSA$_\alpha$. *More specifically, let* H *be an elegant ultrafilter containing an opremum element of* $\mathfrak{A}$. $\mathfrak{A}$ *is embeddable in the full set algebra with base* $\mathfrak{F}_H := \langle \alpha^+/H, R_H \rangle$.

Proof: Just like in Section 4.3, associate with elements of $\mathfrak{A}$ formulas of the language whose propositional variables are $\{p_a \mid a \in A^{[0]}\}$, so that every formula ψ corresponds to a term Ψ in the extended language and every element $a \in \mathfrak{A}$ is named by such a term. Define a valuation V_H of propositional variables in $\mathfrak{F}_H$ by

$$V_H(p_a) := \{[k] \mid @_k a \in H\}.$$

Let $\mathfrak{M} = \langle \mathfrak{F}_H, V_H \rangle$. We are going to show that $\mathfrak{A}$ is isomorphic to **Ss**$\mathfrak{M}$. By Corollary 5, we can restrict our attention only to those τ's which are retractions. Thus, by Lemma 10 it is enough to formulate all claims and proofs only for replacements. For a mapping $\tau : \alpha \mapsto \alpha^+$, let $\tau^+ := \tau|_{\alpha^+}$. For arbitrary term Ψ, define auxiliary mapping g' as

$$g'(\Psi) := \{\tau' : \alpha \mapsto \alpha^+ \mid @_{\tau'(0)} \Psi^{\tau'^+} \in H\},$$

and then

$$g(\Psi) := \{\tau : \alpha \mapsto \alpha^+/H \mid \exists \tau' \in g'(\Psi). \forall i \in \alpha. \tau'(i) \in \tau(i)\}$$

Now, for arbitrary a choose ψ_a to be arbitrary formula s.t. $\Psi_a = a$ and define $f(a) := g(\Psi_a)$. We have to show that this is a correct definition, i.e., that f

is independent of the choice of ψ. It is enough to prove that for arbitrary Ψ, Φ, $\Psi \rightarrow \Phi = \top$ implies $g(\Psi) \leq g(\Phi)$. Assume $g(\Psi) \not\leq g(\Phi)$. Let τ be such that $@_{\tau(0)}\Psi^{\tau^+} \in H$ and $@_{\tau(0)}\Phi^{\tau^+} \notin H$. It means that $@_{\tau(0)}(\Psi \wedge \neg\Phi)^{\tau^+} \in H$, i.e, $@_{\tau(0)}(\Psi \wedge \neg\Phi)^{\tau^+} \neq \bot$ and hence $(\Psi \wedge \neg\Phi)^{\tau^+} \neq \bot$. Choose arbitrary $k \in \Delta(\Psi \wedge \neg\Phi)$ in the range of τ. By Lemma 6, $@_k(\Psi \wedge \neg\Phi)^{\tau^+} \neq \bot$ and hence by Corollary 4, $\Psi \not\leq \Phi$.

Let us prove that f is a homomorphism. We don't need the assumption of subdirect irreducibility here, only the fact that H is an elegant ultrafilter. Subdirect irreducibility will be used only to show f is an embedding.

Given $\tau : \alpha \mapsto \alpha^+$, let $[\tau] : \alpha \mapsto \alpha^+/H$ be a mapping defined as $[\tau](i) = [\tau(i)]$. Thus, g can be redefined as $g(\Psi) = \{[\tau'] | \tau' \in g'(\Psi)\}$. With every $\sigma : \alpha^+ \mapsto \alpha^+$, we can associate a nominal assignment v^σ in $\mathfrak{M}$ defined as $v^\sigma(i_k) := [\sigma(k)]$. $v^\sigma(l/k)(i_l) := [k]$ and $v^\sigma(l/k)(i_j) := v^\sigma(i_j)$ for $j \neq l$. Sometimes, we denote by v the mapping $v(i_k) = [k]$, i.e., v^{id}. As we are interested only in finitary retractions, every v^σ is of the form $v(l_1/k_1) \ldots (l_n/k_n)$ for some $l_1, \ldots, l_n, k_1, \ldots, k_n$. By $[k] \in v^\sigma(\psi)$ we mean that ψ holds at $[k]$ in $\mathfrak{M}$ under v^σ.

Claim 1: $[k] \in v^\sigma(\downarrow^l \psi)$ iff $[k] \in v^\sigma(l/k)(\psi)$. In fact, this is just a clause from definition of satisfaction, as $v^\sigma(l/k) = (v^\sigma)^l_{[k]}$. We just use more elegant notation to avoid clumsiness.

Claim 2: $v(l/k)(\psi) = v(\psi^{(l/k)})$. Thus, for every σ, $v^\sigma(\psi) = v(\psi^\sigma)$ and $v^\sigma(l/k)(\psi) = v(\psi^{\sigma^l_k})$.

Claim 3: $[j] \in v(\psi^\sigma)$ iff $@_j\Psi^\sigma \in H$.

Proof of claim: For $\psi = p_a$ it follows from the definition of V_H. For $\psi = i_k$ — from the definition of $\sim_H$. For booleans: from distributivity of $@_k$ over boolean connectives and the fact that H is an ultrafilter. The clause for $\Diamond$ is the one where we use the fact that H is elegant: $[j] \in v(\Diamond\phi^\sigma)$ iff (by definition of valuation in hybrid model) exists $[k]$ s.t. $[j]R_H[k]$ and $k \in v(\phi^\sigma)$ iff (by definition of R_H and IH) there is $[k]$ s.t. $@_j\Diamond\mathbf{d}_k \in H$ and $@_k\Phi^\sigma \in H$ iff (by assumption on H) $@_j\Diamond\Phi^\sigma \in H$.

Assume now $\psi = @_k\phi$. Then $[j] \in v((@_k\phi)^\sigma)$ iff $[j] \in v(@_{\sigma(k)}\phi^\sigma)$ iff $[\sigma(k)] \in v(\phi^\sigma)$ iff (by IH) $@_{\sigma(k)}\Phi^\sigma \in H$ iff $\Psi^\sigma \in H$ iff (by Ax3d) $@_j\Psi^\sigma \in H$.

Finally, assume $\psi = \downarrow^k \phi$. Then $[j] \in v((\downarrow^k \phi)^\sigma)$ iff $[j] \in v(\downarrow^k \phi^{\sigma-k})$ iff (by Claim 1) $[j] \in v(k/j)(\phi^{\sigma-k})$ iff (by Claim 2) $[j] \in v(\phi^{\sigma^k_j})$ iff (by IH) $@_j\Phi^{\sigma^k_j} \in H$ iff (by Lemma 12) $@_j \downarrow^k \Phi^{\sigma-k} \in H$ iff $@_j\Psi^\sigma \in H$. $\dashv$

Claim 3 immediately implies that f is a homomorphism: for every $\tau : \alpha \mapsto \alpha^+$ and every $a \in \mathfrak{A}$, $\tau \in f(a)$ iff $[\tau(0)] \in v^{\tau^+}(\psi_a)$.

In order to show f is an embedding we finally use assumption of subdirect irreducibility and the fact that H contains opremum. We want to show that $a \not\leq b$ implies $f(a) \not\leq f(b)$. By Lemma 3 $\Psi_a \rightarrow \Psi_b \neq \top$ implies there is $\blacklozenge_1, \ldots, \blacklozenge_n \in \mathbf{Mod}_\alpha$ s.t. $@_k\blacklozenge_1(\Psi_a \wedge \neg\Psi_b) \vee \cdots \vee @_k\blacklozenge_n(\Psi_a \wedge \neg\Psi_b) \in H$. By the fact that H is an ultrafilter, we obtain that there is $\blacklozenge \in \mathbf{Mod}_\alpha$ s.t. $@_k\blacklozenge(\Psi_a \wedge \neg\Psi_b) \in H$. Because of Ax3d and the fact that H is elegant, if $@_k \blacklozenge (\Psi_a \wedge \neg \Psi_b) = @_k \blacklozenge_1 \blacklozenge_2 (\Psi_a \wedge \neg \Psi_b)$ for some $\blacklozenge_1$ consisting only of diamonds and satisfaction operators, then for some $l \in \alpha^+$, $@_l \blacklozenge_2 (\Psi_a \wedge \neg \Psi_b) \in H$. In other words, we can get rid of initial diamonds and satisfaction operators. $@_l \downarrow^m (\blacklozenge_2 (\Psi_a \wedge \neg \Psi_b)) \in H$ can be rewritten as $@_l (\blacklozenge_2 (\Psi_a \wedge \neg \Psi_b))^{(m/l)} \in H$. Proceeding in this way, we finally obtain that for some j and some σ, $@_j (\Psi_a^\sigma \wedge \neg \Psi_b^\sigma) \in H$. Reasoning the same way as in the proof that f is correctly defined, we finally obtain that $\Psi_a \rightarrow \Psi_b \neq \top$, i.e., $a \not\leq b$.

f is in fact an isomorphism onto $\mathbf{Ss}\mathfrak{M}$, cf. the proof of Theorem 3. $\quad\dashv$

Theorem 7 (Representation). *For a countably infinite α, every subdirectly irreducible $\mathfrak{A} \in \mathbf{Prop}_\alpha$ is isomorphic to a* FSSA$_\alpha$. *Consequently,* $\mathbf{Prop}_\alpha = $ RSSA$_\alpha$.

Proof: For countable algebras, this already follows from Theorem 6. For uncountable $\mathfrak{A}$, we can prove it very similarly to Lemma 3 in [6]. Namely, let $\{\mathfrak{B}_l\}_{l \in \beta}$ be a directed system of s.i. algebras in $I(\text{CSSA}_\alpha)$ sharing a common opremum element. Then $\bigcup_{l \in \beta} \mathfrak{B}_l \in I(\text{CSSA}_\alpha)$. Lack of space (i.e., LNCS 15 pages limit) prevents us from proving the theorem in detail. $\quad\dashv$

5   Open Problems and Further Developments

The algebraic axiomatization of Section 3 suggests there should exist a $\mathcal{H}(\downarrow, @)$ analogue of Tarski's axiomatization of first-order logic which uses neither the notion of a free variable nor the notion of proper substitution of a variable in a formula. [10] Also, our Andréka-Németi style proof of The Representation Theorem employs a slightly different strategy from the one used by Henkin, based on the notion of *thin elements* and *rich algebras* — cf. [11] or [12]. It could be interesting to prove the representation theorem for SSA's also in this way.

Another path open for exploration: Monk [13] shows that significant part of algebraic model theory can be presented by focusing on set-theoretical algebras, without any axiomatic definition of abstract cylindric algebras. By analogy, it could be tempting to develop a part of algebraic $\mathcal{H}(\downarrow, @)$-model theory or algebraic bounded model theory by means of set SSA's.

The referee of the present paper posed two interesting questions, which are currently investigated by the author. First, the bounded fragment is known to be a *conservative reduction class* for first-order logic. That suggests that known results about cylindric algebras may be derivable from theorems concerning SSA's. Second, both bounded fragment and hybrid logic with binders are model-theoretically well-behaved: one example is the interpolation property. This should translate into nice algebraic characteristics of SSA's (e.g., the amalgamation property).

Finally, the possible connection with computer science, which was in fact a motivation to present these results to the computer science community. It is known that cylindric algebras capture exactly those database queries which are first-order expressible, cf. [14] for details. Is there an a related interpretation for

SSA's — for example, in terms of databases where the user is allowed to ask questions concerning only *accessible* entries?

References

1. Areces, C., Blackburn, P., Marx, M.: Hybrid logic is the bounded fragment of first order logic. In de Queiroz, R., Carnielli, W., eds.: Proceedings of 6th Workshop on Logic, Language , Information and Computation, WOLLIC99, Rio de Janeiro, Brazil (1999) 33–50
2. Feferman, S., Kreisel, G.: Persistent and invariant formulas relative to theories of higher order. Bulletin of the American Mathematical Society **72** (1966) 480–485 Research Announcement.
3. Feferman, S.: Persistent and invariant formulas for outer extensions. Compositio Mathematica **20** (1968) 29–52
4. Halmos, P.: Algebraic Logic. Chelsea Publishing Company (1962)
5. Pinter, C.: A simple algebra of first order logic. Notre Dame Journal of Formal Logic **1** (1973) 361–366
6. Andréka, H., Németi, I.: A simple, purely algebraic proof of the completeness of some first order logics. Algebra Universalis **5** (1975) 8–15
7. ten Cate, B.: Model theory for extended modal languages. PhD thesis, University of Amsterdam (2005) ILLC Dissertation Series DS-2005-01.
8. Blackburn, P., Cate, B.: Pure extensions, proof rules, and hybrid axiomatics. In Schmidt, R., Pratt-Hartmann, I., Reynolds, M., Wansing, H., eds.: Preliminary proceedings of Advances in Modal Logic (AiML 2004), Manchester (2004)
9. Koppelberg, S.: Handbook of boolean algebras. Volume I. Elsevier, North-Holland (1989)
10. Tarski, A.: A simplified formalization of predicate logic with identity. Archiv für Mathematische Logik und Grundlagenforschung **7** (1965)
11. Henkin, L., Monk, J., Tarski, A.: Cylindric algebras, Part II. North Holland, Amsterdam (1985)
12. Andréka, H., Givant, S., Mikulás, S., Németi, I., Simon, A.: Notions of density that imply representability in algebraic logic. Annals of Pure and Applied Logic **91** (1998) 93–190
13. Monk, D.: An introduction to cylindric set algebras (with an appendix by H. Andréka). Logic Journal of the IGPL **8** (2000) 451–506
14. Van den Bussche, J.: Applications of Alfred Tarski's ideas in database theory. Lecture Notes in Computer Science **2142** (2001) 20–37

Using Probabilistic Kleene Algebra
for Protocol Verification

AK McIver[1], E Cohen[2], and CC Morgan[3]

[1] Dept. Computer Science, Macquarie University, NSW 2109 Australia
anabel@ics.mq.edu.au
[2] Microsoft, US
Ernie.Cohen@microsoft.com
[3] School of Engineering and Computer Science, University of New South Wales,
NSW 2052 Australia
carrollm@ecs.unsw.edu.au

Abstract. We describe *pKA*, a probabilistic Kleene-style algebra, based on a well known model of probabilistic/demonic computation [3,16,10]. Our technical aim is to express probabilistic versions of Cohen's *separation theorems*[1].

Separation theorems simplify reasoning about distributed systems, where with purely algebraic reasoning they can reduce complicated interleaving behaviour to "separated" behaviours each of which can be analysed on its own. Until now that has not been possible for *probabilistic* distributed systems.

Algebraic reasoning in general is very robust, and easy to check: thus an algebraic approach to probabilistic distributed systems is attractive because in that "doubly hostile" environment (probability *and* interleaving) the opportunities for subtle error abound. Especially tricky is the interaction of probability and the demonic or "adversarial" scheduling implied by concurrency.

Our case study — based on Rabin's *Mutual exclusion with bounded waiting* [6] — is one where just such problems have already occurred: the original presentation was later shown to have subtle flaws [15]. It motivates our interest in algebras, where assumptions relating probability and secrecy are clearly exposed and, in some cases, can be given simple characterisations in spite of their intricacy.

Keywords: Kleene algebra, probabilistic systems, probabilistic verification.

1 Introduction

The verification of probabilistic systems creates significant challenges for formal proof techniques. The challenge is particularly severe in the distributed context where quantitative system-wide effects must be assembled from a collection of disparate localised behaviours. Here carefully prepared probabilities may become inadvertently skewed by the interaction of so-called adversarial scheduling, the well-known abstraction of unpredictable execution order.

R.A. Schmidt (Ed.): RelMiCS /AKA 2006, LNCS 4136, pp. 296–310, 2006.
© Springer-Verlag Berlin Heidelberg 2006

One approach is probabilistic model checking, but it may quickly become over-whelmed by state-space explosion, and so verification is often possible only for small problem instances. On the other hand quantitative proof-based approaches [10,4], though in principle independent of state-space issues, may similarly fail due to the difficulties of calculating complicated probabilities, effectively "by hand".

In this paper we propose a third way, in which we apply proof as a "pre-processing" stage that simplifies a distributed architecture *without the need to do any numerical calculations whatsoever*, bringing the problem within range of quantitative model-based analysis after all. It uses *reduction*, the well-known technique allowing simplification of distributed algorithms, *but applied in the probabilistic context*.

We describe a program algebra *pKA* introduced elsewhere [8] in which stan-dard *Kleene algebra* [5] has been adapted to reflect the interaction of proba-bilistic assignments with nondeterminism, a typical phenomenon in distributed algorithms. Standard (i.e. non-probabilistic) Kleene algebra his been used ef-fectively to verify some non-trivial distributed protocols [1], and we will argue that the benefits carry over to the probabilistic setting as well. The main dif-ference between *pKA* and standard Kleene Algebra is that *pKA* prevents cer-tain distributions of nondeterminism +, just in those cases where whether that nondeterminism can "see" probabilistic choices is important [16,3,10]. That dis-tribution failure however removes some conventional axioms on which familiar techniques depend: and so we must replace those axioms with adjusted (weaker) probabilistic versions.

Our case study is inspired by Rabin's solution to the mutual exclusion problem with bounded waiting [14,6], whose original formulation was found to contain some subtle flaws [15] due precisely to the combination of adversarial and prob-abilistic choice we address. Later it became clear that the assumptions required for the correctness of Rabin's probabilistic protocol — that the outcome of some probabilistic choices are invisible to the adversary — cannot be supported by the usual model for probabilistic systems. We investigate the implications on the model and algebra of adopting those assumptions which, we argue, have wider applications for secrecy and probabilities.

Our specific contributions are as follows.

1. A summary of *pKA*'s characteristics (Sec. 2), including a generalisation of Cohen's work on separation [1] for probabilistic distributed systems using *pKA* (Sec. 4);
2. Application of the general separation results to Rabin's solution to distrib-uted mutual exclusion with bounded waiting (Sec. 5);
3. Introduction of a model which supports the algebraic characterisation of secrecy in a context of probability (Sec. 6).

The notational conventions used are as follows. Function application is repre-sented by a dot, as in $f.x$. If K is a set then $\overline{K}$ is the set of discrete probability distributions over K, that is the normalised functions from K into the real in-terval $[0, 1]$. A point distribution centered at a point k is denoted by δ_k. The

$(p, 1-p)$-weighted average of distributions d and d' is denoted $d \, _p\oplus d'$. If K is a subset, and d a distribution, we write $d.K$ for $\sum_{s \in K} d.s$. The power set of K is denoted $\mathbb{P}K$. We use early letters a, b, c for general Kleene expressions, late letters x, y for variables, and t for tests.

2   Probabilistic Kleene Algebra

Given a (discrete) state space S, the set of functions $S \to \mathbb{P}\overline{S}$, from (initial) states to subsets of distributions over (final) states has now been thoroughly worked out as a basis for the transition-system style model now generally accepted for probabilistic systems [10] though, depending on the particular application, the conditions imposed on the subsets of (final) probability distributions can vary [12,3]. Briefly the idea is that probabilistic systems comprise both *quantifiable* arbitrary behaviour (such as the chance of winning an automated lottery) together with *un*-quantifiable arbitrary behaviour (such as the precise order of interleaved events in a distributed system). The functions $S \to \mathbb{P}\overline{S}$ model the unquantifiable aspects with powersets ($\mathbb{P}(\cdot)$) and the quantifiable aspects with distributions ($\overline{S}$).

For example, a program that simulates a fair coin is modelled by a function that maps an arbitrary state s to (the singleton set containing only) the distribution weighted evenly between states 0 and 1; we write it

$$flip \quad \hat{=} \quad s := 0 \;\; _{1/2}\oplus \;\; s := 1 \,. \tag{1}$$

In contrast a program that simulates a possible bias favouring 0 of at most $2/3$ is modelled by a nondeterministic choice delimiting a range of behaviours:

$$biasFlip \quad \hat{=} \quad s := 0 \;\; _{1/2}\oplus \;\; s := 1 \quad \sqcap \quad s := 0 \;\; _{2/3}\oplus \;\; s := 1 \,, \tag{2}$$

and in the semantics (given below) its result set is represented by the *set* of distributions defined by the two specified probabilistic choices at (2).

In setting out the details, we follow Morgan et al. [12] and take a domain theoretical approach, restricting the result sets of the semantic functions according to an underlying order on the state space. We take a flat domain $(S^\top, \sqsubseteq)$, where $S^\top$ is $S \cup \{\top\}$ (in which $\top$ is a special state used to model miraculous behaviour) and the order $\sqsubseteq$ is constructed so that $\top$ dominates all (proper) states in S, which are otherwise unrelated.

Definition 1. *Our probabilistic power domain is a pair $(\overline{S^\top}, \sqsubseteq)$, where $\overline{S^\top}$ is the set of normalised functions from $S^\top$ into the real interval $[0, 1]$, and $\sqsubseteq$ is induced from the underlying $\sqsubseteq$ on $S^\top$ so that*

$$d \sqsubseteq d' \quad \text{iff} \quad (\forall K \subseteq S \cdot d.K + d.\top \;\; \leq \;\; d'.K + d'.\top) \,.$$

Probabilistic programs are now modelled as the set of functions from initial state in $S^\top$ to sets of final distributions over $S^\top$, where the result sets are restricted by so-called *healthiness conditions* characterising viable probabilistic behaviour, motivated in detail elsewhere [10]. By doing so the semantics accounts for specific features of probabilistic programs. In this case we impose *up-closure* (the inclusion of all $\sqsubseteq$-dominating distributions), *convex closure* (the inclusion of all convex combinations of distributions), and *Cauchy closure* (the inclusion of all limits of distributions according to the standard Cauchy metric on real-valued functions [12]). Thus, by construction, viable computations are those in which miracles dominate (refine) all other behaviours (implied by up-closure), nondeterministic choice is refined by probabilistic choice (implied by convex closure), and classic limiting behaviour of probabilistic events (such as so-called "zero-one laws" [1]) is also accounted for (implied by Cauchy closure). A further bonus is that (as usual) program refinement is simply defined as reverse set-inclusion. We observe that probabilistic properties are preserved with increase in this order.

Definition 2. *The space of probabilistic programs is given by* $(\mathcal{LS}, \sqsubseteq)$ *where* $\mathcal{LS}$ *is the set of functions from* $S^\top$ *to the power set of* $\overline{S^\top}$*, restricted to subsets which are* Cauchy- *, convex- and up-closed with respect to* $\sqsubseteq$*. All programs are* $\top$*-preserving (mapping* $\top$ *to* $\{\delta_\top\}$*). The order between programs is defined*

$$Prog \sqsubseteq Prog' \quad \text{iff} \quad (\forall s \colon S \cdot Prog.s \supseteq Prog'.s) \ .$$

For example the healthiness conditions mean that the semantics of the program at (2) contains all mappings of the form

$$s \to \delta_0 \ _q\!\oplus \delta_1 \ , \quad \text{for} \quad 2/3 \geq q \geq 1/2 \ ,$$

where respectively δ_0 and δ_1 are the point distributions on the states $s = 0$ and $s = 1$.

In Fig.1 we define some mathematical operators on the space of programs: they will be used to interpret our language of Kleene terms. Informally composition *Prog; Prog'* corresponds to a program *Prog* being executed followed by *Prog'*, so that from initial state s, any result distribution d of *Prog.s* can be followed by an arbitrary distribution of *Prog'*. The probabilistic operator takes the weighted average of the distributions of its operands, and the nondeterminism operator takes their union (with closure).

Iteration is the most intricate of the operations — operationally *Prog** represents the program that can execute *Prog* an arbitrary finite number of times. In the probabilistic context, as well as generating the results of all "finite iterations" of (*Prog* $\sqcap$ skip) (*viz*, a finite number of compositions of (*Prog* $\sqcap$ skip)), imposition of Cauchy closure acts as usual on metric spaces, in that it also generates all *limiting* distributions — i.e. if d_0, $d_1, \ldots$ are distributions contained in a result set U, and they converge to d, then d is contained in U as well. To illustrate, we consider

$$halfFlip \quad \hat{=} \quad \text{if } (s = 0) \text{ then } flip \text{ else skip} \ , \tag{3}$$

---

[1] An easy consequence of a zero-one law is that if a fair coin is flipped repeatedly, then with probability 1 a head is observed eventually. See the program '*flip*' inside an iteration, which is discussed below.

Skip	$\text{skip}.s$	$\triangleq$	$\lceil\{\delta_s\}\rceil$,
Miracle	$\text{magic }.s$	$\triangleq$	$\{\delta_\top\}$,
Chaos	$\text{chaos}_K.s$	$\triangleq$	$\mathbb{P}K^\top$
Composition	$(Prog;\ Prog').s$	$\triangleq$	$\{\sum_{u:\ s\top}(d.u)\times d'_u \mid d \in Prog.s;\ d'_u \in Prog'.u\}$,
Choice	$(if\ B\ then\ Prog\ else\ Prog').s$	$\triangleq$	$if\ B.s,\ then\ Prog.s,\ otherwise\ Prog'.s$
Probability	$(Prog\ _p\oplus Prog').s$	$\triangleq$	$\{d\ _p\oplus d' \mid d \in r.s;\ d' \in r'.s\}$,
Nondeterminism	$(Prog\ \sqcap\ Prog').s$	$\triangleq$	$\lceil\{d \mid d \in (Prog.s \cup Prog'.s)\}\rceil$,
Iteration	$Prog^*$	$\triangleq$	$(\nu X \cdot Prog;\ X\ \sqcap\ 1)$.

In the above definitions s is a state in S and $\lceil K\rceil$ is the smallest up-, convex- and Cauchy-closed subset of distributions containing K. Programs are denoted by $Prog$ and $Prog'$, and the expression $(\nu X \cdot f.X)$ denotes the greatest fixed point of the function f — in the case of iteration the function is the monotone $\sqsubseteq$-program-to-program function $\lambda X \cdot (Prog;\ X\ \sqcap\ 1)$. All programs map $\top$ to $\{\delta_\top\}$.

Fig. 1. Mathematical operators on the space of programs [10]

where *flip* was defined at (1). It is easy to see that the iteration $halfFlip^*$ corresponds to a transition system which can (but does not have to) flip the state from $s = 0$ an arbitrary number of times. Thus after n iterations of $halfFlip$, the result set contains the distribution $\delta_0/2^n + (1-1/2^n)\delta_1$. Cauchy Closure implies the result distribution must contain δ_1 as well, because $\delta_0/2^n + (1-1/2^n)\delta_1$ converges to that point distribution as n approaches infinity.

We shall repeatedly make use of *tests*, defined as follows. Given a predicate B over the state s, we write $[B]$ for the test

$$(if\ B\ then\ skip\ else\ magic) ,\qquad(4)$$

viz. the program which skips if the initial state satisfies B, and behaves like a miracle otherwise. We use $[\neg B]$ for the *complement* of $[B]$. Tests are standard (non-probabilistic) programs which satisfy the following properties.

- $\text{skip} \sqsubseteq [B]$, meaning that the identity is refined by a test.
- $Prog\ ;\ [B]$ determines the *greatest probability* that $Prog$ may establish B. For example if $Prog$ is the program $biasFlip$ at (2), then $biasFlip\ ;\ [s = 0]$ is

$$s := 0\ _{1/2}\oplus \text{magic}\ \ \sqcap\ \ s := 0\ _{2/3}\oplus \text{magic}\ \ =\ \ s := 0\ _{2/3}\oplus \text{magic} ,$$

  a program whose probability of not blocking $(2/3)$ is the maximum probability that $biasFlip$ establishes $s = 0$.
- Similarly, $Prog\ ;\ [B]\ ;\ \text{chaos}_K\ =\ \text{magic}\ _{p_s}\oplus \text{chaos}_K$, where $(1-p_s)$ is the greatest probability that $Prog$ may establish B from initial state s, because chaos_K masks all information except for the probability that the test is successful.
- If $Prog$ contains no probabilistic choice, then $Prog$ distributes $\sqcap$, i.e. for any $Prog'$ and $Prog''$, we have $Prog; (Prog'\ \sqcap\ Prog'') = Prog; Prog'\ \sqcap\ Prog; Prog''$.

Now we have introduced a model for general probabilistic contexts, our next task is to investigate its program algebra. That is the topic of the next section.

2.1   Mapping *pKA* into $\mathcal{LS}$

Kleene algebra consists of a sequential composition operator (with a distinguished identity (1) and zero (0)); a binary plus (+) and unary star (∗). Terms are ordered by $\leq$ defined by + (see Fig.2), and both binary as well as the unary operators are monotone with respect to it. Sequential composition is indicated by the sequencing of terms in an expression so that ab means the program denoted by a is executed first, and then b. The expression $a + b$ means that either a or b is executed, and the Kleene star a^* represents an arbitrary number of executions of the program a.

In Fig.2 we set out the rules for the *probabilistic Kleene algebra, pKA*. We shall also use *tests*, whose denotations are programs of the kind (4). We normally denote a test by t, and for us its complement is $\neg t$.

The next definition gives an interpretation of *pKA* in $\mathcal{LS}$.

Definition 3. *Assume that for all simple variables x, the denotation $[\![x]\!] \in \mathcal{LS}$ as a program (including tests) is given explicitly. We interpret the Kleene operators over terms as follows:*

$$[\![1]\!] \; \hat{=} \; \text{skip} \; , \quad [\![0]\!] \; \hat{=} \; \text{magic} \; ,$$
$$[\![ab]\!] \; \hat{=} \; [\![a]\!]; [\![b]\!] \; , \quad [\![a + b]\!] \; \hat{=} \; [\![a]\!] \sqcap [\![b]\!] \; , \quad [\![a^*]\!] \; \hat{=} \; [\![a]\!]^* \; .$$

Here a and b stand for other terms, including simple variables.

We use $\geq$ for the order in *pKA*, which we identify with $\sqsubseteq$ from Def. 2; the next result shows that Def. 3 is a valid interpretation for the rules in 1, in that theorems in *pKA* apply in general to probabilistic programs.

Theorem 1. *([8]) Let $[\![\cdot]\!]$ be an interpretation as set out at Def. 3. The rules at Fig.2 are all satisfied, namely if $a \leq b$ is a theorem of pKA set out at Fig.2, then $[\![b]\!] \sqsubseteq [\![a]\!]$.*

To see why we cannot have equality at (†) in Fig.2, consider the expressions $a(b + c)$ and $ab + ac$, and an interpretation where a is *flip* at (1), and b is skip and c is $s := 1-s$. In this case in the interpretation of $a(b+c)$, the demon (at +) is free to make his selection after the probabilistic choice in a has been resolved, and for example could arrange to set the final state to $s = 0$ with probability 1, since if a sets it to 0 then the demon chooses to execute b, and if a sets it to 1, the demon may reverse it by executing c. On the other hand, in $ab + ac$, the demon must choose which of ab or ac to execute before the probability in a has been resolved, and either way there is a chance of at least $1/2$ that the final state is 1. (The fact that distribution fails says that there is more information available to the demon after execution of a than before.)

Similarly the rule at Fig.2 (‡) is not the usual one for Kleene-algebra. Normally this induction rule only requires a weaker hypothesis, but that rule, $ab \leq a \Rightarrow ab^* = a$, is unsound for the interpretation in $\mathcal{LS}$, again due to the interaction of probability and nondeterminism. Consider, for example, the interpretation where each of a, b and c represent the *flip* defined at (1) above. We may prove directly that $flip\ ;\ flip^* = s := 0 \sqcap s := 1$, i.e. $flip\ ;\ flip^* \neq flip$ in spite of the fact that $flip\ ;\ flip = flip$. To see why, we note that from Def. 3 the Kleene-star is interpreted as an iteration which may stop at any time. In this case, if a result $s = 1$ is required, then $flip$ executes for as long as necessary (probability theory ensures that $s = 1$ will eventually be satisfied). On the other hand if $s = 0$ is required then that result too may be guaranteed eventually by executing $flip$ long enough. To prevent an incorrect conclusion in this case, we use instead the sound rule (‡) (for which the antecedent fails). Indeed the effect of the $(1 + \cdot)$ in rule (‡) is to capture explicitly the action of the demon, and the hypothesis is satisfied only if the demon cannot skew the probabilistic results in the way illustrated above.

$(i)\ 0 + a = a$ $\qquad\qquad (viii)\ ab + ac \le a(b + c)\ \ (\dagger)$

$(ii)\ a + b = b + a$ $\qquad\quad (ix)\ (a + b)c = ac + bc$

$(iii)\ a + a = a$ $\qquad\qquad (x)\ a \le b\quad iff\quad a + b = b$

$(iv)\ a + (b + c) = (a + b) + c$

$(v)\ a(bc) = (ab)c$ $\qquad\qquad (xi)\ a^* = 1 + aa^*$

$(vi)\ 0a = a0 = 0$ $\qquad\qquad (xii)\ a(b + 1) \le a\quad \Rightarrow\quad ab^* = a\ \ (\ddagger)$

$(vii)\ 1a = a1 = a$ $\qquad\qquad (xiii)\ ab \le b\quad \Rightarrow\quad a^*b = b$

Fig. 2. Rules of Probabilistic Kleene algebra, pKA[8]

---

pKA purposefully treats probabilistic choice implicitly, and it is only the failure of the equality at (†) which suggests that the interpretation may include probability: in fact it is this property that characterises probabilistic-like models, separating them from those which contain only pure demonic nondeterminism. Note in the case that the interpretation is standard — where probabilities are not present in a — then the distribution goes through as usual. The use of implicit probabilities fits in well with our applications, where probability is usually confined to code residing at individual processors within a distributed protocol and nondeterminism refers to the arbitrary sequencing of actions that is controlled by a so-called *adversarial scheduler* [16]. For example, if a and b correspond to atomic program fragments (containing probability), then the expression $(a + b)^*$ means that either a or b (possibly containing probability) is executed an arbitrary number of times (according to the scheduler), and in any order — in other words it corresponds to the concurrent execution of a and b.

Typically a two-stage verification of a probabilistic distributed protocol might involve first the transformation a distributed implementation architecture, such as $(a + b)^*$, to a simple, separated specification architecture, such as a^*b^* (first a executes for an arbitrary number of times, and then b does), using general hypotheses, such as $ab = ba$ (program fragments a and b commute). The second stage would then involve a model-based analysis in which the hypotheses postulated to make the separation go through would be individually validated by examining the semantics in $\mathcal{LS}$ of the precise code for each. We do not deal with that stage here: indeed our purpose is precisely to make that stage a separate concern, not further complicated by the algorithm under study.

In the following sections we introduce our case study and illustrate how *pKA* may be used to simplify the overall analysis.

3   Mutual Exclusion with Bounded Waiting

In this section we describe the mutual exclusion protocol, and discuss how to apply the algebraic approach to it.

Let $P_1, \ldots P_N$ be N processes that from time to time need to have exclusive access to a shared resource.

The *mutual exclusion problem* is to define a protocol which will ensure both the exclusive access, and the "lockout free" property, namely that any process needing to access the shared resource will eventually be allowed to access it.

A protocol is said to satisfy the *bounded waiting condition* if, whenever no more than k processes are actively competing for the resource, each has probability at least α/k of obtaining it, for some fixed α (independent of N). [2]

The randomised solution we consider is based on one proposed by Rabin [6]. Processes can coordinate their activities by use of a shared "test-and-set" variable, so that "testing and setting" is an atomic action. The solution assumes an "adversarial scheduler", the mechanism which controls the otherwise autonomous executions of the individual P_i. The scheduler chooses nondeterministically between the P_i, and the chosen process then may perform a single atomic action, which might include the test and set of the shared variable together with some updates of its own local variables. Whilst the scheduler is not restricted in its choice, it must treat the processes fairly in the sense that it must always eventually schedule any particular process.

The broad outline of the protocol is as follows — more details are set out at Fig.3. Each process executes a program which is split into two phases, one voting, and one notifying. In the voting phase, processes participate in a lottery; the current winner's lottery number is recorded as part of the shared variable. Processes draw at most once in a competition, and the winner is notified when it executes its notification phase. The notification phase may only begin when the critical section becomes free.

Our aim is to show that when processes follow the above protocol, the bounded waiting condition is satisfied. Rabin observed [6] that in a lottery with k participants in which tickets are drawn according to (independent) exponential distributions, there is a probability of at least $1/3$ of a unique winner. However that model-based proof cannot be applied directly here, since it assumes (a) that

---

[2] Note that this is a much stronger condition than a probability α/N for some constant α, since it is supposed that in practice $k \ll N$.

there is no scheduler/probability interaction; (b) that the voting is unbiased between processes, and (c) that the voting may be separated from the notification. In Rabin's original solution, (c) was false (which led to the protocol's overall incorrectness); in fact both (a) and (b) are also not true, although the model-based argument still applies provided that the voting may be (almost) separated. We shall use an algebraic approach to do exactly that.

- *Voting phase.* P_i checks if it is eligible to vote, then draws a number randomly; if that number is strictly greater than the largest value drawn so far, it sets the shared variable to that value, and keeps a record. If P_i is ineligible to vote, it skips.
- *Notification phase.* P_i checks if it is eligible to test, and if it is, then checks whether its recorded vote is the same as the maximum drawn (by examining the shared variable); if it is, it sets itself to be the winner. If P is ineligible, then it just skips.
- *Release of the critical section.* When this is executed, the critical section becomes free, and processes may begin notification.

These events occur in a single round of the protocol; the verifier of the protocol must ensure that when these program fragments are implemented, they satisfy the algebraic properties set out at Fig.4.

Fig. 3. The key events in a single round of the mutual exclusion protocol

---

4   Separation Theorems and Their Applications

In this section we extend some standard separation theorems of Cohen [1] to the probabilistic context, so that we may apply them to the mutual exclusion problem set out above. Although the lemmas are somewhat intricate we stress their generality: proved once, they can be used in many applications.

Our first results at Lem. 1 consider a basic iteration, generalising loop-invariant rules to allow the body of an iteration to be transformed by passage of a program a.

Lemma 1.

$$a(b+1) \leq ca + d \quad \Rightarrow \quad ab^* \leq c^*(a + db^*) \tag{5}$$

$$ac \leq cb \quad \Rightarrow \quad a^*c \leq cb^* \tag{6}$$

Proof. The proof of (5) is set out elsewhere [8]. For (6) we have the following inequalities, justified by the hypothesis and monotonicity.

$$acb^* \quad \leq \quad cbb^* \quad \leq \quad cb^* .$$

Now applying (xiii), we deduce that $a^*cb^* = cb^*$, and the result now follows since $a^*c \leq a^*cb^*$.

Note that the weaker commutativity condition of $ab \leq ca + d$ will not do at (5), as the example with $a, b, c \mathrel{\hat{=}} flip$ and $d \mathrel{\hat{=}} magic$ illustrate. In this case we see, that a^* and b^* both correspond to the program $(s := 0 \sqcap s := 1)$, and this is not the same as the corresponding interpretation for c^*a which corresponds to $flip$ again.

Lem. 1 implies that with suitable commutativity between phases a and b of a program, an iteration involving the interleaving of the phases may be thought of as executing in two separated pieces. Note that again we need to use a hypothesis $b(1 + a) \leq (1 + a)b$, rather than a weaker $ba \leq ab$.

Lemma 2. $b(1 + a) \leq (1 + a)b^* \quad \Rightarrow \quad (a + b)^* \leq a^*b^*$.

Proof. We reason as follows

$$
\begin{aligned}
& (a + b)^* \\
\leq\ & (a + b)^*a^*b^* && 1 \leq a^*b^* \\
\leq\ & a^*b^* . && (a + b)a^*b^* \leq a^*b^*,\ see\ below;\ (xiii)
\end{aligned}
$$

For the "see below", we argue

$$
\begin{aligned}
& (a + b)a^*b^* \\
=\ & aa^*b^* + ba^*b^* \\
\leq\ & a^*b^* + b(1 + a)^*b^* && a \leq a^* \leq a^* \leq (1 + a)^* \\
\leq\ & a^*b^* + (1 + a)^*b^*b^* && hyp;\ (5) \Rightarrow b(1 + a)^* \leq (1 + a)^*b^* \\
=\ & a^*b^* .
\end{aligned}
$$

5   The Probability That a Participating Process Loses

We now show how the lemmas of Sec. 4 can be applied to the example of Sec. 3: we show how to compute the probability that a particular process P (one of the P_i's) participating in the lottery loses. Writing V and T for the two phases of P, respectively *vote* and *notify* (recall Fig.3) and representing scheduler choice by "+", we see that the chance that P loses can be expressed as

$$(V + T + \widetilde{V} + \widetilde{T} + C)^* \, A \ ,$$

where $\widetilde{V} \mathrel{\hat{=}} +_{P_i \neq P} V_i$, and $\widetilde{T} \mathrel{\hat{=}} +_{P_i \neq P} T_i$ are the two phases (overall) of the remaining processes, and A tests for whether P has lost. Thus A is a test of the form "skip if P has not drawn yet, or has lost, otherwise magic ", followed by an abstraction function which forgets the values of all variables except those required to decide whether P has lost or not.

The crucial algebraic properties of the program fragments are set out at Fig.4, and as a separate analysis the verifier must ensure that the actual code fragments implementing the various phases of the protocol satisfy them. This task however is considerably easier than analysing an explicit model of the distributed architecture, because the code fragments can be treated one-by-one, in isolation.

The next lemma uses separation to show show that we can separate the voting from the notification within a single round, with the round effectively ending the voting phase with the execution of the critical section.

1. *Voting and notification commute:* $V_i T_j = T_j V_i$.
2. *Notification occurs when the critical section is free:* $T_j(C+1) \leq (C+1)T_j$.
3. *Voting occurs when the critical section is busy:* $C(V_j+1) \leq (V_j+1)C$.
4. *It's more likely to lose, the later the vote:* $VA(\widetilde{V}A+1) \leq (\widetilde{V}A+1)VA$.

Here V corresponds to a distinguished process P's voting phase, V_i to P_i's voting phase, and $\widetilde{V}$ to the nondeterministic choice of all the voting phases (not P's). Similarly T and T_j are the various notification phases. A essentially tests for whether P has lost or not.

Fig. 4. Algebraic properties of the system

Lemma 3. $(V + T + \widetilde{V} + \widetilde{T} + C)^* \quad \leq \quad (V + \widetilde{V})^* \, C^* \, (T + \widetilde{T})^*$.

Proof. We use Lem. 2 twice, first to pull $(T + \widetilde{T})$ to the right of everything else, and then to pull $(V + \widetilde{V})$ to the left of C. In detail, we can verify from Fig.4, that

$$(T + \widetilde{T}) \, (V + \widetilde{V} + C + 1) \quad \leq \quad (V + \widetilde{V} + C + 1) \, (T + \widetilde{T})^* \, ,$$

(since $T + \widetilde{T}$ has a standard denotation, so distributes +) to deduce from Lem. 2 that $(V + T + \widetilde{V} + \widetilde{T} + C)^* \leq (V + \widetilde{V} + C)^* \, (T + \widetilde{T})^*$. Similarly $C(V + \widetilde{V} + 1) \leq (V + \widetilde{V} + 1)C^*$, so that $(V + \widetilde{V} + C)^* \leq (V + \widetilde{V})^* \, C^*$.

Next we may consider the voting to occur in an orderly manner in which the selected processor P votes last, with the other processors effectively acting as a "pool" of anonymous opponents who collectively "attempt" to lower the chance that P will win — this is the fact allowing us to use the model-based observation of Rabin to compute a lower bound on the chance that P wins.

Lemma 4. $(V + \widetilde{V})^* A \leq \widetilde{V}^*(VA)^*$.

Proof. We reason as follows.

	$(V + \widetilde{V})^* A$	
$\leq$	$A(VA + \widetilde{V}A)^*$	*see below*
$\leq$	$A(\widetilde{V}A)^*(VA)^*$	*Fig.4 (4); Lem. 2*
$\leq$	$\widetilde{V}^*(VA)^*$.	*P not voted, implies $A\widetilde{V}A = \widetilde{V}$*

For the "see below" we note that

	$(V + \widetilde{V})A(VA + \widetilde{V}A)^*$	
$=$	$(VA + \widetilde{V}A)(VA + \widetilde{V}A)^*A$	$A(V + \widetilde{V}) = (V + \widetilde{V})A$, then (6), (5)
$\leq$	$(VA + \widetilde{V}A)^*(VA + \widetilde{V}A)^*A$	
$=$	$A(VA + \widetilde{V}A)^*$,	

so that $(V + \widetilde{V})^* A(VA + \widetilde{V}A)^* \leq A(VA + \widetilde{V}A)^*$ by (xiii), and therefore that $(V + \widetilde{V})^* A \leq A(VA + \widetilde{V}A)^*$.

The calculation above is based on the assumption that P is eligible to vote when it is first scheduled in a round. The mechanism for testing eligibility uses a round number as part of the shared variable, and after a process votes, it sets a local variable to the same value as the round number recorded by the shared variable. By this means the process is prevented from voting more than once in any round. In the case that the round number is unbounded, P will indeed be eligible to vote the first time it is scheduled. However one of Rabin's intentions was to restrict the size of the shared variable, and in particular the round number. His observation was that round numbers may be reused provided they are chosen *randomly* at the start of the round, and that the *scheduler cannot see the result* when it decides which process to schedule. In the next section we discuss the implications of this assumption on $\mathcal{L}$ and pKA.

6   Secrecy and Its Algebraic Characterisation

The actual behaviour of Rabin's protocol includes *probabilistically* setting the round number, which we denote R and which makes the protocol in fact

$$R(V + T + \widetilde{V} + \widetilde{T} + C)^* \ . \tag{7}$$

The problem is that the interpretation in $\mathcal{L}S$ assumes that the value chosen by R is observable by all, in particular by the adversarial scheduler, that latter implying that the scheduler can use the value during voting to determine whether to schedule P. In a multi-round scenario, that would in turn allow the policy that P is scheduled *only* when its just-selected round variable is (accidentally) the same as the current global round: while satisfying fairness (since that equality happens infinitely often with probability one), it would nevertheless allow P to be scheduled only when it cannot possibly win (in fact will not even be allowed to vote).

   Clearly that strategy must be prevented (if the algorithm is to be correct!) — and it is prevented *provided* the scheduler cannot see the value set by R. Thus we need a model to support algebraic characterisations for "cannot see".

   The following (sketched) description of a model $\mathcal{Q}S$ [9, Key QMSRM] — necessarily more detailed than QS — is able to model cases where probabilistic outcomes cannot be seen by subsequent demonic choice. The idea (based on "non-interference" in security) is to separate the state into *visible* and *hidden* parts, the latter not accessible directly by demonic choice. The state s is now a pair (v, h) where v, like s, is given some conventional type but h now has type *distribution* over some conventional type. The $\mathcal{Q}S$ model is effectively the $\mathcal{L}S$ model built over this more detailed foundation.[3]

   For example, if a sets the hidden h probabilistically to 0 or 1 then (for some p) in the $\mathcal{Q}S$ model a denotes

Hidden resolution of probability. $(v, h) \quad \overset{a}{\mapsto} \quad \{ (v, (0 \ {}_p{\oplus}\ 1)) \} \ .$[4]

---

[3] Thus we have "distributions over values-and-distributions" so that the type of a program in $\mathcal{Q}S$ is $(V \times \overline{H}) \to \mathbb{P}(\overline{V \times \overline{H}})^{\top}$, that is $\mathcal{L}S$ where $S = V \times \overline{H}$.

[4] Strictly speaking we should write $\delta_0 \ {}_p{\oplus}\ \delta_1$.

In contrast, if b sets the visible v similarly we'd have b denoting

Visible resolution of probability. $(v, h) \quad \overset{b}{\mapsto} \quad \{\, (0, h)\,_{1/2} \oplus (1, h) \,\}$.

The crucial difference between a and b above is in their respective interactions with subsequent nondeterminism; for we find

$$a(c + d) \quad = \quad ac + ad$$
$$\text{but in general} \quad b(c + d) \quad \neq \quad bc + bd \ ,$$

because in the a case the nondeterminism between c and d "cannot see" the probability hidden in h. In the b case, the probability (in v) is not hidden.

A second effect of hidden probability is that tests are no longer necessarily "read-only". For example if t denotes the test $[h = 0]$ then we would have (after a say)

$$(v, \ (0 \ _p \oplus 1) \ ) \quad \overset{t}{\mapsto} \quad \{(v, 0) \ _p \oplus \ \mathsf{magic} \,\}$$

where the test, by its access to h, has revealed the probability that was formerly hidden and, in doing so, has changed the state (in what could be called a particularly subtle way — which is precisely the problem when dealing with these issues informally!)

In fact this state-changing property gives us an algebraic characterisation of observability.

Definition 4. *Observability; resolution.*
For any program a and test t we say that "t is known after a" just when

$$a(t + \neg t) \quad = \quad a \ . \tag{8}$$

As a special case, we say that "t is known" just when $t + \neg t = 1$.

Say that Program a "contains no visible probability" just when for all programs b, c we have

$$a(b + c) \quad = \quad ab + ac \ .$$

Thus the distributivity through $+$ in Def. 4 expresses the adversary's ignorance in the case that a contains hidden probabilistic choice. If instead the choice were visible, then the $+$-distribution would fail: if occurring first it could not see the probabilistic choice[5] whereas, if occurring second, it could.

Secrecy for the Randomised Round Number
We illustrate the above by returning to mutual exclusion. Interpret R as the random selection of a local round number (as suggested above), and consider the probability that the adversarial scheduler can guess the outcome. For example, if the adversary may guess the round number with probability 1 during the voting phase, according to Def. 4 we would have

$$R \ (V + \widetilde{V})^*([rn = 0] + [rn = 1]) \ \mathsf{chaos} \quad = \quad \mathsf{chaos} \ ,$$

---

[5] Here it cannot see it because it has not yet happened, not because it is hidden.

(because $[rn = 0] + [rn = 1]$ would be skip).[6] But since $(V + \widetilde{V} + 1)([rn = 0] + [rn = 1]) = ([rn = 0] + [rn = 1])(V + \widetilde{V} + 1)$ we may reason otherwise:

$$
\begin{aligned}
& R(V + \widetilde{V})^*([rn = 0] + [rn = 1])\text{chaos} \\
= \ & R([rn = 0] + [rn = 1])(V + \widetilde{V})^*\text{chaos} && \text{Lem. 1}\\
= \ & R[rn = 0]\text{chaos} + R[rn = 1]\text{chaos} , && \text{Def. 4 and (8)}
\end{aligned}
$$

which, now back in the model we can compute easily to be magic $_{1/2}\oplus$ chaos, deducing that the chance that the scheduler may guess the round number is at most $1/2$, and not 1 at all.

7 Conclusions and Other Work

Rabin's probabilistic solution to the mutual exclusion problem with bounded waiting is particularly apt for demonstrating the difficulties of verifying probabilistic protocols, as the original solution contained a particularly subtle flaw [15]. The use of pKA makes it clear what assumptions need to be checked relating to the individual process code and the interaction with the scheduler, and moreover a model-based verification of a complex distributed architecture is reduced to checking the appropriate hypotheses are satisfied. Our decision to introduce the models separately stems from QSH's complexity to $\mathcal{LS}$, and the fact that in many protocols $\mathcal{LS}$ is enough. The nice algebraic characterisations of hidden and visible state, may suggest that QSH may support a logic for probabilities and ignorance in the refinement context, though that remains an interesting topic for research.

Others have investigated instances of Rabin's algorithm using model checking [13]; there are also logics for "probability-one properties" [7], and models for investigating the interaction of probability, knowledge and adversaries [2].

There are other variations on Kleene Algebra which allow for the relaxation of distributivity laws [11], including those which are equivalent to pKA, except for the left 0-annihilation [17].

References

1. E. Cohen. Separation and reduction. In *Mathematics of Program Construction, 5th Intern. Conference*, volume 1837 of *LNCS*, pages 45–59. Springer, July 2000.
2. J. Halpern and M. Tuttle. Knowledge, probabilities and adversaries. *J. ACM*, 40(4):917–962, 1993.
3. Jifeng He, K. Seidel, and A.K. McIver. Probabilistic models for the guarded command language. *Science of Computer Programming*, 28:171–92, 1997. Earlier appeared in Proc. FMTA '95, Warsaw, May 1995. Available at [9, key HSM95].

---

[6] Here we are abusing notation, by using program syntax directly in algebraic expressions.

4. Joe Hurd. A formal approach to probabilistic termination. In Víctor A. Carreño, César A. Muñoz, and Sofiène Tahar, editors, *15th International Conference on Theorem Proving in Higher Order Logics: TPHOLs 2002*, volume 2410 of *LNCS*, pages 230–45, Hampton, VA., August 2002. Springer. www.cl.cam.ac.uk/~jeh1004/research/papers.
5. D. Kozen. Kleene algebra with tests. *ACM Transactions on Programming Languages and Systems (TOPLAS)*, 19(3):427–443, 1997.
6. Eyal Kushilevitz and M.O. Rabin. Randomized mutual exclusion algorithms revisited. In *Proc. 11th Annual ACM Symp. on Principles of Distributed Computing*, 1992.
7. D. Lehmann and S. Shelah. Reasoning with time and chance. *Information and Control*, 53(3):165–98, 1982.
8. A. McIver and T. Weber. Towards automated proof support for probabilistic distributed systems. In *Proceedings of Logic for Programming and Automated Reasoning*, volume 3835 of *LNAI*, pages 534–548. Springer, 2005.
9. A.K. McIver, C.C. Morgan, J.W. Sanders, and K. Seidel. Probabilistic Systems Group: Collected reports. web.comlab.ox.ac.uk/oucl/research/areas/probs.
10. Annabelle McIver and Carroll Morgan. *Abstraction, Refinement and Proof for Probabilistic Systems*. Technical Monographs in Computer Science. Springer, 2004.
11. B. Moeller. Lazy Kleene Algebra. In D. Kozen and C. Shankland, editors, *MPC*, volume 3125 of *Lecture Notes in Computer Science*, pages 252–273. Springer, 2004.
12. C.C. Morgan, A.K. McIver, and K. Seidel. Probabilistic predicate transformers. *ACM Transactions on Programming Languages and Systems*, 18(3):325–53, 1996. doi.acm.org/10.1145/229542.229547.
13. PRISM. Probabilistic symbolic model checker. www.cs.bham.ac.uk/~dxp/prism.
14. M.O. Rabin. N-process mutual exclusion with bounded waiting by 4 log 2n-valued shared variable. *Journal of Computer and System Sciences*, 25(1):66–75, 1982.
15. I. Saias. Proving probabilistic correctness statements: the case of Rabin's algorithm for mutual exclusion. In *Proc. 11th Annual ACM Symp. on Principles of Distributed Computing*, 1992.
16. Roberto Segala. *Modeling and Verification of Randomized Distributed Real-Time Systems*. PhD thesis, MIT, 1995.
17. T. Takai and H. Furusawa. Monadic tree Kleene Algebra. This proceedings: *Relations and Kleene Algebras in Computer Science* LNCS 4136.

A   Equational Identities That Still Apply in *pKA*

$$a^*a^* = a^* \tag{9}$$
$$a^*(b+c) = a^*(a^*b + a^*c) \tag{10}$$

Monotone Predicate Transformers as Up-Closed Multirelations

Ingrid Rewitzky and Chris Brink

Department of Mathematical Sciences, University of Stellenbosch, South Africa

Abstract. In the study of semantic models for computations two independent views predominate: relational models and predicate transformer semantics. Recently the traditional relational view of computations as binary relations between states has been generalised to multirelations between states and properties allowing the simultaneous treatment of angelic and demonic nondeterminism. In this paper the two-level nature of multirelations is exploited to provide a factorisation of up-closed multirelations which clarifies exactly how multirelations model nondeterminism. Moreover, monotone predicate transformers are, in the precise sense of duality, up-closed multirelations. As such they are shown to provide a notion of effectivity of a specification for achieving a given postcondition.

1 Introduction

Until recently it was commonly accepted that if programs and specifications are to be modelled in a single framework then a predicate transformer semantics could be defined in terms of monotone predicate transformers but there is no traditional relational model. However, game theoretic descriptions [16,3] of a specification computation with both angelic and demonic nondeterminism have suggested that there is indeed a relational representation in terms of binary multirelations as introduced in [24]. The basic idea is that binary multirelations, being relations from states to sets of states, specify computations at the level of properties that an atomic step has to satisfy, while binary relations, being relations between states, specify computations at the level of states that an atomic step may reach.

Within the multirelational model we may still define the traditional relational model capturing angelic- or demonic nondeterminism. But in addition multirelations can model two kinds of nondeterminism: demonic nondeterminism in terms of states at the level of the computations specified, and angelic nondeterminism in terms of properties at the level of the specifications. This multirelational model is more than just an empty generalisation of the traditional relational model. It in fact corresponds, in the precise sense of a Jónsson/Tarski [17] duality, to monotone predicate transformer semantics. This and lattice-theoretic properties of families of multirelations have been studied in [24]. A subsequent paper [18] demonstrates how multirelations can be used for the specification of multi-agent systems involving human-information interactions including resource sharing protocols and games.

R.A. Schmidt (Ed.): RelMiCS /AKA 2006, LNCS 4136, pp. 311–327, 2006.
© Springer-Verlag Berlin Heidelberg 2006

This paper begins with a description of multirelations and their operations and properties useful for modelling angelic and demonic nondeterminism. Here the presentation differs from that in [24] in that it emphasises the two levels at which nondeterminism is being modelled. In Section 3 monotone predicate transformers are shown to correspond in the precise sense of duality to upclosed multirelations. The topological perspective of multirelations introduced in Section 4 leads to a new and illuminating topological characterisation of strongest postconditions. Section 5 considers two factorisations: a factorisation of binary multirelations which reveals exactly how multirelations can model two kinds of nondeterminism, and a generalisation of the known [12] factorisation of monotone predicate transformers which provides a notion of effectivity of a computation for achieving a given postcondition.

2   Relations and Multirelations

In standard relational models for programs a binary relation specifies the input-output behaviour of a program in terms of the states it may reach from a given state. Lifting this description from the level of states to the level of properties, and using an idea that goes back to Hoare's [15] seminal paper of 1969, the behaviour of a program α may be specified in terms of the postconditions (or the properties) that it has to satisfy, that is, if α has demonic choice then

$$sR_\alpha Q \quad \text{iff} \quad \text{every execution of program } \alpha, \text{ from state } s,$$
$$\text{reaches a state in which } Q \text{ is true.}$$

or, if α has angelic choice then

$$sR_\alpha Q \quad \text{iff} \quad \text{some execution of program } \alpha, \text{ from state } s,$$
$$\text{reaches a state in which } Q \text{ is true.}$$

We now use these ideas to move towards a relational representation of specification computations with demonic and angelic nondeterminism. One way to think of such a specification α is as a two-player two-step game of choice. We may refer colloquially to the players as 'the user' and 'the machine'. The user makes the first move by selecting a set of possible winning positions, and the machine makes the second move by selecting the actual final position from this set. The game is lost by the player who is faced with the choice from an empty set. It is assumed that the user will take any available opportunity to win, while the machine assumes the role of a devil's advocate who is trying to make sure the user loses. In this context the angelic choices are interpreted as those made by the user, while the demonic choices are those made by the machine. A loss for the user is the empty family, and the user has the opportunity of winning the game whenever the family of choices contains the empty set. From each starting position s the user may have the choice of more than one set of possible winning positions. Therefore, the set

$$\{ Q \mid sR_\alpha Q \}$$

captures the angelic choices available to the user (or angel). If this set includes the empty set then it is a winning strategy for the angel. While for each Q with $sR_\alpha Q$, the set

$$\{t \mid t \in Q\}$$

captures the choices available to the machine (or demon). Therefore, a specification computation α may be represented as a relation R_α, the idea being that α, when started in state s is guaranteed to achieve postcondition Q for some angelic choice regardless of the demonic choices. We may formalise this representation of a specification computation α in terms of a binary multirelation R_α relating states and postconditions.

Definition 1. Let X and Y be sets. A binary multirelation is a subset of the Cartesian product $X \times \mathcal{P}(Y)$, that is, a set of ordered pairs (x, Q) where $x \in X$ and $Q \subseteq Y$. The image under a multirelation R of any $x \in X$ is denoted $R(x)$ and defined to be the set $\{Q \subseteq Y \mid xRQ\}$. Mostly we will deal with the case of $X = Y = S$.

For any binary multirelations $R, T \subseteq S \times \mathcal{P}(S)$, their composition may be defined as follows:

$$R \, {}_9^\circ \, T = \{(s, Q) \mid (\exists Q')[sRQ' \text{ and } Q' \subseteq \{y \mid yTQ\}]\}.$$

So, given input value s, the angel can only guarantee that $R \, {}_9^\circ \, T$ will achieve postcondition Q if s/he can ensure that R will establish some intermediate postcondition Q' and if s/he can also guarantee that T will establish Q given any value in Q'.

Multirelations model two kinds of nondeterminism: angelic nondeterminism captured in terms of sets $\{Q \mid sRQ\}$ of properties or postconditions at the level of the specifications, and demonic nondeterminism in terms of sets $\{t \mid t \in Q\}$ of states at the level of the computations specified. So there are two levels at which multirelations may be compared. At the level of specifications a comparison of multirelations may be based on the number of sets Q each relates to a given state s. That is, for multirelations $R, T \subseteq S \times \mathcal{P}(S)$,

$$R \sqsubseteq_a T \quad \text{iff} \quad R \supseteq T \quad \text{iff} \quad \forall s \in S, \ \{Q \mid sRQ\} \supseteq \{Q \mid sTQ\},$$

with the intuition that T is 'better' from the demon's perspective than R if T has less angelic choice (and possible more demonic choice) than R. This provides a notion of *angelic refinement* of R by T since the angelic nondeterminism is reduced. At the level of computations a comparison of multirelations may be based on the size of the sets Q related to a given state s. That is, for multirelations $R, T \subseteq S \times \mathcal{P}(S)$,

$$R \sqsubseteq_d T \quad \text{iff} \quad (\forall s \in S)(\forall Q \in R(s))(\exists Q' \in T(s))[Q' \subseteq Q]$$

with the intuition that T is 'better' from the angel's perspective than R if T has less demonic choice (and possible more angelic choice) than R. This provides a notion of *demonic refinement* of R by T since the demonic nondeterminism is reduced.

We distinguish two extreme multirelations, namely the *universal multirelation* defined by

$$\top^+ = \{(s, Q) \mid s \in S \text{ and } Q \subseteq S\}$$

and the *empty multirelation* $\bot^+ = \emptyset$. Then for all multirelations $R \subseteq S \times \mathcal{P}(S)$,

$$\top^+ \sqsubseteq_a R \qquad \text{and} \qquad R \sqsubseteq_d \top^+,$$

that is, $\top^+$ has the most angelic choice and the least demonic choice (since $sR\emptyset$, for each $s \in S$). Dually, for all multirelations $R \subseteq S \times \mathcal{P}(S)$,

$$R \sqsubseteq_a \bot^+ \qquad \text{and} \qquad \bot^+ \sqsubseteq_d R,$$

that is, $\bot^+$ has the least angelic choice (since $sR\emptyset$, for each $s \in S$) and the most demonic choice .

With respect to the refinement orderings $\sqsubseteq_a$ and $\sqsubseteq_d$, notions of lub and glb may be defined: $\sqcup_a$ and $\sqcap_a$ are defined in terms of intersection and union of subsets of $\mathcal{P}(S)$ respectively, and hence at the level of the specifications; while $\sqcup_d$ and $\sqcap_d$ are defined as union or intersection of subsets of S respectively, and hence at the level of the computations specified.

Definition 2. *Let R and T be binary multirelations over a set S.*

Angelic intersection	$R \sqcap_a T = \{(s, Q) \mid sRQ \text{ or } sTQ\}.$
Angelic union	$R \sqcup_a T = \{(s, Q) \mid sRQ \text{ and } sTQ\}.$
Demonic intersection	$R \sqcap_d T = \{(s, P_1 \cap P_2) \mid sRP_1 \text{ and } sTP_2\}.$
Demonic union	$R \sqcup_d T = \{(s, P_1 \cup P_2) \mid sRP_1 \text{ and } sTP_2\}.$

Since the image set of any $s \in S$ under any binary multirelation is an element of $\mathcal{P}(\mathcal{P}(S))$ the demonic union and intersection may be viewed as pointwise extensions of the *power operations* [7,8] of union and intersection on the powerset Boolean algebra $\mathcal{P}(\mathcal{S}) = (\mathcal{P}(S), \cup, \cap, ^-, \emptyset, S)$. The order $\sqsubseteq_d$ is a pre-order given by the *upper power order*, in the sense of [7], of set inclusion.

With these operations we may give a semantics of program and specification constructs in extensions [19,21,2] of Dijkstra's guarded command language [9,10]. For example,

Definition 3.

No op	$R_{\text{skip}} = \{(s, Q) \mid s \in S \text{ and } s \in Q\}$
divergence	$R_{\text{abort}} = \bot^+$
miracle	$R_{\text{magic}} = \top^+$
sequential composition	$R_{\alpha;\beta} = R_\alpha \, \mathbin{\S} \, R_\beta$
angelic choice	$R_{\alpha \sqcap \beta} = R_\alpha \sqcap_a R_\beta$
demonic choice	$R_{\alpha \sqcup \beta} = R_\alpha \sqcup_d R_\beta.$

Binary multirelations have many interesting and useful properties. Here are some of them.

Definition 4. *Let $R \subseteq S \times P(S)$ be a binary multirelation. Then*

(a) R *is* proper *if, for each $s \in S$, $R(s) \neq \emptyset$.*
(b) R *is* total *if, for each $s \in S$, $\emptyset \notin R(s)$.*
(c) R *is* up-closed *if, for each $s \in S$ and any $Q \subseteq S$,*

$$sRQ \quad \text{iff} \quad \forall Q' \subseteq S,\ Q \subseteq Q' \Rightarrow sRQ'.$$

(d) R *is* multiplicative *if, for each $s \in S$ and any non-empty $\mathbf{Q}$ of subsets of S,*

$$sR(\cap \mathbf{Q}) \quad \text{iff} \quad \forall Q \in \mathbf{Q},\ sRQ.$$

(e) R *is* additive *if, for each $s \in S$ and any non-empty set $\mathbf{Q}$ of subsets of S*

$$sR(\cup \mathbf{Q}) \quad \text{iff} \quad \exists Q \in \mathbf{Q},\ sRQ.$$

For the remainder of the paper we will consider only up-closed multirelations since these are the multirelations that are, in the sense of the duality of Section 3, monotone predicate transformers. As shown in [24], the family of up-closed binary multirelations over S has a very rich lattice-theoretic structure inherited from the lattice of up-closed sets of the powerset Boolean algebra $P(S) = (P(S), \cup, \cap, ^-, \emptyset, P(S))$. Formally,

Theorem 1. *The family of up-closed binary multirelations is a complete ring of sets in which*

$$\bigwedge \{R_i \mid i \in I\} \quad = \quad \cap \{R_i \mid i \in I\} \quad = \quad \sqcup_d \{R_i \mid i \in I\}$$
$$\bigvee \{R_i \mid i \in I\} \quad = \quad \cup \{R_i \mid i \in I\} \quad = \quad \sqcap_a \{R_i \mid i \in I\}.$$

The bottom element is $\perp^+$ and the top element is $\top^+$. The finite elements are the finite joins of proper multiplicative multirelations. The completely join-irreducible elements are the proper multiplicative multirelations and the completely meet-irreducible elements are the total additive multirelations. □

Since a proper multiplicative multirelation is completely join-irreducible it cannot be expressed as a union of distinct non-empty multirelations, that is, it contains no angelic choice and hence will be referred to as a *demonic multirelation*. Dually, a total additive multirelation contains no demonic choice and will be referred to as an *angelic multirelation*.

The next theorem provides some distributivity properties of multirelational composition; the proofs are easy and left for the reader.

Theorem 2. *For any up-closed multirelations $R, T_1, T_1 \subseteq S \times P(S)$,*

(a) $\top^+ \mathbin{_9^o} R = \top^+$
    $R \mathbin{_9^o} \top^+ = \top^+$, *if R is proper.*
(b) $\perp^+ \mathbin{_9^o} R = \perp^+$
    $R \mathbin{_9^o} \perp^+ = \perp^+$, *if R is total.*

(c) $T_1 \subseteq T_2$ *implies* $T_1 \cap R \subseteq T_2 \cap R$ *and* $T_1 \cup R \subseteq T_2 \cup R$.
$T_1 \subseteq T_2$ *implies* $T_1 \, \mathring{\S} \, R \subseteq T_2 \, \mathring{\S} \, R$ *and* $R \, \mathring{\S} \, T_1 \subseteq R \, \mathring{\S} \, T_2$.

(d) $(T_1 \cap T_2) \, \mathring{\S} \, R \subseteq T_1 \, \mathring{\S} \, R \cap T_2 \, \mathring{\S} \, R$
$R \, \mathring{\S} \, (T_1 \cap T_2) = R \, \mathring{\S} \, T_1 \cap R \, \mathring{\S} \, T_2$, *if R is multiplicative.*

(e) $(T_1 \cup T_2) \, \mathring{\S} \, R = T_1 \, \mathring{\S} \, R \cup T_2 \, \mathring{\S} \, R$
$R \, \mathring{\S} \, (T_1 \cup T_2) = R \, \mathring{\S} \, T_1 \cup R \, \mathring{\S} \, T_2$, *if R is additive.*   □

We conclude this description of multirelations with the correspondence between demonic multirelations and binary relations, and between angelic multirelations and binary relations. Any binary relation r over the state space S may be viewed as a demonic multirelation $R_d \subseteq S \times \mathcal{P}(S)$ given by:

$$s R_r^d Q \quad \text{iff} \quad (\forall t)[srt \Rightarrow t \in Q], \qquad \text{for any } s \in S \text{ and any } Q \subseteq S,$$

and as an angelic binary multirelation $R_a \subseteq S \times \mathcal{P}(S)$ given by:

$$s R_r^a Q \quad \text{iff} \quad (\exists t)[srt \wedge t \in Q], \qquad \text{for any } s \in S \text{ and any } Q \subseteq S.$$

It is easy to check that R_r^d is a demonic multirelation and R_r^a an angelic multirelation. To each demonic multirelation $R \subseteq S \times \mathcal{P}(S)$ there corresponds some binary relation $r_R \subseteq S \times S$ given by $r_R(s) = \bigcap \{Q \mid Q \in R(s)\}$ for $s \in S$. Dually, if R is an angelic multirelation then the corresponding binary relation $r_R \subseteq S \times S$ is given by $r_R(s) = \bigcap \{Q \mid \overline{Q} \notin R(s)\}$ for $s \in S$. It is easy to check that for any demonic multirelation R over S, $R = R_{r_R}^d$; for any angelic multirelation R over S, $R = R_{r_R}^a$; and for any binary relation r over S, $r = r_{R_r^a} = r_{R_r^d}$.

3   Up-Closed Multirelations Are Monotone Predicate Transformers

In this section we establish a bijective correspondence between up-closed binary multirelations and monotone predicate transformers which is a generalisation of Jónsson/Tarski duality [17] for binary relations and certain monotone unary operators. For monotone predicate transformers $g : \mathcal{P}(S) \to \mathcal{P}(S)$ the correspondence is somewhat trivial, and involves finding the states from which a specification will achieve a given postcondition. The correspondence for monotone predicate transformers g over a (not necessarily powerset) Boolean algebra B is more interesting in that it invokes the canonical extension[13,14] of g.

For an up-closed binary multirelation $R \subseteq S \times \mathcal{P}(S)$, a predicate transformer g_R over $\mathcal{P}(S)$ is defined by

$$g_R(Q) = \{s \in S \mid sRQ\}, \qquad \text{for any } Q \subseteq S.$$

Since R is up-closed, g_R is monotone. Conversely, for a monotone predicate transformer $g : \mathcal{P}(S) \to \mathcal{P}(S)$, a multirelation $R_g \subseteq S \times \mathcal{P}(S)$ is defined by

$$s R_g Q \quad \text{iff} \quad s \in g(Q), \qquad \text{for any } s \in S \text{ and any } Q \subseteq S.$$

Since g is monotone, R_g is up-closed. As a trivial consequence of these definitions, we have

$$R_{g_R} = R \qquad \text{and} \qquad g_{R_g} = g.$$

Therefore, up-closed binary multirelations are monotone predicate transformers over a powerset Boolean algebra. In what follows we show that the restriction to powerset Boolean algebras can be dropped.

Definition 5. *A binary multirelational structure* $\mathbf{S} = (S, \{R_i \mid i \in I\})$ *is such that S is a set and $\{R_i \mid i \in I\}$ is a collection of up-closed binary multirelations over S.*

By translating each up-closed binary multirelation R in a binary multirelational structure into a monotone predicate transformer g_R we obtain a certain kind of Boolean algebra with operators. Since we wish to establish a bijective correspondence we define these in general.

Definition 6. *A Boolean algebra with monotone predicate transformers*

$$\mathbf{B} = (B, \vee, \wedge, {}^-, 0, 1, \{g_i \mid i \in I\})$$

is such that $\mathbf{B} = (B, \vee, \wedge, {}^-, 0, 1)$ *is a Boolean algebra and* $\{g_i \mid i \in I\}$ *is a collection of monotone predicate transformers over B.*

In order to establish a bijective correspondence between Boolean algebras with monotone predicate transformers and binary multirelational structures, we must show how each gives rise to and can be recovered from the other. From above we have,

Theorem 3. *Given any binary multirelational structure* $\mathbf{S} = (S, \{R_i \mid i \in I\})$ *its power algebra* $\mathcal{P}(\mathbf{S}) = (\mathcal{P}(S), \cup, \cap, {}^-, \emptyset, S, \{g_{R_i} \mid i \in I\})$ *is a Boolean algebra with monotone predicate transformers.*

Next we show that any Boolean algebra with monotone predicate transformers in turn gives rise to a binary multirelational structure by invoking the basic Stone representation [25]. That is, we represent the elements of the Boolean algebra as subsets of some universal set, namely the set of all prime filters, and then define binary multirelations over this universe.

Let $\mathbf{B} = (B, \vee, \wedge, {}^-, 0, 1, \{g_i \mid i \in I\})$ be a Boolean algebra with monotone predicate transformers, and let $\mathcal{F}(B)$ be the set of all prime filters in $\mathbf{B}$ considered as a Boolean algebra. For each monotone operator $g : B \to B$, we may define a mapping g^σ over $\mathcal{P}(\mathcal{F}(B))$, called the *canonical extension* [13,14] of g, by

$$g^\sigma(\mathsf{Y}) \quad = \quad \bigcup\{g^\sigma(N_\mathsf{Y}) \mid N_\mathsf{Y} \subseteq \mathsf{Y}\}, \qquad \text{for } \mathsf{Y} \subseteq \mathcal{F}(B),$$

$$\text{where} \quad \begin{aligned} N_Y &= \{F \in \mathcal{F}(B) \mid Y \subseteq F\}, & \text{for } Y \subseteq B \\ g^\sigma(N_Y) &= \bigcap\{g^\sigma(N_y) \mid y \in Y\}, & \text{for } Y \subseteq B \\ g^\sigma(N_y) &= \{F \in \mathcal{F}(B) \mid g(y) \in F\} & \text{for } y \in B. \end{aligned}$$

Using these mappings we may define, for each monotone operator $g : B \to B$, a binary multirelation $R_g \subseteq \mathcal{F}(B) \times \mathcal{P}(\mathcal{F}(B))$ by

$$X R_g \mathsf{Y} \quad \text{iff} \quad X \in g^\sigma(\mathsf{Y}) \quad \text{iff} \quad \exists N_Y \subseteq \mathsf{Y} \; \forall y \in Y \; g(y) \in X,$$

for any $X \in \mathcal{F}(B)$ and any $\mathsf{Y} \subseteq \mathcal{F}(B)$.

If elements of B are viewed as properties then prime filters in $\mathcal{F}(B)$ may be viewed as states. A subset Y of $\mathcal{F}(B)$ may then be viewed as an angelic/demonic joint strategy, as follows. The demon chooses a postcondition N_Y from the set

$$\{N_Y \subseteq \mathcal{F}(B) \mid Y \subseteq B \text{ and } N_Y \subseteq \mathsf{Y}\}$$

of postconditions stronger that postcondition Y, and the angel chooses from the set

$$\{g(y) \mid y \in Y \text{ and } g(y) \in X\}$$

of preconditions a property of the state X from which the chosen postcondition N_Y will be achieved.

In the case of a complete atomic Boolean algebra with monotone predicate transformers, g and g^σ coincide so the above translation reduces to

$$x R_g Q \quad \text{iff} \quad x \in g(Q), \qquad \text{for any } x \in S \text{ and any } Q \subseteq S.$$

If Y is a set $\{F \in \mathcal{F}(B) \mid y \in F\}$ (for $y \in B$), then

$$X R_g \mathsf{Y} \quad \text{iff} \quad g(y) \in X.$$

If Y is a set $\{F \in \mathcal{F}(B) \mid Y \subseteq F\}$ (for $Y \subseteq B$), then

$$X R_g \mathsf{Y} \quad \text{iff} \quad Y \subseteq g^{-1}(X).$$

Theorem 4. *Given any Boolean algebra with monotone predicate transformers*

$$\mathbf{B} = (B, \vee, \wedge, ^-, 0, 1, \{g_i \mid i \in I\})$$

its prime filter structure $(\mathcal{F}(B), \{R_{g_i} \mid i \in I\})$ is a binary multirelational structure.

Proof. For each monotone operator $g : B \to B$, the mapping g^σ over $\mathcal{P}(\mathcal{F}(B))$ is monotone. Hence the binary multirelation $R_g \subseteq \mathcal{F}(B) \times \mathcal{P}(\mathcal{F}(B))$ is up-closed. $\qquad \square$

Thus every binary multirelational structure gives rise to a Boolean algebra with monotone predicate transformers, and conversely. The next two theorems show that each can also be recovered from the other. Let $\mathbf{B} = (B, \vee, \wedge, ^-, 0, 1, \{g_i \mid i \in I\})$ be a Boolean algebra with monotone predicate transformers. Then the Stone [25] mapping $h : B \to \mathcal{P}(\mathcal{F}(B))$, given by $h(a) = \{F \in \mathcal{F}(B) \mid a \in F\}$, is an embedding of the Boolean algebra $\mathbf{B} = (B, \vee, \wedge, ^-, 0, 1)$ into the powerset Boolean algebra

$$\mathcal{P}(\mathcal{F}(\mathbf{B})) = (\mathcal{P}(\mathcal{F}(B)), \cup, \cap, ^-, \emptyset, \mathcal{F}(B)).$$

Now we need to show that h preserves monotone predicate transformers over B.

Theorem 5. *Any Boolean algebra with monotone predicate transformers is isomorphic to a subalgebra of the Boolean algebra with predicate transformers of its underlying binary multirelational structure.*

Proof. Given any monotone operator $g : B \to B$ and any $a \in B$, we have to show that $h(g(a)) = g_{R_g}(h(a))$.

$$g_{R_g}(h(a))$$
$$= \{F \in \mathcal{F}(B) \mid F R_g h(a)\}$$

by definition on page 316 of g_{R_g} from R_g

$$= \{F \in \mathcal{F}(B) \mid g(a) \in F\}$$

by a special case of definition on page 318 of R_g from g

$$= h(g(a))$$

by definition of h. □

Consider any binary multirelational structure $\mathbf{S} = (S, \{R_i \mid i \in I\})$. The power-set $\mathcal{P}(S)$ of S endowed with the mappings g_R yields the Boolean algebra with monotone predicate transformers $\mathcal{P}(\mathbf{S}) = (\mathcal{P}(S), \cup, \cap, ^-, \emptyset, S, \{g_{R_i} \mid i \in I\})$. Forming the prime filter structure of this yields a binary multirelational structure which contains an isomorphic copy of the original binary multirelational structure. Each of the original up-closed binary multirelations R over S gives rise to a monotone predicate transformer $g_R : \mathcal{P}(S) \to \mathcal{P}(S)$, which in turn gives rise to an up-closed binary multirelation R_{g_R} over $\mathcal{F}(\mathcal{P}(S))$. There is a bijective correspondence between the elements of S and certain prime filters in $\mathcal{P}(S)$, namely the principal prime filters under the mapping $a \mapsto k(a) = \{A \subseteq S \mid a \in A\}$. An extension of this mapping provides a bijective correspondence between subsets of S and principal filters, namely $Y \mapsto k(Y) = \{A \subseteq S \mid Y \subseteq A\}$. We need to show that this mapping preserves structure.

Theorem 6. *Any binary multirelational structure is isomorphic to a substructure of the prime filter binary multirelational structure of its Boolean algebra with monotone predicate transformers.*

Proof. Consider any up-closed binary multirelation $R \subseteq S \times \mathcal{P}(S)$. For $x \in S$ and $Y \subseteq S$, we show that

$$k(x) R_{g_R} N_{k(Y)} \quad \text{iff} \quad xRY, \quad \text{where } N_{k(Y)} = \{Q \in \mathcal{F}(\mathcal{P}(S)) \mid k(Y) \subseteq Q\}.$$

$$k(x) R_{g_R} N_{k(Y)}$$

iff $k(Y) \subseteq (g_R)^{-1}(k(x))$

by special case of definition on page 318 of R_{g_R} from g_R

iff $\{Z \subseteq S \mid Y \subseteq Z\} \subseteq (g_R)^{-1}(k(x))$

by definition of $k(Y)$

iff $(\forall Z \subseteq S)[Y \subseteq Z \Rightarrow g_R(Z) \in k(x)]$

by definition of $\subseteq$

iff $(\forall Z \subseteq S)[Y \subseteq Z \Rightarrow x \in g_R(Z)]$

by definition of $k(x)$

iff $(\forall Z \subseteq S)[Y \subseteq Z \Rightarrow xRZ]$

by definition on page 316 of g_R from R

iff xRY

since $R(x)$ is up–closed   □

Therefore, the relational counterpart of a monotone predicate transformers are binary multirelations.

As a consequence of these bijective correspondences, there is a translation between properties of up-closed binary multirelations and properties of monotone predicate transformers.

Theorem 7. *Let $g : B \to B$ be a monotone predicate transformer. Then properties of g translate into properties of R_g as (i) and (ii) below. Conversely, let $R \subseteq S \times \mathcal{P}(S)$ be an up-closed binary multirelation. Then the properties of R translate into properties of g_R as (ii) to (i) below.*

(a) (i) *g is normal*   (ii) *R is total;*
(b) (i) *g is full*   (ii) *R is proper;*
(c) (i) *g is multiplicative* (ii) *R is multiplicative;*
(d) (i) *g is additive*   (ii) *R is additive.*   □

It may also be shown (as in [24] p269) that sequential composition of up-closed multirelations corresponds to composition of monotone predicate transformers.

4   Strongest Postconditions as Closed Sets

Up-closed multirelations $R \subseteq S \times \mathcal{P}(S)$ satisfy the property that for any state $s \in S$ and postcondition $Q \subseteq S$,

$$sRQ \quad \text{iff} \quad \forall Q' \subseteq S, \ Q \subset Q' \Rightarrow sRQ'.$$

This suggests that it suffices to consider the strongest postconditions Q of R with respect to s, that is, the postconditions $Q \subseteq S$, with the property that

$$sRQ \quad \text{and} \quad \forall Q' \subseteq S, \ sRQ' \Rightarrow Q' \not\subseteq Q.$$

However, the set of postconditions for an up-closed multirelation with respect to a state s does not necessarily contain a strongest postcondition. For example, consider the multirelational structure $(\mathbb{R}, \nu)$ where $\mathbb{R}$ is the set of real numbers and

$$\nu(0) = \uparrow\{(0, x_0) \mid x_0 > 0\} = \{X \subseteq \mathbb{R} \mid \{x \mid 0 < x < x_0\} \subseteq X \text{ for some } x_0 > 0\}.$$

Then $\{Q \mid 0\nu Q \land \forall Q_0 \subset Q \; Q_0 \not\in \nu(0)\} = \emptyset$. This kind of descending chain of postconditions does not occur in up-closed multirelations $R \subseteq S \times \mathcal{P}(S)$ satisfying the property that for any state $s \in S$ and any postcondition $Q \subseteq S$,

$$sRQ \quad \text{iff} \quad \exists Q' \subseteq S, \; sRQ' \land Q' \subset Q.$$

In order to distinguish between postconditions and strongest postconditions, we invoke some ideas from topology.

Definition 7. *A general multirelational structure* $(S, \{R_i \mid i \in I\}, \mathcal{A})$ *is a multirelational structure* $(S, \{R_i \mid i \in I\})$ *together with a set* $\mathcal{A}$ *of basic open sets for a topology* Ω_S *on* S. *In such a structure the interaction of the multirelations* R *with the topology is captured by the following properties:*

(a) *For any set* $Q \subseteq S$, sRQ *iff* $\exists$ *closed set* $C \subseteq S$, $sRC \land C \subseteq Q$.
(b) *For any closed set* $C \subseteq S$, sRC *iff* $\forall Q \subseteq S$, $C \subseteq Q \Rightarrow sRQ$.

As a consequence of the properties (a) and (b) of multirelations we have the following characterisation of strongest postconditions as closed sets of the topological space $(S, \Omega_S, \{R_i\}_{i \in I})$.

Theorem 8. *For any multirelation* $R \subseteq S \times \mathcal{P}(S)$ *of a general multirelational structure, the set* $\{Q \subseteq S \mid sRQ \land \forall Q' \subseteq S, \; sRQ' \Rightarrow Q' \not\subset Q\}$ *of strongest postconditions of* R *with respect to* s *is a set of closed sets.*

Proof. For any $s \in S$ and $Q \subseteq S$, assume sRQ and $\forall Q', \; sRQ' \Rightarrow Q' \not\subset Q$. Then, by Definition 7 (a), for some closed set $C \subseteq S$, sRC and $C \subseteq Q$ and $\forall Q' \subseteq S, \; sRQ' \Rightarrow Q' \not\subset Q$. Thus, by predicate calculus, $Q = C$ and hence Q is a closed set. $\qquad \Box$

This result may be seen as the multirelational analogue of a result in [6] showing that relations of a differentiated and compact general relational structure are point closed, i.e. the image set of each point is closed.

It turns out that if a general multirelational structure $(S, \{R_i \mid i \in I\}, \mathcal{A})$ is

(a) *differentiated* (i.e., $w = v$ iff $\forall A \in \mathcal{A}, (w \in A \Leftrightarrow v \in A)$)
(b) *compact* (i.e., $\forall \mathcal{A}' \subseteq \mathcal{A}, \bigcap \mathcal{A}' \neq \emptyset$ if $\mathcal{A}'$ has the finite intersection property)

then the topology Ω_S is the Stone topology on S with $\mathcal{A}$ as clopen basis.

In this context, we obtain alternative characterisations of the binary multirelation R_g defined from a monotone operator $g : B \to B$. For this we note that given a Boolean algebra with monotone predicate transformers $(B, \{g_i\}_{i \in I})$, the set $\mathcal{F}(B)$ of prime filters has a natural topology called the Stone topology generated by a subbasis of sets of form

$$N_a = \{F \in \mathcal{F}(B) \mid a \in F\} \qquad \text{for } a \in B.$$

Then subsets of $\mathcal{F}(B)$ of the form

$$O_A = \{F \in \mathcal{F}(B) \mid A \cap F \neq \emptyset\} \qquad \text{for } A \subseteq B$$

are called open sets while those of them form

$$C_A = \{F \in \mathcal{F}(B) \mid A \subseteq F\} \qquad \text{for } A \subseteq B$$

are called closed sets. Let $\mathcal{C}(\mathcal{F}(B))$ denote the set of all closed sets. For any $X \in \mathcal{F}(B)$ and any $\mathsf{Y} \subseteq \mathcal{F}(B)$, if Y is a closed set C_Y then

$$X R_g \mathsf{Y} \quad \text{iff} \quad \mathsf{Y} \subseteq g^{-1}(X) \quad \text{iff} \quad \forall a \in B, \quad a \in Y \Rightarrow g(a) \in X.$$

Also, for any $X \in \mathcal{F}(B)$ and any $\mathsf{Y} \subseteq \mathcal{F}(B)$,

$$X R_g \mathsf{Y} \quad \text{iff} \quad X \in g^\sigma(\mathsf{Y}) \quad \text{iff} \quad \exists \mathsf{C}_Y \in \mathcal{C}(\mathcal{F}(B)), \ \mathsf{C}_Y \subseteq \mathsf{Y} \wedge \mathsf{Y} \subseteq g^{-1}(X).$$

Thus R_g satisfies property (a) of Definition 7. Since R_g is up-closed, property (b) is satisfied. Hence $(\mathcal{F}(B), \{R_{g_i}\}_{i \in I}, \{N_a \mid a \in B\})$ is a general multirelational structure. Moreover, the interpretation in Section 3 of a subset Y of $\mathcal{F}(B)$ as an angelic/demonic joint strategy can be refined to one where the demon chooses a strongest postcondition.

The topological perspective of this section may be used to establish a full topological duality between Boolean algebras with monotone predicate transformers and general multirelational structures.

5   Angelic/Demonic Factorisation

It is known that the standard epi/monic factorisation [1] can be obtained uniquely for meet- and join operators but for monotone operators in general there is no unique such factorisation. However, in the case of monotone operators over a power set Boolean algebra (as shown in [12]) we have a meet/join factorisation which was used in [11] for proving the completeness of Morgan's [20] refinement laws. In this section we give a factorisation for up-closed multirelations and for monotone operators over a (not necessarily power set) Boolean algebra. The multirelational factorisation will be used to justify the intuition of Section 2 that multirelations can model both angelic and demonic nondeterminism; the monotone predicate transformer factorisation will be used to refine the notion of winning strategy for the game theoretic interpretation of specifications with angelic and demonic nondeterminism.

First we recall a well-known fact from category theory used in [4] for a categorical description of the power construction [7] lifting structure from states (or individuals) to postconditions (or sets of individuals). Namely, the power set functor $\mathcal{P}$ induces a monad $(\mathcal{P}, \eta, \mu)$ on the category SET of sets, where $\mathcal{P}(S)$ is the set of all subsets of S, $\eta : \mathrm{id}_{SET} \to \mathcal{P}$ is given by $\eta_S : S \to \mathcal{P}(S)$ defined by $\eta_S(s) = \{s\}$ (for $s \in S$) and $\mu : \mathcal{P} \circ \mathcal{P} \to \mathcal{P}$ is given by $\mu_S : \mathcal{P}(\mathcal{P}(S)) \to \mathcal{P}(S)$ defined by $\mu_A(\mathsf{X}) = \cup \mathsf{X}$ (for $\mathsf{X} \subset \mathcal{P}(S)$). Any monotone map $g : \mathcal{P}(S) \to \mathcal{P}(S)$ gives rise to a monotone map $g^+ : \mathcal{P}(\mathcal{P}(S)) \to \mathcal{P}(\mathcal{P}(S))$ by $g^+(\mathsf{Q}) = \{g(Q) \mid Q \in \mathsf{Q}\}$ (for $\mathsf{Q} \subseteq \mathcal{P}(S)$).

Given an up-closed multirelation $R \subseteq S \times \mathcal{P}(S)$ it may be viewed as a monotone map $R : S \to \mathcal{P}(\mathcal{P}(S))$. Then, it is easy to show that

$$R = \mu_{\mathcal{P}(S)} \circ R^+ \circ \eta_S.$$

Observe that, for any $s \in S$,

$$R(s) = \cup R^+(\{s\}) = \cup\{Q \subseteq \mathcal{P}(S) \mid \exists Q \in \mathbf{Q}, \ sR\mathbf{Q}\}$$

and

$$Q \in \cup\{\mathbf{Q} \subseteq \mathcal{P}(S) \mid \exists Q \in \mathbf{Q} \ sR\mathbf{Q}\} \quad \text{iff} \quad \exists \mathbf{Q} \subseteq \mathcal{P}(S), \ Q \in \mathbf{Q} \ \wedge \ \mathbf{Q} \subseteq \{Q \mid sR\mathbf{Q}\}.$$

Hence, for any $s \in S$ and any $Q \subseteq S$,

$$sRQ \quad \text{iff} \quad \exists \mathbf{Q} \subseteq \mathcal{P}(S), \ Q \in \mathbf{Q} \ \wedge \ \mathbf{Q} \subseteq \{Q' \subseteq S \mid sRQ'\}.$$

Thus a factorisation of an up-closed multirelation $R \subseteq S \times \mathcal{P}(S)$ is given by

$$R = A_R \mathbin{\overset{\circ}{,}} D_R,$$

where the relation $A_R \subseteq S \times \mathcal{P}(\mathcal{P}(S))$ defined by

$$sA_R\mathbf{Q} \quad \text{iff} \quad \mathbf{Q} \subseteq \{Q' \subseteq S \mid sRQ'\}, \qquad \text{for any } s \in S \text{ and any } \mathbf{Q} \subseteq \mathcal{P}(S)$$

is an angelic multirelation, and the relation $D_R \subseteq \mathcal{P}(\mathcal{P}(S)) \times \mathcal{P}(S)$ defined by

$$\mathbf{Q}D_RQ \quad \text{iff} \quad Q \in \mathbf{Q}, \qquad \text{for any } \mathbf{Q} \subseteq \mathcal{P}(S) \text{ and any } Q \subseteq S$$

is a demonic multirelation. Now suppose that there is another factorisation of R given by $R = A' \mathbin{\overset{\circ}{,}} D'$ where A' is an angelic multirelation and D' is a demonic multirelation. Then $A_R = A' \mathbin{\overset{\circ}{,}} A''$ for some angelic multirelation A''. Define $A'' : \mathcal{P}(\mathcal{P}(S)) \times \mathcal{P}(\mathcal{P}(S))$ by

$$\mathbf{Q}'A''\mathbf{Q} \quad \text{iff} \quad \mathbf{Q}'D'(\cap\mathbf{Q}), \qquad \text{for any } \mathbf{Q}', \mathbf{Q} \subseteq \mathcal{P}(S).$$

Then A'' is angelic, and for any $s \in S$ and any $\mathbf{Q} \subseteq \mathcal{P}(S)$,

$$
\begin{array}{llll}
sA_R\mathbf{Q} & \text{iff} & \forall Q \in \mathbf{Q}, \ sRQ & \text{by definition of } A_R \\
& \text{iff} & \forall Q \in \mathbf{Q}, \ sA' \mathbin{\overset{\circ}{,}} D'Q & \text{since } R = A' \mathbin{\overset{\circ}{,}} D' \\
& \text{iff} & sA' \mathbin{\overset{\circ}{,}} D'(\cap\mathbf{Q}) & \text{since } D' \text{ is multiplicative} \\
& \text{iff} & \exists \mathbf{Q}', \ sA'\mathbf{Q}' \ \wedge \ \mathbf{Q}'D'(\cap\mathbf{Q})\} & \text{definition of } \mathbin{\overset{\circ}{,}} \\
& \text{iff} & sA' \mathbin{\overset{\circ}{,}} A''(\mathbf{Q}) & \text{by definition of } A''
\end{array}
$$

Therefore, the angelic/demonic factorisation of an up-closed multirelation has a uniqueness property similar to that for the standard epi/monic factorisation.

This factorisation of an up-closed multirelation may seem one-sided with only A_R being dependent on R but it reflects how multirelations capture angelic and demonic nondeterminism. If $\{Q \mid sRQ\}$ is a singleton, then there is only one computation starting from s and hence there is no angelic nondeterminism. If each Q in $\{Q \mid sRQ\}$ is a singleton, there may be more than one computation starting from s but in each case the postcondition has a singleton and hence there is no demonic nondeterminism. Thus demonic nondeterminism depends on the size of the sets Q in $\{Q \mid sRQ\}$ and is captured by the relation D_R, while angelic nondeterminism depends on the number of sets in $\{Q \mid sRQ\}$ and is captured by the relation A_R.

Given a monotone predicate transformer $g : \mathcal{P}(S) \to \mathcal{P}(S)$,

$$g \circ \mu_S = \mu_S \circ g^+.$$

So a factorisation of a monotone predicate transformer $g : \mathcal{P}(S) \to \mathcal{P}(S)$ is given by

$$g = j \circ m$$

where $j : \mathcal{P}(\mathcal{P}(S)) \to \mathcal{P}(S)$ defined by

$$j(\mathsf{Q}) \quad = \quad (\mu_S \circ g^+)(\mathsf{Q}) \quad = \quad \bigcup\{g(Q) \mid Q \in \mathsf{Q}\}, \qquad \text{for any } \mathsf{Q} \subseteq \mathcal{P}(S)$$

is an additive predicate transformer, and $m : \mathcal{P}(S) \to \mathcal{P}(\mathcal{P}(S))$ defined by

$$m(Q) \quad = \quad (\mu_S)^{-1}(Q) \quad = \quad \{Q' \mid Q' \subseteq Q\}, \qquad \text{for any } Q \subseteq S$$

is a multiplicative predicate transformer. Now suppose that there is another factorisation of $g : \mathcal{P}(S) \to \mathcal{P}(S)$ given by $g = j' \circ m'$ where j' is an additive predicate transformer j' and m' is a multiplicative predicate transformer. Then there is an additive predicate transformer, j'' such that $j = j' \circ j''$. Define $j'' : \mathcal{P}(\mathcal{P}(S)) \to \mathcal{P}(\mathcal{P}(S))$ by

$$j''(\mathsf{Q}) = \cup\{m'(Q) \mid Q \in \mathsf{Q}\}, \qquad \text{for any } \mathsf{Q} \subseteq \mathcal{P}(S).$$

Then j'' is additive and

$$
\begin{aligned}
j(\mathsf{Q}) &= \mu_S \circ g^+(\mathsf{Q}) && \text{by definition of } j \\
&= \cup\{g(Q) \mid Q \in \mathsf{Q}\} && \text{by definition of } \mu_S, g^+ \\
&= \cup\{j' \circ m'(Q) \mid Q \in \mathsf{Q}\} && \text{since } g = j' \circ m' \\
&= j'(\cup\{m'(Q) \mid Q \in \mathsf{Q}\}) && \text{since } j' \text{ additive} \\
&= j' \circ j''(\mathsf{Q}) && \text{by definition of } j''
\end{aligned}
$$

Therefore, the additive/multiplicative factorisation of a monotone predicate transformer has a uniqueness property similar to that for the epi/monic factorisation.

This factorisation may be viewed as providing an angelic/demonic joint strategy for achieving a postcondition Q. Namely, the demon chooses a (strongest) postcondition $Q' \subseteq Q$ and for the angel chooses a property $g(Q')$ of states from which the chosen (strongest) postcondition Q' can be achieved. Thus demonic nondeterminism depends on the size of the postcondition Q and is captured by predicate transformer m, while angelic nondeterminism depends on the number of sets in $\{g(Q) \mid Q \in \mathsf{Q}\}$ and is captured by predicate transformer j.

For a monotone predicate transformer $g : B \to B$ over a (not necessarily power set) Boolean algebra B we combine the above unique factorisation (with $S = \mathcal{F}(B)$) and the fact that the Stone mapping $h : B \to \mathcal{P}(\mathcal{F}(B))$ preserves monotone operators g over B, that is, $h \circ g = g_{R_g} \circ h$. This yields a factorisation of a monotone predicate transformer $g : B \to B$ given by

$$g = j \circ m$$

where $j : \mathcal{P}(\mathcal{P}(\mathcal{F}(B))) \to B$ defined by

$$
\begin{aligned}
j(\mathcal{A}) \;&=\; h^{-1}(\cup\{g_{R_g}(\mathsf{Y}) \mid \mathsf{Y} \in \mathcal{A}\}), \qquad \text{for any } \mathcal{A} \subseteq \mathcal{P}(\mathcal{F}(B)) \\
&=\; \{b \in B \mid \exists \mathsf{Y} \in \mathcal{A},\ \forall X \in \mathcal{F}(B),\quad b \in X \quad \text{iff}\quad X \in g^\sigma(\mathsf{Y})\}
\end{aligned}
$$

is an additive predicate transformer and $m : B \to \mathcal{P}(\mathcal{P}(\mathcal{F}(B)))$ defined by

$$
m(a) \;=\; (\mu_{\mathcal{F}(B)})^{-1} \circ h \;=\; \{\mathcal{A} \subseteq \mathcal{P}(\mathcal{F}(B)) \mid a \in \bigcap_{F \in \mathcal{A}} F\}, \qquad \text{for any } a \in B
$$

is a multiplicative predicate transformer. Uniqueness of this factorisation follows from the uniqueness of the factorisation of g_{R_g}. The factorisation may be viewed as providing an angelic/demonic joint strategy for effectively achieving a postcondition a. Namely, the demon chooses a (strongest) postcondition A from the set

$$
\{\mathsf{A} \subseteq \mathcal{F}(B) \mid \forall F \in \mathsf{A}\ a \in F\}
$$

of postconditions stronger than postcondition a, and the angel chooses from the set

$$
\{b \in B \mid \exists \mathsf{Y} \in \mathcal{A},\ \forall X \in \mathcal{F}(B),\quad b \in X \quad \text{iff}\quad X \in g^\sigma(\mathsf{Y})\}
$$

of preconditions a property b of the states from which the chosen (strongest) postcondition will be achieved. Thus demonic nondeterminism depends on the postcondition $a \in B$ and is captured by the predicate transformer m, while angelic nondeterminism depends on the number of sets in $\{g_{R_g}(\mathsf{Y}) \mid \mathsf{Y} \in \mathcal{A}\}$ and is captured by the predicate transformer j.

6   Conclusion

The aim of this paper has been to stimulate interest in multirelations as a model for simultaneously reasoning about angelic and demonic nondeterminism, and in duality as a tool for unifying semantic models.

Multirelations were introduced in [24] as an alternative to monotone predicate transformers, and their expressivity for modelling nondeterminism has been explored in [18]. In the book [8], the classical dualities of Stone [25], Jónsson/Tarski [17] and Priestley [22] are invoked to compare semantic models, notwithstanding their differences in formulation, and provide a surprisingly uniform picture of program semantics.

This paper builds on the earlier work in a number of ways. First, the factorisation of up-closed multirelations is new and reveals the two levels at which multirelations capture angelic and demonic nondeterminism. Second, the topological perspective of multirelations introduced here provides a natural characterisation of strongest postconditions as closed sets. Third, the translation between monotone predicate transformers and up-closed multirelations formulated in terms of canonical extensions is equivalent to that in [24], but is more useful since it suggests a natural interpretation of monotone predicate transformers in terms of angelic/demonic joint strategies for effectively achieving given postconditions. Fourth, the framework of [23,8] for program semantics has been extended and applied to semantic models for specification computations.

A number of challenges remain. For example: to develop an approach based on binary multirelations for deriving strategies of games, to extend the relational calculus for program derivation of [5], and to use multirelations for proving completeness of data refinement in the relational model. Perhaps further questions will occur to the reader.

References

1. Adámek, J., Herrlich, H., Strecker, G.E.: Abstract and Concrete Categories. John Wiley and Sons, Inc (1991).
2. Back, R.J.R., von Wright, J.: Combining angels, demons and miracles in program specifications. Theoretical Computer Science **100** (1992) 365–383.
3. Back, R.J.R., J. von Wright, J.:. Refinement Calclulus: A Systematic Introduction. Graduate Texts in Computer Science. Springer-Verlag, New York (1998).
4. Bargenda, H.W., Brink, C., Vajner, V.: Categorical aspects of power algebras. Quaestiones Mathematica **16** (1993) 133–147.
5. Bird, R., de Moor, O.: Algebra of Programming. Prentice Hall (1997).
6. Blackburn, P., De Rijke, M., Venema, Y.: Modal Logic. Cambridge Tracts in Theoretical Computer Science 53. Cambridge University Press, Cambridge (2001).
7. Brink, C.: Power structures. Algebra Universalis **30** (1993) 177–216.
8. Brink, C., Rewitzky, I.: A Paradigm for Program Semantics: Power Structures and Duality. CSLI Publications, Stanford (2001).
9. Dijkstra, E.W.: Guarded commands, nondeterminacy and formal derivation of programs. Communications of the ACM **18** (8) (1975) 453–458.
10. Dijkstra, E.W.: A Discipline of Programming. Englewood Cliffs, New Jersey: Prentice-Hall (1976).
11. Gardiner, P.H., Morgan, C.C.: Data refinement of predicate transformers. Theoretical Computer Science **87** (1) (1991) 143–162.
12. Gardiner, P.H., Martin, C.E., de Moor, O.: An algebraic construction of predicate transformers. Science of Computer Programming **22** (1-2) (1994) 21–44.
13. Gehrke, M., Jónsson, B.: Bounded distributive lattices with operators. Mathematica Japonica **40** (2) (1994) 207–215.
14. Gehrke, M., Jónsson, B.: Monotone bounded distributive lattice expansions. Mathematica Japonica **52** (2) (2000) 197–213.
15. Hoare, C.A.R.: An axiomatic basis for computer programming. Communications of the ACM **12**(10) (1969) 576–583.
16. Hoare, C.A.R.: An algebra of games of choice. Unpublished manuscript, 4 pages (1996).
17. Jónsson, B., Tarski, A.: Boolean algebras with operators I. American Journal of Mathematics **73** (1951) 891–939.
18. Martin, C., Curtis, S., Rewitzky, I.: Modelling nondeterminism. In Proceedings of the 7th International Conference on Mathematics of Program Construction. Lecture Notes in Computer Science Vol 3125. Spinger-Verlag, Berlin Heidelberg New York (2004) 228–251.
19. Morgan, C.C.: The specification statement. Transactions of Programming Language Systems **10** (3) (1998) 403–491.

20. Morgan, C.C., Robertson, K.A.: Specification statements and refinement. IBM Journal of Research and Development **31** (5) (1987) 546–555.
21. Nelson, G.: A generalisation of Dijkstra's calculus. ACM Transactions on Programming Languages and Systems **11** (4) (1989) 517–562.
22. Priestley, H.A.: Representation of distributive lattices by means of ordered Stone spaces. Bulletin of the London Mathematical Society **2** (1970) 186–190.
23. Rewitzky, I., Brink, C.: Predicate transformers as power operations. Formal Aspects of Computing **7** (1995) 169–182.
24. Rewitzky, I.: Binary multirelations. In *Theory and Application of Relational Structures as Knowledge Instruments*. (eds: H de Swart, E Orlowska, G Schmidt, M Roubens). Lecture Notes in Computer Science Vol 2929. Spinger-Verlag, Berlin Heidelberg New York (2003) 259–274.
25. Stone, M.H.: Topological representations of distributive lattices and Brouwerian logics. Casopis Pro Potování Mathematiky **67** (1937) 1–25.

Homomorphism and Isomorphism Theorems Generalized from a Relational Perspective

Gunther Schmidt*

Institute for Software Technology, Department of Computing Science
Universität der Bundeswehr München, 85577 Neubiberg, Germany
Gunther.Schmidt@unibw.de

Abstract. The homomorphism and isomorphism theorems traditionally taught to students in a group theory or linear algebra lecture are by no means theorems of group theory. They are for a long time seen as general concepts of universal algebra. This article goes even further and identifies them as relational properties which to study does not even require the concept of an algebra. In addition it is shown how the homomorphism and isomorphism theorems generalize to not necessarily algebraic and thus relational structures.

Keywords: homomorphism theorem, isomorphism theorem, relation algebra, congruence, multi-covering.

1 Introduction

Relation algebra has received increasing interest during the last years. Many areas have been reconsidered from the relational point of view, which often provided additional insight. Here, the classical homomorphism and isomorphism theorems (see [1], e.g.) are reviewed from a relational perspective, thereby simplifying and slightly generalizing them.

The paper is organized as follows. First we recall the notion of a heterogeneous relation algebra and some of the very basic rules governing work with relations. With these, function and equivalence properties may be formulated concisely. The relational concept of homomorphism is defined as well as the concept of a congruence which is related with the concept of a multi-covering, which have connections with topology, complex analysis, and with the equivalence problem for flow-chart programs. We deal with the relationship between mappings and equivalence relations. The topics include the so-called substitution property and the forming of quotients.

Homomorphisms may be used to give universal characterizations of domain constructions. Starting from sets, further sets may be obtained by construction, as pair sets (direct product), as variant sets (direct sum), as power sets (direct power), or as the quotient of a set modulo some equivalence. Another

---

* Cooperation and communication around this research was partly sponsored by the European COST Action 274: TARSKI (*Theory and Applications of Relational Structures as Knowledge Instruments*), which is gratefully acknowledged.

R.A. Schmidt (Ed.): RelMiCS /AKA 2006, LNCS 4136, pp. 328–342, 2006.

construction that is not so easily identified as such is subset extrusion. It serves to promote a subset of a set, which needs the larger one to exist, to a set of its own right.

Using the so-called dependent types, quotient set and subset extrusion, we then formulate the homomorphism and isomorphism theorems and prove them in a fully algebraic style. The paper ends with hints on coverings with locally univalent outgoing fans.

2   Homogeneous and Heterogeneous Relation Algebras

A **homogeneous relation algebra** $(\mathcal{R}, \cup, \cap, {}^{-}, {}_{;}, {}^{\mathsf{T}})$ consists of a set $\mathcal{R} \neq \emptyset$, whose elements are called relations, such that $(\mathcal{R}, \cup, \cap, {}^{-})$ is a complete, atomic boolean algebra with zero element $\mathbb{L}$, universal element $\mathbb{T}$, and ordering $\subseteq$, that $(\mathcal{R}, {}_{;})$ is a semigroup with precisely one unit element $\mathbb{I}$, and, finally, the Schröder equivalences $Q_{;}R \subseteq S \iff Q^{\mathsf{T}}_{;}\overline{S} \subseteq \overline{R} \iff \overline{S}_{;}R^{\mathsf{T}} \subseteq \overline{Q}$ are satisfied.

One may switch to heterogeneous relation algebra, which has been proposed in, e.g., [2,3]. A **heterogeneous relation algebra** is a category $\mathcal{R}$ consisting of a set $\mathcal{O}$ of objects and sets $\mathtt{Mor}(A, B)$ of morphisms, where $A, B \in \mathcal{O}$. Composition is denoted by ${}_{;}$ while identities are denoted by $\mathbb{I}_A \in \mathtt{Mor}(A, A)$. In addition, there is a totally defined unary operation ${}^{\mathsf{T}}_{A,B} : \mathtt{Mor}(A, B) \longrightarrow \mathtt{Mor}(B, A)$ between morphism sets. Every set $\mathtt{Mor}(A, B)$ carries the structure of a complete, atomic boolean algebra with operations $\cup, \cap, {}^{-}$, zero element $\mathbb{L}_{A,B}$, universal element $\mathbb{T}_{A,B}$ (the latter two non-equal), and inclusion ordering $\subseteq$. The Schröder equivalences—where the definedness of one of the three formulae implies that of the other two—are postulated to hold.

Most of the indices of elements and operations are usually omitted for brevity and can easily be reinvented. For the purpose of self-containedness, we recall the following computational rules; see, e.g., [4,5].

2.1 Proposition.
  i) $\mathbb{L}_{;}R = R_{;}\mathbb{L} = \mathbb{L}$;
  ii) $R \subseteq S \implies Q_{;}R \subseteq Q_{;}S, \; R_{;}Q \subseteq S_{;}Q$;
  iii) $Q_{;}(R \cap S) \subseteq Q_{;}R \cap Q_{;}S, \quad (R \cap S)_{;}Q \subseteq R_{;}Q \cap S_{;}Q$
        $Q_{;}(R \cup S) = Q_{;}R \cup Q_{;}S, \quad (R \cup S)_{;}Q = R_{;}Q \cup S_{;}Q$
  iv) $(R^{\mathsf{T}})^{\mathsf{T}} = R$;
  v) $(R_{;}S)^{\mathsf{T}} = S^{\mathsf{T}}_{;}R^{\mathsf{T}}$;
  vi) $R \subseteq S \iff R^{\mathsf{T}} \subseteq S^{\mathsf{T}}$;
  vii) $\overline{R}^{\mathsf{T}} = \overline{R^{\mathsf{T}}}$;
  viii) $(R \cup S)^{\mathsf{T}} = R^{\mathsf{T}} \cup S^{\mathsf{T}}$;
        $(R \cap S)^{\mathsf{T}} = R^{\mathsf{T}} \cap S^{\mathsf{T}}$;
  ix) $Q_{;}R \cap S \subseteq (Q \cap S_{;}R^{\mathsf{T}})_{;}(R \cap Q^{\mathsf{T}}_{;}S)$.   (Dedekind rule)

A relation R is called **univalent** (or a partial function) if $R^{\mathsf{T}}_{;}R \subseteq \mathbb{I}$. When R satisfies $\mathbb{I} \subseteq R_{;}R^{\mathsf{T}}$ (or equivalently if $\mathbb{T} \subseteq R_{;}\mathbb{T}$), then R is said to be **total**. If both these requirements are satisfied, i.e., if R resembles a total and univalent function, we shall often speak of a **mapping**. A relation R is called injective, surjective and bijective, if R^T is univalent, total, or both, respectively. Furthermore

$$R \subseteq Q,\, Q \text{ univalent},\, R_{;}\mathbb{T} \supseteq Q_{;}\mathbb{T} \implies R = Q \tag{*}$$

The following basic properties are mainly recalled from [4,5].

2.2 Proposition (*Row and column masks*).   The following formulae hold for arbitrary relations $P : V \longrightarrow W, Q : U \longrightarrow V, R : U \longrightarrow W, S : V \longrightarrow W$, provided the constructs are defined.

i) $(Q \cap R_{;}\mathbb{T}_{WV})_{;}S = Q_{;}S \cap R_{;}\mathbb{T}_{WW}$;

ii) $(Q \cap (P_{;}\mathbb{T}_{WU})^\mathsf{T})_{;}S = Q_{;}(S \cap P_{;}\mathbb{T}_{WW})$.   □

We now recall a rule which is useful for calculations involving equivalence relations; it deals with the effect of composition with an equivalence relation with regard to intersection. For a proof see [4,5].

2.3 Proposition. Let Θ be an equivalence and let A, B be arbitrary relations.

$$(A_{;}\Theta \cap B)_{;}\Theta = A_{;}\Theta \cap B_{;}\Theta = (A \cap B_{;}\Theta)_{;}\Theta \qquad\qquad □$$

It is sometimes useful to consider a vector, which is what has at other occasions been called a right ideal element. It is characterized by $U = U_{;}\mathbb{T}$ and thus corresponds to a subset or a predicate. One may, however, also use a partial diagonal to characterize a subset. There is a one-to-one correspondence between the two concepts. Of course, $p \subseteq \mathbb{I} \implies p^2 = p^\mathsf{T} = p$. The **symmetric quotient** has been applied in various applications:

$$\mathsf{syq}\,(A, B) := \overline{A^\mathsf{T}_{;}\overline{B}} \cap \overline{\overline{A}^\mathsf{T}_{;}B}$$

3   Homomorphisms

We recall the concept of homomorphism for relational structures with Fig. 3.1. Structure and mappings shall commute, however, not as an equality but just as containment.

Fig. 3.1. Relational homomorphism

3.1 Definition. Given two relations R, S, we call the pair (Φ, Ψ) of relations a **homomorphism** from R to S, if Φ, Ψ are mappings satisfying

$$R_{;}\Psi \subseteq \Phi_{;}S. \qquad\qquad □$$

The homomorphism condition has four variants

$$R_{;}\Psi \subseteq \Phi_{;}S \iff R \subseteq \Phi_{;}S_{;}\Psi^{\mathsf{T}} \iff \Phi^{\mathsf{T}}{}_{;}R \subseteq S_{;}\Psi^{\mathsf{T}} \iff \Phi^{\mathsf{T}}{}_{;}R_{;}\Psi \subseteq S$$

which may be used interchangeably. This is easily recognized applying the mapping properties

$$\Phi^{\mathsf{T}}{}_{;}\Phi \subseteq \mathbb{I}, \quad \mathbb{I} \subseteq \Phi_{;}\Phi^{\mathsf{T}}, \quad \Psi^{\mathsf{T}}{}_{;}\Psi \subseteq \mathbb{I}, \quad \mathbb{I} \subseteq \Psi_{;}\Psi^{\mathsf{T}}$$

As usual, also isomorphisms are introduced.

3.2 Definition. We call (Φ, Ψ) an **isomorphism** between the two relations R, S, if it is a homomorphism from R to S and if $(\Phi^{\mathsf{T}}, \Psi^{\mathsf{T}})$ is a homomorphism from S to R. □

The following lemma will sometimes help in identifying an isomorphism.

3.3 Lemma. Let relations R, S be given together with a homomorphism (Φ, Ψ) from R to S such that

Φ, Ψ are bijective mappings and $R_{;}\Psi = \Phi_{;}S$.

Then (Φ, Ψ) is an isomorphism.

Proof. $S_{;}\Psi^{\mathsf{T}} = \Phi^{\mathsf{T}}{}_{;}\Phi_{;}S_{;}\Psi^{\mathsf{T}} = \Phi^{\mathsf{T}}{}_{;}R_{;}\Psi_{;}\Psi^{\mathsf{T}} = \Phi^{\mathsf{T}}{}_{;}R.$ □

4 Universal Characterizations

Given a mathematical structure, one is immediately interested in homomorphisms, substructures, and congruences. When handling these, there is a characteristic difference between algebraic and relational structures.

Algebraic structures are defined by composition laws such as a binary multiplication $mult: A \times A \longrightarrow A$ or the unary operation of forming the inverse $inv: A \longrightarrow A$. These operations can, of course, be interpreted as relations. The first example furnishes a "ternary" relation $R_{mult} : (A \times A) \longrightarrow A$, the second, a binary relation $R_{inv} : A \longrightarrow A$, and both are univalent and total.

Relational structures are also defined by certain relations, but these need no longer be univalent or total. Purely relational structures are orders, strictorders, equivalences, and graphs. Typically, however, mixed structures with both, algebraic and relational, features occur, such as ordered fields, for example.

4.1 Standard Domain Constructions

The *direct product* resembling the pair set construction is given via two generic relations π, ρ, the left and the right projection, satisfying

$$\pi^{\mathsf{T}}{}_{;}\pi = \mathbb{I}, \quad \rho^{\mathsf{T}}{}_{;}\rho = \mathbb{I}, \quad \pi_{;}\pi^{\mathsf{T}} \cap \rho_{;}\rho^{\mathsf{T}} = \mathbb{I}, \quad \pi^{\mathsf{T}}{}_{;}\rho = \mathbb{T}$$

Whenever a second pair π_1, ρ_1 of relations with these properties should be presented, one may construct the isomorphism $\Phi := \pi_{;}\pi_1^{\mathsf{T}} \cap \rho_{;}\rho_1^{\mathsf{T}}$, thus showing that the direct product is defined uniquely up to isomorphism.

The *direct sum* resembling variant set forming (disjoint union) is given via two generic relations ι, κ, the left and the right injection, satisfying

$$\iota_;\iota^{\mathsf{T}} = \mathbb{I}, \quad \kappa_;\kappa^{\mathsf{T}} = \mathbb{I}, \quad \iota^{\mathsf{T}}{}_;\iota \cup \kappa^{\mathsf{T}}{}_;\kappa = \mathbb{I}, \quad \iota_;\kappa^{\mathsf{T}} = \mathbb{\bot}$$

Whenever a second pair ι_1, κ_1 of relations with these properties should be presented, one may construct the isomorphism $\varPhi := \iota^{\mathsf{T}}{}_;\iota_1 \cup \kappa^{\mathsf{T}}{}_;\kappa_1$, thus showing that the direct sum is defined uniquely up to isomorphism.

The *direct power* resembling powerset construction is given via a generic relation ε, the membership relation, satisfying

$$\mathsf{syq}(\varepsilon, \varepsilon) \subseteq \mathbb{I} \text{ and that } \mathsf{syq}(\varepsilon, X) \text{ is surjective for every relation } X$$

Should a second membership relation ε_1 with these properties be presented, one may construct the isomorphism $\varPhi := \mathsf{syq}(\varepsilon, \varepsilon_1)$, thus showing that the direct power is defined uniquely up to isomorphism. These constructions are by now standard; proofs may be found in [4,5].

4.2 Quotient Forming and Subset Extrusion

In addition to these, other domain constructions are possible which are usually not handled as such. Although relatively simple, they need a bit of care. Known as *dependent types* they do not just start with a domain or two, but with an additional construct, namely an equivalence or a subset.

4.1 Proposition (*Quotient set*). Given an equivalence $\varXi$ on the set V, one may generically define the quotient set $V_{\varXi}$ together with the natural projection $\eta : V \longrightarrow V_{\varXi}$ postulating both to satisfy

$$\varXi = \eta_;\eta^{\mathsf{T}}, \qquad \eta^{\mathsf{T}}{}_;\eta = \mathbb{I}_{V_{\varXi}}.$$

The natural projection η is uniquely determined up to isomorphism: should a second natural projection η_1 be presented, the isomorphism is $(\mathbb{I}, \eta^{\mathsf{T}}{}_;\eta_1)$.

Proof. Assume two such projections $V_{\varXi} \xleftarrow{\eta} V \xrightarrow{\eta_1} W_{\varXi}$, for which therefore

$$\varXi = \eta_1{}_;\eta_1^{\mathsf{T}}, \qquad \eta_1^{\mathsf{T}}{}_;\eta_1 = \mathbb{I}_{W_{\varXi}}.$$

Looking at this setting, the only way to relate $V_{\varXi}$ with $W_{\varXi}$ is to define $\varPhi := \eta^{\mathsf{T}}{}_;\eta_1$ and proceed showing

$$
\begin{aligned}
\varPhi^{\mathsf{T}}{}_;\varPhi &= (\eta_1^{\mathsf{T}}{}_;\eta)_;(\eta^{\mathsf{T}}{}_;\eta_1) && \text{by definition of } \varPhi \\
&= \eta_1^{\mathsf{T}}{}_;(\eta_;\eta^{\mathsf{T}})_;\eta_1 && \text{associative} \\
&= \eta_1^{\mathsf{T}}{}_;\varXi_;\eta_1 && \text{as } \varXi = \eta_;\eta^{\mathsf{T}} \\
&= \eta_1^{\mathsf{T}}{}_;(\eta_1{}_;\eta_1^{\mathsf{T}})_;\eta_1 && \text{as } \varXi = \eta_1{}_;\eta_1^{\mathsf{T}} \\
&= (\eta_1^{\mathsf{T}}{}_;\eta_1)_;(\eta_1^{\mathsf{T}}{}_;\eta_1) && \text{associative} \\
&= \mathbb{I}_{W_{\varXi}}{}_;\mathbb{I}_{W_{\varXi}} && \text{since } \eta_1^{\mathsf{T}}{}_;\eta_1 = \mathbb{I}_{W_{\varXi}} \\
&= \mathbb{I}_{W_{\varXi}} && \text{since } \mathbb{I}_{W_{\varXi}}{}_;\mathbb{I}_{W_{\varXi}} = \mathbb{I}_{W_{\varXi}}
\end{aligned}
$$

$\varPhi_;\varPhi^{\mathsf{T}} = \mathbb{I}_{V_{\varXi}}$ is shown analogously. Furthermore, $(\mathbb{I}, \varPhi)$ satisfies the property of an isomorphism between η and η_1 following Lemma 3.3:

$$\eta_;\varPhi = \eta_;\eta^{\mathsf{T}}{}_;\eta_1 = \varXi_;\eta_1 = \eta_1{}_;\eta_1^{\mathsf{T}}{}_;\eta_1 = \eta_1{}_;\mathbb{I}_{W_{\varXi}} = \eta_1 \qquad\qquad \square$$

Not least when working on a computer, one is interested in such quotients as the quotient set is usually smaller and may be handled more efficiently. The same reason leads us to consider subset extrusion in a very formal way.

A subset is assumed to exist *relatively to some other set* so that it is not a first-class citizen in our realm of domains. With a bit of formalism, however, it can be managed to convert a subset so as to have it as a set of its own right, a process which one might call a *subset extrusion*.

4.2 Proposition (*Extruded subset*). Given a subset U of some set V, one may generically define the extruded set D_U together with the natural injection $\chi : D_U \longrightarrow V$ postulating both to satisfy

$$\chi \,\chi^\mathsf{T} = \mathbb{I}_{D_U}, \quad \chi^\mathsf{T} \,\chi = \mathbb{I}_V \cap U \,\mathbb{T}_{V,V}.$$

The natural injection χ is uniquely determined up to isomorphism: should a second natural injection χ_1 be presented, the isomorphism is $(\chi \,\chi_1^\mathsf{T}, \mathbb{I})$.

Proof. We have $D_U \xrightarrow{\chi} V \xleftarrow{\chi_1} D$ with the corresponding properties:

$$\chi_1 \,\chi_1^\mathsf{T} \subseteq \mathbb{I}_D, \quad \chi_1^\mathsf{T} \,\chi_1 = \mathbb{I}_V \cap U \,\mathbb{T}_{V,V}$$

and show

$$\Phi^\mathsf{T} \,\Phi = \chi_1 \,\chi^\mathsf{T} \,\chi \,\chi_1^\mathsf{T} = \chi_1 \,(\mathbb{I}_V \cap U \,\mathbb{T}) \,\chi_1^\mathsf{T} = \chi_1 \,\chi_1^\mathsf{T} \,\chi_1 \,\chi_1^\mathsf{T} = \mathbb{I}_D \,\mathbb{I}_D = \mathbb{I}_D$$

and analogously also $\Phi \,\Phi^\mathsf{T} = \mathbb{I}_{D_U}$. Furthermore, $(\Phi, \mathbb{I})$ satisfies the property of an isomorphism between χ and χ_1 using Lemma 3.3:

$$\chi \,\mathbb{I}_V = \chi = \mathbb{I}_{D_U} \,\chi = \chi \,\chi^\mathsf{T} \,\chi = \chi \,(\mathbb{I}_V \cap U \,\mathbb{T}) = \chi \,\chi_1^\mathsf{T} \,\chi_1 = \Phi \,\chi_1 \qquad \square$$

A point to mention is that subset extrusion allows to switch from set-theoretic consideration to an algebraic one. When using a computer and a formula manipulation system or a theorem prover, this means a considerable restriction in expressivity which is honored with much better efficiency.

An important application of extrusion is the concept of *tabulation* introduced by Roger Maddux. It now turns out to be a composite construction; see [6,7], e.g. An arbitrary relation $R : X \longrightarrow Y$ is said to be tabulated by relations (due to the following characterization, they turn out to be mappings) P, Q if

$$P^\mathsf{T} \,Q = R, \quad P^\mathsf{T} \,P = \mathbb{I}_X \cap R \,\mathbb{T}_{YX}, \quad Q^\mathsf{T} \,Q = \mathbb{I}_Y \cap R^\mathsf{T} \,\mathbb{T}_{XY}, \quad P \,P^\mathsf{T} \cap Q \,Q^\mathsf{T} = \mathbb{I}_{X \times Y}$$

This may indeed be composed of extruding with $\chi : D_U \longrightarrow X \times Y$ the subset of related pairs out of a direct product

$$
\begin{aligned}
U &:= (\pi \,R \cap \rho) \,\mathbb{T}_{Y,X \times Y} = (\pi \,R \cap \rho) \,\mathbb{T}_{YX} \,\pi^\mathsf{T} = (\pi \,R \,\rho^\mathsf{T} \cap \mathbb{I}) \,\rho \,\mathbb{T}_{YX} \,\pi^\mathsf{T} \\
&= (\pi \,R \,\rho^\mathsf{T} \cap \mathbb{I}) \,\mathbb{T}_{X \times Y, X \times Y} = (\rho \,R^\mathsf{T} \,\pi^\mathsf{T} \cap \mathbb{I}) \,\mathbb{T}_{X \times Y, X \times Y} \\
&= (\rho \,R^\mathsf{T} \,\pi^\mathsf{T} \cap \mathbb{I}) \,\pi \,\mathbb{T}_{XY} \,\rho^\mathsf{T} = (\rho \,R^\mathsf{T} \cap \pi) \,\mathbb{T}_{XY} \,\rho^\mathsf{T} = (\rho \,R^\mathsf{T} \cap \pi) \,\mathbb{T}_{X, X \times Y}
\end{aligned}
$$

and defining $P := \chi \,\pi$ and $Q := \chi \,\rho$. This is proved quite easily as follows.

$$
\begin{aligned}
P^\mathsf{T} \,Q &= \pi^\mathsf{T} \,\chi^\mathsf{T} \,\chi \,\rho = \pi^\mathsf{T} \,(\pi \,R \,\rho^\mathsf{T} \cap \mathbb{I}) \,\rho \\
&= \pi^\mathsf{T} \,(\pi \,R \cap \rho) \\
&= R \cap \pi^\mathsf{T} \,\rho \\
&= R \cap \mathbb{T} = R
\end{aligned}
$$

$$P^\mathsf{T} \,P = \pi^\mathsf{T} \,\chi^\mathsf{T} \,\chi \,\pi = \pi^\mathsf{T} \,(\mathbb{I} \cap (\pi \,R \cap \rho) \,\rho^\mathsf{T} \,\pi \,\pi^\mathsf{T}) \,\pi$$

$$= \pi^{\mathsf{T}};(\pi \cap (\pi;R \cap \rho));\rho^{\mathsf{T}};\pi)$$
$$= \mathbb{I} \cap \pi^{\mathsf{T}};(\pi;R \cap \rho);\rho^{\mathsf{T}};\pi$$
$$= \mathbb{I} \cap (R \cap \pi^{\mathsf{T}};\rho);\rho^{\mathsf{T}};\pi = \mathbb{I} \cap R;\mathbb{T}_{YX}$$

$Q^{\mathsf{T}};Q$ is handled analogously

$$P;P^{\mathsf{T}} \cap Q;Q^{\mathsf{T}} = \chi;\pi;\pi^{\mathsf{T}};\chi^{\mathsf{T}} \cap \chi;\rho;\rho^{\mathsf{T}};\chi^{\mathsf{T}} = \chi;(\pi;\pi^{\mathsf{T}} \cap \rho;\rho^{\mathsf{T}});\chi^{\mathsf{T}} = \chi;\mathbb{I};\chi^{\mathsf{T}} = \mathbb{I}$$

5   Congruences and Multi-coverings

Whenever equivalences behave well with regard to some other structure, we are accustomed to call them congruences. This is well-known for algebraic structures, i.e., those defined by mappings on some set. We define it correspondingly for the non-algebraic case, including heterogeneous relations; i.e., possibly neither univalent nor total. While the basic idea is known from many application fields, the following general concepts may be a novel abstraction.

5.1 Definition. Let B be a relation and Ξ, Θ equivalences. The pair (Ξ, Θ) is called a B-**congruence** if $\Xi;B \subseteq B;\Theta$.   ☐

If B were an operation on a given set and we had $\Xi = \Theta$, we would say that B "has the substitution property with regard to Ξ". The concept of congruence is related to the concept of a multi-covering.

5.2 Definition. A homomorphism (Φ, Ψ) from B to B' is called a **multi-covering**, provided the functions are surjective and satisfy $\Phi;B' \subseteq B;\Psi$ in addition to being a homomorphism.   ☐

The relationship between congruences and multi-coverings is close and seems not to have been pointed out yet.

5.3 Theorem.

  i) If (Φ, Ψ) is a multi-covering from B to B', then $(\Xi, \Theta) := (\Phi;\Phi^{\mathsf{T}}, \Psi;\Psi^{\mathsf{T}})$ is a B-congruence.
 ii) If the pair (Ξ, Θ) is a B-congruence, then there exists up to isomorphism at most one multi-covering (Φ, Ψ) satisfying $\Xi = \Phi;\Phi^{\mathsf{T}}$ and $\Theta = \Psi;\Psi^{\mathsf{T}}$.

Proof. i) Ξ is certainly reflexive and transitive, as Φ is total and univalent. In the same way, Θ is reflexive and transitive. The relation $\Xi = \Phi;\Phi^{\mathsf{T}}$ is symmetric by construction and so is Θ. Now we prove

$$\Xi;B = \Phi;\Phi^{\mathsf{T}};B \subseteq \Phi;B';\Psi^{\mathsf{T}} \subseteq B;\Psi;\Psi^{\mathsf{T}} = B;\Theta$$

applying one after the other the definition of Ξ, one of the four homomorphism definitions, the multi-covering condition, and the definition of Θ.

ii) Let (Φ_i, Ψ_i) be a multi-covering from B to B_i, $i = 1, 2$. Then

$$B_i \subseteq \Phi_i^{\mathsf{T}};\Phi_i;B_i \subseteq \Phi_i^{\mathsf{T}};B;\Psi_i \subseteq B_i, \text{ and therefore everywhere “=”,}$$

applying surjectivity, the multi-covering property and one of the homomorphism conditions. Now we indicate how to prove that $(\xi, \vartheta) := (\Phi_1^\mathsf{T}{}_,\Phi_2, \Psi_1^\mathsf{T}{}_,\Psi_2)$ is a homomorphism from B_1 onto B_2 — which is then of course also an isomorphism.

$$\xi^\mathsf{T}{}_,\xi = \Phi_2^\mathsf{T}{}_,\Phi_1{}_,\Phi_1^\mathsf{T}{}_,\Phi_2 = \Phi_2^\mathsf{T}{}_,\Xi{}_,\Phi_2 = \Phi_2^\mathsf{T}{}_,\Phi_2{}_,\Phi_2^\mathsf{T}{}_,\Phi_2 = \mathbb{I}{}_,\mathbb{I} = \mathbb{I}$$
$$B_1{}_,\vartheta = \Phi_1^\mathsf{T}{}_,B{}_,\Psi_1{}_,\Psi_1^\mathsf{T}{}_,\Psi_2 = \Phi_1^\mathsf{T}{}_,B{}_,\Theta{}_,\Psi_2 = \Phi_1^\mathsf{T}{}_,B{}_,\Psi_2{}_,\Psi_2^\mathsf{T}{}_,\Psi_2$$
$$\subseteq \Phi_1^\mathsf{T}{}_,\Phi_2{}_,B_2{}_,\mathbb{I} = \xi{}_,B_2 \qquad\qquad \square$$

The multi-covering (Φ, Ψ) for some given congruences Ξ, Θ need not exist in the given relation algebra. It may, however, be constructed by setting Φ, Ψ to be the quotient mappings according to the two equivalences Ξ, Θ together with $R' := \Phi^\mathsf{T}{}_,R{}_,\Psi$.

A multi-covering between relational structures most closely resembles a homomorphism on algebraic structures:

5.4 Proposition. A homomorphism between algebraic structures is necessarily a multi-covering.

Proof. Assume two mappings B, B' and the homomorphism (Ψ, Φ) from B to B', so that $B{}_,\Psi \subseteq \Phi{}_,B'$. The relation $\Phi{}_,B'$ is univalent, since Φ and B' are mappings. The domains $B{}_,\Psi{}_,\mathbb{T} = \mathbb{T} = \Phi{}_,B'{}_,\mathbb{T}$ of $B{}_,\Psi$ and $\Phi{}_,B'$ coincide, because all the relations are mappings and, therefore, are total. So we may use (*) and obtain $B{}_,\Psi = \Phi{}_,B'$. $\square$

6   Homomorphism and Isomorphism Theorems

Now we study the homomorphism and isomorphism theorems (see [1], e.g.) traditionally offered in a course on group theory or on universal algebra from the relational point of view. In the courses mentioned, R, S are often n-ary mappings such as addition and multiplication. In Fig. 6.1, we are more general allowing them to be relations, i.e., not necessarily mappings. The algebraic laws they satisfy in the algebra are completely irrelevant.

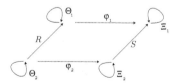

Fig. 6.1. Basic situation of the homomorphism theorem

6.1 Proposition (*Homomorphism Theorem*). Let a relation R be given with an R-congruence (Θ_2, Θ_1) as well as a relation S together with an S-congruence (Ξ_2, Ξ_1). Assume a multi-covering (φ_2, φ_1) from R to S such that at the same time we have $\Theta_i = \varphi_i{}_,\Xi_i{}_,\varphi_i^\mathsf{T}$ for $i = 1, 2$; see Fig. 6.1. Introducing the natural projections η_i for Θ_i as well as δ_i for Ξ_i, one has that $\psi_i := \eta_i^\mathsf{T}{;}\varphi_i{;}\delta_i$, $i = 1, 2$, establish an isomorphism from $R' := \eta_2^\mathsf{T}{;}R{;}\eta_1$ to $S' := \delta_2^\mathsf{T}{;}S{;}\delta_1$.

Proof. The equivalences (Θ_2, Θ_1) satisfy $\Theta_2{;}R \subseteq R{;}\Theta_1$ while (Ξ_2, Ξ_1) satisfy $\Xi_2{;}S \subseteq S{;}\Xi_1$. Furthermore, we have that (φ_2, φ_1) are surjective mappings satisfying $R{;}\varphi_1 \subseteq \varphi_2{;}S$ for homomorphism and $R{;}\varphi_1 \supseteq \varphi_2{;}S$ for multi-covering.

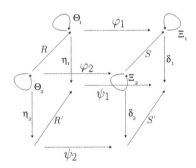

Fig. 6.2. Natural projections added to Fig. 6.1

The ψ_i are bijective mappings, which we prove omitting indices:

$$
\begin{aligned}
\psi^\mathsf{T}{;}\psi &= (\eta^\mathsf{T}{;}\varphi{;}\delta)^\mathsf{T}{;}\eta^\mathsf{T}{;}\varphi{;}\delta && \text{by definition} \\
&= \delta^\mathsf{T}{;}\varphi^\mathsf{T}{;}\eta{;}\eta^\mathsf{T}{;}\varphi{;}\delta && \text{executing transposition} \\
&= \delta^\mathsf{T}{;}\varphi^\mathsf{T}{;}\Theta{;}\varphi{;}\delta && \text{natural projection } \eta \\
&= \delta^\mathsf{T}{;}\varphi^\mathsf{T}{;}\varphi{;}\Xi{;}\delta && \text{multi-covering} \\
&= \delta^\mathsf{T}{;}\Xi{;}\delta && \text{as } \varphi \text{ is surjective and univalent} \\
&= \delta^\mathsf{T}{;}\delta{;}\delta^\mathsf{T}{;}\delta = \mathbb{I}{;}\mathbb{I} = \mathbb{I} && \text{natural projection } \delta
\end{aligned}
$$

and

$$
\begin{aligned}
\psi{;}\psi^\mathsf{T} &= \eta^\mathsf{T}{;}\varphi{;}\delta{;}(\eta^\mathsf{T}{;}\varphi{;}\delta)^\mathsf{T} && \text{by definition} \\
&= \eta^\mathsf{T}{;}\varphi{;}\delta{;}\delta^\mathsf{T}{;}\varphi^\mathsf{T}{;}\eta && \text{transposing} \\
&= \eta^\mathsf{T}{;}\varphi{;}\Xi{;}\varphi^\mathsf{T}{;}\eta && \text{natural projection } \delta \\
&= \eta^\mathsf{T}{;}\Theta{;}\eta && \text{property of } \varphi \text{ wrt. } \Theta, \Xi \\
&= \eta^\mathsf{T}{;}\eta{;}\eta^\mathsf{T}{;}\eta = \mathbb{I}{;}\mathbb{I} = \mathbb{I} && \text{natural projection } \eta
\end{aligned}
$$

Proof of the isomorphism property:

$$
\begin{aligned}
R'{;}\psi_1 &= \eta_2^\mathsf{T}{;}R{;}\eta_1{;}\eta_1^\mathsf{T}{;}\varphi_1{;}\delta_1 && \text{by definition} \\
&= \eta_2^\mathsf{T}{;}R{;}\Theta_1{;}\varphi_1{;}\delta_1 && \text{natural projection } \eta_1 \\
&= \eta_2^\mathsf{T}{;}R{;}\varphi_1{;}\Xi_1{;}\varphi_1^\mathsf{T}{;}\varphi_1{;}\delta_1 && \text{property of } \varphi \text{ wrt. } \Theta, \Xi \\
&= \eta_2^\mathsf{T}{;}R{;}\varphi_1{;}\Xi_1{;}\delta_1 && \text{as } \varphi_1 \text{ is surjective and univalent} \\
&= \eta_2^\mathsf{T}{;}\varphi_2{;}S{;}\Xi_1{;}\delta_1 && \text{multi-covering} \\
&= \eta_2^\mathsf{T}{;}\varphi_2{;}\Xi_2{;}S{;}\Xi_1{;}\delta_1 && S{;}\Xi_1 \subseteq \Xi_2{;}S{;}\Xi_1 \subseteq S{;}\Xi_1{;}\Xi_1 = S{;}\Xi_1 \\
&= \eta_2^\mathsf{T}{;}\varphi_2{;}\delta_2{;}\delta_2^\mathsf{T}{;}S{;}\delta_1{;}\delta_1^\mathsf{T}{;}\delta_1 && \text{natural projections} \\
&= \eta_2^\mathsf{T}{;}\varphi_2{;}\delta_2{;}S'{;}\delta_1^\mathsf{T}{;}\delta_1 && \text{definition of } S' \\
&= \eta_2^\mathsf{T}{;}\varphi_2{;}\delta_2{;}S' && \text{as } \delta_1 \text{ is surjective and univalent} \\
&= \psi_2{;}S' && \text{definition of } \psi_2
\end{aligned}
$$

According to Lemma 3.3, this suffices for an isomorphism. $\qquad\square$

One should bear in mind that this proposition was in several respects slightly more general than the classical homomorphism theorem: R, S need not be mappings, nor need they be homogeneous relations, Ξ was not confined to be the identity congruence, and not least does relation algebra admit non-standard models.

6.2 Proposition (*First Isomorphism Theorem*). Let a homogeneous relation R on X together with an equivalence Ξ and a non-empty subset U. Assume that U is contracted by R and that Ξ is an R-congruence:

$$R^{\mathsf{T}}{}_{;}U \subseteq U \quad \text{and} \quad \Xi{}_{;}R \subseteq R{}_{;}\Xi.$$

Now extrude both, U and its Ξ-saturation $\Xi{}_{;}U$ so as to obtain natural injections

$$\iota : Y \longrightarrow X \text{ and } \lambda : Z \longrightarrow X,$$

universally characterized by (see Fig. 6.3)

$$\iota^{\mathsf{T}}{}_{;}\iota = \mathbb{I}_X \cap U{}_{;}\mathbb{T}, \qquad \iota{}_{;}\iota^{\mathsf{T}} = \mathbb{I}_Y,$$
$$\lambda^{\mathsf{T}}{}_{;}\lambda = \mathbb{I}_X \cap \Xi{}_{;}U{}_{;}\mathbb{T}, \quad \lambda{}_{;}\lambda^{\mathsf{T}} = \mathbb{I}_Z.$$

On Y and Z, we consider the derived equivalences $\Xi_Y := \iota{}_{;}\Xi{}_{;}\iota^{\mathsf{T}}$ and $\Xi_Z := \lambda{}_{;}\Xi{}_{;}\lambda^{\mathsf{T}}$ and in addition their natural projections $\eta : Y \longrightarrow Y_\Xi$ and $\delta : Z \longrightarrow Z_\Xi$. In a standard way, restrictions of R may be defined, namely

$$S := \eta^{\mathsf{T}}{}_{;}\iota{}_{;}R{}_{;}\iota^{\mathsf{T}}{}_{;}\eta \text{ and } T := \delta^{\mathsf{T}}{}_{;}\lambda{}_{;}R{}_{;}\lambda^{\mathsf{T}}{}_{;}\delta.$$

In this setting, $\varphi := \delta^{\mathsf{T}}{}_{;}\lambda{}_{;}\iota^{\mathsf{T}}{}_{;}\eta$ gives an isomorphism (φ, φ) between S and T.

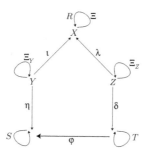

Fig. 6.3. Situation of the First Isomorphism Theorem

Proof. We prove several results in advance, namely

$$\Xi{}_{;}\iota^{\mathsf{T}}{}_{;}\iota{}_{;}\Xi = \Xi{}_{;}\lambda^{\mathsf{T}}{}_{;}\lambda{}_{;}\Xi, \tag{1}$$

proved using rules for composition of equivalences:

$$
\begin{aligned}
\Xi{}_{;}\iota^{\mathsf{T}}{}_{;}\iota{}_{;}\Xi &= \Xi{}_{;}(\mathbb{I} \cap U{}_{;}\mathbb{T}){}_{;}\Xi & \text{definition of natural injection } \iota \\
&= \Xi{}_{;}\Xi{}_{;}(\mathbb{I} \cap U{}_{;}\mathbb{T}{}_{;}\Xi){}_{;}\Xi{}_{;}\Xi & \Xi \text{ surjective and an equivalence} \\
&= \Xi{}_{;}(\mathbb{I} \cap \Xi{}_{;}U{}_{;}\mathbb{T}){}_{;}\Xi & \text{several applications of Prop. 2.3} \\
&= \Xi{}_{;}\lambda^{\mathsf{T}}{}_{;}\lambda{}_{;}\Xi & \text{definition of natural injection } \lambda
\end{aligned}
$$

In a similar way follow

$$\iota{}_{;}\lambda^{\mathsf{T}}{}_{;}\lambda = \iota \qquad \qquad \iota{}_{;}R{}_{;}\iota^{\mathsf{T}}{}_{;}\iota = \iota{}_{;}R \tag{2}$$

The left identity is proved with

$$
\begin{aligned}
\iota;\lambda^{\mathsf{T}};\lambda &= \iota;\iota^{\mathsf{T}};\iota;\lambda^{\mathsf{T}};\lambda && \iota \text{ is injective and total}\\
&= \iota;(\mathbb{I}\cap U;\mathbb{T});(\mathbb{I}\cap \varXi;U;\mathbb{T}) && \text{definition of natural injections}\\
&= \iota;(\mathbb{I}\cap U;\mathbb{T}\cap \varXi;U;\mathbb{T}) && \text{intersecting partial identities}\\
&= \iota;(\mathbb{I}\cap U;\mathbb{T}) = \iota;\iota^{\mathsf{T}} = \iota
\end{aligned}
$$

The contraction condition $R^{\mathsf{T}};U \subseteq U$ and $\varXi;R \subseteq R;\varXi$ allows to prove the right
one for which " $\subseteq$ " is obvious. For " $\supseteq$ ", we apply $\iota;\iota^{\mathsf{T}} = \mathbb{I}$ after having shown

$$
\begin{aligned}
\iota^{\mathsf{T}};\iota;R &= (\mathbb{I}\cap U;\mathbb{T});R = U;\mathbb{T};\mathbb{I}\cap R && \text{according to Prop. 2.2}\\
&\subseteq (U;\mathbb{T}\cap R;\mathbb{I}^{\mathsf{T}});(\mathbb{I}\cap (U;\mathbb{T})^{\mathsf{T}};R) && \text{Dedekind}\\
&\subseteq (R\cap U;\mathbb{T});(\mathbb{I}\cap \mathbb{T};U^{\mathsf{T}}) && \text{since } R^{\mathsf{T}};U \subseteq U\\
&= (R\cap U;\mathbb{T});(\mathbb{I}\cap U;\mathbb{T}) && \text{as } Q \subseteq \mathbb{I} \text{ implies } Q = Q^{\mathsf{T}}\\
&= (\mathbb{I}\cap U;\mathbb{T});R;(\mathbb{I}\cap U;\mathbb{T}) && \text{according to Prop. 2.2 again}\\
&= \iota^{\mathsf{T}};\iota;R;\iota^{\mathsf{T}};\iota && \text{definition of natural injection}
\end{aligned}
$$

With $R^{\mathsf{T}};\varXi;U \subseteq \varXi;R^{\mathsf{T}};U \subseteq \varXi;U$, we get in a completely similar way

$$
\lambda;R;\lambda^{\mathsf{T}};\lambda = \lambda;R \tag{3}
$$

We show that φ is univalent and surjective:

$$
\begin{aligned}
\varphi^{\mathsf{T}};\varphi &= \eta^{\mathsf{T}};\iota;\lambda^{\mathsf{T}};\delta;\delta^{\mathsf{T}};\lambda;\iota^{\mathsf{T}};\eta && \text{by definition}\\
&= \eta^{\mathsf{T}};\iota;\lambda^{\mathsf{T}};\varXi_Z;\lambda;\iota^{\mathsf{T}};\eta && \text{natural projection}\\
&= \eta^{\mathsf{T}};\iota;\lambda^{\mathsf{T}};\lambda;\varXi;\lambda^{\mathsf{T}};\lambda;\iota^{\mathsf{T}};\eta && \text{definition of } \varXi_Z\\
&= \eta^{\mathsf{T}};\iota;\varXi;\iota^{\mathsf{T}};\eta && \text{as proved initially}\\
&= \eta^{\mathsf{T}};\varXi_Y;\eta && \text{definition of } \varXi_Y\\
&= \eta^{\mathsf{T}};\eta;\eta^{\mathsf{T}};\eta = \mathbb{I};\mathbb{I} = \mathbb{I} && \text{natural projection}
\end{aligned}
$$

To show that φ is injective and total, we start

$$
\begin{aligned}
\delta;\varphi;\varphi^{\mathsf{T}};\delta^{\mathsf{T}} &= \delta;\delta^{\mathsf{T}};\lambda;\iota^{\mathsf{T}};\eta;\eta^{\mathsf{T}};\iota;\lambda^{\mathsf{T}};\delta;\delta^{\mathsf{T}} && \text{by definition}\\
&= \varXi_Z;\lambda;\iota^{\mathsf{T}};\varXi_Y;\iota;\lambda^{\mathsf{T}};\varXi_Z && \text{natural projections}\\
&= \lambda;\varXi;\lambda^{\mathsf{T}};\lambda;\iota^{\mathsf{T}};\iota;\varXi;\iota^{\mathsf{T}};\iota;\lambda^{\mathsf{T}};\lambda;\varXi;\lambda^{\mathsf{T}} && \text{by definition of } \varXi_Y, \varXi_Z\\
&= \lambda;\varXi;\iota^{\mathsf{T}};\iota;\varXi;\iota^{\mathsf{T}};\iota;\varXi;\lambda^{\mathsf{T}} && \text{as } \iota;\lambda^{\mathsf{T}};\lambda = \iota\\
&= \lambda;\varXi;\lambda^{\mathsf{T}};\lambda;\varXi;\lambda^{\mathsf{T}};\lambda;\varXi;\lambda^{\mathsf{T}} && \text{see above}\\
&= \varXi_Z;\varXi_Z;\varXi_Z = \varXi_Z && \text{by definition of } \varXi_Z
\end{aligned}
$$

so that we may go on with

$$
\begin{aligned}
\varphi;\varphi^{\mathsf{T}} &= \delta^{\mathsf{T}};\delta;\varphi;\varphi^{\mathsf{T}};\delta^{\mathsf{T}};\delta && \text{as } \delta \text{ is univalent and surjective}\\
&= \delta^{\mathsf{T}};\varXi_Z;\delta && \text{as before}\\
&= \delta^{\mathsf{T}};\delta;\delta^{\mathsf{T}};\delta = \mathbb{I};\mathbb{I} = \mathbb{I} && \text{natural projection}
\end{aligned}
$$

The interplay of subset forming and equivalence classes is visualized in Fig. 6.4.

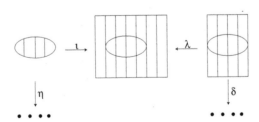

Fig. 6.4. Visualization of the First Isomorphism Theorem

It turns out that $\varXi_Y$ is an R_Y-congruence for $R_Y := \iota\,;R\,;\iota^\mathsf{T}$:

$$
\begin{aligned}
\varXi_Y\,;R_Y &= \iota\,;\varXi\,;\iota^\mathsf{T}\,;\iota\,;R\,;\iota^\mathsf{T} && \text{by definition} \\
&\subseteq \iota\,;\varXi\,;R\,;\iota^\mathsf{T} && \iota \text{ is univalent} \\
&\subseteq \iota\,;R\,;\varXi\,;\iota^\mathsf{T} && \text{congruence} \\
&\subseteq \iota\,;R\,;\iota^\mathsf{T}\,;\iota\,;\varXi\,;\iota^\mathsf{T} && (2) \\
&\subseteq R_Y\,;\varXi_Y && \text{definition of } R_Y, \varXi_Y
\end{aligned}
$$

The construct $\alpha := \iota\,;\varXi\,;\lambda^\mathsf{T}\,;\delta$ is a surjective mapping:

$$
\begin{aligned}
\alpha^\mathsf{T}\,;\alpha &= \delta^\mathsf{T}\,;\lambda\,;\varXi\,;\iota^\mathsf{T}\,;\iota\,;\varXi\,;\lambda^\mathsf{T}\,;\delta && \text{by the definition just given} \\
&= \delta^\mathsf{T}\,;\lambda\,;\varXi\,;\lambda^\mathsf{T}\,;\lambda\,;\varXi\,;\lambda^\mathsf{T}\,;\delta && (1) \\
&= \delta^\mathsf{T}\,;\varXi_Z\,;\varXi_Z\,;\delta && \text{definition of } \varXi_Z \\
&= \delta^\mathsf{T}\,;\varXi_Z\,;\delta && \varXi_Z \text{ is indeed an equivalence} \\
&= \delta^\mathsf{T}\,;\delta\,;\delta^\mathsf{T}\,;\delta = \mathbb{I}\,;\mathbb{I} = \mathbb{I} && \delta \text{ is natural projection for } \varXi_Z \\[4pt]
\alpha\,;\alpha^\mathsf{T} &= \iota\,;\varXi\,;\lambda^\mathsf{T}\,;\delta\,;\delta^\mathsf{T}\,;\lambda\,;\varXi\,;\iota^\mathsf{T} && \text{by definition} \\
&= \iota\,;\varXi\,;\lambda^\mathsf{T}\,;\varXi_Z\,;\lambda\,;\varXi\,;\iota^\mathsf{T} && \delta \text{ is natural projection for } \varXi_Z \\
&= \iota\,;\varXi\,;\lambda^\mathsf{T}\,;\lambda\,;\varXi\,;\lambda^\mathsf{T}\,;\lambda\,;\varXi\,;\iota^\mathsf{T} && \text{definition of } \varXi_Z \\
&= \iota\,;\varXi\,;\iota^\mathsf{T}\,;\iota\,;\varXi\,;\iota^\mathsf{T}\,;\iota\,;\varXi\,;\iota^\mathsf{T} && (1) \\
&= \varXi_Y\,;\varXi_Y\,;\varXi_Y = \varXi_Y \supseteq \mathbb{I} && \text{definition of equivalence } \varXi_Y
\end{aligned}
$$

With α, we may express S, T in a shorter way:

$$
\begin{aligned}
\alpha^\mathsf{T}\,;R_Y\,;\alpha &= \delta^\mathsf{T}\,;\lambda\,;\varXi\,;\iota^\mathsf{T}\,;R_Y\,;\iota\,;\varXi\,;\lambda^\mathsf{T}\,;\delta && \text{definition of } \alpha \\
&= \delta^\mathsf{T}\,;\lambda\,;\varXi\,;\iota^\mathsf{T}\,;\iota\,;R\,;\iota^\mathsf{T}\,;\iota\,;\varXi\,;\lambda^\mathsf{T}\,;\delta && \text{definition of } R_Y \\
&= \delta^\mathsf{T}\,;\lambda\,;\varXi\,;\iota^\mathsf{T}\,;\iota\,;R\,;\varXi\,;\lambda^\mathsf{T}\,;\delta && (2) \\
&= \delta^\mathsf{T}\,;\lambda\,;\varXi\,;\iota^\mathsf{T}\,;\iota\,;\varXi\,;R\,;\varXi\,;\lambda^\mathsf{T}\,;\delta && \varXi\,;R\,;\varXi \subseteq R\,;\varXi\,;\varXi = R\,;\varXi \subseteq \varXi\,;R\,;\varXi \\
&= \delta^\mathsf{T}\,;\lambda\,;\varXi\,;\lambda^\mathsf{T}\,;\lambda\,;\varXi\,;R\,;\varXi\,;\lambda^\mathsf{T}\,;\delta && (1) \\
&= \delta^\mathsf{T}\,;\varXi_Z\,;\lambda\,;R\,;\varXi\,;\lambda^\mathsf{T}\,;\delta && \text{as before, definition of } \varXi_Z \\
&= \delta^\mathsf{T}\,;\varXi_Z\,;\lambda\,;R\,;\lambda^\mathsf{T}\,;\lambda\,;\varXi\,;\lambda^\mathsf{T}\,;\delta && (3) \\
&= \delta^\mathsf{T}\,;\varXi_Z\,;\lambda\,;R\,;\lambda^\mathsf{T}\,;\varXi_Z\,;\delta && \text{definition of } \varXi_Z \\
&= \delta^\mathsf{T}\,;\delta\,;\delta^\mathsf{T}\,;\lambda\,;R\,;\lambda^\mathsf{T}\,;\delta\,;\delta^\mathsf{T}\,;\delta && \delta \text{ is natural projection for } \varXi_Z \\
&= \delta^\mathsf{T}\,;\lambda\,;R\,;\lambda^\mathsf{T}\,;\delta = T && \delta \text{ is a surjective mapping} \\[4pt]
\eta^\mathsf{T}\,;R_Y\,;\eta &= \eta^\mathsf{T}\,;\iota\,;R\,;\iota^\mathsf{T}\,;\eta && \text{definition of } R_Y \\
&= S && \text{definition of } S
\end{aligned}
$$

Relations α and φ are closely related:

$$
\begin{aligned}
\alpha\,;\varphi &= \iota\,;\varXi\,;\lambda^\mathsf{T}\,;\delta\,;\delta^\mathsf{T}\,;\lambda\,;\iota^\mathsf{T}\,;\eta && \text{definition of } \alpha, \varphi \\
&= \iota\,;\varXi\,;\lambda^\mathsf{T}\,;\varXi_Z\,;\lambda\,;\iota^\mathsf{T}\,;\eta && \delta \text{ is natural projection for } \varXi_Z \\
&= \iota\,;\varXi\,;\lambda^\mathsf{T}\,;\lambda\,;\varXi\,;\lambda^\mathsf{T}\,;\lambda\,;\iota^\mathsf{T}\,;\eta && \text{definition of } \varXi_Z \\
&= \iota\,;\varXi\,;\lambda^\mathsf{T}\,;\lambda\,;\varXi\,;\iota^\mathsf{T}\,;\eta && (2) \\
&= \iota\,;\varXi\,;\iota^\mathsf{T}\,;\iota\,;\varXi\,;\iota^\mathsf{T}\,;\eta && (1) \\
&= \varXi_Y\,;\varXi_Y\,;\eta && \text{definition of } \varXi_Y \\
&= \eta\,;\eta^\mathsf{T}\,;\eta\,;\eta^\mathsf{T}\,;\eta = \eta && \eta \text{ is natural projection for } \varXi_Y \\[4pt]
\alpha^\mathsf{T}\,;\eta &= \alpha^\mathsf{T}\,;\alpha\,;\varphi && \text{see before} \\
&= \varphi && \alpha \text{ is univalent and surjective}
\end{aligned}
$$

This enables us already to prove the homomorphism condition:

$$
\begin{aligned}
T\,;\varphi &= \alpha^\mathsf{T}\,;R_Y\,;\alpha\,;\alpha^\mathsf{T}\,;\eta && \text{above results on } T, \varphi \\
&= \alpha^\mathsf{T}\,;R_Y\,;\varXi_Y\,;\eta && \alpha\,;\alpha^\mathsf{T} = \varXi_Y, \text{ see above} \\
&= \alpha^\mathsf{T}\,;\varXi_Y\,;R_Y\,;\varXi_Y\,;\eta && \varXi_Y \text{ is an } R_Y\text{-congruence}
\end{aligned}
$$

$$= \alpha^\mathsf{T}{}_;\eta{}_;\eta^\mathsf{T}{}_;R_Y{}_;\eta{}_;\eta^\mathsf{T}{}_;\eta \qquad \eta \text{ is natural projection for } \Xi_Y$$
$$= \varphi{}_;\eta^\mathsf{T}{}_;R_Y{}_;\eta \qquad\qquad \eta \text{ is univalent and surjective}$$
$$= \varphi{}_;S \qquad\qquad\qquad\quad \text{see above}$$

This was an equality, so that it suffices according to Lemma 3.3.   □

It will have become clear, that these proofs completely rely on generic constructions and their algebraic laws. When elaborated they seem lengthy. With a supporting system, however, they reduce considerably to a sequence of rules to be applied.

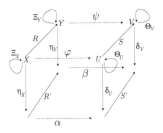

Fig. 6.5. Situation of the Second Isomorphism Theorem

6.3 Proposition (*Second Isomorphism Theorem*). Let a multi-covering (φ, ψ) between any two relations $R : X \longrightarrow Y$ and $S : U \longrightarrow V$ be given as well as an R-congruence (Ξ_X, Ξ_Y) and an S-congruence (Θ_U, Θ_V). Let also the equivalences be related through φ, ψ as $\Xi_Y = \psi{}_;\Theta_V{}_;\psi^\mathsf{T}$ and $\Xi_X = \varphi{}_;\Theta_U{}_;\varphi^\mathsf{T}$. Given this situation, introduce the natural projections $\eta_X, \eta_Y, \delta_U, \delta_V$ for the equivalences and proceed to relations $R' := \eta_X^\mathsf{T}{}_;R{}_;\eta_Y$ and $S' := \delta_U^\mathsf{T}{}_;S{}_;\delta_V$. Then $\alpha := \eta_X^\mathsf{T}{}_;\varphi{}_;\delta_U$ and $\beta := \eta_Y^\mathsf{T}{}_;\psi{}_;\delta_V$ constitute an isomorphism from R' to S' (see Fig. 6.5).

Proof. α is univalent and surjective (β follows completely analogous)

$$\begin{aligned}
\alpha^\mathsf{T}{}_;\alpha &= (\eta_X^\mathsf{T}{}_;\varphi{}_;\delta_U)^\mathsf{T}{}_;\eta_X^\mathsf{T}{}_;\varphi{}_;\delta_U &&\text{by definition}\\
&= \delta_U^\mathsf{T}{}_;\varphi^\mathsf{T}{}_;\eta_X{}_;\eta_X^\mathsf{T}{}_;\varphi{}_;\delta_U &&\text{transposing}\\
&= \delta_U^\mathsf{T}{}_;\varphi^\mathsf{T}{}_;\Xi_X{}_;\varphi{}_;\delta_U &&\text{natural projection}\\
&= \delta_U^\mathsf{T}{}_;\varphi^\mathsf{T}{}_;\varphi{}_;\Theta_U{}_;\varphi^\mathsf{T}{}_;\varphi{}_;\delta_U &&\text{condition on mapping equivalences}\\
&= \delta_U^\mathsf{T}{}_;\Theta_U{}_;\delta_U &&\text{as } \varphi \text{ is a surjective mapping}\\
&= \delta_U^\mathsf{T}{}_;\delta_U{}_;\delta_U^\mathsf{T}{}_;\delta_U &&\text{natural projection}\\
&= \mathbb{I}{}_;\mathbb{I} = \mathbb{I}
\end{aligned}$$

We show that α is total and injective (β follows completely analogous)

$$\begin{aligned}
\alpha{}_;\alpha^\mathsf{T} &= \eta_X^\mathsf{T}{}_;\varphi{}_;\delta_U{}_;(\eta_X^\mathsf{T}{}_;\varphi{}_;\delta_U)^\mathsf{T} &&\text{by definition}\\
&= \eta_X^\mathsf{T}{}_;\varphi{}_;\delta_U{}_;\delta_U^\mathsf{T}{}_;\varphi^\mathsf{T}{}_;\eta_X &&\text{transposing}\\
&= \eta_X^\mathsf{T}{}_;\varphi{}_;\Theta_U{}_;\varphi^\mathsf{T}{}_;\eta_X &&\text{natural projection}\\
&= \eta_X^\mathsf{T}{}_;\Xi_X{}_;\eta_X &&\text{condition on mapping equivalences}\\
&= \eta_X^\mathsf{T}{}_;\eta_X{}_;\eta_X^\mathsf{T}{}_;\eta_X &&\text{natural projection}\\
&= \mathbb{I}{}_;\mathbb{I} = \mathbb{I}
\end{aligned}$$

We show that $\alpha_;\beta$ is a homomorphism:

$$
\begin{aligned}
R'_;\beta &= \eta_X^\mathsf{T}{}_;R_;\eta_Y{}_;\eta_Y^\mathsf{T}{}_;\psi_;\delta_V && \text{by definition}\\
&= \eta_X^\mathsf{T}{}_;R_;\Xi_Y{}_;\psi_;\delta_V && \text{natural projection}\\
&= \eta_X^\mathsf{T}{}_;R_;\psi_;\Theta_V{}_;\psi^\mathsf{T}{}_;\psi_;\delta_V && \text{condition on mapping equivalences}\\
&= \eta_X^\mathsf{T}{}_;R_;\psi_;\Theta_V{}_;\delta_V && \text{as } \psi \text{ is surjective and univalent}\\
&= \eta_X^\mathsf{T}{}_;\varphi_;S_;\Theta_V{}_;\delta_V && \text{multi-covering}\\
&= \eta_X^\mathsf{T}{}_;\varphi_;\Theta_U{}_;S_;\Theta_V{}_;\delta_V && S_;\Theta_V \subseteq \Theta_U{}_;S_;\Theta_V \subseteq S_;\Theta_V{}_;\Theta_V = S_;\Theta_V\\
&= \eta_X^\mathsf{T}{}_;\varphi_;\Theta_U{}_;S_;\delta_V{}_;\delta_V^\mathsf{T}{}_;\delta_V && \text{natural projection}\\
&= \eta_X^\mathsf{T}{}_;\varphi_;\Theta_U{}_;S_;\delta_V && \text{as } \delta \text{ is a surjective mapping}\\
&= \eta_X^\mathsf{T}{}_;\varphi_;\delta_U{}_;\delta_U^\mathsf{T}{}_;S_;\delta_V && \text{natural projection}\\
&= \alpha_;S' && \text{by definition}
\end{aligned}
$$

This was an equality, so that it suffices according to Lemma 3.3. □

7 Covering of Graphs and Path Equivalence

There is another point to mention here which has gained considerable interest in an algebraic or topological context, not least for Riemann surfaces.

7.1 Proposition (*Lifting property*). Let a homogeneous relation B be given together with a multi-covering (Φ, Φ) on the relation B'. Let furthermore some rooted graph B_0 with root a_0, i.e., satisfying and $B_0^{\mathsf{T}*}{}_;a_0 = \mathbb{T}$, be given together with a homomorphism Φ_0 that sends the root a_0 to $a' := \Phi_0^\mathsf{T}{}_;a_0$. If $a \subseteq \Phi^\mathsf{T}{}_;a'$ is some point mapped by Φ to a', there exists always a relation Ψ — not necessarily a mapping — satisfying the properties

$$\Psi^\mathsf{T}{}_;a_0 = a \quad \text{and} \quad B_0{}_;\Psi \subseteq \Psi_;B.$$

Idea of proof: Define $\Psi := \inf\{X \mid a_0{}_;a^\mathsf{T} \cup (B_0^\mathsf{T}{}_;X_;B \cap \Phi_0{}_;\Phi^\mathsf{T}) \subseteq X\}$. □

The relation Ψ enjoys the homomorphism property but fails to be a mapping in general. In order to make it a mapping, one will choose one of the following two possibilities:

- Firstly, one might follow the recursive definition starting from a_0 and at every stage make an arbitrary choice among the relational images offered, thus choosing a fiber.
- Secondly, one may further restrict the multi-covering condition to "locally univalent" fans in Φ, requiring $B_0^\mathsf{T}{}_;\Psi_;B \cap \Phi_0{}_;\Phi^\mathsf{T} \subseteq \mathbb{I}$ to hold for it, which leads to a well-developed theory, see [2,3,8].

In both cases, one will find a homomorphism from B_0 to B. The effect of a flow chart diagram is particularly easy to understand when the underlying rooted graph is also a rooted *tree*, so that the view is not blocked by nested circuits which can be traveled several times. When dealing with a rooted graph that does contain such circuits one has to keep track of the possibly infinite number of ways in which the graph can be traversed from its root. To this end there exists a theory of coverings which is based on the notion of homomorphy.

The idea is to unfold circuits. We want to characterize those homomorphisms of a graph that preserve to a certain extent the possibilities of traversal. We shall see that such a homomorphism is surjective and that it carries the successor relation at any point onto that at the image point.

7.2 Definition. A surjective homomorphism $\Phi\colon G \longrightarrow G'$ is called a **covering**, provided that it is a multi-covering satisfying $B^\mathsf{T}{,}B \cap \Phi{,}\Phi^\mathsf{T} \subseteq \mathbb{I}$. □

The multi-covering Φ compares two relations between the points of G and of G' and ensures that for any inverse image point x of some point x' and successor y' of x' there is *at least* one successor y of x which is mapped onto y'. The new condition guarantees that there is *at most* one such y since it requires that the relation "have a common predecessor according to B, and have a common image under Φ" is contained in the identity.

8 Concluding Remark

We have reworked mathematical basics from a relational perspective. First the step from an algebraic to a relational structure has been made. This is so serious a generalization, that one would not expect much of the idea of homomorphism and isomorphism theorems to survive. With the concept of a multi-covering, however, a new and adequate concept seems to have been found. Prop. 5.4 shows that it reduces completely to homomorphisms when going back to the algebraic case. For relational structures, a multi-covering behaves nicely with respect to quotient forming. This relates to earlier papers (see [2,3,8]) where semantics of programs (partial correctness, total correctness, and flow equivalence, even for systems of recursive procedures) has first been given a componentfree relational form.

References

1. Grätzer, G.: Universal Algebra, 2nd Ed. Springer-Verlag (1978)
2. Schmidt, G.: Programme als partielle Graphen. Habil. Thesis 1977 und Bericht 7813, Fachbereich Mathematik der Techn. Univ. München (1977) English as [3,8].
3. Schmidt, G.: Programs as partial graphs I: Flow equivalence and correctness. Theoret. Comput. Sci. **15** (1981) 1–25
4. Schmidt, G., Ströhlein, T.: Relationen und Graphen. Mathematik für Informatiker. Springer-Verlag (1989) ISBN 3-540-50304-8, ISBN 0-387-50304-8.
5. Schmidt, G., Ströhlein, T.: Relations and Graphs — Discrete Mathematics for Computer Scientists. EATCS Monographs on Theoretical Computer Science. Springer-Verlag (1993) ISBN 3-540-56254-0, ISBN 0-387-56254-0.
6. Freyd, P.J., Scedrov, A.: Categories, Allegories. Volume 39 of North-Holland Mathematical Library. North-Holland, Amsterdam (1990)
7. Kahl, W.: A Relation-Algebraic Approach to Graph Structure Transformation. Technical Report 2002/03, Fakultät für Informatik, Universität der Bundeswehr München (2002) http://ist.unibw-muenchen.de/Publications/TR/2002-03/.
8. Schmidt, G.: Programs as partial graphs II: Recursion. Theoret. Comput. Sci. **15** (1981) 159–179

Relational Measures and Integration

Gunther Schmidt*

Institute for Software Technology, Department of Computing Science
Universität der Bundeswehr München, 85577 Neubiberg, Germany
Gunther.Schmidt@UniBw.DE

Abstract. Work in fuzzy modeling has recently made its way from the interval $[0, 1] \subseteq \mathbb{R}$ to the ordinal or even to the qualitative level. We proceed further and introduce relational measures and relational integration. First ideas of this kind, but for the real-valued linear orderings stem from Choquet (1950s) and Sugeno (1970s). We generalize to not necessarily linear order and handle it algebraically and in a componentfree manner. We thus open this area of research for treatment with theorem provers which would be extremely difficult for the classical presentation of Choquet and Sugeno integrals.

Keywords: Sugeno integral, Choquet integral, relation algebra, evidence and belief, plausibility, necessity, and possibility measures, relational measure.

1 Introduction

Mankind has developed a multitude of concepts to reason about something that is *better than* or is *more attractive than* something else or *similar to* something else. Such concepts lead to an enormous bulk of formulae and interdependencies.

We start from the concept of an *order* and a *strictorder*, defined as a transitive, antisymmetric, reflexive relation or as a transitive and asymmetric, respectively. In earlier times it was not at all clear that orderings need not be *linear* orderings. But since the development of lattice theory in the 1930s it became more and more evident that most of our reasoning with orderings was also possible when they failed to be linear ones. So the people studied fuzziness mainly along the linear order of $\mathbb{R}$ and began only later to generalize to the ordinal level: Numbers indicate the relative position of items, but no longer the magnitude of difference. Then they moved to the interval level: Numbers indicate the magnitude of difference between items, but there is no absolute zero point. Examples are attitude scales and opinion scales. We proceed even further and introduce relational measures with values in a lattice. Measures traditionally provide a basis for integration. Astonishingly, this holds true for these relational measures so that it becomes possible to introduce a concept of relational integration.

---

* Cooperation and communication around this research was partly sponsored by the European COST Action 274: TARSKI (*Theory and Applications of Relational Structures as Knowledge Instruments*), which is gratefully acknowledged.

2   Modelling Preferences

Who is about to make severe decisions will usually base these on carefully selected basic information and clean lines of reasoning. It is in general not too
difficult to apply *just one* criterion and to operate according to this criterion.
If several criteria must be taken into consideration, one has also to consider the
all too often occurring situation that these provide contradictory information:
"This car looks nicer, but it is much more expensive". Social and economical
sciences have developed techniques to model what takes place when decisions
are to be made in an environment with a multitude of diverging criteria. Preference is assumed to represent the degree to which one alternative is preferred to
another. Often it takes the form of expressing that alternative A is considered
being "not worse than" alternative B. Sometimes a linear ranking of the set of
alternatives is assumed, which we avoid.

So finding decisions became abstracted to a scientific task. We may observe
two lines of development. The Anglo-Saxon countries, in particular, formulated
utility theory, in which numerical values shall indicate the intensity of some
preference. Mainly in continental Europe, on the other hand side, binary relations
were used to model pairwise preference; see [1], e.g. While the former idea allows
to easily relate to statistics, the latter is based on evidence via direct comparison.
In earlier years indeed, basic information was quite often statistical in nature
and expressed in real numbers. Today we have more often fuzzy, vague, rough,
etc. forms of qualification.

3   Introductory Example

We first give an example of relational integration deciding for a car to be bought
out of several offers. We intend to follow a set $\mathcal{C}$ of three criteria, namely color,
price, and speed. They are, of course, not of equal importance for us; price will
most certainly outweigh the color of the car, e.g. Nevertheless let the valuation
with these criteria be given on an ordinal scale $\mathcal{L}$ with 5 linearly ordered values
as indicated on the left side of (1). (Here for simplicity, the ordering is linear,
but it need not.) We name these values 1,2,3,4,5, but do not combine this with
any arithmetic; i.e., value 4 is not intended to mean two times as good as value
2. Rather they might be described with linguistic variables as *bad, not totally
bad, medium, outstanding, absolutely outstanding*; purposefully these example
qualifications have not been chosen "equidistant".

$$
\begin{matrix}
\text{color} \\
\text{price} \\
\text{speed}
\end{matrix}
\begin{pmatrix}
0 & 0 & 0 & 1 & 0 \\
0 & 0 & 0 & 1 & 0 \\
0 & 1 & 0 & 0 & 0
\end{pmatrix}
\qquad
\begin{aligned}
4 = \mathtt{lub}\,\big[\,&\mathtt{glb}\big(4_{v(color)}, 4_{\mu\{c,p\}}\big), \\
&\mathtt{glb}\big(4_{v(price)}, 4_{\mu\{c,p\}}\big), \\
&\mathtt{glb}\big(2_{v(speed)}, 5_{\mu\{c,p,s\}}\big)\,\big]
\end{aligned}
\qquad (1)
$$

First we concentrate on the left side of (1). The task is to arrive at *one* overall
valuation of the car out of these three. In a simple-minded approach, we might

indeed conceive numbers $1, 2, 3, 4, 5 \in \mathbb{R}$ and then evaluate in a classical way the average value as $\frac{1}{3}(4 + 4 + 2) = 3.3333\ldots$, which is a value not expressible in the given scale. When considering the second example (2), we would arrive at the same average value although the switch from (1) to (2) between price and speed would trigger most people to decide differently.

$$
\begin{array}{l}
\text{color} \\
\text{price} \\
\text{speed}
\end{array}
\begin{pmatrix}
0 & 0 & 0 & 1 & 0 \\
0 & 1 & 0 & 0 & 0 \\
0 & 0 & 0 & 1 & 0
\end{pmatrix}
\qquad
\begin{aligned}
3 = \mathtt{lub}\,\big[&\mathtt{glb}\,(4_{v(color)}, 3_{\mu\{c,s\}}), \\
&\mathtt{glb}\,(2_{v(price)}, 5_{\mu\{c,p,s\}}), \qquad (2) \\
&\mathtt{glb}\,(4_{v(speed)}, 3_{\mu\{c,s\}}) \big]
\end{aligned}
$$

With relational integration, we learn to make explicit which set of criteria to apply with which weight. It is conceivable that criteria c_1, c_2 are given a low weight but the criteria set $\{c_1, c_2\}$ in conjunction a high one. This means that we introduce a **relational measure** assigning values in $\mathcal{L}$ to subsets of $\mathcal{C}$.

$$
\mu =
\begin{array}{l}
\{\} \\
\{color\} \\
\{price\} \\
\{color,price\} \\
\{speed\} \\
\{color,speed\} \\
\{price,speed\} \\
\{color,price,speed\}
\end{array}
\begin{pmatrix}
1 & 0 & 0 & 0 & 0 \\
1 & 0 & 0 & 0 & 0 \\
0 & 0 & 1 & 0 & 0 \\
0 & 0 & 0 & 1 & 0 \\
0 & 1 & 0 & 0 & 0 \\
0 & 0 & 1 & 0 & 0 \\
0 & 0 & 1 & 0 & 0 \\
0 & 0 & 0 & 0 & 1
\end{pmatrix}
$$

For gauging purposes we demand that the empty criteria set gets assigned the least value in $\mathcal{L}$ and the full criteria set the greatest. A point to stress is that we assume the criteria themselves as well as the measuring of subsets of criteria as *commensurable*.

The relational measure μ should obviously be monotonic with respect to the ordering Ω on the powerset of $\mathcal{C}$ and the ordering E on $\mathcal{L}$. We do not demand continuity (additivity), however. The price alone is ranked of medium importance 3, higher than speed alone, while color alone is considered completely unimportant and ranks 1. However, color and price together are ranked 4, i.e., higher than the supremum of ranks for color alone and for price alone, etc.

As now the valuations according to the criteria as well as the valuation according to the relative measuring of the criteria are given, we may proceed as visualized on the right sides of (1) and (2). We run through the criteria and always look for two items: their corresponding value and in addition for the value of that subset of criteria *assigning equal or higher values*. Then we determine the greatest lower bound for the two values. From the list thus obtained, the least upper bound is taken. The two examples above show how by simple evaluation along this concept, one will arrive at the overall values 4 or 3, respectively. This results from the fact that in the second case only such rather unimportant criteria as color and speed assign the higher values.

The effect is counterrunning: Low values of criteria as for s in (1) are intersected with rather high μ's as many criteria give higher scores and μ is monotonic. Highest values of criteria as for color or speed in (2) are intersected with the μ of a small or even one-element criteria set; i.e., with a rather small one. In total we find that here are two operations applied in a way we already know from matrix multiplication: a "sum" operator, $\mathtt{lub}$ or $\vee$, following application a "product" operator, $\mathtt{glb}$ or $\wedge$.

This example gave a first idea of how relational integration works and how it may be useful. Introducing a relational measure and using it for integration serves an important purpose: Concerns are now separated. One may design the criteria and the measure in a design phase prior to polling. Only then shall the questionnaire be filled, or the voters be polled. The procedure of coming to an overall valuation is now just computation and should no longer lead to quarrels.

4   Order-Theoretic Functionals

Given the page limit, we cannot present all the prerequisites on relation algebra and give [2,3] as a general reference for handling relations as boolean matrices and subsets of a set as boolean vectors. Let an order relation E be given on a set V. An element e is called an *upper bound* (also: *majorant*) of the subset of V characterized by the vector u of V provided $\forall x \in u : E_{xe}$. From the predicate logic version, we easily derive a relation-algebraic formulation as $e \subseteq \overline{\overline{E}^{\mathsf{T}} {}_{;} u}$, so that we introduce the order-theoretic functional $\mathtt{ubd}_E(u) := \overline{\overline{E}^{\mathsf{T}} {}_{;} u}$ to return the possibly empty vector of all upper bounds. Analogously, we have the set of lower bounds $\mathtt{lbd}_E(u) := \overline{\overline{E} {}_{;} u}$.

Starting herefrom, also the other traditional functionals may be obtained, as the least upper bound u, (also: *supremum*), the at most 1-element set of least elements among the set of all upper bounds of u

$$\mathtt{lub}_E(u) = \mathtt{ubd}_E(u) \cap \mathtt{lbd}_E(\mathtt{ubd}_E(u))$$

In contrast to our expectation that a least upper bound may exist or not, it will here always exist as a vector; it may, however be the null vector resembling that there is none.

As a tradition, a vector is often a column vector. In many cases, however, a row vector would be more convenient. We decided to introduce a variant denotation for order-theoretic functionals working on row vectors:

$$\mathtt{lubR}_E(X) := [\mathtt{lub}_E(X^{\mathsf{T}})]^{\mathsf{T}}, \text{ etc.}$$

We are here concerned with lattice orderings E only. For convenience we introduce notation for least and greatest elements as

$$0_E = \mathtt{glb}_E(\mathbb{T}), \quad 1_E = \mathtt{lub}_E(\mathbb{T})$$

5   Relational Measures

Assume the following basic setting with a set $\mathcal{C}$ of so-called criteria and a measuring lattice $\mathcal{L}$. Depending on the application envisaged, the set $\mathcal{C}$ may also be interpreted as one of players in a cooperative game, of attributes, of experts, or of voters in an opinion polling problem. This includes the setting with $\mathcal{L}$

the interval $[0,1] \subseteq \mathbb{R}$ or a linear ordering for measuring. We consider a (relational) measure generalizing the concept of a fuzzy measure (or capacité in French origin) assigning measures in $\mathcal{L}$ for subsets of $\mathcal{C}$.

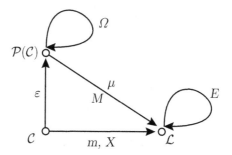

Fig. 5.1. Basic situation for relational integration

The relation ε is the membership relation between $\mathcal{C}$ and its powerset $\mathcal{P}(\mathcal{C})$. The measures envisaged will be called μ, other relations will be denoted as M. Valuations according to the criteria will be X or m depending on the context.

For a running example assume the task to assess persons of the staff according to their intellectual abilities as well as according to the workload they achieve to master.

$$
E = \begin{array}{r}
\text{(low,lazy)} \\
\text{(medium,lazy)} \\
\text{(low,fair)} \\
\text{(high,lazy)} \\
\text{(medium,fair)} \\
\text{(low,good)} \\
\text{(high,fair)} \\
\text{(medium,good)} \\
\text{(low,bulldozer)} \\
\text{(high,good)} \\
\text{(medium,bulldozer)} \\
\text{(high,bulldozer)}
\end{array}
\left(\begin{array}{cccccccccccc}
1 & 1 & 1 & 1 & 1 & 1 & 1 & 1 & 1 & 1 & 1 & 1 \\
0 & 1 & 0 & 1 & 1 & 0 & 1 & 1 & 0 & 1 & 1 & 1 \\
0 & 0 & 1 & 0 & 1 & 1 & 1 & 1 & 1 & 1 & 1 & 1 \\
0 & 0 & 0 & 1 & 0 & 0 & 1 & 0 & 0 & 1 & 0 & 1 \\
0 & 0 & 0 & 0 & 1 & 0 & 1 & 1 & 0 & 1 & 1 & 1 \\
0 & 0 & 0 & 0 & 0 & 1 & 0 & 1 & 1 & 1 & 1 & 1 \\
0 & 0 & 0 & 0 & 0 & 0 & 1 & 0 & 0 & 1 & 0 & 1 \\
0 & 0 & 0 & 0 & 0 & 0 & 0 & 1 & 0 & 1 & 1 & 1 \\
0 & 0 & 0 & 0 & 0 & 0 & 0 & 0 & 1 & 0 & 1 & 1 \\
0 & 0 & 0 & 0 & 0 & 0 & 0 & 0 & 0 & 1 & 0 & 1 \\
0 & 0 & 0 & 0 & 0 & 0 & 0 & 0 & 0 & 0 & 1 & 1 \\
0 & 0 & 0 & 0 & 0 & 0 & 0 & 0 & 0 & 0 & 0 & 1
\end{array}\right)
$$

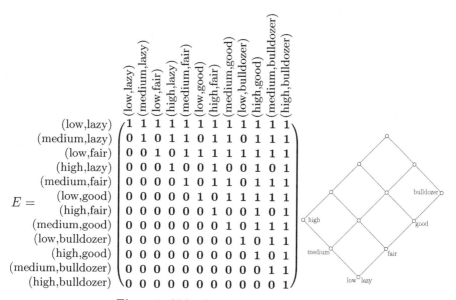

Fig. 5.2. Value lattice $\mathcal{L}$ ordered with E

5.1 Definition. Suppose a set of criteria $\mathcal{C}$ to be given together with some lattice $\mathcal{L}$, ordered by E, in which subsets of these criteria shall be given a measure

$\mu : \mathcal{P}(\mathcal{C}) \longrightarrow \mathcal{L}$. Let Ω be the ordering on $\mathcal{P}(\mathcal{C})$. We call a mapping $\mu : \mathcal{P}(\mathcal{C}) \to \mathcal{L}$ a (**relational**) **measure** provided

- $\Omega_{;}\mu \subseteq \mu_{;}E$,   meaning that μ is isotonic wrt. to the orderings Ω and E.
- $\mu^{\mathsf{T}}{}_{;}0_{\Omega} = 0_E$,   meaning that the empty subset of $\mathcal{P}(\mathcal{C})$ is mapped to the least element of $\mathcal{L}$.
- $\mu^{\mathsf{T}}{}_{;}1_{\Omega} = 1_E$,   meaning that the full subset of $\mathcal{P}(\mathcal{C})$ is mapped to the greatest element of $\mathcal{L}$.   □

A (relational) measure for $s \in \mathcal{P}(\mathcal{C})$, i.e., $\mu(s)$ when written as a mapping or $\mu^{\mathsf{T}}{}_{;}s$ when written in relation form, may be interpreted as the weight of importance we attribute to the combination s of criteria. It should not be mixed up with a probability. The latter would require the setting $\mathcal{L} = [0,1] \subseteq \mathbb{R}$ and in addition that μ be continuous.

Many ideas of this type have been collected by Glenn Shafer under the heading *theory of evidence*, calling μ a *belief function*. Using it, he explained a basis of rational behaviour. We attribute certain weights to evidence, but do not explain in which way. These weights shall in our case be lattice-ordered. This alone gives us reason to rationally decide this or that way. Real-valued belief functions have numerous applications in artificial intelligence, expert systems, approximate reasoning, knowledge extraction from data, and Bayesian Networks.

Concerning additivity, the example of Glenn Shafer [4] is when one is wondering whether a Ming vase is a genuine one or a fake. We have to put the full amount of our belief on the disjunction "*genuine* or *fake*" as one of the alternatives will certainly be the case. But the amount of trust we are willing to put on the alternatives may in both cases be very small as we have only tiny hints for being genuine, but also very tiny hints for being a fake.

With the idea of probability, we could not so easily cope with the ignorance just mentioned. Probability does not allow one to withhold belief from a proposition without *according the withheld amount of belief to the negation*. When thinking on the Ming vase in terms of probability we would have to attribute p to *genuine* and $1 - p$ to *fake*.

In the extreme case, we have complete ignorance expressed by the so-called **vacuous belief mapping**

$$\mu_0(s) = \begin{cases} 0_E & \text{if } \mathcal{C} \neq s \\ 1_E & \text{if } \mathcal{C} = s \end{cases}$$

On the other side, we may completely overspoil our trust expressed by what we may call a **light-minded belief mapping**

$$\mu_1(s) = \begin{cases} 0_E & \text{if } 0_{\Omega} = s \\ 1_E & \text{otherwise} \end{cases}$$

To an arbitrary non-empty set of criteria, the light-minded belief mapping attributes all the components of trust or belief.

5.2 Definition. Given this setting, we call μ

i) a **Bayesian measure** if it is lattice-continuous, i.e.,

$$\mathtt{lub}_E(\mu^\mathsf{T};s) = \mu^\mathsf{T};\mathtt{lub}_\Omega(s)$$

for a subset $s \subseteq \mathcal{P}(\mathcal{C})$, or expressed differently, a set of subsets of $\mathcal{C}$.

ii) a **simple support mapping** focused on U valued with v, if U is a non-empty subset $U \subseteq \mathcal{C}$ and $v \in \mathcal{L}$ an element such that

$$\mu(s) = \begin{cases} 0_E & \text{if } s \not\supseteq U \\ v & \text{if } \mathcal{C} \neq s \supseteq U \\ 1_E & \text{if } \mathcal{C} = s \end{cases} \qquad \square$$

In particular, μ_1 is Bayesian while μ_0 is not. In the real-valued environment, the condition for a Bayesian measure is: additive when non-overlapping. Lattice-continuity incorporates two concepts, namely additivity

$$\mu^\mathsf{T};(s_1 \cup s_2) = \mu^\mathsf{T};s_1 \cup_\mathcal{L} \mu^\mathsf{T};s_2$$

and sending 0_Ω to 0_E.

Combining measures

Dempster [5] found for the real-valued case a way of combining measures in a form closely related to conditional probability. It shows a way of adjusting opinion in the light of new evidence. We have re-modeled this for the relational case. One should be aware of how a measure behaves on upper and lower cones:

$$\mu = \mathtt{lubR}_E(\Omega^\mathsf{T};\mu) \quad \mu = \mathtt{glbR}_E(\Omega;\mu)$$

When one has in addition to μ got further evidence from a second measure μ', one will intersect the upper cones resulting in a possibly smaller cone positioned higher up and take its greatest lower bound:

$$\mu \oplus \mu' := \mathtt{glbR}_E(\mu;E \cap \mu';E)$$

One might, however, also look where μ and μ' agree, and thus intersect the lower bound cones resulting in a possibly smaller cone positioned deeper down and take its least upper bound:

$$\mu \otimes \mu' := \mathtt{lubR}_E(\mu;E^\mathsf{T} \cap \mu';E^\mathsf{T})$$

5.3 Proposition. If the measures μ, μ' are given, $\mu \oplus \mu'$ as well as $\mu \otimes \mu'$ are measures again. Both operations are commutative and associative. The vacuous belief mapping μ_0 is the null element while the light-minded belief mapping μ_1 is the unit element among measures:

$$\mu \oplus \mu_0 = \mu, \quad \mu \otimes \mu_1 = \mu, \quad \text{and} \quad \mu \otimes \mu_0 = \mu_0$$

Proof: The least element must be sent to the least element. This result is prepared observing that 0_Ω is a transposed mapping, in

$$\mathtt{lbd}_E([\mu;E \cap \mu';E]^\mathsf{T});0_\Omega$$
$$= \overline{\overline{E};[\mu;E \cap \mu';E]^\mathsf{T};0_\Omega}$$
$$= \overline{E};[\mu;E \cap \mu';E]^\mathsf{T};0_\Omega \quad \text{a mapping may slip under a negation from the left}$$

$$= \overline{\overline{E_i\,[E^{\mathsf{T}}_i\,\mu^{\mathsf{T}} \cap E^{\mathsf{T}}_i\,\mu'^{\mathsf{T}}]_i 0_\Omega}}$$
$$= \overline{\overline{E_i\,[E^{\mathsf{T}}_i\,\mu^{\mathsf{T}}_i 0_\Omega \cap E^{\mathsf{T}}_i\,\mu'^{\mathsf{T}}_i 0_\Omega]}} \text{ multiplying an injective relation from the right}$$
$$= \overline{\overline{E_i\,[E^{\mathsf{T}}_i\,0_E \cap E^{\mathsf{T}}_i\,0_E]}} \text{ definition of measure}$$
$$= \overline{\overline{E_i\,E^{\mathsf{T}}_i\,0_E}}$$
$$= \overline{\overline{E_i\,\mathbb{T}}} \text{ in the complete lattice } E$$
$$= \mathtt{lbd}\,(\mathbb{T}) = 0_E \text{ in the complete lattice } E$$

Now
$$(\mu \oplus \mu')^{\mathsf{T}}_i 0_\Omega = \mathtt{glb}_E([\mu_i\,E \cap \mu'_i\,E]^{\mathsf{T}})_i 0_\Omega$$
$$= \big(\mathtt{lbd}_E([\mu_i\,E \cap \mu'_i\,E]^{\mathsf{T}}) \cap \mathtt{ubd}\,(\mathtt{lbd}_E([\mu_i\,E \cap \mu'_i\,E]^{\mathsf{T}}))\big)_i 0_\Omega$$
$$= \mathtt{lbd}_E([\mu_i\,E \cap \mu'_i\,E]^{\mathsf{T}})_i 0_\Omega \cap \overline{\overline{E}}^{\mathsf{T}}_i\mathtt{lbd}_E([\mu_i\,E \cap \mu'_i\,E]^{\mathsf{T}})_i 0_\Omega$$
$$= 0_E \cap \overline{\overline{E}}^{\mathsf{T}}_i\mathtt{lbd}_E([\mu_i\,E \cap \mu'_i\,E]^{\mathsf{T}})_i 0_\Omega$$
$$= 0_E \cap \overline{\overline{E}}^{\mathsf{T}}_i 0_E$$
$$= 0_E \cap \mathtt{ubd}\,(0_E)$$
$$= 0_E \cap \mathbb{T} = 0_E$$

For reasons of space, the other parts of the proof are left to the reader.   □

6   Relational Integration

Assume now that for all the criteria $\mathcal{C}$ a valuation has taken place resulting in a mapping $X : \mathcal{C} \longrightarrow \mathcal{L}$. The question is how to arrive at an overall valuation by rational means, for which μ shall be the guideline.

6.1 Definition. Given a relational measure μ and a mapping X indicating the values given by the criteria, we define the **relational integral**

$$(R) \int X \circ \mu := \mathtt{lubR}_E(\mathbb{T}_i\,\mathtt{glbR}_E[(X \cup \mathtt{syq}(X_i\,E_i\,X^{\mathsf{T}}, \varepsilon)_i\,\mu)]) \qquad □$$

As already mentioned, we apply a sum operator $\mathtt{lub}$ after applying the product operator $\mathtt{glb}$. When values are assigned with X, we look with E for those greater or equal, then with X^{T} for the criteria so valuated. Now comes a technically difficult step, namely proceeding to the union of the resulting sets with the symmetric quotient $\mathtt{syq}$ and the membership relation ε. The μ-score of this set is then taken.

The tables in Fig. 6.1 show a measure, a valuation and then the relational integral computed with the TITUREL system.

We are now in a position to understand why gauging $\mu^{\mathsf{T}}_i 1_\Omega = 1_E$ is necessary for μ, or "greatest element is sent to greatest element". Consider, e.g., the special case of an X with all criteria assigning the same value. We certainly expect the relational integral to precisely deliver this value regardless of the measure chosen. But this might not be the case if a measure should assign too small a value to the full set.

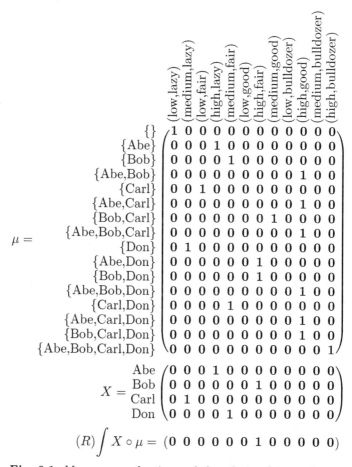

$$(R)\int X \circ \mu = (0\ 0\ 0\ 0\ 0\ 0\ 1\ 0\ 0\ 0\ 0\ 0)$$

Fig. 6.1. Measure, a valuation and the relational integral

These considerations originate from a free re-interpretation of the following concepts for work in $[0,1] \subseteq \mathbb{R}$. The **Sugeno integral** operator is in the literature defined as

$$M_{S,\mu}(x_1.\ldots,x_m) = (S)\int x \circ \mu = \bigvee_{i=1}^{m} [x_i \wedge \mu(A_i)]$$

and the **Choquet integral** operator as

$$M_{C,\mu}(x_1,\ldots,x_m) = (C)\int x \circ \mu = \sum_{i=1}^{m} [(x_i - x_{i-1}) \cdot \mu(A_i)]$$

In both cases the elements of vector $(x_1,\ldots,x_m)$, and parallel to this, the criteria set $\mathcal{C} = \{C_1,\ldots,C_m\}$ have each time been reordered such that

$$0 = x_0 \leq x_1 \leq x_2 \leq \cdots \leq x_m \leq x_{m+1} = 1 \text{ and } \mu(A_i) = \mu(C_i,\ldots,C_m).$$

The concept of Choquet integral was first introduced for a real-valued context in [6] and later used by Michio Sugeno [7]. This integral has nice properties for aggregation: It is continuous, non-decreasing, and stable under certain interval preserving transformations. Not least reduces it to the weighted arithmetic mean as soon as it becomes additive.

7 Defining Relational Measures

Such measures may be given directly, which is, however, a costly task as a power-set is involved all of whose elements need values. Therefore, they mainly originate in some other way.

Measures originating from direct valuation of criteria

Let a **direct valuation** of the criteria be given as any relation m between $\mathcal{C}$ and $\mathcal{L}$. Although it is allowed to be contradictory and non-univalent, we provide for a way of defining a relational measure based on it. This will happen via the following constructs

$$\sigma(m) := \overline{\varepsilon^{\mathsf{T}} ; m ; \overline{E}} \qquad \pi(\mu) := \overline{\varepsilon ; \mu ; \overline{E}^{\mathsf{T}}}, \tag{3}$$

which very obviously satisfy the Galois correspondence requirement

$$m \subseteq \pi(\mu) \quad \Longleftrightarrow \quad \mu \subseteq \sigma(m).$$

They satisfy $\sigma(m ; E^{\mathsf{T}}) = \sigma(m)$ and $\pi(\mu ; E) = \pi(\mu)$, so that in principle only lower, respectively upper, cones occur as arguments. Applying $\overline{W ; E} = \overline{W ; E ; E^{\mathsf{T}}}$, we get

$$\sigma(m) ; E = \overline{\varepsilon^{\mathsf{T}} ; m ; \overline{E}} ; E = \overline{\varepsilon^{\mathsf{T}} ; m ; \overline{E} ; E^{\mathsf{T}}} ; E = \overline{\varepsilon^{\mathsf{T}} ; m ; \overline{E}} = \sigma(m),$$

so that images of σ are always upper cones — and thus best described by their greatest lower bound $\mathtt{glbR}_E(\sigma(m))$.

7.1 Proposition. Given any relation $m : \mathcal{C} \to \mathcal{L}$, the construct

$$\mu_m := \mu_0 \oplus \mathtt{glbR}_E(\sigma(m))$$

forms a relational measure, the so-called **possibility measure**. □

Addition of the vacuous belief mapping μ_0 is again necessary for gauging purposes. In case m is a mapping, the situation becomes even nicer. From

$$\pi(\sigma(m ; E^{\mathsf{T}})) = \pi(\sigma(m)) = \overline{\varepsilon ; \overline{\varepsilon^{\mathsf{T}} ; m ; \overline{E}} ; \overline{E}^{\mathsf{T}}}$$

$$= \overline{m ; \overline{E} ; \overline{E}^{\mathsf{T}}} \text{ as it can be shown that in general } \varepsilon ; \overline{\varepsilon^{\mathsf{T}} ; X} = \overline{X} \text{ for all } X$$

$$= \overline{m ; \overline{E} ; \overline{E}^{\mathsf{T}}} \text{ as } m \text{ was assumed to be a mapping}$$

$$= \overline{m ; E ; \overline{E}^{\mathsf{T}}}$$

$$= m ; E^{\mathsf{T}}$$

we see that this is an adjunction on cones. The lower cones $m ; E^{\mathsf{T}}$ in turn are $1 : 1$ represented by their least upper bounds $\mathtt{lubR}_E(m ; E)$.

The following proposition exhibits that a Bayesian measure is a rather special case, namely more or less directly determined as a possibility measure for a direct valuation via a mapping m. Fig. 7.1 shows an example. One may proceed from m to the measure according to Prop. 7.1 or vice versa according to Prop. 7.2.

7.2 Proposition. Let μ be a Bayesian measure. Then $m_\mu := \texttt{lubR}_E(\pi(\mu))$ is that direct valuation for which $\mu = \mu_{m_\mu}$. $\qquad\square$

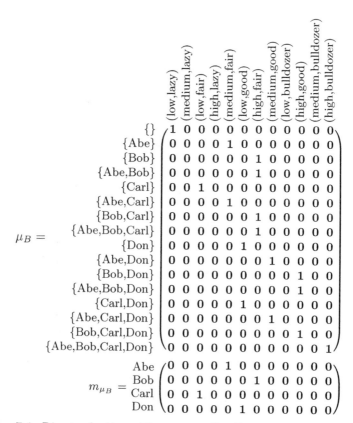

Fig. 7.1. Direct valuation with corresponding Bayesian measure

With this method just a few of the many relational measures will be found. By construction they are all continuous (or additive).

Measures originating from a body of evidence

We may also derive relational measures out of some relation between $\mathcal{P}(\mathcal{C})$ and $\mathcal{L}$. Although it is allowed to be non-univalent, we provide for a way of defining two measures based on it — which may coincide.

7.3 Definition. Let our general setting be given.

i) A **body of evidence** is an arbitrary relation $M : \mathcal{P}(\mathcal{C}) \longrightarrow \mathcal{L}$, restricted by the requirement that $M^\mathsf{T};0_\Omega \subseteq 0_E$.

ii) When the body of evidence M is in addition a mapping, we speak — following [4] — of a **basic probability assignment**.   □

If I dare saying that occurrence of $A \subseteq C$ deserves my trust to the amount $M(A)$, then $A' \subseteq A \subseteq C$ deserves at least this amount of trusting as it occurs whenever A occurs. I might, however, not be willing to consider that $A'' \subseteq C$ with $A \subseteq A''$ deserves to be trusted with the same amount as there is a chance that it occurs not so often.

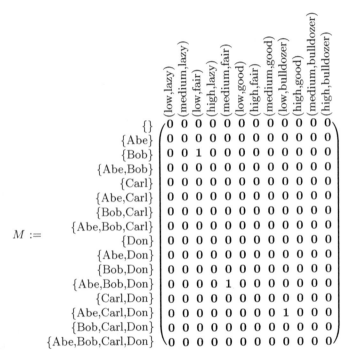

Fig. 7.2. A body of evidence

We should be aware that the basic probability assignment is meant to assign something to a set regardless of what is assigned to its proper subsets. The condition $M^\mathsf{T}\!{\,;\,}0_\Omega \subseteq 0_E$ expresses that M either does not assign any belief to the empty set or assigns it just 0_E.

Now a construction similar to that in (3) becomes possible, introducing

$$\sigma'(M) := \overline{\Omega^\mathsf{T}\!{\,;\,}M{\,;\,}\overline{E}} \qquad \pi'(\mu) := \overline{\Omega{\,;\,}\mu{\,;\,}\overline{E}^\mathsf{T}}, \tag{4}$$

which again satisfies the Galois correspondence requirement

$$M \subseteq \pi'(\mu) \quad\Longleftrightarrow\quad \mu \subseteq \sigma'(M).$$

Obviously $\sigma'(M{\,;\,}E^\mathsf{T}) = \sigma'(M)$ and $\pi'(\mu{\,;\,}E) = \pi'(\mu)$, so that in principle only upper (E) and lower (E^T), respectively, cones are set into relation. But again applying $\overline{W{\,;\,}E} = \overline{W{\,;\,}E}{\,;\,}E^\mathsf{T}$, we get

$$\sigma'(M){\,;\,}E = \overline{\Omega^{\mathsf{T}}{\,;\,}M{\,;\,}\overline{E}}{\,;\,}E = \overline{\Omega^{\mathsf{T}}{\,;\,}M{\,;\,}\overline{E}{\,;\,}E^{\mathsf{T}}}{\,;\,}E = \overline{\Omega^{\mathsf{T}}{\,;\,}M{\,;\,}\overline{E}} = \sigma'(M),$$

so that images of σ' are always upper cones — and thus best described by their greatest lower bound $\mathtt{glbR}_E(\sigma'(M))$.

7.4 Proposition. Should some body of evidence M be given, there exist two relational measures closely resembling M,

i) the **belief measure** $\mu_{\text{belief}}(M) := \mu_0 \oplus \mathtt{lubR}_E(\Omega^{\mathsf{T}}{\,;\,}M)$
ii) the **plausibility measure** $\mu_{\text{plausibility}}(M) := \mu_0 \oplus \mathtt{lubR}_E((\Omega \cap \overline{\Omega}{\,;\,}\mathbb{T})^{\mathsf{T}}{\,;\,}\Omega{\,;\,}M)$
iii) In general, the belief measure assigns values not exceeding those of the plausibility measure, i.e., $\mu_{\text{belief}}(M) \subseteq \mu_{\text{plausibility}}(M){\,;\,}E^{\mathsf{T}}$. □

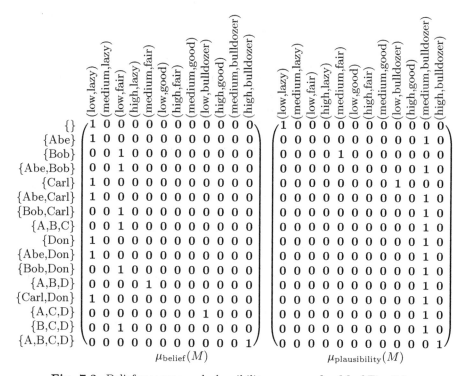

Fig. 7.3. Belief measure and plausibility measure for M of Fig. 7.2

The belief measure adds information to the extent that all evidence of subsets with an evidence attached is incorporated. Another idea is followed by the plausibility measure. One asks which sets have a non-empty intersection with some set with an evidence attached and determines the least upper bound of all these.

The plausibility measure collects those pieces of evidence that do *not* indicate trust against occurrence of the event or non-void parts of it. The belief as well as the plausibility measure more or less precisely determine their original body of evidence.

7.5 Proposition. Should the body of evidence be concentrated on singleton sets only, the belief and the plausibility measure will coincide.

Proof: That M is concentrated on arguments which are singleton sets means that $M = a\,;M$ with a the partial diagonal relation describing the atoms of the ordering Ω. For Ω and a one can prove $(\Omega \cap \overline{\Omega}\,;\mathbb{T})\,;a = a$ as the only other element less or equal to an atom, namely the least one, has been cut out via $\overline{\Omega}$. Then

$$
\begin{aligned}
(\Omega \cap \overline{\Omega}\,;\mathbb{T})^{\mathsf{T}}\,;\Omega\,;M &= (\Omega^{\mathsf{T}} \cap \mathbb{T}\,;\overline{\Omega}^{\mathsf{T}})\,;\Omega\,;a\,;M && M = a\,;M \text{ and transposing} \\
&= \Omega^{\mathsf{T}}\,;(\Omega \cap \overline{\Omega}\,;\mathbb{T})\,;a\,;M && \text{mask shifting} \\
&= \Omega^{\mathsf{T}}\,;a\,;M && \text{see above} \\
&= \Omega^{\mathsf{T}}\,;M && \text{again since } M = a\,;M \qquad \square
\end{aligned}
$$

One should compare this result with the former one assuming m to be a mapping putting $m := \varepsilon\,;M$. One may also try to go in reverse direction, namely from a measure back to a body of evidence.

7.6 Definition. Let some measure μ be given and define strict subset containment $C := \overline{\mathbb{I}} \cap \Omega$. We introduce two basic probability assignments, namely

i) $A_\mu := \text{lubR}_E(C^{\mathsf{T}}\,;\mu)$, its **purely additive part**,

ii) $J_\mu := \mu_1 \otimes (\mu \cap \overline{\text{lubR}_E(C^{\mathsf{T}}\,;\mu)})$, its **jump part**. $\square$

As an example, the purely additive part A_μ of the μ of Fig. 6.1 would assign in line {Abe,Bob} the value {high,fair} only as $\mu(\{\text{Abe}\}) = \{\text{high,lazy}\}$ and $\mu(\{\text{Bob}\}) = \{\text{medium,fair}\}$. In excess to this, μ assigns {high,good}, and is, thus, not additive or Bayesian. We have for A_μ taken only what could have been computed already by summing up the values attached to *strictly* smaller subsets. In J_μ the excess of μ to A_μ is collected. In the procedure for J_μ all the values attached to atoms of the lattice will be saved as from an atom only one step down according to C is possible. The value for the least element is, however, the least element of $\mathcal{L}$. Multiplication with μ_1 serves the purpose that rows full of **0**'s be converted to rows with the least element 0_E attached as a value.

Now some arithmetic on these parts is possible, not least providing the insight that a measure decomposes into an additive part and a jump part.

7.7 Proposition. Given the present setting, we have

i) $A_\mu \oplus J_\mu = \mu$.

ii) $\mu_{\text{belief}}(J_\mu) = \mu$. $\square$

In the real-valued case, this result is not surprising at all as one may always decompose into a part continuous from the left and a jump part.

In view of these results it seems promising to investigate in which way also concepts such as commonality, consonance, necessity measures, focal sets, and cores may be found in the relational approach. This seems particularly interesting as also the concepts of De Morgan triples have been transferred to the componentfree relational side. We leave this to future research.

8   Concluding Remark

There exists a bulk of literature around the topic of Dempster-Shafer belief. It concentrates mostly on work with real numbers and their linear order and applies traditional free-hand mathematics. This makes it sometimes difficult to follow the basic ideas, not least as authors are all too often falling back to probability considerations.

We feel that the componentfree relational reformulation of this field and the important generalization accompanying it is a clarification — at least for the strictly growing community of those who do not fear to use relations. Proofs may now be supported via proof systems. The results of this paper have been formulated also in the relational language TiTuRel [8,9], for which some system support is available making it immediately operational. Not least has it provided computation and representation of the example matrices.

Acknowledgement. My thanks go to the unknown referees for their detailed comments which considerably improved the exposition.

References

1. Fodor, J., Roubens, M.: Fuzzy Preference Modelling and Multicriteria Decision Support. Volume 14 of Theory and Decision Library, Series D: System Theory, Knowledge Engineering and Problem Solving. Kluwer Academic Publishers (1994)
2. Schmidt, G., Ströhlein, T.: Relationen und Graphen. Mathematik für Informatiker. Springer-Verlag (1989) ISBN 3-540-50304-8, ISBN 0-387-50304-8.
3. Schmidt, G., Ströhlein, T.: Relations and Graphs — Discrete Mathematics for Computer Scientists. EATCS Monographs on Theoretical Computer Science. Springer-Verlag (1993) ISBN 3-540-56254-0, ISBN 0-387-56254-0.
4. Shafer, G.: A Mathematical Theory of Evidence. Princeton University Press (1976)
5. Dempster, A.P.: Upper and lower probabilities induced by a multivalued mapping. Annals of Math. Statistics **38** (1967) 325–339
6. Choquet, G.: Theory of capacities. Annales de l'Institut Fourier **5** (1953) 131–295
7. Sugeno, M., ed.: Industrial Applications fo Fuzzy Control. North-Holland (1985)
8. Schmidt, G.: Relational Language. Technical Report 2003-05, Fakultät für Informatik, Universität der Bundeswehr München (2003) 101 pages, `http://homepage.mac.com/titurel/Papers/LanguageProposal.html`.
9. Schmidt, G.: The Relational Language TiTuRel: Revised Version (2005) In preparation; see `http://homepage.mac.com/titurel/TituRel/LanguageProposal2.pdf`.

A Relational View of Recurrence and Attractors in State Transition Dynamics

Giuseppe Scollo[1], Giuditta Franco[2], and Vincenzo Manca[2]

[1] University of Catania, Department of Mathematics and Computer Science,
Viale A. Doria, 6, I-95125 Catania, Italy
scollo@dmi.unict.it
http://www.dmi.unict.it/~scollo

[2] University of Verona, Department of Computer Science, Strada Le Grazie, 15,
I-37134 Verona, Italy
franco@sci.univr.it, vincenzo.manca@univr.it
http://www.sci.univr.it/~manca

Abstract. The classical dynamics concepts of recurrence and attractor are analysed in the basic mathematical setting of state transition systems, where both time and space are discrete, and no structure is assumed on the state space besides a binary transition relation. This framework proves useful to the dynamical analysis of computations and biomolecular processes. Here a relational formulation of this framework is presented, where the concepts of attractor and recurrence surface in two variants, respectively relating to the two fundamental modalities. A strong link between recurrence and both existence and extent of attractors, in either variant, is established by a novel characterization theorem.

1 Introduction

Analyses of dynamical systems represent the main application of mathematical sciences to the study of natural phenomena. Three constituents are essential in a dynamical system: space, collecting all the possible states of the system; time, collecting the different instants at which the system is considered; and dynamics, which associates, to each instant, the system state at that instant. The various kinds of dynamical systems are essentially determined by the structure of the space, by the nature of the time, and the way dynamics is characterized [5,9].

The classical approach to study dynamical systems is focused on differential equations, that impose local (infinitesimal) relations on quantity variations, from which, under suitable hypotheses, one can analytically reconstruct the global dynamical behaviour of the system. Recent developments of discrete models to analyse biological processes motivate the revisitation of typical concepts of classical dynamics in a completely discrete context. A couple of discrete models already applied with remarkable success are cellular automata [16], having the Lindenmayer systems as a special case, and Kauffman networks [7]. In these systems, viewed as dynamical systems, typical properties that are relevant in computation models, such as termination, confluence, and reducibility, are replaced

R.A. Schmidt (Ed.): RelMiCS /AKA 2006, LNCS 4136, pp. 358–372, 2006.

by other properties, which share nontermination as their outstanding prominent feature: periodicity, recurrence, emergence, propagation, stability, evolution. Indeed, the dynamics of biological adaptive systems may be viewed as a computation where the "result", which a system searches for, is not a state but a stable pattern, that is an attractor which fulfils certain desirable conditions.

Dynamics of discrete systems, besides being instrumental to their direct algorithmic simulation, often proves most natural in the representation of biomolecular dynamics, where the symbolic entities which come into play are easily amenable to strings or closely related structures, or to dynamical networks [3]. The book [1] is a fundamental pioneering work where state transition graphs, called kinetic graphs, are introduced in the context of dynamical concepts. However, despite its strong biological relevance, the book does not cope with the mathematical aspects of its conceptual apparatus. Most interesting results on dynamical indicators have been identified in [16], that discriminate classes of behaviour which appear to be different, but with no formal definition of the specific characteristics of a given behaviour. Important experimental results in this perspective may be found in [7] and [17].

In [9], we addressed the problem of considering, in general terms, dynamical systems that are completely discrete in that, not only the instants are natural or integer numbers, but also the space is a discrete entity. We introduced state transition dynamics, and we showed a few applications to computational and biomolecular dynamics. We argued that string manipulation systems from formal language theory may be naturally expressed as string transition systems. In this perspective, notable concepts from grammars and automata theory take up a purely dynamic character; for instance, the language generated by a grammar is an attractor in the computation dynamics. Concepts of periodicity and quasi-periodicity of state transition dynamics were applied to P systems with boundary rules and dynamical environment (PBE systems) [4], and to Metabolic P systems [10]. P systems are formalisms based on distributed rewriting rules [12,13] simulating the space distributed action of different agents that rule system global patterns such as oscillations and synchronization phenomena [11].

In this paper we cast the conceptual framework developed in [9] in purely relational terms. The generality of this formulation seems apt to present definitions of typically discrete dynamical concepts at a convenient abstraction level, which supports reasoning and insight. Here, we actually move beyond the first results presented in [9], to build a (fairly complex) full proof of a characterization of deep connections between both existence and extent of recurrence and attractors in discrete dynamics, that were only conjectured in [9].

The rest of the paper is organized as follows. In Sect. 3 we shall recast basic concepts and definitions relating to state transition dynamics, as proposed in [9], in terms of relation algebra concepts [15]. Some of these are preliminarily recalled in Sect. 2, where we also introduce the notation that will serve our purpose. In Sect. 4 a characterization theorem is achieved which links recurrence, eternal recurrence and attractors. Possible applications and a future extension of this work are outlined in Sect. 5, which concludes the article.

2   Notational Preliminaries

Let us designate the *universal* and *identity* binary relations on an arbitrary set S, with 1_S and $1'_S$, respectively. We shall omit the subscript whenever it is clear from the context. Let q, r be binary relations on S. Boolean difference is defined by $q \backslash r = q \cdot r^{-1}$, with standard notation for the Boolean product and complement operations, while 0 denotes the empty binary relation in any relation algebra. The standard Boolean ordering $\leq$ is defined in any relation algebra just like in its Boolean algebra reduct. This reduct is actually a *complete* Boolean algebra, hence the binary Boolean sum operation is extended to summation over arbitrary sets of binary relations.

Let $r^\vee$ denote the relation-algebraic converse of r. Consistently, we let $f^\vee$ denote the inverse of an invertible function f, as well as the inverse image relation of any function f. The binary relation $q \mathbin{;} r$ is the relation-algebraic composition of binary relations q and r. That is, $x \; q \mathbin{;} r \; y$ if and only if there exists an element z (in S) such that both $x \, q \, z$ and $z \, r \, y$.

If q^i denotes the i-fold iterated composition of q with itself, then the *reflexive-transitive closure of q*, the *transitive closure of q*, and the *at least n-fold iterated composition of q with itself*, are respectively defined as follows.

$$q^* = \sum\nolimits_{i \in \mathbb{N}} q^i$$

$$q^+ = \sum\nolimits_{i > 0} q^i$$

$$q^{\geq n} = \sum\nolimits_{i \geq n} q^i$$

One may note that $q^* = q^{\geq 0}$ and $q^+ = q^{\geq 1}$.

We recall here that a *monotype* binary relation x is a subrelation of the identity: $x \leq 1'_S$, for more details see [2]. The *domain* and the *image* of a binary relation q may be expressed as monotypes, respectively defined by the equations: $\mathsf{dom} \; q = 1' \cdot (q \mathbin{;} 1)$, and $\mathsf{img} \; q = 1' \cdot (1 \mathbin{;} q)$. A few higher-order binary relations on monotypes will prove useful.

First, if q is a binary relation and x, y are monotypes on S, then we define

$$x \leq_q^{(n)} y \quad \text{iff} \quad \mathsf{img}(x \mathbin{;} q^{\geq n}) \leq y \;.$$

Then we define when a monotype is *eventually below* another one, under the iterated composition with a binary relation q, by a simple summation in the higher-order relation algebra on such monotypes, as follows:

$$\preceq_q = \sum_{n \in \mathbb{N}} \leq_q^{(n)}.$$

Two useful operations will allow us to *lift* binary relations to (atomic) monotypes of a higher-order relation algebra and, conversely, to *flatten* higher-order monotypes to lower-order binary relations. These operations are respectively defined as follows.

If $r \leq 1_S$, then $\uparrow r$ is the *higher-order one-element monotype generated by* r, that is $\uparrow r \leq 1'_{2^{S \times S}}$ and it is defined by

$$p \uparrow r\, q \quad \Leftrightarrow \quad p = q = r.$$

Clearly, the binary relation $\uparrow r$ on $2^{S \times S}$ has only one element, that is (r, r).

If $x \leq 1'_{2^{S \times S}}$, that is x is a higher-order monotype, then $\downarrow x$ is the lower-order binary relation, $\downarrow x \leq 1_S$, defined by:

$$\downarrow x \;=\; \sum_{r\,x\,r} r.$$

At times we shall need to refer to individual elements of binary relations in terms of relation algebra. We shall do so by letting them be represented by the singleton monotypes they generate, which are *atoms* of the relation algebra. Atomicity may be formalized quasiequationally in relation algebra, by requiring that the monotype be the only nonempty subrelation of itself, so we get three factors to this purpose, where the equation $1 \,;\, x \,;\, 1 = 1$ specifies nonemptiness (by Tarski rule, cf. [14]): x is an *atomic monotype* if

$$x \leq 1', \quad 1 \,;\, x \,;\, 1 = 1, \quad \text{and} \quad y \leq x \wedge 1 \,;\, y \,;\, 1 = 1 \Rightarrow y = x.$$

If x is a monotype, the notation $x \leq y$ means that x is atomic and $x \leq y$.

3   A Relational View of State Transition Dynamics

Definition 1. *A* state transition dynamics *is a pair* $\langle S, q \rangle$, *where* S *is a set of states and* q *is a binary relation on* S, *the* transition relation *on states.*

As in [9], we call *quasistate* any subset of S, and often we shall actually be concerned with subsets rather than elements of S. For this reason, we shall refer to quasistates as states, whereas the lengthier term *individual state* will refer to elements of S. In terms of binary relations, states may be represented by monotypes in the relation algebra over S; in this respect, we adopt the notational convention of letting lowercase letters a, b, r, s, x, y, z, possibly with subscripts, denote monotypes that represent subsets of S, and we shall often confuse such monotypes with the subsets they represent. We may thus say that the notation $x \leq a$ means that "x is an individual state within state a", rather than the more accurate "x represents an individual state within the state represented by a".

A state transition dynamics is *eternal* if the transition relation q is total. This condition is easily expressed in the language of relation algebra by the equation $\mathrm{dom}\, q = 1'_S$. This condition is not met iff there is some *final* state in S, viz. there is a nonzero monotype x such that $x \,;\, q = 0$. Every final state may be easily turned into a nonfinal one by suitably extending the transition relation; thus, for example, one can extend every dynamics to become an eternal one by turning each final state into a fixed point of the transition relation, also called *q-dynamics*, according to the following definition.

Definition 2. *An individual state x is a* fixed point *of the q-dynamics if it turns out that* $\mathrm{img}(x\,;q)\!=\!x$.

The previous definition actually applies to any state as well, not just to individual ones. We shall henceforth consider eternal dynamics only. With reference to such a q-dynamics, the following definitions prove straightforward.

Definition 3. *An* orbit *is an infinite sequence of states* $(x_i \mid i \in \mathbb{N})$ *such that* $x_{i+1} = \mathrm{img}(x_i\,;q)$. *State* x_0 *is the* origin *of the orbit.*

Definition 4. *(Periodicity, eventual periodicity).*
An orbit $(x_i \mid i \in \mathbb{N})$ *is* periodic *if* $\exists n\!>\!0 : x_n = x_0$, *that is* $\mathrm{img}(x_0\,;q^n) = x_0$.
An orbit is eventually periodic *if, for some* $k\!\geq\!0$, *it evolves into a periodic one after a k-step transient, that is, if* $\exists n\!>\!0\colon \mathrm{img}(x_0\,;q^{k+n}) = \mathrm{img}(x_0\,;q^k)$.

Definition 5. *(Orbits' inclusion, eventual inclusion).*
An orbit $(x_i \mid i \in \mathbb{N})$ *is* included *in the orbit* $(y_i \mid i \in \mathbb{N})$ *if* $x_i\!\leq\!y_i \;\forall i \in \mathbb{N}$.
An orbit $(x_i \mid i \in \mathbb{N})$ *is* eventually included *in the orbit* $(y_i \mid i \in \mathbb{N})$ *if, for some* $j \in \mathbb{N}$, $x_i\!\leq\!y_i$ *holds* $\forall i\!\geq\!j$.

We may also use this terminology for any infinite sequences of states, rather than just for orbits. So, for example, we may say that orbit $(x_i \mid i \in \mathbb{N})$ is eventually included in $a^\omega = (a_i \mid i \in \mathbb{N})$, with $a_i\!=\!a$ for all $i\!\in\!\mathbb{N}$, without thereby implying $a\!=\!\mathrm{img}(a\,;q)$.

Definition 6. *A* basin *b is a nonempty state, $b \neq 0$, that is* closed under q, *that is* $\mathrm{img}(b\,;q) \leq b$.

In the terminology of [14], basins are nonempty states that are "contracted" by the q-dynamics. The higher-order binary relations defined in Sect. 2, as well as the lifting and flattening operations, find their first application in the following formalization of the two notions of *attracting set* introduced in [9].

Definition 7. *(Attracting sets). Let a and b be a nonempty state and a basin, respectively, such that $a\!\leq\!b$.*
(i) *a is an* unavoidable attracting set *of b if* $b \leq \downarrow\!\mathrm{dom}(\preceq_q\,;\uparrow\!a)$.
(ii) *a is a* potential attracting set *of b if* $b \leq \mathrm{dom}(q^*\,;\downarrow\!\mathrm{dom}(\preceq_q\,;\uparrow\!a))$.

The previous definition deserves a few explanations.
   An *unavoidable* attracting set a of basin b is characterized by the property that every orbit taking its origin inside b is *eventually* included in a^ω. For a correct formalization of this property, one must take into account that

- the higher-order *eventually below* relation $\preceq_q$ on monotypes fits nicely, but it lives in the higher-order relation algebra on the $2^{S\times S}$ universe, whence the lifting of the a monotype proves necessary to composition with $\preceq_q$;
- it would *not* be correct to require $b\preceq_q a$, since orbits starting at different origins inside b may have transients of different length, before getting included in a^ω, and the cardinality of b may well be infinite, hence the set of the transients' lengths may well be unbounded.

The following couple of facts may further elucidate the matter.

Proposition 1. *The definitions of $\preceq_q$ and of the lifting operation entail*

$$\uparrow r \leq \text{dom}(\preceq_q \,;\, \uparrow a) \Leftrightarrow r \preceq_q a \,.$$

Proposition 2. *Prop. 1 and the definition of the flattening operation entail*

$$\downarrow \text{dom}(\preceq_q \,;\, \uparrow a) \;=\; \sum_{r \preceq_q a} r \,.$$

This should be sufficient to see the correctness of the formalization of the concept of unavoidable attracting set. Unlike this, a *potential* attracting set may be infinitely often escaped from by orbits starting inside the basin; however, it can never be *definitely* escaped from, since it is always reachable (in a finite number of steps) from every individual state in the basin, and furthermore the basin is closed under q transitions. It should not be difficult to recognize such a "persistent q^*-reachability of unavoidable eventual inclusion" character in the formalization proposed in Def. 7.(ii), where the composition with q^* makes the essential difference with the preceding definition of unavoidable attracting set.

The two notions of attracting set may be expressed as binary relations on monotypes, in the relation algebra on $2^{S \times S}$, where they are designated by $\square$-attracts, $\lozenge$-attracts. That is, given a q-dynamics on S, for any monotypes a, b in the relation algebra on S we define:

- $a \,\square$-attracts b if b is a basin and a is an unavoidable attracting set of b;
- $a \,\lozenge$-attracts b if b is a basin and a is a potential attracting set of b.

For a given basin, the search for minimal attracting sets is supported by the following definition, which has to do with removability of states from an attracting set while preserving its attractiveness, in either form.

Definition 8. *(Removable states).* Let b be a basin in the q-dynamics.

(i) $x \,\square_b$-removable from a *(read: x is "must-removable" from a w.r.t. b) if*
$x \leq a$, $a \,\square$-attracts b *and* $a \backslash x \,\square$-attracts b ;

(ii) $x \,\lozenge_b$-removable from a *(read: x is "may-removable" from a w.r.t. b) if*
$x \leq a$, $a \,\lozenge$-attracts b *and* $a \backslash x \,\lozenge$-attracts b .

Minimal attracting sets are called *attractors*. This is formalized as follows.

Definition 9. *(Attractors).* Let b be a basin in the q-dynamics.

(i) *An* unavoidable attractor *of b is an unavoidable attracting set a of b that is minimal in the standard Boolean ordering, viz. no nonempty subset of a is must-removable from a w.r.t. b :*
$a \,\square$-attractor of b *if* $a \,\square$-attracts b *and* $x \,\square_b$-removable from $a \Rightarrow x = 0$.

(ii) *A* potential attractor *of b is a potential attracting set a of b that is minimal in the standard Boolean ordering, viz. no nonempty subset of a is may-removable from a w.r.t. b :*
$a \,\lozenge$-attractor of b *if* $a \,\lozenge$-attracts b *and* $x \,\lozenge_b$-removable from $a \Rightarrow x = 0$.

When we write "attractor" or "attracting set" without qualification, then "potential" is implicitly understood.

We recall a few useful facts from [9], while recasting them in our present notation. Those collected in Prop. 3 below easily follow from the definitions, whereas Prop. 4, which is proven in [9], offers a first characterization of either form of removability of individual states from attracting sets (of corresponding form, of course).

Proposition 3. *Let b be a basin in the q-dynamics.*

(i) $b\,\square$-attracts b .

(ii) $a\,\square$-attracts $b \Rightarrow a\,\Diamond$-attracts b .

(iii) $a\,\square$-attracts $b \wedge a \leq a' \leq b \Rightarrow a'\,\square$-attracts b .

(iv) $a\,\Diamond$-attracts $b \wedge a \leq a' \leq b \Rightarrow a'\,\Diamond$-attracts b .

(v) $x\,\square_b$-removable from $a \Rightarrow x\,\Diamond_b$-removable from a .

(vi) *An attractor of b is unique, if it exists. We then speak of the attractor $a_\Diamond$ of b, whereas the notation $a_\Diamond = 0$ means that b has no attractor.*

(vii) *An unavoidable attractor of b is unique, if it exists. We then speak of the unavoidable attractor $a_\square$ of b, whereas the notation $a_\square = 0$ means that b has no unavoidable attractor.*

(viii) *The (unavoidable) attractor of b is also the (unavoidable) attractor of any $x \leq b$ that is a basin and is above the (unavoidable) attractor of b in the standard Boolean ordering. In particular, every (unavoidable) attractor is also its own (unavoidable) attractor.*

Proposition 4. *Let b be a basin in the q-dynamics.*

(i) *If $a\,\Diamond$-attracts b and $y \leq a$, then $y\,\Diamond_b$-removable from a iff $y \leq \text{img}(x\,;q^*\,;q^{*-1})$ for all $x \leq b$.*

(ii) *If $a\,\square$-attracts b and $y \leq a$, then $y\,\square_b$-removable from a iff for no $x \leq b$ does y occur infinitely often in the x-orbit.*

Recurrence is defined in [9] for individual states in a given basin. Individual states may be represented as atomic monotypes in the relation algebra on the state set S. Two modal shapes of recurrence are defined: *recurrence* as occurrence of an individual state in its own orbit, and *eternal recurrence* as occurrence of an individual state in all orbits of individual states that fall in the orbit of the given individual state. This is formalized as follows.

Definition 10. *(Recurrence). Let x be an individual state in a basin in the q-dynamics.*

(i) x *is recurrent if* $x \leq \text{img}(x\,;q^+)$

(ii) x *is eternally recurrent if* $x\,;q^* \leq x\,;q^{\vee*}$

We write $x\,\Diamond\text{-rec}\,b$ to mean that b is a basin where x is recurrent, while $x\,\square\text{-rec}\,b$ means that x is eternally recurrent in basin b, with $x \leq b$ in both cases. Henceforth, with reference to a fixed basin b, $r_\Diamond$ denotes the monotype of recurrent states in b, while $r_\square$ denotes the monotype of eternally recurrent states in b, that is we have the following concepts.

Definition 11. *(Recurrence sets). Let b be a basin in the q-dynamics, then with respect to b, $r_\diamond$ and $r_\square$ are defined by the following equations:*

$$r_\diamond = \sum_{x \diamond\text{-rec } b} x \,, \qquad r_\square = \sum_{x \square\text{-rec } b} x \,.$$

The following facts easily follow from the definitions.

Proposition 5. *Let b be a basin in the q-dynamics.*

(i) *Every eternally recurrent state is recurrent, thus $r_\square \leq r_\diamond$.*
(ii) *The set of eternally recurrent states is closed under transitions, i.e. it is either a basin or empty, and in both cases $\mathrm{img}(r_\square\,; q) \leq r_\square$, whence $\mathrm{img}(r_\square\,; q^*) = r_\square$.*

4   A Characterization of Recurrence and Attractors

A first link between recurrence and attractors surfaces as a cross-connection between eternal recurrence and $\diamond$-nonremovability of individual states. We speak of a *cross*-connection because of the difference in the modalities involved—this turns out to be a recurrent phenomenon in this section.

Prop. 4(i) and Def. 10(ii) entail that an individual state is $\diamond$-nonremovable from any attracting set of a basin b, hence from b itself (cf. Prop. 3(i),(ii)) iff it is eternally recurrent. Thus, $r_\square$ is *the* monotype of $\diamond$-nonremovable individual states in b. One may formalize this, for a fixed basin b, by letting

$$\diamond\text{-removable} \;=\; \sum_{y \,\diamond_b\text{-removable from } b} y \,,$$

in the following equation.

Proposition 6. *For any basin b in the q-dynamics, $r_\square = b \setminus \diamond\text{-removable}$.*

A modal dual of Prop. 6 does not hold in the general case, but only under certain assumptions. In order to state them precisely, a few concepts are needed, that are phrased in celestial terminology as in [9].

Definition 12. *(Trajectory, flight, antiflight, blackhole).*
Let x represent an individual state in the q-dynamics on the state space S.

(i) *A trajectory of origin x, briefly an x-trajectory, is a function $\xi : \mathbb{N} \to 1'_S$ such that, with subscript argument, $\xi_0 = x$ and $\xi_{n+1} \leq \mathrm{img}(\xi_n\,; q)$. $\xi_\mathbb{N}$ denotes the image of this function.*
(ii) *A flight of origin x, or x-flight, is an injective x-trajectory.*
(iii) *An antiflight of target x, or x-antiflight, in the q-dynamics is an x-flight in the converse $q^\vee$-dynamics.*
(iv) *A flight ξ is antiflight-free if no individual state in $\xi_\mathbb{N}$ is the target of an antiflight.*
(v) *An x-flight ξ is an x-blackhole if $\mathrm{img}(x\,; q^*) \leq \xi_\mathbb{N}$.*

So, in summary: a trajectory develops through an infinite sequence of transitions between individual states; a flight is a trajectory where every individual state occurs at most once; an antiflight is a backward flight; an antiflight-free flight has no backward flight starting at any of its individual states; and a blackhole is a flight that is closed under transitions.

In absence of flights and antiflights, a modal dual of Prop. 6 holds, where the dual of $r_\square$ actually is $\mathrm{img}(r_\lozenge\,;q^*)$. This is duality proper, thanks to Prop. 5(ii). Flights and antiflights introduce further possibilities of $\square$-nonremovability, in agreement with the next proposition. For a fixed basin b, let

$$\square\text{-removable} = \sum_{y\,\square_b\text{-removable from }b} y\,.$$

We then have the following situation.

Proposition 7. *For any basin b in the q-dynamics:*

(i) $\mathrm{img}(r_\lozenge\,;q^*) \le b \setminus \square\text{-removable}$,
(ii) $\mathrm{img}(r_\lozenge\,;q^*) = b \setminus \square\text{-removable}$, *if the basin has neither flights nor antiflights.*

Flights affect not only the aforementioned duality but also the existence of attractors, of either kind. As shown in [9], in the presence of certain kinds of flights in a basin, it may happen that $a_\lozenge = 0$ while $a_\square \ne 0$, as well as that $a_\lozenge \ne 0$ while $a_\square = 0$. The examples presented there reveal that the presence of a flight in the basin may, but need not, hamper the validity of either or both of the dual equations $a_\lozenge = r_\square$ and $a_\square = \mathrm{img}(r_\lozenge\,;q^*)$. Purpose of the rest of this section is to establish necessary and sufficient conditions for the validity of each of these equations, thereby characterizing both the existence and the extent of attractors of either kind. The following definitions prove purposeful.

Definition 13. *(Recurrent flights).*
Let b be a basin and ξ be a flight in b, with the q-dynamics.

(i) *ξ is recurrent in b if $\xi_n \le \mathrm{img}(r_\lozenge\,;q^*)$ for some $n \in \mathbb{N}$.*
(ii) *ξ is eternally recurrent in b if $\xi_{\mathbb{N}} \le \mathrm{img}(r_\square\,;q^{\vee *})$.*

Definition 14. *(Finitary dynamics).*
The q-dynamics is finitary if the q relation is image-finite on individual states, i.e. $\mathrm{img}\,(x\,;q)$ represents a finite state whenever x represents an individual state.

A few preliminary lemmas will shorten part of the proof of the subsequent theorem. The first one relates to existence of nonrecurrent flights in a basin.

Lemma 1. *(Existence of nonrecurrent flights).*
Let b be a basin in a finitary q-dynamics. If there exists $x \le b$ such that for no $n \in \mathbb{N}$ $\mathrm{img}\,(x\,;q^{\ge n}) \le \mathrm{img}\,(r_\lozenge\,;q^*)$, then there is a nonrecurrent x-flight in b.

Proof. Let's arrange the x-orbit in a tree, with the root labeled by x and where the children of node labeled by individual state y are labeled by the individual states in $\text{img}(y;q)$. In this tree, which is finitely branching since the q-dynamics is finitary, if a node is labeled by some individual state in $b\backslash\text{img}(r_\diamond;q^*)$, then so are all nodes in the path leading from the root to that node—otherwise one would get a nonempty intersection of disjoint monotypes, which is clearly absurd. Now, let's prune the tree by removing those nodes that are labeled by individual states in $\text{img}(r_\diamond;q^*)$, so all remaining nodes are labeled by individual states in $b\backslash\text{img}(r_\diamond;q^*)$. It is fairly immediate to see that the hypothesis on x entails that the pruned tree is infinite, but since it is the outcome of pruning a finitely branching tree, it is finitely branching as well, therefore it must have an infinite path, by König's Lemma. Since all nodes in this path are labelled by states in $b\backslash\text{img}(r_\diamond;q^*)$, none of these is recurrent, hence each of them occurs only once in the path, thus the path corresponds to an x-flight in b, indeed a nonrecurrent one, since no state in the path may ever be found in $\text{img}(r_\diamond;q^*)$. $\square$

Remark 1. The hypothesis that the q-dynamics is finitary is fairly essential. This is apparent in the use of König's Lemma in the proof, and is further corroborated by the following counterexample to validity of the statement in a case where that hypothesis does not hold.

Let the q-dynamics consist of antiflight ξ, with ξ_0 the only fixed point in basin b, and an additional individual state $x \leqq b$ with $\text{img}(x;q)=\xi_{\mathbb{N}}$. Clearly, this dynamics is not finitary. However, state x does fulfil the condition required by Lemma 1, since the x-orbit is eventually periodic, with transient x and period $\xi_{\mathbb{N}}$, whereas $r_\diamond = \text{img}(r_\diamond;q^*)=\xi_0$. Nonetheless, there's no x-flight in b, a *fortiori* no nonrecurrent x-flight.

The next lemma provides a sufficient condition for nonexistence of the unavoidable attractor.

Lemma 2. *(Nonexistence of the unavoidable attractor).*
For any basin b in the q-dynamics, $a_\square = 0$ if

(i) *the converse $q^\smallsmile$-dynamics is finitary, and*
(ii) *there is a nonrecurrent antiflight-free flight in b, under the q-dynamics.*

Proof. Let ξ be a nonrecurrent antiflight-free flight in b. We claim that for no individual state $x \leqq b$ may ξ_0 occur infinitely often in the x-orbit. This will entail $\xi_0\,\square_b$-removable from b by Prop. 4(ii), and much the same for any individual state ξ_n in $\xi_{\mathbb{N}}$, since the ξ_n-flight in ξ meets the same conditions stated above for the ξ_0-flight. Then it will follow that $\xi_{\mathbb{N}} \leqq \square$-removable, but no infinite subset of $\xi_{\mathbb{N}}$ is $\square$-removable from the basin b, whence $a_\square = 0$. Here is the proof of the claim.

First, if ξ_0 occurs in the x-orbit, then x is not recurrent, indeed not even $x \leqq \text{img}(r_\diamond;q^*)$, by nonrecurrence of the ξ_0-flight.

Second, hypothesis (i) entails that $\text{img}(\xi_0;q^{\smallsmile n})$ is finite for all $n>0$, hence these images may be displayed in a finitely branching tree, with the root labeled by ξ_0, where the children of node labeled by individual state y are labeled by the individual states in $\text{img}(y;q^\smallsmile)$.

Again, by the nonrecurrence of the ξ_0-flight, one observes that no individual state may occur more than once as a node label in any given path from the root in the aforementioned tree.

Now, by contradiction, assume ξ_0 occurs infinitely often in the x-orbit, then the set of lengths of paths from the root in the tree is unbounded, hence the set of individual states that label the nodes of the aforementioned tree must be infinite, by the previous observation. Thus, the tree itself must have an infinite number of nodes, hence it has an infinite path from the root, by König's Lemma. It follows that ξ_0 is the target of an antiflight, again by the previous observation now relating to the infinite path, but this outcome is against the hypothesis that ξ is antiflight-free.   □

Our final lemma provides a sufficient condition for the existence of flights in the basin of any q-dynamics. It tells something more, viz. in the absence of eternal recurrence, flights start everywhere in the basin.

Lemma 3. *(Flights in absence of eternal recurrence).*
If b is a basin in the q-dynamics with no eternally recurrent states, then every $x \leq b$ is the origin of a flight.

Proof. Since no state is eternally recurrent, by Def. 10(ii) one gets immediately nonemptiness of $\mathrm{img}(x\,;q^+) \setminus \mathrm{img}(x\,;q^{\vee *})$ for every $x \leq b$. Furthermore, one may always find an individual state x' in this set such that there exists a finite sequence of $n{+}2$ individual states $(\xi_i \mid 0 \leq i \leq n+1)$, for some $n \geq 0$, that satisfies the following requirements:

1. $\xi_0 = x$, $\xi_{n+1} = x'$, $\xi_{i+1} \leq \mathrm{img}(\xi_i\,;q)$, for $0 \leq i \leq n$;
2. $\xi_i \leq \mathrm{img}(x\,;q^{\vee *})$, for $0 < i \leq n$;
3. $\xi_i = \xi_j \Leftrightarrow i = j$, for $0 \leq i,j \leq n+1$.

Satisfiability of the third requirement comes from the simple observation that, if there is a path that links a given pair of distinct source and target nodes, through a set of nodes in a directed graph, then there is a cycle-free path which links the given pair through the same set of nodes.

The construction of an x-flight takes place by iterating the procedure specified above to x', then to x'', and so on. More precisely, the mapping $\xi : \mathbb{N} \to \mathrm{img}(x\,;q^*)$ is defined as follows. Let $x_0 = x$, $x_{k+1} = x_k{}'$, n_k the (possibly 0) number of intermediate states in the chosen finite sequence $(x_{kj} \mid 0 \leq j \leq n_k+1)$ linking the source state $x_k = x_{k_0}$ to the target state $x_{k_{n_k+1}} = x_{k+1}$. By convening that summation is 0-valued when the upper bound index is negative, we define for all $k \in \mathbb{N}$, $0 \leq j \leq n_k$:

$$\xi \; : j + k + \sum_{h=0}^{k-1} n_h \; \mapsto \; x_{kj} \, .$$

It is easy to see that the mapping ξ is indeed defined for all $n \in \mathbb{N}$. To see that it is injective, it is enough to observe that

- x_{k+1} lies outside of $\text{img}(x_k ; q^{\vee*})$ by construction;
- more generally, it is a fact that if $m < k$, then x_k lies outside of $\text{img}(x_m ; q^{\vee*})$, for otherwise we should have $x_k \leq \text{img}(x_{k-1} ; q^{\vee*})$, against the previous observation;
- $x_{k_i} = x_{k_j} \Leftrightarrow i = j$, for $0 \leq i, j \leq n_k+1$ by construction as well (third requirement above);
- if $m < k$ and $n_k > 0$, then for $0 < i \leq n_k$ it must be $x_{k_i} \neq x_m$, otherwise it would be $x_k \leq \text{img}(x_m ; q^{\vee*})$ with $m < k$, against the aforementioned fact;
- and, finally, that if there were some $m < k$, with $n_m > 0$, $n_k > 0$, and some i, j, such that $0 < i \leq n_m$, $0 < j \leq n_k$, and $x_{k_j} = x_{m_i}$, then by construction (second requirement above) it would follow that $x_{k_j} \leq \text{img}(x_m ; q^{\vee*})$, and since $x_k \leq \text{img}(x_{k_j} ; q^{\vee*})$, it would turn out that $x_k \leq \text{img}(x_m ; q^{\vee*})$ with $m < k$, against the aforementioned fact.

□

We now have all ingredients to state and prove the desired characterization.

Theorem 1. *(Recurrence and Attractors).*
In any basin b with the q-dynamics:

(i) $a_\square = \text{img}(r_\diamond ; q^*)$ *if the q-dynamics is finitary and every flight is recurrent, otherwise $a_\square = 0$ if the converse $q^\vee$-dynamics is finitary and if there is a nonrecurrent antiflight-free flight, under the q-dynamics.*
(ii) $a_\diamond = r_\square$ *if every flight is eternally recurrent, otherwise $a_\diamond = 0$.*

Proof. The basic fact is that, in the presence of flights that do not meet the conditions stated for the existence of attractors, one may find nonremovable (infinite) subsets of the basin that consist of removable individual states, for either modality. Typically, if y represents such a set, it so happens that any finite $z \leq y$ is removable, while no infinite $z \leq y$ so is. Clearly, whenever such a situation occurs, the modally corresponding attractor does not exist.
Proof of (i).
First, if a □-attracts b, then $b \setminus$ □-removable $\leq a$ by Def. 8(i) and Prop. 3(iii) with contraposition, hence $\text{img}(r_\diamond ; q^*) \leq a$ by Prop. 7(i).

Next, we show that, provided the q-dynamics is finitary, if $r_\diamond \neq 0$ and all flights are recurrent, then $\text{img}(r_\diamond ; q^*)$ □-attracts b. According to Def. 7(i) and Prop. 2, we have got to show that

$$b \leq \sum_{r \preceq_q \text{img}(r_\diamond ; q^*)} r .$$

To this purpose, it suffices to show that for every $x \leq b$ the eventual inclusion $x \preceq_q \text{img}(r_\diamond ; q^*)$ holds, i.e., $\exists n \in \mathbb{N} : \text{img}(x ; q^{\geq n}) \leq \text{img}(r_\diamond ; q^*)$. By contradiction, let's assume the existence of $x \leq b$ such that for no $n \in \mathbb{N}$ $\text{img}(x ; q^{\geq n}) \leq \text{img}(r_\diamond ; q^*)$. Since the q-dynamics is finitary, by Lemma 1 a nonrecurrent x-flight exists in b, against the hypothesis that all flights in b are recurrent.

Putting together what is proven so far, we get that, in finitary q-dynamics, $a_\square = \text{img}(r_\diamond ; q^*)$ if $r_\diamond \neq 0$ and every flight in b is recurrent. For the case $r_\diamond = 0$, the condition that every flight in b be recurrent would only be met if there were no flights in b, since there are no recurrent states. However, the assumption of eternal dynamics (made throughout this paper) entails that all trajectories in the basin are nonrecurrent flights, whereby the second part of statement (i) applies. The first part of statement (i) is thus proven, while its second part is Lemma 2.

Proof of (ii).

By Prop. 6, the only $\Diamond$-nonremovable individual states are the eternally recurrent ones, viz. those in $r_\square$. So, whenever all sets consisting of $\Diamond$-removable individual states are $\Diamond$-removable themselves, then $a_\Diamond = r_\square$ holds. This is immediate for $r_\square \neq 0$, while the case $r_\square = 0$ deserves special treatment. In such a case, *all* individual states in the basin are $\Diamond$-removable, but the basin itself cannot be so (since no attracting set may be empty), and the basin must be infinite (by a corollary of Lemma 3), hence the attractor does not exist, or $a_\Diamond = 0$, in this case—formally, $a_\Diamond = r_\square$ holds in this case, too. We shall thus prove two facts:

1. $y \leq \Diamond$-removable $\Rightarrow y \, \Diamond_b$-removable from b if every flight is eternally recurrent.
2. If there is a flight that is not eternally recurrent, then $a_\Diamond = 0$.

We prove fact 1 by contraposition. Assume $y \, \Diamond_b$-removable from b does not hold, while for every $x \leq y$ $x \, \Diamond_b$-removable from b holds, that is $y \leq \Diamond$-removable, we then show the existence of a flight that is not eternally recurrent.

First, $y \leq \Diamond$-removable by Prop. 6 entails $y \leq b \backslash r_\square$ (†).

Second, for all individual states $z \leq b$ we have $\mathrm{img}(z\,;q^*) \cdot y \neq 0 \Rightarrow z \leq b \backslash r_\square$, by Prop. 5(ii), therefore $\mathrm{img}(y\,;q^*) \leq b \backslash r_\square$ (‡).

Now, the first assumption just means that $b \backslash y$ is not an attracting set. We have two cases where this may happen:

• $y = b$, thus $r_\square = 0$. By Lemma 3 there exist flights in the basin; none of them is eternally recurrent, since there is no eternally recurrent state in the basin.

• $b \backslash y \neq 0$, actually $r_\square \neq 0$ and $r_\square \leq b \backslash y$, by (†) above. Since $b \backslash y$ is not an attracting set, by Def. 7(ii) there is an $x \leq b$ such that for all $z \leq \mathrm{img}(x\,;q^*)$ one has $\mathrm{img}(z\,;q^*) \cdot 1' \backslash (b \backslash y) \neq 0$, and since basin b is closed under transitions, this is equivalent to $\mathrm{img}(z\,;q^*) \cdot y \neq 0$ for every $z \leq \mathrm{img}(x\,;q^*)$. For such an x it must hold that $\mathrm{img}(x\,;q^*) \cdot r_\square = 0$, by (‡) above. If we can show the existence of an x-flight, this surely would not be eternally recurrent, according to Def. 13(ii), since $x \leq b \backslash \mathrm{img}(r_\square\,;q^*)$, by the previously inferred equation. The existence of such a flight is a consequence of the absence of eternally recurrent states in $\mathrm{img}(x\,;q^*)$, according to Lemma 3, since $\mathrm{img}(x\,;q^*)$ is a basin.

Finally, here is a proof of fact 2 stated above. Suppose ξ is a flight such that for some $k \in \mathbb{N}$ $\mathrm{img}(\xi_k\,;q^*) \cdot r_\square = 0$. Then $\mathrm{img}(\xi_n\,;q^*) \cdot r_\square = 0$ $\forall n \geq k$, so for the ξ_k-flight ξ' defined by $\xi'_i = \xi_{k+i}$ we have $\mathrm{img}(\xi'_\mathbb{N}\,;q^*) \cdot r_\square = 0$, thus by Prop. 6 $\mathrm{img}(\xi'_\mathbb{N}\,;q^*) \leq \Diamond$-removable. We now have to show that there exists an infinite subset of $\mathrm{img}(\xi'_\mathbb{N}\,;q^*)$ that is not $\Diamond_b$-removable from b. This may well be $\mathrm{img}(\xi'_\mathbb{N}\,;q^*)$ itself. This set is infinite, since ξ' is a flight, and furthermore $\mathrm{img}(\xi'_\mathbb{N}\,;q^*) \, \Diamond_b$-removable from b does not hold because $\mathrm{img}(\xi'_\mathbb{N}\,;q^*)$ is closed under transitions; to see this, consider that $b \backslash \mathrm{img}(\xi'_\mathbb{N}\,;q^*)$ cannot be an attracting set of b since, $\forall x \leq \mathrm{img}(\xi'_\mathbb{N}\,;q^*) \leq b$, $y \leq \mathrm{img}(x\,;q^*) \Rightarrow \mathrm{img}(y\,;q^*) \cdot (b \backslash \mathrm{img}(\xi'_\mathbb{N}\,;q^*)) = 0$, therefore $y \preceq_q b \backslash \mathrm{img}(\xi'_\mathbb{N}\,;q^*)$ cannot hold, by Def. 7(ii) and Prop. 2. □

Remark 2. Finitarity assumptions are only needed for the characterization of the unavoidable attractor. The modal difference between the two forms of attractor, and of recurrence alike, obviously disappears in deterministic dynamics, yet it is not easy to translate the content of Theorem 1 in terms of classical dynamical systems, not even those of symbolic dynamics [8]. These are deterministic systems but rely on a metric structure of the state space, enabling seriously different concepts of attraction and recurrence, that are based on *approximate* transition through states, *i.e.* transition at arbitrarily small distance from the given state. While no easy translation of our theorem can be given in so different a setting, a certain analogy with Poincaré Recurrence Theorem surfaces, with boundedness and invariance replaced by finitarity and flight recurrence hypotheses.

5   Applications and Future Perspectives

Generally speaking, discrete models of system dynamics prove especially useful when the locality structure has a spatial rather than temporal profile and the evolution of the system is "computed" by some sort of algorithms. In such contexts, a few properties may surface in the discrete approach that are not apparent in the differential theory.

An application area of the relational formulation of state transition dynamics could be the formal description of biomolecular processes in the cell, as they are the result of many individual reactions, each of these being formed by subprocesses whose extension is limited in space and in time. These subprocesses only get information on what is happening in the whole organism from their respective neighbourhood. Nonetheless, they often exhibit a surprising overall co-ordination, and in this sense biomolecular processes are by all means asynchronous. We expect that the development of an analysis of the local reactions in terms of binary transition relations would take place at a convenient abstraction level, to represent the independent subprocesses and their interaction dynamics.

Even the assumption to work on a state space with no metric structure seems suitable for such an application area. If one thinks of chemical reactions that involve molecules inside the cytoplasm or the nucleus of a cell, topological features of a model seem to be important to simulate these processes. The concept of distance is not necessary, though, at least at a certain level of abstraction. Let us explain this view by way of example; we suppose to have two molecules that interact (namely by transforming each other) just by contact, rather than by sending signals to each other. In this case, having a distance would be important; as a matter of fact, one could evaluate their relative distance at any computational step of the system, and assume that they interact when this distance is zero. Such an approach even allows one to estimate in how many steps do the two molecules meet each other in the system. On the other hand, however, in this way one is forced to dive very much into the gory details of processes, which fact may clutter one's ability to observe the global properties of the system. Indeed, the only relevant information for the dynamics of the system may just be that those specific molecules interact *eventually*, given that they stay 'close enough', no matter when and how exactly [6].

We aim at continuing the relational formulation and analysis undertaken in the present paper on other dynamical concepts of actual biological interest, such as "creods", "centers", "focuses", "saddles", and of "weak" forms of chaos, which could be defined by combining some of the features defined here.

We think that the work outlined above could suggest definitions of other forms of attractors, more directly connected to the relational formulation of state transition dynamics, in that suitable concepts of stability, control, and randomness could be analyzed by associating information sources to relational dynamical systems. In this perspective, informational and entropic concepts could point out interesting characterizations of fundamental dynamical concepts for complex biological dynamics.

References

1. W.R. Ashby, *An Introduction to Cybernetics*, Chapman and Hall (1956)
2. R. Backhouse and J. van der Woude, Demonic Operators and Monotype Factors, *Mathematical Structures in Computer Science*, **3**:4 (1993) 417–433
3. C. Bonanno and V. Manca, Discrete dynamics in biological models, Gh. Păun, C. Calude (Eds.), *Romanian Journal of Information Science and Technology*, **1-2**:5 (2002) 45–67
4. L. Bianco, F. Fontana, G. Franco and V. Manca, P Systems for Biological Dynamics, G. Ciobanu, Gh. Păun, M. J. Perez-Jimenez (Eds.), *Applications of Membrane Computing*, Natural Computing Series, Springer (2006) 81–126
5. R.L. Devaney, *Introduction to chaotic dynamical systems*, Addison-Wesley (1989)
6. G. Franco, *Biomolecular Computing — Combinatorial Algorithms and Laboratory Experiments*, PhD thesis, University of Verona, Italy (2006)
7. S. Kauffman, *Investigations*, Oxford University Press (2000)
8. P. Kůrka, *Topological and Symbolic Dynamics*, Cours Spécialisés **11**, Société Mathématique de France (2003)
9. V. Manca, G. Franco and G. Scollo, State transition dynamics: basic concepts and molecular computing perspectives, M. Gheorghe (Ed.), *Molecular Computational Models: Unconventional Approaches*, Idea Group, Hershey, PA, USA (2005) 32–55
10. V. Manca and L. Bianco, Biological Networks in Metabolic P Systems, (2006) submitted.
11. V. Manca, L. Bianco and F. Fontana, Evolutions and oscillations of P systems: Theoretical considerations and applications to biochemical phenomena, G. Mauri, Gh. Păun, M.J. Pérez-Jiménez, G. Rozenberg, A. Salomaa (Eds.), *Membrane Computing*, LNCS **3365**, Springer (2005) 63–84
12. Gh. Păun, Computing with membranes, *J. Comput. System Sci.*, **61**:1 (2000) 108–143
13. Gh. Păun, *Membrane Computing. An Introduction*, Springer (2002)
14. G. Schmidt, Th. Ströhlein, *Relations and Graphs*, Springer (1993)
15. A. Tarski, On the calculus of relations, *Journal of Symbolic Logic*, **6** (1941) 73–89
16. S. Wolfram *Theory and Application of Cellular Automata*, Addison-Wesley (1986)
17. A. Wuensche, Basins of Attraction in Network Dynamics: A Conceptual Framework for Biomolecular Networks, G. Schlosser, G.P. Wagner (Eds.), *Modularity in Development and Evolution*, Chicago University Press (2002)

On Two Dually Nondeterministic Refinement Algebras

Kim Solin[*]

Turku Centre for Computer Science
Lemminkäinengatan 14 A, FIN-20520 Åbo, Finland
kim.solin@utu.fi

Abstract. A dually nondeterministic refinement algebra with a negation operator is proposed. The algebra facilitates reasoning about total-correctness preserving program transformations and nondeterministic programs. The negation operator is used to express enabledness and termination operators through a useful explicit definition. As a small application, a property of action systems is proved employing the algebra. A dually nondeterministic refinement algebra without the negation operator is also discussed.

1 Introduction

Refinement algebras are abstract algebras for reasoning about program refinement [18,21,22,20]. Axiomatic reasoning can, in a certain sense, provide a simpler reasoning tool than the classical set and order-theoretic frameworks [1,5,16]. Different classes of predicate transformers over a fixed state space form the motivating models, but should not be seen as exclusive.

The first papers on refinement algebras, in our sense of the term, were von Wright's initial paper [21], followed by [22], which builds on the aforementioned. In these papers von Wright outlines an axiomatisation with the set of isotone predicate transformers as an intended model. He also proposes the introduction of angelic choice as a separate operator in the algebra.

This paper proposes a refinement algebra that extends the original framework with three operators: the angelic choice (as suggested by von Wright), strong angelic iteration and a negation operator. Looking at the predicate-transformer models, the negation operator demands that the set of all predicate transformers be a model, whereas the iteration operator demands that the predicate transformers be isotone (as a consequence of needing to establish the existence of fixpoints via the Knaster-Tarski theorem). To solve this conflict, we let the carrier set be the set of all predicate transformers over a fixed state space and impose isotony conditions on elements of axioms involving iteration. Taking one step further, we also add ways of imposing conjunctivity and disjunctivity conditions

---

[*] Currently visiting Institut für Informatik, Universität Augsburg.

R.A. Schmidt (Ed.): RelMiCS /AKA 2006, LNCS 4136, pp. 373–387, 2006.

on elements. Thus one could say that the algebra we propose is an algebra intended for reasoning about isotone predicate transformers, but having the whole class of predicate transformers as a model.

In the earlier frameworks, assertions were always defined in terms of guards. In the framework we propose here, guards can also be defined in terms of assertions. The guards and the assertions thus have equal status. Together with von Wright we have investigated an enabledness and a termination operator in refinement algebra [20]. The enabledness operator applied to a program denotes those states from which the program is enabled, that is those states from which the program will not terminate miraculously. The termination operator, on the other hand, yields those states from which the program is guaranteed to terminate in some state, that is, the program will not abort. In this paper, these operators are defined in terms of the other operators as opposed to our earlier work where they were introduced with an implicit axiomatisation. Thus, the framework of this paper subsumes the one of [20].

Action systems comprise a formalism for reasoning about parallel programs [2,4]. The intuition is that an action system is an iteration of a fixed number of demonic choices that terminates when none of the conjunctive *actions* are enabled any longer. In the refinement algebra, an action system can be expressed using the enabledness operator. An action system can be decomposed so that the order of execution is clearly expressed; this has been shown by Back and von Wright using predicate transformer reasoning [6]. In the axiomatisation of [20] we were able to prove one direction of action system decomposition, but the other direction seems to be harder. Using the framework we present here, both directions can be derived quite easily.

When the negation operator is left out we obtain a dually nondeterministic refinement algebra for which the isotone predicate transformers constitute a model. This means that no special conditions need to be imposed on the elements to guarantee the existence of fixpoints. Also in this framework guards and assertions can be defined in terms of each other. On the other hand, explicit definitions of the enabledness and termination operators, upon which the proof of action-system decomposition relies, seem not to be possible.

The following work can be traced in the history of this paper. Kozen's axiomatisation of Kleene algebra and his introduction of tests into the algebra has been a very significant inspiration for us [12,14]. Von Wright's non-conservative extension of Kleene algebra with tests was the first abstract algebra that was genuinely an algebra for total correctness (it drops right-annihilation) [21]. It rests upon previous work on algebraic program reasoning by Back and von Wright [6]. Desharnais, Möller, and Struth extended Kleene algebra with a domain operator [8], upon which Möller relaxed Kleene algebra by giving up right-annihilation (as in [21]) *and* right-distributivity of composition. These two papers laid a firm ground to the the developments in [20], where the enabledness and termination operators were introduced. Angelic nondeterminism takes off in the theory of nondeterministic automata and Floyd's nondeterministic programs [10]. In the context of program refinement, Broy and Nelson [15,7], Back and von Wright [3], and Gardiner and Morgan [11] are early names. The present paper extends an earlier workshop version [19].

The paper is set up as follows. First the abstract algebra is proposed and a program intuition is given. Then a predicate-transformer model for the algebra is provided, which serves as a program-semantical justification. After the model, basic properties of the algebra are discussed. The third section extends the algebra by guards and assertions. After this the termination and enabledness operators are introduced. Action systems are considered under the abstract-algebraic view in Section 4. The final section before the concluding one, remarks on a dually nondetereministic refinement algebra without the negation operator.

The purpose of this paper is not to provide more grandiose applications nor to give a complete algebraic treatment; the purpose is to lay down the first strokes of the brush, the purpose is to get started.

2   A Dually Nondeterministic Refinement Algebra with Negation

In this section we propose a dually nondeterministic refinement algebra with negation, give a predicate transformer model, and have a glance at some basic properties that should be taken into account.

2.1   Axiomatisation

A *dually nondeterministic refinement algebra with negation* (dndRAn) is a structure over the signature $(\sqcap, \sqcup, \neg, ; ,^{\omega},^{\dagger}, \bot, \top, 1)$ such that $(\sqcap, \sqcup, \neg, \bot, \top)$ is a Boolean algebra, $(;, 1)$ is a monoid, and the following equations hold (; left implicit, $x \sqsubseteq y \stackrel{\text{def}}{\Leftrightarrow} x \sqcap y = x$):

$$\neg xy = \neg(xy),$$
$$\top x = \top, \qquad \bot x = \bot,$$
$$(x \sqcap y)z = xz \sqcap yz \quad \text{and} \quad (x \sqcup y)z = xz \sqcup yz.$$

Moreover, if an element x satisfies $y \sqsubseteq z \Rightarrow xy \sqsubseteq xz$ we say that x is *isotone* and if x and y are isotone, then

$$x^{\omega} = xx^{\omega} \sqcap 1, \qquad xz \sqcap y \sqsubseteq z \Rightarrow x^{\omega}y \sqsubseteq z,$$
$$x^{\dagger} = xx^{\dagger} \sqcup 1 \quad \text{and} \quad z \sqsubseteq xz \sqcup y \Rightarrow z \sqsubseteq x^{\dagger}y$$

hold. If x satisfies $x(y \sqcap z) = xy \sqcap xz$ and $x(y \sqcup z) = xy \sqcup xz$ we say that x is *conjunctive* and *disjunctive*, respectively. Of course, conjunctivity or disjunctivity implies isotony.

The operator $\neg$ binds stronger than the equally strong $^{\omega}$ and †, which in turn bind stronger than ;, which, finally, binds stronger than the equally strong $\sqcap$ and $\sqcup$.

Let us remark that the signature could be reduced to $(\sqcap, \neg, ; ,^{\omega}, 1)$, since the other operators can be defined in terms of these. Some of the axioms could also be left out, since they can be derived as theorems. For clarity, we choose to have the more spelled-out axiomatisation.

As a rough intuition, the elements of the carrier set can be seen as program statements. The operators should be understood so that $\sqcap$ is *demonic choice* (a choice we cannot affect), $\sqcup$ is *angelic choice* (a choice we can affect), ; is *sequential composition*, $\neg x$ terminates in any state where x would not terminate and the other way around, $^\omega$, the *strong (demonic) iteration*, is an iteration that either terminates *or* goes on infinitely, in which case it aborts; and †, the *strong angelic iteration*, is an iteration that terminates *or* goes on infinitely, in which case a miracle occurs. If y establishes anything that x does and possibly more, then x is *refined* by y: $x \sqsubseteq y$. The constant $\bot$ is abort, an always aborting program statement; $\top$ is magic, a program statement that establishes any postcondition; and 1 is skip. A conjunctive element can be seen as facilitating demonic non-determinism, but not angelic, whereas a disjunctive element can have angelic nondeterminism, but not demonic. An isotone element permits both kinds of nondeterminism.

2.2 A Model

A predicate transformer S is a function $S : \wp(\Sigma) \to \wp(\Sigma)$, where Σ is any set. Programs can be modelled by predicate transformers according to a weakest precondition semantics [9,5]: $S.q$ denotes those sets of states from which the execution of S is bound to terminate in q.

If $p, q \in \wp(\Sigma)$ and S satisfies $p \subseteq q \Rightarrow S.p \subseteq S.q$ then S is *isotone*. If S for any set I satisfies $S.(\bigcap_{i \in I} p_i) = \bigcap_{i \in I} S.p_i$ and $S.(\bigcup_{i \in I} p_i) = \bigcup_{i \in I} S.p_i$ it is *conjunctive* and *disjunctive*, respectively. There are three named predicate transformers abort $= (\lambda q \bullet \emptyset)$, magic $= (\lambda q \bullet \Sigma)$, and skip $= (\lambda q \bullet q)$. A predicate transformer S *is refined by* T, written $S \sqsubseteq T$, if $(\forall q \in \wp(\Sigma) \bullet S.q \subseteq T.q)$. This paper deals with six operations on predicate transformers defined by

$$(S \sqcap T).q = S.q \cap T.q,$$
$$(S \sqcup T).q = S.q \cup T.q,$$
$$\neg S.q = (S.q)^{\mathbf{C}},$$
$$(S;T).q = S.(T.q),$$
$$S^\omega = \mu.(\lambda X \bullet S; X \sqcap \mathsf{skip}) \text{ and}$$
$$S^\dagger = \nu.(\lambda X \bullet S; X \sqcup \mathsf{skip}),$$

where $^{\mathbf{C}}$ is set complement, μ denotes the least fixpoint with respect to $\sqsubseteq$, and ν denotes the greatest.

That our isotony condition of the axiomatisation actually singles out the isotone predicate transformers is settled by the next lemma. Similarly it can be proved that our conjunctivity and disjunctivity conditions single out the conjunctive and disjunctive predicate transformers, respectively.

Lemma 1. *Let* $S : \wp(\Sigma) \to \wp(\Sigma)$. *Then* S *is isotone if and only if for all predicate transformers* $T, U : \wp(\Sigma) \to \wp(\Sigma)$, *if* $T \sqsubseteq U$ *then* $S;T \sqsubseteq S;U$.

Proof. If S is isotone, then clearly $T \sqsubseteq U \Rightarrow S; T \sqsubseteq S; U$. Assume now that for all predicate transformers U and T it holds that $T \sqsubseteq U \Rightarrow S; T \sqsubseteq S; U$. We show that this implies that S is isotone. Indeed, suppose that $p, q \in \wp(\Sigma)$ and $p \subseteq q$. Then construct two predicate transformers I and J such that for any $r \in \wp(\Sigma)$ it holds that $I.r = p$ and $J.r = q$. Since $p \subseteq q$, we then have that $I \sqsubseteq J$. By the assumption, this means that $S; I \sqsubseteq S; J$, that is $(\forall r \in \wp(\Sigma) \bullet S.I.r \subseteq S.J.r)$, or in other words $(\forall r \in \wp(\Sigma) \bullet S.p \subseteq S.q)$. Removing the idle quantifier, this says exactly that $S.p \subseteq S.q$.   □

With the aid of the above lemma and the Knaster-Tarski theorem, it is easily verified that the set of all predicate transformers forms a model for the dndRAn with the interpretation of the operators given as above.

Proposition 1. *Let* Ptran_Σ *be the set of all predicate transformers over a set* Σ*. Then*

$$(\mathsf{Ptran}_\Sigma, \neg, \sqcap, \sqcup, ;, ^\omega, ^\dagger, \mathsf{magic}, \mathsf{abort}, \mathsf{skip})$$

is a dndRAn*, when the interpretation of the operators is given according to the above.*

2.3 What Is Going on?

The basic properties of the algebra differ from the algebras in [21,20] in the fact that not all operators are isotone any longer and that some propositions are weakened.

The ; is not right isotone for all elements, but for isotone *elements* it is. In fact, the isotony condition on elements says exactly this. All other operators are isotone, except the negation operator which is antitone

$$x \sqsubseteq y \Rightarrow \neg y \sqsubseteq \neg x.$$

This is to be kept in mind when doing derivations. The leapfrog property of strong iteration (strong angelic iteration is dual) is weakened, but the decomposition property is not. That is, if x and y are isotone, then

$$x(yx)^\omega \sqsubseteq (xy)^\omega x \tag{1}$$

is in general only a refinement, whereas

$$(x \sqcap y)^\omega = x^\omega (yx^\omega)^\omega \tag{2}$$

is always an equality. If x and y are conjunctive, then (1) can be strengthened to an equality [21].

3   Guards and Assertions

This section extends the algebra with guards and assertions, shows that they can be defined in terms of each other, and provides an interpretation in the predicate-transformer model.

3.1   Definitions and Properties

First, some notation. If an element of a dndRAn is both conjunctive and disjunctive, then we say that it is *functional*. A functional element thus permits no kind of nondeterminism.

Guards should be thought of as programs that check if some predicate holds, skip if that is the case, and otherwise a miracle occurs.

Definition 1. *An element g of a* dndRAn *is a* guard *if*

(1g) *g is functional,*

(2g) *g has a complement $\bar{g}$ satisfying*
$$g\bar{g} = \top \quad and \quad g \sqcap \bar{g} = 1, \quad and$$

(3g) *for any g' also satisfying (1g) and (2g) it holds that*
$$gg' = g \sqcup g'.$$

Assertions are similar to guards, but instead of performing a miracle when the predicate does not hold, they abort. That is to say, an assertion that is executed in a state where the predicate does not hold establishes no postcondition.

Definition 2. *An element p is an* assertion *if*

(1a) *p is functional,*

(2a) *p has a complement $\bar{p}$ satisfying*
$$p\bar{p} = \bot \quad and \quad p \sqcup \bar{p} = 1, \quad and$$

(3a) *for any p' also satisfying (1a) and (2a) it holds that*
$$pp' = p \sqcap p'.$$

It is easily established that the guards and the assertions form Boolean algebras, since guards and assertions are closed under the operators $\sqcap, \sqcup$, and $;$.

Proposition 2. *Let G be the set of guards and let A be the set of assertions of a* dndRAn. *Then*

$$(G, \sqcap, ;, \bar{\ }, 1, \top) \qquad and \qquad (A, ;, \sqcup, \bar{\ }, \bot, 1)$$

are Boolean algebras.

From this we get the following useful fact, by verifying that $g\bot \sqcup 1$ is the unique complement of $\bar{g}$ in the sense of (2g).

Corollary 1. *For any guard g of a* dndRAn, *we have that $g = g\bot \sqcup 1$.*

We will also need the following lemma.

Lemma 2. *For any x in the carrier set of an* dndRAn *it holds that $x\bot$ and $x\top$ are functional.*

Proof. The first case is proved by

$$x\bot(y \sqcap z) = x\bot = x\bot \sqcap x\bot = x\bot y \sqcap x\bot z$$

and the other three cases are similar.   □

With the aid of the previous lemma, the guard and assertion conditions yielding the following proposition are easily verified.

Proposition 3. *Let g be a guard and let p be an assertion in a* dndRAn. *Then*

$\bar{g}\bot \sqcap 1$	*is an assertion with the complement*	$g\bot \sqcap 1$, *and*
$\bar{p}\top \sqcup 1$	*is a guard with the complement*	$p\top \sqcup 1$.

We can now prove the following the following theorem.

Theorem 1. *Any guard/assertion can be defined in terms of an assertion/guard.*

Proof. Let G be the set of guards and let A be the set of assertions in a dndRAn. We establish a bijection between the set of guards and the set of assertions. First define $^\circ : G \to A$ by

$$g^\circ = \bar{g}\bot \sqcap 1$$

and $^\diamond : A \to G$ by

$$p^\diamond = \bar{p}\top \sqcup 1.$$

Clearly, the mappings are well-defined by Proposition 3. Now, we show that they are surjective and each other's inverses, thus bijections. Take any $g \in G$. Then $(g^\circ)^\diamond = g\bot \sqcup 1 = g$, by Proposition 3 and Corollary 1. Thus $^\diamond$ is surjective and is the inverse function of $^\circ$. The case for $^\circ$ is analogous.   □

This means that the set of guards and the set of assertions can be defined in terms of each other.

3.2 A Predicate-Transformer Model

Consider the function $[\cdot] : \wp(\Sigma) \to (\wp(\Sigma) \to \wp(\Sigma))$ such that $[p].q = p^{\mathbf{C}} \cup q$, when $p, q \in \wp(\Sigma)$. For every element $p \in \wp(\Sigma)$ there is thus a predicate transformer $S_p : \wp(\Sigma) \to \wp(\Sigma)$, $q \mapsto p^{\mathbf{C}} \cup q$. These predicate transformers are called *guards*. There is also a dual, an *assertion* and it is defined by $\{p\}.q = p \cap q$. Complement $^-$ is defined on guards and assertions by $\overline{[p]} = [p^{\mathbf{C}}]$ and $\overline{\{p\}} = \{p^{\mathbf{C}}\}$.

It follows directly from the definitions that the complement of any guard is also a guard, and moreover, that the guards are closed under the operators $\sqcap, \sqcup$, and ; defined in Section 2.2. If $[p]$ is any guard, it holds that

$$[p].(q_1 \cap q_2) = p^{\mathbf{C}} \cup (q_1 \cap q_2) = (p^{\mathbf{C}} \cup q_1) \cap (p^{\mathbf{C}} \cup q_2) = [p].q_1 \cap [p].q_2$$

for any $q_1, q_2 \in \wp(\Sigma)$. Similarly one can show that $[p].(q_1 \cup q_2) = [p].q_1 \cup [p].q_2$, so any guard is functional. Finally, it is easily verified that the axioms (g2) and (g3) also hold when the guards are interpreted in the predicate-transformer sense above. This means that guards in the predicate-transformer sense constitute a model for the guards in the abstract-algebraic sense. A similar argumentation shows that assertions in the predicate-transformer sense are a model for assertions in the abstract-algebraic sense.

4   Enabledness and Termination

We here introduce explicit definitions of the enabledness and the termination operators and show that, in this framework, the explicit definitions are equivalent to the implicit ones of [20].

4.1   Definitions

The enabledness operator ϵ is an operator that maps any program to a guard that skips in those states in which the program is enabled, that is, in those states from which the program will not terminate miraculously. It binds stronger than all the other operators and is a mapping from the set of isotone elements to the set of guards defined by

$$\epsilon x = x \bot \sqcup 1. \tag{3}$$

To see that the operator is well-defined, note that $x \bot \sqcup 1$ can be shown to be a guard with

$$\neg x \bot \sqcup 1 \tag{4}$$

as the complement.

In [20] the enabledness operator was defined implicitly similarly to the domain operator of Kleene algebra with domain (KAD) [8]. The next theorem shows that, in this framework, the implicit definition found in [20] is equivalent to the explicit definition above (in fact, as shown below only the two first axioms of the implicit axiomatisation are needed). Note that a similar move could not be done in KAD, since the explicit definition (3) relies on the lack of the right annihilation axiom for $\top$.

Theorem 2. *For any guard g and any isotone x in the carrier set of an* dndRAn, *ϵx satisfies*

$$\epsilon x x = x \qquad and \tag{5}$$
$$g \sqsubseteq \epsilon(gx) \tag{6}$$

if and only if

$$\epsilon x = x \bot \sqcup 1. \tag{7}$$

Proof. The first two axioms of ϵ can be replaced by the equivalence

$$gx \sqsubseteq x \Leftrightarrow g \sqsubseteq \epsilon x. \tag{8}$$

This can proved by reusing the proofs from [8]. Uniqueness of ϵx then follows from the principle of indirect equality and (8). Then it suffices to show that the right hand side of the explicit definition satisfies (5–6). This is verified by

$$(x \bot \sqcup 1)x \sqsubseteq x$$
$$\Leftrightarrow \{\text{axiom}\}$$
$$x \bot x \sqcup x \sqsubseteq x$$
$$\Leftrightarrow \{\text{axioms}\}$$
$$x \bot \sqcup x \sqsubseteq x 1 \sqcup x$$
$$\Leftarrow \{\text{isotony}\}$$
$$\bot \sqsubseteq 1$$
$$\Leftrightarrow \{\bot \text{ bottom element}\}$$
$$\textbf{True}$$

and

$$g \sqsubseteq g x \bot \sqcup 1$$
$$\Leftrightarrow \{\text{Corollary 1}\}$$
$$g \bot \sqcup 1 \sqsubseteq g x \bot \sqcup 1$$
$$\Leftrightarrow \{\text{axiom}\}$$
$$g \bot \bot \sqcup 1 \sqsubseteq g x \bot \sqcup 1$$
$$\Leftarrow \{\text{isotony}\}$$
$$\bot \sqsubseteq x$$
$$\Leftrightarrow \{\bot \text{ bottom element and left annihilator}\}$$
$$\textbf{True}$$

which proves the proposition.   □

Moreover, in contrast to the domain operator of [8], the compositionality property

$$\epsilon(xy) = \epsilon(x\epsilon y) \tag{9}$$

can be shown to always hold for the enabledness operator in a dndRAn (in [20] this was taken as an axiom of ϵ):

$$\epsilon(xy) = \epsilon(x\epsilon y)$$
$$\Leftrightarrow \{\text{definitions}\}$$
$$xy \bot \sqcup 1 = x(y \bot \sqcup 1) \bot \sqcup 1$$
$$\Leftrightarrow \{\text{axiom}\}$$
$$xy \bot \sqcup 1 = x(y \bot \bot \sqcup \bot) \sqcup 1$$
$$\Leftrightarrow \{\bot \text{ bottom element}\}$$
$$xy \bot \sqcup 1 = xy \bot \sqcup 1$$
$$\Leftrightarrow \{\text{reflexivity}\}$$
$$\textbf{True}.$$

Using the explicit definition, the properties

$$\epsilon(x \sqcap y) = \epsilon x \sqcap \epsilon y \text{ and} \tag{10}$$
$$\epsilon(x \sqcup y) = \epsilon x \sqcup \epsilon y \tag{11}$$

can be proved by the calculations

$$\epsilon(x \sqcap y) = (x \sqcap y)\bot \sqcup 1 = (x\bot \sqcap y\bot) \sqcup 1 = (x\bot \sqcup 1) \sqcap (y\bot \sqcup 1) = \epsilon x \sqcap \epsilon y$$

and

$$\epsilon(x \sqcup y) = (x \sqcup y)\bot \sqcup 1 = x\bot \sqcup y\bot \sqcup 1 \sqcup 1 = x\bot \sqcup 1 \sqcup y\bot \sqcup 1 = \epsilon x \sqcup \epsilon y$$

respectively. From this, isotony of enabledness

$$x \sqsubseteq y \Rightarrow \epsilon x \sqsubseteq \epsilon y \tag{12}$$

easily follows.

The termination operator τ is a mapping from isotone elements to the set of assertions defined by

$$\tau x = x \top \sqcap 1.$$

It binds equally strong as ϵ. The intuition is that the operator τ applied to a program denotes those states from which the program is guaranteed to terminate, that is, states from which it will not abort. Analogously to the enabledness operator, it can be shown that $x \top \sqcap 1$ is an assertion with complement $\neg x \top \sqcap 1$, so τ is well-defined. Moreover, using similar reasoning as above, it can also be shown that τ can equivalently be defined by

$$\tau x x = x \text{ and} \tag{13}$$
$$\tau(px) \sqsubseteq p. \tag{14}$$

That τ satisfies the properties

$$\tau(x \sqcap y) = \tau \sqcap \tau y, \tag{15}$$
$$\tau(x \sqcup y) = \tau \sqcup \tau y \text{ and} \tag{16}$$
$$x \sqsubseteq y \Rightarrow \tau x \sqsubseteq \tau y \tag{17}$$

can be proved as for enabledness.

4.2   A Predicate-Transformer Model and a Digression

In [5] the *miracle guard* is defined by $\neg(\bigsqcap_{q \in \wp(\Sigma)} S.q)$ and the *abortion guard* by $\bigsqcup_{q \in \wp(\Sigma)} S.q$. Intuitively, the miracle guard is a predicate that holds in a state $\sigma \in \Sigma$ if and only if the program S is guaranteed not to perform a miracle, that is S does not establish every postcondition starting in σ. The abortion guard holds in a state $\sigma \in \Sigma$ if and only if the program S will always terminate starting in σ, it will establish some postcondition when starting in σ. When S is isotone the least $S.q$ is $S.\emptyset$ and the greatest $S.\Sigma$, so the miracle guard can be written $\neg(S.\emptyset)$ and the abortion guard $S.\Sigma$.

A predicate-transformer interpretation of ϵx is $[\neg S.\emptyset]$, when x is interpreted as the predicate transformer S. The termination operator τx is interpreted as $\{S.\Sigma\}$. The enabledness operator and the termination operator thus correspond to the miracle guard and the abortion guard of [5], respectively, but lifted to predicate-transformer level. That the interpretation is sound is seen by the fact that $[\neg S.\emptyset] = S; \mathsf{abort} \sqcup 1$ and $\{S.\Sigma\} = S; \mathsf{magic} \sqcap 1$.

The following is a slight digression. What we did above was to turn the miracle and abortion guards into a guard and an assertion (in the predicate-transformer and the abstract-algebraic sense), respectively, since predicate transformers make up our concrete carrier set in this model. There is, however, another way of lifting the miracle and the abortion guard to the predicate-transformer level which is closer to their original definition. This is done by setting the miracle guard to be $\neg S \bot$ and the termination guard to be $S \top$. This interpretation does not, however, satisfy the the respective axioms of enabledness and termination, so the connection to KAD is lost. Nonetheless, in certain applications the possibility to work with miracle and abortion guard without turning them into a guard and an assertion could prove useful.

4.3 Expressing Relations Between Programs

The enabledness and the termination operator can be used to express relations between programs [20]. We list here some examples of the use of the first-mentioned operator. First note that $\overline{\epsilon x}$ is a guard that skips in those states where x is *disabled*.

A program x *excludes* a program y if whenever x is enabled y is not. This can be formalised by saying that x is equal to first executing a guard that checks that y is disabled and then executing x: $x = \overline{\epsilon y}x$. A program x *enables* y if y is enabled after having executed x: $x = x\epsilon y$. Similarly as above x *disables* y if $x = x\overline{\epsilon y}$. The exclusion condition will be used in the application of the next section.

5 A Small Application: Action-System Decomposition

Action systems comprise a formalism for reasoning about parallel programs [2,4]. The intuition is that an *action system* $\mathsf{do}\, x_1 [\!] \ldots [\!] x_n \,\mathsf{od}$ is an iteration of a demonic choice $x_0 \sqcap \cdots \sqcap x_n$ between a fixed number of demonically nondeterministic *actions*, $x_0, \ldots, x_n$, that terminates when none of them are any longer enabled. In dndRAn, an action system can be expressed as $(x_0 \sqcap \cdots \sqcap x_n)^\omega \overline{\epsilon(x_0)} \ldots \overline{\epsilon(x_n)}$ where $x_0, \ldots, x_n$ are conjunctive. The actions are thus iterated, expressed with the strong iteration operator, until none of them is any longer enabled, expressed with the enabledness operator.

An action system can be decomposed so that the order of execution is clearly expressed: if x excludes y, i.e. $x = \overline{\epsilon y}x$, then

$$(x \sqcap y)^\omega \overline{\epsilon x}\ \overline{\epsilon y} = y^\omega \overline{\epsilon y}(xy^\omega \overline{\epsilon y})^\omega \overline{\epsilon(xy^\omega \overline{\epsilon y})}.$$

Note that $(xy^\omega \overline{\epsilon y})^\omega \overline{\epsilon(xy^\omega \overline{\epsilon y})}$ is of the form $z^\omega \overline{\epsilon z}$.

Action-system decomposition has been shown by Back and von Wright using predicate transformer reasoning [6]. We now prove this axiomatically. We begin by an outer derivation, collecting assumptions as needed:

$$(x \sqcap y)^\omega \overline{\epsilon x} \ \overline{\epsilon y}$$
$$= \{(2)\}$$
$$y^\omega (xy^\omega)^\omega \overline{\epsilon x} \ \overline{\epsilon y}$$
$$= \{\text{assumption}\}$$
$$y^\omega (\overline{\epsilon y} xy^\omega)^\omega \overline{\epsilon x} \ \overline{\epsilon y}$$
$$= \{\text{guards Boolean algebra}\}$$
$$y^\omega (\overline{\epsilon y} xy^\omega)^\omega \overline{\epsilon y} \ \overline{\epsilon x}$$
$$= \{\text{leapfrog, conjunctivity}\}$$
$$y^\omega \overline{\epsilon y} (xy^\omega \overline{\epsilon y})^\omega \ \overline{\epsilon x}$$
$$= \{\textbf{collect: if } \epsilon x = \epsilon(xy^\omega \overline{\epsilon y})\}$$
$$y^\omega \overline{\epsilon y} (xy^\omega \overline{\epsilon y})^\omega \ \overline{\epsilon(xy^\omega \overline{\epsilon y})}.$$

The collected assumption is then, in turn, proved by two refinements. First we refine the left term into the right by (setting $z = y^\omega \overline{\epsilon y}$)

$$\epsilon x \sqsubseteq \epsilon(xz)$$
$$\Leftrightarrow \{\text{definitions}\}$$
$$x \bot \sqcup 1 \sqsubseteq xz \bot \sqcup 1$$
$$\Leftarrow \{\text{isotony}\}$$
$$\bot \sqsubseteq z \bot$$
$$\Leftrightarrow \{\bot \text{ bottom element}\}$$
$$\text{True}$$

and then the right into the left by

$$\epsilon(xy^\omega \overline{\epsilon y}) \sqsubseteq \epsilon x$$
$$\Leftrightarrow \{\text{definition}\}$$
$$xy^\omega \overline{\epsilon y} \bot \sqcup 1 \sqsubseteq x \bot \sqcup 1$$
$$\Leftarrow \{\text{isotony}\}$$
$$y^\omega \overline{\epsilon y} \bot \sqsubseteq \bot$$
$$\Leftarrow \{\text{induction}\}$$
$$y \bot \sqcap \overline{\epsilon y} \bot \sqsubseteq \bot$$
$$\Leftrightarrow \{\text{definition and (4)}\}$$
$$y \bot \sqcap (\neg y \bot \sqcup 1) \bot \sqsubseteq \bot$$
$$\Leftrightarrow \{\text{axioms}\}$$
$$y \bot \sqcap (\neg y \bot \sqcup \bot) \sqsubseteq \bot$$
$$\Leftrightarrow \{\bot \text{ bottom element}\}$$
$$y \bot \sqcap \neg y \bot \sqsubseteq \bot$$
$$\Leftrightarrow \{\text{axiom}\}$$
$$(y \sqcap \neg y) \bot \sqsubseteq \bot$$
$$\Leftrightarrow \{\text{axioms}\}$$
$$\text{True}.$$

Using the implicit definition without the negation operator, the first refinement of the assumption can also easily be proved [20], but the second refinement seems to require some additional axioms for the enabledness operator (see below).

6 Leaving Out the Negation

This section contains some remarks on a dually nondeterministic refinement algebra wihtout negation. The algebra was suggested by von Wright in [21,22], but without the strong angelic iteration. The negation operator with its axiom is dropped and we strenghten the axioms so that all elements are isotone by adding the axioms $x(y \sqcap z) \sqsubseteq xy \sqcap xz$ and $xy \sqcup xz \sqsubseteq x(y \sqcup z)$. The structure over $(\sqcap, ;, \top, 1)$ is thus a left semiring [17]. Dropping the negation means that we no longer have a Boolean algebra, however we have a complete bounded distributive lattice over $(\sqcap, \sqcup, \bot, \top)$. The spelled out axiomatisation over the signature $(\sqcap, \sqcup, ;, ^{\omega}, ^{\dagger}, \bot, \top, 1)$ is thus given by the following:

$$x \sqcap (y \sqcap z) = (x \sqcap y) \sqcap z \qquad\qquad x \sqcup (y \sqcup z) = (x \sqcup y) \sqcup z$$
$$x \sqcap y = y \sqcap x \qquad\qquad\qquad x \sqcup y = y \sqcup x$$
$$x \sqcap \top = x \qquad\qquad\qquad\qquad x \sqcup \bot = x$$
$$x \sqcap x = x \qquad\qquad\qquad\qquad x \sqcup x = x$$
$$x \sqcap (y \sqcup z) = (x \sqcap y) \sqcup (x \sqcap z) \qquad x \sqcup (y \sqcap z) = (x \sqcup y) \sqcap (x \sqcup z)$$
$$x(yz) = (xy)z$$
$$1x = x$$
$$x1 = x$$
$$\top x = \top$$
$$\bot x = \bot$$
$$x(y \sqcap z) \sqsubseteq xy \sqcap xz \qquad\qquad x(y \sqcup z) \sqsupseteq xy \sqcup xz$$
$$(x \sqcap y)z = xz \sqcap yz \qquad\qquad (x \sqcup y)z = xz \sqcup yz$$
$$x^{\omega} = xx^{\omega} \sqcap 1 \qquad\qquad\qquad x^{\dagger} = xx^{\dagger} \sqcup 1$$
$$xz \sqcap y \sqsubseteq z \;\Rightarrow\; x^{\omega}y \sqsubseteq z \qquad z \sqsubseteq xz \sqcup y \;\Rightarrow\; z \sqsubseteq x^{\dagger}y$$

Since the isotone predicate transformers are closed under union, they constitute a predicate-transfomer model for the algebra under the interpretation given in Section 2.2.

By examining the proofs of Section 3, it is clear that the results regarding guards and assertions can be re-proved without the negation operator. On the other hand, it seems to us that the enabledness operator cannot be cast in the explicit form, since we cannot express the complement $\neg x\bot \sqcup 1$ of $x\bot \sqcup 1$ and this is needed for showing that $x\bot \sqcup 1$ actually is a guard. Analogously, the termination operator cannot be given an explicit definition either. Thus, the operators have to be axiomatised along the lines of [20]. The termination operator is axiomatized by

$$x = \tau xx, \tag{18}$$
$$\tau(g^{\circ}x) \sqsubseteq g^{\circ}, \tag{19}$$

$$\tau(x\tau y) = \tau(xy) \text{ and} \tag{20}$$

$$\tau(x \sqcap y) = \tau x \sqcap \tau y, \tag{21}$$

and the enabledness operator by

$$\epsilon x x = x, \tag{22}$$

$$g \sqsubseteq \epsilon(gx), \tag{23}$$

$$\epsilon(xy) = \epsilon(x\epsilon y) \text{ and} \tag{24}$$

$$\epsilon(x \sqcup y) = \epsilon x \sqcup \epsilon y. \tag{25}$$

The last axiom for ϵ was not given in [20], since in that framework angelic choice is not even present. We conjecture that (21) and (25) are independent from the other axioms of their respective operators.

In [20] it is noted that to prove action-system decomposition an additional axiom for the enabledeness operator seems to be required: $\epsilon x \bot = x \bot$. The addition of the angelic choice operator does not seem to facilitate a proof. Due to this, the proof of action-system decomposition does not follow as neatly as when the negation operator is at hand. Dually, it is possible that $\tau x \top = x \top$ needs to be postulated for some specific purpose.

The dually nondeterminisitic refinement algebra without negation thus gives a cleaner treatment of iteration, but as a drawback the enabledness and the termination operators start crackling.

7   Concluding Remarks

We have proposed a dually nondeterministic refinement algebra with a negation operator for reasoning about program refinement and applied it to proving a rather humble property of action systems. The negation operator facilitates useful explicit definitions of the enabledness and the termination operators and it is a powerful technical tool. It is, however, antitone, which perhaps makes the reasoning a bit more subtle. When dropping the negation operator, but keeping the angelic choice, guards and assertions can still be defined in terms of each other, whereas the enabledness and termination operators no longer can be given explicit definitions.

Finding more application areas of this refinement algebra is one of our intents. Applications that genuinely include angelic nondeterminism (here it only comes into play indirectly via the definition of enabledness) is a field where the algebra could be put to use. The strong angelic iteration and the termination operator beg for application. A systematic investigation striving towards a collection of calculational rules is yet to be done.

Acknowledgements. Thanks are due to Orieta Celiku, Peter Höfner, Linas Laibinis, Bernhard Möller, Ville Piirainen, Joakim von Wright and the anonymous referees for stimulating discussions, helpful suggestions, and careful scrutiny.

References

1. R.-J. Back. *Correctness Preserving Program Refinements: Proof Theory and Applications*, volume 131 of *Mathematical Centre Tracts*. Mathematical Centre, Amsterdam, 1980.
2. R.-J. Back and R. Kurki-Suonio. Decentralization of Process Nets with Centralized Control. In *2nd ACM SIGACT-SIGOPS Symp. on Principles of Distributed Computing*, ACM, Montreal, Quebec, Canada, 1983.
3. R.-J. Back and J. von Wright. Duality in specification languages: A lattice theoretical approach. Acta Informatica, 27(7), 1990.
4. R.-J. Back and K. Sere. Stepwise refinement of action systems. Structured Programming, 12, 1991.
5. R.-J. Back and J. von Wright. *Refinement Calculus: A Systematic Introduction.* Springer-Verlag, 1998.
6. R.-J. Back and J. von Wright. Reasoning algebraically about loops. Acta Informatica, 36, 1999.
7. M. Broy and G. Nelson. Adding Fair Choice to Dijkstra's Calculus. ACM Transactions on Programming Languages and Systems, Vol 16, NO 3, 1994.
8. J. Desharnais, B. Möller and G. Struth. Kleene algebra with domain. Technical Report 2003-7, Universität Augsburg, Institut für Informatik, 2003.
9. E.W. Dijkstra. *A Discipline of Programming.* Prentice-Hall International, 1976.
10. R.W. Floyd. Nondeterministic algorithms. Journal of the ACM, 14(4), 1967.
11. P.H. Gardiner and C.C. Morgan. Data refinement of predicate transformers. Theoretical Computer Science, 87(1), 1991.
12. D. Kozen. A Completeness Theorem for Kleene Algebras and the Algebra of Regular Events. Inf. Comput. 110(2), 1994.
13. D. Kozen. *Automata and Computability.* Springer-Verlag, 1997.
14. D. Kozen. Kleene algebra with tests. ACM Transactions on Programming Languages and Systems, 19(3), 1997.
15. G. Nelson. A Generalization of Dijkstra's Calculus. ACM Transactions on Programming Languages and Systems, 11(4), 1989.
16. C.C. Morgan. *Programming from Specifications* (2nd edition). Prentice-Hall, 1994.
17. B. Möller. Lazy Kleene algebra. In D. Kozen (ed.): Mathematics of Program Construction, LNCS 3125, Springer, 2004.
18. Sampaio, A.C.A. *An Algebraic Approach To Compiler Design.* World Scientific, 1997.
19. K. Solin. An Outline of a Dually Nondeterministic Refinement Algebra with Negation. In Peter Mosses, John Power, and Monika Seisenberger (eds.): CALCO Young Researchers Workshop 2005, Selected Papers. Univ. of Wales, Swansea, Technical Report CSR 18-2005, 2005.
20. K. Solin and J. von Wright. Refinement Algebra with Operators for Enabledness and Termination. Accepted to MPC 2006.
21. J. von Wright. From Kleene algebra to refinement algebra. In E.A. Boiten and B. Möller (eds.): *Mathematics of Program Construction,* volume 2386 of *Lecture Notes in Computer Science*, Germany, Springer-Verlag, 2002.
22. J. von Wright. Towards a refinement algebra. Science of Computer Programming, 51, 2004.

On the Fixpoint Theory of Equality and Its Applications*

Andrzej Szałas[1,2] and Jerzy Tyszkiewicz[3]

[1] Dept. of Computer and Information Science, Linköping University
SE-581 83 Linköping, Sweden
andsz@ida.liu.se
[2] The University of Economics and Computer Science, Olsztyn, Poland
[3] Institute of Informatics, University of Warsaw, ul. Banacha 2, 02-097 Warsaw, Poland
jty@mimuw.edu.pl

Abstract. In the current paper we first show that the fixpoint theory of equality is decidable. The motivation behind considering this theory is that second-order quantifier elimination techniques based on a theorem given in [16], when successful, often result in such formulas. This opens many applications, including automated theorem proving, static verification of integrity constraints in databases as well as reasoning with weakest sufficient and strongest necessary conditions.

1 Introduction

In this paper we investigate the fixpoint theory of equality, FEQ, i.e., the classical first-order theory with equality as the only relation symbol, extended by allowing least and greatest fixpoints. We show that FEQ is decidable.

The motivation behind considering this theory follows from important applications naturally appearing in artificial intelligence and databases. Namely, we propose a technique, which basically depends on expressing some interesting properties as second-order formulas with all relation symbols appearing in the scope of second-order quantifiers, then on eliminating second-order quantifiers, if possible, and obtaining formulas expressed in the theory FEQ and finally, on reasoning in FEQ.

Second-order formalisms are frequent in knowledge representation. On the other hand, second-order logic is too complex[1] to be directly applied in practical reasoning. The proposed technique allows one to reduce second-order reasoning to fixpoint calculus for a large class of formulas and then to apply the decision procedure for FEQ.

To achieve our goal we first introduce a logic with simultaneous least fixpoints (Section 2) and then define the theory FEQ, prove its decidability and estimate complexity of reasoning (see Section 3). Next, in Section 4, we recall the fixpoint theorem of [16]. Then we discuss some applications of the proposed technique in automated theorem proving (Section 5.1), static verification of integrity constraints in deductive databases

---

* Supported in part by the grants 3 T11C 023 29 and 4 T11C 042 25 of the Polish Ministry of Science and Information Society Technologies.

[1] It is totally undecidable over arbitrary models and PSPACE-complete over finite models.

R.A. Schmidt (Ed.): RelMiCS /AKA 2006, LNCS 4136, pp. 388–401, 2006.

(Section 5.2) and reasoning with weakest sufficient and strongest necessary conditions as considered in [12,7] (Section 5.3).

To our best knowledge, the method proposed in Section 5.1 is original. The method discussed in Section 5.2 substantially extends the method of [9] by allowing recursive rules in addition to relational databases considered in [9]. The method presented in Section 5.3 shows a uniform approach to various forms of reasoning important in many artificial intelligence applications.

2   Fixpoint Logic

In this paper we deal with classical first-order logic (FOL) and the simultaneous least fixpoint logic (SLFP) with equality as a logical symbol, i.e., whenever we refer to the empty signature, we still allow the equality symbol within formulas.

We assume that the reader is familiar with FOL and define below syntax and semantics of SLFP.

Many of the notions of interest for us are syntax independent, so the choice of a syntactical representation of a particular semantics of fixpoints is immaterial. In this semantical sense the logic we consider has been introduced by Chandra and Harel in [5,4]. However, here we use a different syntax. A number of different definitions of SLFP, though of the same expressive power, can be found in the literature. All of them allow iterating a FOL formula up to a fixpoint. The difference is in the form of iteration.

Definition 2.1. *A relation symbol R occurs* positively *(respectively* negatively*) in a formula A if it appears under an even (respectively odd) number of negations.*[2]

A formula A is positive *w.r.t. relation symbol R iff all occurrences of R in A are positive. A formula A is* negative *w.r.t. relation symbol R iff all occurrences of R in A are negative.*   ◁

Definition 2.2. *Let $\varphi_i(R_1, \ldots, R_\ell, \bar{x}_i, \bar{y}_i)$, for $i = 1, \ldots, \ell$, be FOL formulas, where $\bar{x}_i$ and $\bar{y}_i$ are all free first-order variables of φ_i, $|\bar{x}_i| = k_i$, none of the x's is among the y's and where, for $i = 1, \ldots, \ell$, R_i are k_i-argument relation symbols, all of whose occurrences in $\varphi_1, \ldots, \varphi_\ell$ are positive. Then the formula*

$$\text{SLFP}\,[R_1(\bar{x}_1) \equiv \varphi_1(R_1, \ldots, R_\ell, \bar{x}_1, \bar{y}_1), \ldots, R_\ell(\bar{x}_\ell) \equiv \varphi_\ell(R_1, \ldots, R_\ell, \bar{x}_\ell, \bar{y}_\ell)]$$

is called a simultaneous fixpoint formula *(with variables $\bar{x}_1, \ldots, \bar{x}_\ell, \bar{y}_1 \ldots, \bar{y}_\ell$ free). In the rest of the paper we often abbreviate the above formula by $\text{SLFP}\,[\bar{R} \equiv \bar{\varphi}]$.*

Let σ be a signature. Then the set of SLFP formulas *over σ is inductively defined as the least set containing formulas of FOL over σ, closed under the usual syntax rules of first-order logic and applications of simultaneous least fixpoints.*   ◁

Note that according to the above rules, the fixpoint operators cannot be nested in SLFP, however, it is permitted to use boolean combinations of fixpoints, as well as to quantify variables outside of them.

---

[2] It is assumed here that all implications of the form $p \to q$ are substituted by $\neg p \vee q$ and all equivalences of the form $p \equiv q$ are substituted by $(\neg p \vee q) \wedge (\neg q \vee p)$.

FOL^k and SLFP^k stand for the sets of those formulas in FOL and SLFP, respectively, in which only at most k distinct first-order variable symbols occur.

For a structure $\mathbb{A}$, by A we denote the domain of $\mathbb{A}$. By A^k we denote the cartesian product $\underbrace{A \times \ldots \times A}_{k-\text{times}}$. By ω we denote the set of natural numbers.

We assume the standard semantics of FOL. For SLFP we need a semantical rule concerning the semantics of the formula SLFP $[\bar{R} \equiv \bar{\varphi}]$.

Further on $\bar{x} : \bar{a}, R_1 : \Phi_1, \ldots, R_\ell : \Phi_\ell$ denotes a valuation assigning $\bar{a}$ to $\bar{x}$ and Φ_i to R_i, for $i = 1, \ldots, \ell$. The values of the remaining variables play the rôle of parameters and are not reflected in the notation.

Given a structure $\mathbb{A}$, we define the sequence $(\bar{\Phi}^\alpha) = (\langle \Phi_1^\alpha, \ldots, \Phi_\ell^\alpha \rangle)$ indexed by ordinals α, by the following rules,

$$
\begin{aligned}
\Phi_i^0 &= \emptyset \quad \text{for } i = 1, \ldots, \ell \\
\Phi_i^{\alpha+1} &= \{\bar{b} \in A^{k_i} \mid \mathbb{A}, \bar{x}_i : \bar{b}, R_1 : \Phi_1^\alpha, \ldots, R_\ell : \Phi_\ell^\alpha \models \varphi_i\} \quad \text{for } i = 1, \ldots, \ell \\
\Phi_i^\alpha &= \bigcup_{\beta < \alpha} \Phi_i^\beta \quad \text{for } i = 1, \ldots, \ell, \text{ when } \alpha \text{ is a limit ordinal.}
\end{aligned}
$$

Since each φ_i is positive in all the R_j's, a simple transfinite induction shows that the sequence $(\bar{\Phi}^\alpha)$ is ascending in each of the coordinates.

Let $\bar{\Phi}^\infty \stackrel{\text{def}}{=} \langle \Phi_1^\infty, \ldots, \Phi_\ell^\infty \rangle = \left\langle \bigcup_\alpha \Phi_1^\alpha, \ldots, \bigcup_\alpha \Phi_\ell^\alpha \right\rangle$. Then we define

$$\mathbb{A}, \bar{x}_i : \bar{a}_i \models \text{SLFP} \, [\bar{R} \equiv \bar{\varphi}] \text{ iff } \bar{a}_i \in \Phi_i^\infty \text{ for } i = 1, \ldots, \ell.$$

3   Fixpoint Theory of Equality

3.1   The Main Results

Before proceeding, we introduce the main tools.

Below by $A(\bar{x})[\bar{t}]$ we mean the application of $A(\bar{x})$ to terms (or, dependently on the context, to domain values) $\bar{t}$.

Definition 3.1. *Let $\mathbb{A}, \mathbb{B}$ be two structures over a common signature. We write $\mathbb{A} \equiv_k \mathbb{B}$ iff $\mathbb{A}$ and $\mathbb{B}$ cannot be distinguished by any FOL^k sentence, i.e., when for every sentence φ of first-order logic with k variables, $\mathbb{A} \models \varphi$ iff $\mathbb{B} \models \varphi$.*

For two tuples $\bar{a} \in A^k$ and $\bar{b} \in B^k$ we write $\mathbb{A}, \bar{a} \equiv_k \mathbb{B}, \bar{b}$ iff those tuples cannot be distinguished by any FOL^k formula in $\mathbb{A}$ and $\mathbb{B}$, i.e., when for every formula $\varphi(\bar{x}) \in$ FOL^k, $\mathbb{A} \models \varphi[\bar{a}]$ iff $\mathbb{B} \models \varphi[\bar{b}]$. ◁

Another fact that we will need is a characterization of the expressive power of FOL^k in terms of an infinitary Ehrenfeucht-Fraïssé-style pebble game. This game characterizes the expressive power of the logic we have introduced in the sense formulated in Theorem 3.3 of [3,8,17].

Definition 3.2 (The Game).

Players, board and pebbles. *The game is played by two players, Spoiler and Dupli-*
cator, on two σ-structures $\mathbb{A}, \mathbb{A}'$ with two distinguished tuples $\bar{a} \in A^k$ and $\bar{a}' \in \mathbb{A}'$.
There are k pairs of pebbles: $(1, 1'), \dots, (k, k')$. Pebbles without primes are in-
tended to be placed on elements of A, while those with primes on elements of A'.
Initial position. *Initially, the pebbles are located as follows: pebble i is located on a_i,*
and pebble i' is located on a'_i, for $i = 1, \dots, k$.
Moves. *In each of the moves of the game, Spoiler is allowed to choose one of the*
structures and one of the pebbles placed on an element of that structure and move
it onto some other element of the same structure. Duplicator must place the other
pebble from that pair on some element in the other structure so that the partial
function from $\mathbb{A}$ to $\mathbb{A}'$ mapping $x \in \mathbb{A}$ on which pebble i is placed onto the element
$x' \in \mathbb{A}'$ on which pebble i' is placed and constants in $\mathbb{A}$ onto the corresponding
constants in $\mathbb{A}'$, is a partial isomorphism. Spoiler is allowed to alternate between
the structures as often as he likes, when choosing elements.
Who wins? *Spoiler wins if Duplicator does not have any move preserving the isomor-*
phism. We say that Duplicator has a winning strategy if he can play forever despite
of the moves of Spoiler, preventing him from winning. ◁

Theorem 3.3. *Let $\mathbb{A}, \mathbb{B}$ be any two structures of a common signature. Then Duplicator*
has a winning strategy in the game on $\mathbb{A}, \bar{a}$ and $\mathbb{B}, \bar{b}$ iff $\mathbb{A}, \bar{a} \equiv_k \mathbb{B}, \bar{b}$. ◁

Henceforth we restrict our attention to the theory and models of pure equality. Let for
a cardinal number $\mathfrak{m}$ the symbol $\mathbb{E}_{\mathfrak{m}}$ stand for the only (up to isomorphism) model of
pure equality of cardinality $\mathfrak{m}$.

The following theorem can easily be proved using Theorem 3.3.

Theorem 3.4. *Let $k \in \omega$. Then for any cardinal numbers $\mathfrak{m}, \mathfrak{n} \geq k$ and any two tuples*
$\bar{a}, \bar{b}$ of length k over $\mathbb{E}_{\mathfrak{m}}$ and $\mathbb{E}_{\mathfrak{n}}$, respectively, $\mathbb{E}_{\mathfrak{m}}, \bar{a} \equiv_k \mathbb{E}_{\mathfrak{n}}, \bar{b}$ if and only if for every
$i, j \leq k$ the equivalence $a_i = a_j \equiv b_i = b_j$ holds. ◁

Proof. By theorem 3.3 it suffices to prove that the Duplicator has a winning strategy in
the game iff for every $i, j \leq k$, $a_i = a_j \equiv b_i = b_j$.

If the equivalence does not hold, then certainly the Duplicator lost already at the
beginning. In turn, if it does, than the initial position has the required isomorphism,
and this can be preserved by the Duplicator, since the structures have at least as many
elements as the number of pebbles, so the Duplicator can mimic any move of the
Spoiler. ◁

Henceforth if the equivalence $a_i = a_j \equiv b_i = b_j$ holds for every $i, j \leq k$ for two tuples
$\bar{a}, \bar{b}$ of length k, we will write $\bar{a} \equiv_k \bar{b}$. Note that already in $\mathbb{E}_k$ there are tuples which
are representatives of all the equivalence classes of $\equiv_k$.

Definition 3.5. *The quantifier rank of a formula α, denoted by $r(\alpha)$, is defined induc-*
tively by setting $r(\alpha) \stackrel{\text{def}}{=} 0$ when α contains no quantifiers, $r(\neg\alpha) \stackrel{\text{def}}{=} r(\alpha)$, for any bi-
nary propositional connective $\circ$, $r(\alpha \circ \beta) \stackrel{\text{def}}{=} \max\{r(\alpha), r(\beta)\}$ and $r(\exists\alpha) \stackrel{\text{def}}{=} r(\forall\alpha) \stackrel{\text{def}}{=}$
$r(\alpha) + 1$. ◁

An important result is the following theorem, provided in [10, Theorem 2.7].

Theorem 3.6. *Let* $0 < n \in \omega$ *and let all the first-order formulas* φ_i *in an* SLFP *formula* SLFP $[\bar{R} \equiv \bar{\varphi}]$ *have at most* k *free variables and be of quantifier rank at most* d. *Then, over the empty signature,[3] each component* Φ_i^n *is definable by a first-order formula with at most* k *variables and of quantifier rank at most* dn, *i.e., for any* $0 < n \in \omega$ *there are formulas* $\varphi_1^n, \ldots, \varphi_\ell^n$ *of* FOL^k *of quantifier rank* $\leq dn$ *such that for any structure* $\mathbb{A}$ *over the empty signature,* $\Phi_i^n = \{\bar{a} \in A^{k_i} \mid \mathbb{A}, \bar{x}_i : \bar{a} \models \varphi(\bar{x}_i)\}$. ◁

Next, an application of Theorem 3.4 and the previous results, yields the following consequence.

Corollary 3.7. *Let* $0 < k \in \omega$. *If* $\mathbb{A}$ *is a model of pure equality of cardinality at least* k, $\bar{a} \in A^k$, *and* $\varphi(\bar{x}) \in$ SLFP^k, *then* $\mathbb{A} \models \varphi[\bar{a}]$ *iff* $\mathbb{E}_k \models \varphi[\bar{a}']$, *where* $\bar{a}' \equiv_k \bar{a}$.

Proof. First, we claim that every subformula of φ of the form SLFP $[\bar{R} \equiv \bar{\varphi}]$ can be substituted by an FOL^k formula, equivalent to the former both in $\mathbb{A}$ and $\mathbb{E}_k$.

Indeed, in $\mathbb{E}_k$ the sequence of stages $(\bar{\Phi}^\alpha) = (\langle \Phi_1^\alpha, \ldots, \Phi_\ell^\alpha \rangle)$ reaches a fixpoint in a finite number of iterations, say K, i.e., $\bar{\Phi}^\infty = \bar{\Phi}^K$. The reason is that this sequence is ascending in each of the coordinates, and each coordinate for each α is a subset of a fixed, finite set. Therefore

$$\mathbb{E}_k \models \bigwedge_{i=1}^{\ell} \forall \bar{x} \big(\varphi_i^K(\bar{x}) \equiv \varphi_i^{K+1}(\bar{x}) \big),$$

where $\varphi_i^K(\bar{x})$ are the formulas from Theorem 3.6. $\mathbb{A}$ is isomorphic to some $\mathbb{E}_m$ for some $m \geq k$, so by Theorem 3.4,

$$\mathbb{A} \models \bigwedge_{i=1}^{\ell} \forall \bar{x} \big(\varphi_i^K(\bar{x}) \equiv \varphi_i^{K+1}(\bar{x}) \big),$$

This sentence asserts that the iteration of SLFP $[\bar{R} \equiv \bar{\varphi}]$ stops in $\mathbb{A}$ after at most K steps, too. It is now routine to use the FOL^k formulas $\varphi_i^K(\bar{x})$ to replace SLFP $[\bar{R} \equiv \bar{\varphi}]$ in φ.

Our claim has been proven. So let $\varphi' \in$ FOL^k be equivalent to φ in both $\mathbb{A}$ and $\mathbb{E}_k$, and obtained by the substitution of all fixpoints of φ by their FOL^k-equivalents.

Now by Theorem 3.4 it follows that $\mathbb{A} \models \varphi'[\bar{a}]$ iff $\mathbb{E}_k \models \varphi'[\bar{a}']$, where $\bar{a}' \equiv_k \bar{a}$, and this carries over to the formula φ, as desired. ◁

3.2   The Complexity

Now we turn to the problem of satisfiability of SLFP formulas over the empty signature. This means that still the only predicate allowed in formulas is the equality.

By the results of the previous section, we have the following equivalence:

Theorem 3.8. *A formula* φ *of* SLFP^k *is satisfiable if and only if it is satisfiable in one of the structures* $\mathbb{E}_1, \ldots, \mathbb{E}_k$.

---

[3] Recall that equality is still allowed, since it is a logical symbol.

Proof. Indeed, any structure over the empty signature is isomorphic to one of the form $\mathbb{E}_m$, and since $\mathbb{E}_m \equiv_k \mathbb{E}_k$, the equivalence follows.   ◁

This suggests the following algorithm for testing satisfiability of fixpoint formulas over the empty signature: for a given formula $\varphi(\bar{x}) \in \mathrm{SLFP}^k$ we test if it is satisfied by $\langle \mathbb{A}, \bar{a} \rangle$, where $\mathbb{A}$ ranges over all (pure equality) structures of cardinality at most k, and $\bar{a}$ ranges over all equality types of vectors of length $|\bar{x}|$ of elements from A.

Concerning the complexity of this procedure, the number of structures to be tested is linear in k. The number of iterations of any fixpoint in SLFP^k is bounded by $O(B(k)^\ell)$, where $B(n)$ is the n-th Bell number and ℓ the maximal number of formulas whose simultaneous fixed point is used. Indeed, $B(k)$ is the number of $\equiv_k$-equivalence classes. Thus computing the fixpoints literally, according to the definition, takes time bounded by a polynomial of $B(k)^\ell$, and computing the first-order constructs increases this by only a polynomial factor.

Therefore the algorithm we obtained is of exponential complexity.

4   The Fixpoint Theorem

Further on we deal with the first- and the second-order classical logic with equality.

Below we recall the theorem for elimination of second-order quantifiers, proved in [16]. This theorem, combined with the decidability result given in Section 3.2, provides us with a powerful tool for deciding many interesting properties, as shown in Section 5. For an overview of the related techniques see [15].

Let $B(X)$ be a second-order formula, where X is a k-argument relational variable and let $C(\bar{x})$ be a first-order formula with free variables $\bar{x} = \langle x_1, \ldots, x_k \rangle$. Then by $B[X(\bar{t}) := C(\bar{x})]$ we mean the formula obtained from $B(X)$ by substituting each occurrence of X of the form $X(\bar{t})$ in $B(X)$ by $C(\bar{t})$, renaming the bound variables in $C(\bar{x})$ with fresh variables.

Example 4.1. Let $B(X) \equiv \forall z[X(y,z) \vee X(f(y), g(x,z))]$, where X is a relational variable and let $C(x,y) \equiv \exists z R(x,y,z)$. Then $B[X(t_1,t_2) := C(x,y)]$ is defined by

$$\forall z[\ \underbrace{\exists z' R(y,z,z')}_{C'(y,z)} \vee \underbrace{\exists z' R(f(y), g(x,z), z')}_{C'(f(y),g(x,z))}\ ],$$

where $C'(x,y)$ is obtained from $C(x,y)$ by renaming the bound variable z with z'.   ◁

Recall that by $A(\bar{x})[\bar{t}]$ we mean the application of $A(\bar{x})$ to terms $\bar{t}$.

The following theorem, substantial for the applications considered in Section 5, has been provided in [16]. Below, for simplicity, we use the standard least and greatest fixpoint operators LFP and GFP rather than simultaneous fixpoints.

Theorem 4.2. *Assume that formula A is a first-order formula positive w.r.t. X.*

– *if B is a first-order formula negative w.r.t. X then*

$$\exists X \forall \bar{y}[A(X) \to X(\bar{y})] \wedge [B(X)] \equiv B[X(\bar{t}) := \mathrm{LFP}\, X(\bar{y}).A(X)[\bar{t}]] \tag{1}$$

– if B is a first-order formula positive w.r.t. X then
$$\exists X \forall \bar{y}[X(\bar{y}) \rightarrow A(X)] \wedge [B(X)] \equiv B[X(\bar{t}) := \text{GFP}\, X(\bar{y}).A(X)[\bar{t}]]. \qquad (2)$$
◁

Remark 4.3. Observe that, whenever formula A in Theorem 4.2 does not contain X, the resulting formula is easily reducible to a first-order formula, as in this case both LFP $X(\bar{y}).A$ and GFP $X(\bar{y}).A$ are equivalent to A. Thus the Ackermann's lemma (see, e.g., [2,15,18]) is subsumed by Theorem 4.2). ◁

An online implementation of the algorithm based on the above theorem is available online (see [13]). Observe that the techniques applied in that algorithm, initiated in [18] and further developed in [6], allow one to transform a large class of formulas to the form required in Theorem 4.2.

Example 4.4. Consider the following second-order formula:
$$\exists X \forall x \forall y[(S(x,y) \vee X(y,x)) \rightarrow X(x,y)] \wedge [\neg X(a,b) \vee \forall z(\neg X(a,z))] \qquad (3)$$
According to Theorem 4.2(1), formula (3) is equivalent to:
$$\neg \text{LFP}\, X(x,y).(S(x,y) \vee X(y,x))[a,b] \vee$$
$$\forall z(\neg \text{LFP}\, X(x,y).(S(x,y) \vee X(y,x))[a,z]). \qquad (4)$$
Observe that the definition of the least fixpoint appearing in (4) is obtained on the basis of the first conjunct of (3). The successive lines of (4) represent substitutions of $\neg X(a,b)$ and $\forall z(\neg X(a,z))$ of (3) by the obtained definition of the fixpoint. ◁

5   Applications

It can easily be observed that, whenever the elimination of all predicate variables in a formula is possible by applications of Theorem 4.2, the resulting formula is a fixpoint formula over the signature containing equality only. Thus the method applied in the next sections depends on first eliminating all relations appearing in respective formulas and then to apply reasoning in the fixpoint theory of equality.

5.1   Automated Theorem Proving

Introduction. Automated theorem proving in the classical first-order logic is considered fundamental in such applications as formal verification of software[4] and properties of data structures, as well as in the whole spectrum of reasoning techniques appearing in AI, etc. The majority of techniques in these fields are based on various proof systems with resolution-based ones and natural deduction supplemented with algebraic methods, like term rewriting systems etc.

Below we propose another method, which seems to be new in the field. It is not based on any particular proof system. Instead, we first introduce second-order quantifiers in an obvious way, then try to eliminate them and, if this is successful, use the decision procedure for the theory FEQ.

---

[4] In particular, verification of logic programs, where the method we propose is applicable directly.

The Method. Let $A(R_1, \ldots, R_n)$ be a first-order formula. It is assumed that all relation symbols appearing in this formula are $R_1, \ldots, R_n, =$. In order to prove that $A(R_1, \ldots, R_n)$ is a tautology, $\models A(R_1, \ldots, R_n)$, we prove instead that the following second-order formula

$$\forall R_1 \ldots \forall R_n A(R_1, \ldots, R_n) \tag{5}$$

is a tautology. Of course, $A(R_1, \ldots, R_n)$ is a tautology iff (5) is a tautology. In general this problem is totally undecidable. However, to prove (5) we negate formula (5), eliminate second-order quantifiers $\exists R_1 \ldots \exists R_n$ applying Theorem 4.2 and, if this is successful, apply the decision procedure of Section 3.2. The result is FALSE iff the original formula is equivalent to TRUE.

It should be emphasized that whenever $A(R_1, \ldots, R_n)$ itself is second-order, we can first try to eliminate second-order quantifiers from $A(R_1, \ldots, R_n)$ and then apply the proposed method to the resulting formula. So, in fact, we have a decision procedure for a fragment of the second-order logic, too. This is important in many AI applications, e.g., in reasoning based on various forms of circumscription (see, e.g., [14,11,6]).

Example. Assume a is a constant and consider formula

$$\forall x, y[R(x, y) \to R(y, x)] \to [\exists z R(a, z) \to \exists u R(u, a)] \tag{6}$$

The proof of validity of (6) involves the following steps[5]:

— introduce second-order quantifiers over relations (here only over R):
$$\forall R\Big\{\forall x, y[R(x, y) \to R(y, x)] \to [\exists z R(a, z) \to \exists u R(u, a)]\Big\}$$
— negate: $\exists R\Big\{\forall x, y[R(x, y) \to R(y, x)] \land \exists z R(a, z) \land \forall u \neg R(u, a)\Big\}$
— transform the formula to the form required in Theorem 4.2:
$$\exists R\Big\{\forall x, y[R(x, y) \to (R(y, x) \land x \neq a)] \land \exists z R(a, z)\Big\}$$
— apply Theorem 4.2(2): $\exists z\Big[\text{GFP } R(x, y).(R(y, x) \land x \neq a)[a, z]\Big]$.

To see that the last formula is FALSE, meaning that the formula (6) is TRUE, we unfold the greatest fixpoint and obtain that

$$\text{GFP } R(x, y).(R(y, x) \land x \neq a) \equiv (y \neq a \land x \neq a).$$

Thus the resulting formula is equivalent to $\exists z\Big[(y \neq a \land x \neq a)[a, z]\Big]$, i.e., to $\exists z\Big[(z \neq a \land a \neq a)\Big]$, being equivalent to FALSE. This proves the validity of formula (6).

5.2 Static Verification of Integrity Constraints in Deductive Databases

Introduction. In [9] a method for static verification of integrity constraints in relational databases has been presented. According to the relational database paradigm, integrity constraints express certain conditions that should be preserved by all instances of

---

[5] These steps can fully be automated, as done in [18,6] and implemented in [13].

a given database, where by an *integrity constraint* we understand a classical first-order formula in the signature of the database. In the existing implementations these conditions are checked dynamically during the database updates. In the case of software systems dealing with rapidly changing environment and reacting in real time, checking integrity constraints after each update is usually unacceptable from the point of view of the required reaction time. Such situations are frequent in many artificial intelligence applications, including autonomous systems.

The Method. In the method of [9] it is assumed that the database can be modified only by well-defined procedures, called *transactions*, supplied by database designers. In such a case the task of verification of integrity constraints reduces to the following two steps:

1. verify that the initial contents of the database satisfies the defined constraints
2. verify that all transactions preserve the constraints.

If both above conditions hold, a simple induction, where the first point is the base step and the second point is the induction step, shows that all possible instances of the database preserve the considered integrity constraints. Of course, the first step can be computed in time polynomial w.r.t. the size of the initial database. In what follows we then concentrate on the second step.

Consider a transaction, which modifies relations $R_1, \ldots, R_n$ giving as a result relations $R'_1, \ldots, R'_n$. The second of the steps mentioned earlier reduces to verification whether the following second-order formula is a tautology:

$$\forall R_1, \ldots, R_n [I(R_1, \ldots, R_n) \rightarrow I(R'_1, \ldots, R'_n)].$$

The method of [9] depends on the application of the Ackermann's lemma of [2], which itself is subsumed by Theorem 4.2 (see Remark 4.3). If the Ackermann's lemma is successful, the resulting formula is expressed in the classical theory of equality, but the requirement is that formulas involved in integrity constraints are, among others, nonrecursive. Therefore [9] considers relational databases rather that deductive ones, which usually require recursion (see, e.g., [1]).

Definition 5.1. *By an* update *of a deductive database DB we shall mean an expression of one of the forms* ADD $\bar{e}$ TO R *or* DELETE $\bar{e}$ FROM R *, where R is an k-ary relation of DB and $\bar{e}$ is a tuple of k elements.*   ◁

The meaning of ADD and DELETE updates is rather obvious. Namely, ADD e TO R denotes adding a new tuple e to the relation R, whereas DELETE e FROM R denotes deleting e from R. From the logical point of view, the above updates are formula transformers defined as follows, where $A(R)$ is a formula:

$$
\begin{aligned}
(\text{ADD } \bar{e} \text{ TO } R)(A(R(\bar{x}))) &\stackrel{\text{def}}{\equiv} A(R(\bar{x}) := (R(\bar{x}) \vee \bar{x} = \bar{e})) \\
(\text{DELETE } \bar{e} \text{ FROM } R(A(R(\bar{x}))) &\stackrel{\text{def}}{\equiv} A(R(\bar{x}) := (R(\bar{x}) \wedge \bar{x} \neq \bar{e})).
\end{aligned}
\tag{7}
$$

Definition 5.2. *By a* transaction *on a deductive database DB we shall mean any finite sequence of* updates *on DB. Transaction T is* correct *with respect to integrity constraint* $I(R_1, \ldots, R_k)$ *iff the following implication:*

$$I(R_1, \ldots, R_k) \to T(I(R_1, \ldots, R_k)) \tag{8}$$

is a tautology.   ◁

Formula (8) is a tautology iff the following second-order formula is a tautology, too:

$$\forall R_1 \ldots \forall R_k [I(R_1, \ldots, R_k) \to T(I(R_1, \ldots, R_k))]. \tag{9}$$

In order to eliminate quantifiers $\forall R_1 \ldots \forall R_k$ we first negate (9), as done in Section 5.1:

$$\exists R_1 \ldots \exists R_k [I(R_1, \ldots, R_k) \wedge \neg T(I(R_1, \ldots, R_k))], \tag{10}$$

then try to transform formula (10) into the form suitable for application of Theorem 4.2. This transformation can be based on those given in [6,18] and considered in Section 5.1. If the decision procedure of Section 3.2, applied to (10) results in FALSE, then the formula (9) and, consequently (8), are equivalent to TRUE.

Example. Let $R(x)$ stand for "x is rich", $C(y, x)$ stand for "y is a child of x", j stand for "John" and m for "Mary". Consider the constraint

$$\forall x, y\{[R(x) \wedge C(y, x)] \to R(y)\} \tag{11}$$

and the transaction ADD $\langle j, m \rangle$ TO C; DELETE $\langle m \rangle$ FROM R.

To prove correctness of the transaction we first consider the formula reflecting (9),

$$\forall C \forall R\Big\{\forall x, y\{[R(x) \wedge C(y, x)] \to R(y)\} \to$$
$$\forall x, y\{[R(x) \wedge (C(y, x) \vee (y = j \wedge x = m))] \to [R(y) \wedge y \neq m]\}\Big\}. \tag{12}$$

After negating (12) and renaming variables we obtain

$$\exists C \exists R\Big\{\forall x, y\{[R(x) \wedge C(y, x)] \to R(y)\} \wedge$$
$$\exists u, v\{[R(u) \wedge (C(v, u) \vee (v = j \wedge u = m))] \wedge [\neg R(v) \vee v = m]\}\Big\}. \tag{13}$$

Some transformations of (13) made in the spirit of algorithms [18,6,13] result in

$$\exists u, v \exists C \exists R\Big\{\forall x, y\{[R(x) \wedge C(y, x)] \to R(y)\} \wedge [\neg R(v) \vee v = m] \wedge$$
$$R(u) \wedge \forall x, y[(x = v \wedge y = u \wedge (v \neq j \vee u \neq m)) \to C(x, y)]\Big\}.$$

We first eliminate $\exists C$ which, according to Remark 4.3, results in the following formula without fixpoints

$$\exists u, v \exists R\Big\{\forall x, y\{[R(x) \wedge y = v \wedge x = u \wedge (v \neq j \vee u \neq m)] \to R(y)\} \wedge$$
$$[\neg R(v) \vee v = m] \wedge R(u)\Big\},$$

equivalent to

$$\exists u, v \exists R \Big\{$$
$$\forall y \{ [\exists x [R(x) \wedge y = v \wedge x = u \wedge (v \neq j \vee u \neq m)] \vee y = u] \rightarrow R(y) \} \wedge$$
$$[\neg R(v) \vee v = m] \Big\}.$$

Now an application of Theorem 4.2(1) results in

$$\exists u, v \Big\{ v = m \vee$$
$$\neg \text{LFP } R(y).[\exists x [R(x) \wedge y = v \wedge x = u \wedge (v \neq j \vee u \neq m)] \vee y = u] \Big\}. \tag{14}$$

Applying the decision procedure of Section 3.2 shows that formula (14) is equivalent to FALSE, which proves correctness of the considered transaction.

5.3   Reasoning with Weakest Sufficient and Strongest Necessary Conditions

Introduction. Weakest sufficient and strongest necessary conditions have been introduced by Lin in [12] in the context of propositional reasoning and extended to the first-order case in [7].

Consider a formula A expressed in some logical language. Assume that one is interested in approximating A in a less expressive language, say L, which allows for more efficient reasoning. A sufficient condition of A, expressed in L, is a formula implying A and a necessary condition of A, expressed in L, is a formula implied A. Thus the weakest sufficient condition provides "the best" approximation of A that guarantees its satisfiability and the strongest necessary condition provides "the best" approximation of A that still cannot exclude A, both expressed in the less expressive language.

Let us emphasize that sufficient and necessary conditions are vital for providing solutions to important problems concerning, e.g., approximate reasoning, abduction and hypotheses generation, building communication interfaces between agents or knowledge compilation.

Below we assume that theories are finite, i.e., can be expressed by finite conjunctions of axioms.

The Method. The following are definitions for necessary and sufficient conditions of a formula A relativized to a subset $\bar{P}$ of relation symbols under a theory T, as introduced in [12].

Definition 5.3. *By a* necessary condition *of a formula A on the set of relation symbols $\bar{P}$ under theory T we shall understand any formula B containing only symbols in $\bar{P}$ such that $T \models A \rightarrow B$. It is the* strongest necessary condition, *denoted by* SNC $(A; T; \bar{P})$ *if, additionally, for any necessary condition C of A on $\bar{P}$ under T, we have $T \models B \rightarrow C$.*

By a sufficient condition *of a formula A on the set of relation symbols $\bar{P}$ under theory T we shall understand any formula B containing only symbols in $\bar{P}$ such that $T \models B \rightarrow A$. It is the* weakest sufficient condition, *denoted by* WSC $(A; T; \bar{P})$ *if, additionally, for any sufficient condition C of A on $\bar{P}$ under T, we have $T \models C \rightarrow B$.*◁

The set $\bar{P}$ in Definition 5.3 is referred to as the *target language*.

According to [7], we have the following characterization of weakest sufficient and strongest necessary conditions.

Lemma 5.4. *For any formula A, any set of relation symbols $\bar{P}$ and a closed theory T:*

1. SNC $\left(A;T;\bar{P}\right)$ *is defined by* $\exists \bar{X}[T \wedge A]$
2. WSC $\left(A;T;\bar{P}\right)$ *is defined by* $\forall \bar{X}[T \to A]$,

where $\bar{X}$ consists of all relation symbols appearing in T or A, but not in $\bar{P}$.   ◁

Thus, reasoning with weakest sufficient and strongest necessary conditions can again, in many cases, be reduced to FEQ. Namely, one can first try to eliminate second-order quantifiers from second-order formulas appearing in characterizations provided in Lemma 5.4 and then to apply the method of Section 5.1.

The method is best visible in the case when we are interested in formulas of the target language implied by SNC $\left(A;T;\bar{P}\right)$ and implying WSC $\left(A;T;\bar{P}\right)$. In these cases we deal with formulas of the form $\forall \bar{R}\{\exists \bar{X}[T \wedge A] \to B\}$ and $\forall \bar{R}\{B \to \forall \bar{X}[T \to A]\}$, where $\bar{R}$ consists of all relation symbols appearing free in the respective formulas and B contains no relation symbols of $\bar{X}$. Of course, these forms are equivalent to $\forall \bar{R}\forall \bar{X}\{[T \wedge A] \to B\}$ and $\forall \bar{R}\forall \bar{X}\{B \to [T \to A]\}$. Thus the proposed method applies here in an obvious manner.

Example. Consider a theory given by formula (11) of Section 5.2 and

$$\text{SNC}\left(\neg R(m) \wedge R(j); (11); \{C\}\right). \tag{15}$$

Suppose we are interested in verifying whether (15) implies $\exists x, y \neg C(y, x)$.

According to Lemma 5.4, (15) is equivalent to

$$\exists R\forall x, y\{[R(x) \wedge C(y, x)] \to R(y)\} \wedge \neg R(m) \wedge R(j).$$

i.e., we are interested in verifying whether

$$\forall R\forall C\{\forall x, y\{[R(x) \wedge C(y, x)] \to R(y)\} \wedge \neg R(m) \wedge R(j)\} \to \\ \exists x, y \neg C(y, x), \tag{16}$$

i.e., whether

$$\neg \exists R \exists C \forall x, y\{[C(y, x) \to [R(x) \to R(y)]\} \wedge \neg R(m) \wedge R(j) \wedge \\ \forall x, y C(y, x). \tag{17}$$

According to Theorem 4.2(2), the elimination of $\exists C$ from (17) results in

$$\neg \exists R\forall x, y[R(x) \to R(y)] \wedge \neg R(m) \wedge R(j),$$

which is equivalent to

$$\neg \exists R\forall y[(y = j \vee \exists x R(x)) \to R(y)] \wedge \neg R(m). \tag{18}$$

According to Theorem 4.2(1), the elimination of $\exists R$ from (18) results in

$$\neg \neg \text{LFP } R(y).[y = j \vee \exists x R(x)][m].$$

Applying the decision procedure of Section 3.2, one can easily verify that the above formula is TRUE only when $m = j$.

6 Conclusions

In the current paper we have investigated the fixpoint theory of equality and have shown its applications in automatizing various forms of reasoning in theorem proving, deductive databases and artificial intelligence.

Since many non-classical logics can be translated into the classical logic, the method is applicable to non-classical logics, too.

References

1. S. Abiteboul, R. Hull, and V. Vianu. *Foundations of Databases*. Addison-Wesley Pub. Co., 1996.
2. W. Ackermann. Untersuchungen über das eliminationsproblem der mathematischen logik. *Mathematische Annalen*, 110:390–413, 1935.
3. J. Barwise. On Moschovakis closure ordinals. *Journal of Symbolic Logic*, 42:292–296, 1977.
4. A. Chandra and D. Harel. Computable queries for relational databases. *Journal of Computer and System Sciences*, 21:156–178, 1980.
5. A. Chandra and D. Harel. Structure and complexity of relational queries. *Journal of Computer and System Sciences*, 25:99–128, 1982.
6. P. Doherty, W. Łukaszewicz, and A. Szałas. Computing circumscription revisited. *Journal of Automated Reasoning*, 18(3):297–336, 1997.
7. P. Doherty, W. Łukaszewicz, and A. Szałas. Computing strongest necessary and weakest sufficient conditions of first-order formulas. *International Joint Conference on AI (IJCAI'2001)*, pages 145 – 151, 2000.
8. N. Immerman. Upper and lower bounds for first-order expressibility. *Journal of Computer and System Sciences*, 25:76–98, 1982.
9. J. Kachniarz and A. Szałas. On a static approach to verification of integrity constraints in relational databases. In E. Orłowska and A. Szałas, editors, *Relational Methods for Computer Science Applications*, pages 97–109. Springer Physica-Verlag, 2001.
10. Phokion Kolaitis and Moshe Vardi. Ph. kolaitis and m. vardi. on the expressive power of variable-confined logics. In *Proc. IEEE Conf. Logic in Computer Science*, pages 348–359, 1996.
11. V. Lifschitz. Circumscription. In D. M. Gabbay, C. J. Hogger, and J. A. Robinson, editors, *Handbook of Artificial Intelligence and Logic Programming*, volume 3, pages 297–352. Oxford University Press, 1991.
12. F. Lin. On strongest necessary and weakest sufficient conditions. In A.G. Cohn, F. Giunchiglia, and B. Selman, editors, *Proc. 7th International Conf. on Principles of Knowledge Representation and Reasoning, KR2000*, pages 167–175. Morgan Kaufmann Pub., Inc., 2000.
13. M. Magnusson. DLS*. *http://www.ida.liu.se/labs/kplab/projects/dlsstar/*, 2005.
14. J. McCarthy. Circumscription: A form of non-monotonic reasoning. *Artificial Intelligence Journal*, 13:27–39, 1980.
15. A. Nonnengart, H.J. Ohlbach, and A. Szałas. Elimination of predicate quantifiers. In H.J. Ohlbach and U. Reyle, editors, *Logic, Language and Reasoning. Essays in Honor of Dov Gabbay, Part I*, pages 159–181. Kluwer, 1999.

16. A. Nonnengart and A. Szałas. A fixpoint approach to second-order quantifier elimination with applications to correspondence theory. In E. Orłowska, editor, *Logic at Work: Essays Dedicated to the Memory of Helena Rasiowa*, volume 24 of *Studies in Fuzziness and Soft Computing*, pages 307–328. Springer Physica-Verlag, 1998.
17. B. Poizat. Deux ou trois choses que je sais de L_n. *Journal of Symbolic Logic*, 47:641–658, 1982.
18. A. Szałas. On the correspondence between modal and classical logic: An automated approach. *Journal of Logic and Computation*, 3:605–620, 1993.

Monodic Tree Kleene Algebra

Toshinori Takai[1] and Hitoshi Furusawa[2]

[1] Research Center for Verification and Semantics (CVS),
National Institute of Advanced Industrial Science and Technology (AIST)
[2] Faculty of Science, Kagoshima University

Abstract. We propose a quasi-equational sound axiomatization of regular tree languages, called monodic tree Kleene algebra. The algebra is weaker than Kleene algebra introduced by Kozen. We find a subclass of regular tree languages, for which monodic tree Kleene algebra is complete. While regular tree expressions may have two or more kinds of place holders, the subclass can be equipped with only one kind of them. Along the lines of the original proof by Kozen, we prove the completeness theorem based on determinization and minimization of tree automata represented by matrices on monodic tree Kleene algebra.

1 Introduction

A tree language is a set of first-order terms and a tree automaton is a natural extension of a finite automaton. Instead of strings, inputs of a tree automaton are first-order terms[1]. The class of regular tree languages, which are recognized by tree automata, inherits some desirable properties including complexity of some decision problems and closeness under boolean operations. The goal of our study is to make clear the algebraic structure of regular tree expressions.

In 1998, Ésik[2] proposed a complete axiomatization of regular tree languages based on terms with μ-operators. A μ-operator is the same one in the μ-calculus and it is not first-order. Our interest is to find a first-order axiomatization of regular tree languages. In the last year, the authors proposed essentially algebraic structure of a certain subclass of tree languages[3]. But the subclass and the class of regular tree languages are incomparable.

Although regular tree expressions and Kleene theorem for trees have been proposed[1,4], they are rarely used in practice because the structure is too complicated. Generally, regular tree expressions have two or more multiplication, Kleene stars, and place holders. In this paper, we propose a subclass of regular tree expressions, called *monodic regular tree expressions*, and also propose a complete first-order axiomatization, called a *monodic tree Kleene algebra*, of the subclass of regular tree languages corresponding to monodic regular tree expressions. A monodic regular tree expression has at most one kind of multiplications, Kleene stars, and place holders. The subclass corresponds to tree automata in which only one kind of states occurs in the left-hand side of each transition rule.

A monodic tree Kleene algebra is a similar to a Kleene algebra by Kozen[5,6] and a Kleene algebra is always a monodic tree Kleene algebra. The essential difference is the lack of the right-distributivity of the multiplication over addition

R.A. Schmidt (Ed.): RelMiCS /AKA 2006, LNCS 4136, pp. 402–416, 2006.

$+$. For example, let f be a binary function symbol, a and b be constants. Then $f(\Box, \Box)$ and $a + b$ are regular tree expressions with place holder $\Box$ and so is $f(\Box, \Box) \cdot (a + b)$. The expression $f(\Box, \Box) \cdot (a + b)$ is interpreted as a set of terms obtained by replacing $\Box$ with a or b, i.e. $\{f(a, a), f(a, b), f(b, a), f(b, b)\}$. On the other hand, the interpretation of $f(\Box, \Box) \cdot a + f(\Box, \Box) \cdot b$ is $\{f(a, a), f(b, b)\}$. Another differences are shapes of right-unfold and right-induction laws. We compare the original laws with the new ones.

After giving preliminaries, we define monodic tree Kleene algebras and show some basic properties of them. In Section 4, a subclass of regular tree expressions and a subclass of tree automata are proposed and show the correspondence of them via matrices. Since the proof flow of the completeness theorem is the same as the original one by Kozen[5], this paper concentrates showing the lemmas used in the proof.

2   Preliminaries

We review some notions of the language theory on trees. After defining the syntax of regular tree expressions, we give some functions on tree languages. Using the functions, we give an interpretation function of regular tree expressions.

For a signature Σ, we denote the set of all first-order terms without variables constructed from Σ by T_Σ. A *tree language* is a set of terms. For a signature Σ and a set Γ of *substitution constants*, the set $\mathsf{RegExp}(\Sigma, \Gamma)$ of *regular tree expressions* is inductively defined as follows. A substitution constant is a constant not included in the signature.

1. The symbol $\mathbf{0}$ is a regular tree expression.
2. A term in $\mathsf{T}_{\Sigma \cup \Gamma}$ is a regular tree expression.
3. If e_1 and e_2 are regular tree expressions and $\Box$ is a substitution constant, so are $e_1 + e_2$, $e_1 \cdot_\Box e_2$ and $e_1^{*\Box}$.

Let $\mathscr{P}(S)$ denote the power set of a set S. We define binary function $\circ_\Box$ and unary function $*\Box$ on $\mathscr{P}(\mathsf{T}_{\Sigma \cup \Gamma})$ for any $\Box \in \Gamma$ and define *tree substitutions*. A tree substitution is given by a finite set of pairs of a substitution constant and a tree language. For a tree substitution $\theta = \{(\Box_1, L_1), \ldots, (\Box_n, L_n)\}$ and a term $t \in \mathsf{T}_{\Sigma \cup \Gamma}$, define $\theta(t)$ as follows.

1. If $t \in \Gamma$ and $t = \Box_i$, then $\theta(t) = L_i$.
2. If $t \in \Gamma \setminus \{\Box_i \mid 1 \le i \le n\}$, then $\theta(t) = \{t\}$.
3. If $t = f(t_1, \ldots, t_n)$, then $\theta(t) = \{f(t'_1, \ldots, t'_n) \mid t'_i \in \theta(t_i), 1 \le i \le n\}$.

For tree languages L_1 and L_2 and a substitution constant $\Box$, define $L_1 \circ_\Box L_2$ as follows. If $L_1 = \{t\}$, then $\{t\} \circ_\Box L_2 = \{(\Box, L_2)\}(t)$. If $|L_1| \ne 1$, then $L_1 \circ_\Box L_2 = \bigcup_{t \in L_1} \{t\} \circ_\Box L_2$. Define $L^{*\Box} = \bigcup_{j \ge 0} L^{j, \Box}$ where

$$L^{0, \Box} = \{\Box\}$$
$$L^{n+1, \Box} = L^{n, \Box} \cup L \circ_\Box L^{n, \Box}$$

In the book by Comon *et al.*[1], languages do not contains symbols in Γ. In the survey by Gécseg[4], substitution constants can be elements of languages. This paper follows the latter one because substitution constants are essential in the algebraic structure as we see later.

Using the above functions, we define an interpretation of regular tree expressions by the following function.

$$[\![\_]\!]\colon \mathsf{RegExp}(\Sigma,\Gamma) \to \mathscr{P}(\mathsf{T}_{\Sigma\cup\Gamma})$$

Let e be a regular tree expression.

1. $[\![\mathbf{0}]\!]$ is the empty set.
2. If $e \in \mathsf{T}_{\Sigma\cup\Gamma}$, then $[\![e]\!] = \{e\}$.
3. If e has the form $e_1 + e_2$, then $[\![e_1 + e_2]\!] = [\![e_1]\!] \cup [\![e_2]\!]$.
4. If e has the form $e_1 \cdot_\square e_2$, then $[\![e_1 \cdot_\square e_2]\!] = [\![e_1]\!] \circ_\square [\![e_2]\!]$.
5. If e has the form $e_0^{*\square}$, then $[\![e_0^{*\square}]\!] = [\![e_0]\!]^{*\square}$.

The image of the interpretation function above coincides with the class of *regular tree languages*[1]. Let $\mathsf{Reg}(\Sigma,\Gamma)$ be the set of regular tree languages on signature Σ and set Γ of substitution constants. The definition of $^{*\square}$ can be changed into $L^{*\square} = \bigcup_{j\geq 0} L^{j,\square}$ where

$$L^{0,\square} = \{\square\}$$
$$L^{n+1,\square} = L^{n,\square} \circ_\square (L \cup \{\square\})$$

The proposition below can be shown by induction on n of $L^{n,\square}$ and $L^{n,\square}$.

Proposition 1. *For a tree language L, $L^{*\square} = L^{*\square}$ holds.*   □

3   Monodic Tree Kleene Algebra

In this section, we give an essentially algebraic structure of the subclass of regular tree languages. The definition of the subclass will be shown in the next section. After giving the axioms, we show some basic properties of the algebra.

Definition 1. *A monodic tree Kleene algebra $(A, +, \cdot, {}^*, 0, 1)$ satisfies the following equations and Horn clauses where $\cdot$ is omitted.*

$$a + (b + c) = (a + b) + c \tag{1}$$
$$a + b = b + a \tag{2}$$
$$a + 0 = a \tag{3}$$
$$a + a = a \tag{4}$$
$$a(bc) = (ab)c \tag{5}$$
$$1a = a \tag{6}$$

$$a1 = a \tag{7}$$
$$ac + bc = (a + b)c \tag{8}$$
$$ab + ac \leq a(b + c) \tag{9}$$
$$0a = 0 \tag{10}$$
$$1 + aa^* \leq a^* \tag{11}$$
$$1 + a^*(a + 1) \leq a^* \tag{12}$$
$$b + ax \leq x \rightarrow a^*b \leq x \tag{13}$$
$$b + x(a + 1) \leq x \rightarrow ba^* \leq x \tag{14}$$

The order is defined as $a \leq b$ if $a + b = b$.

Operators $+$, $\cdot$ and * are respectively called an addition, a multiplication and a Kleene star. Axioms (11) and (12) are sometimes called unfold laws. Axioms (13) and (14), which are called induction laws, can be replaced by the following axioms, respectively.

$$ax \leq x \rightarrow a^*x \leq x \tag{15}$$
$$x(a + 1) \leq x \rightarrow xa^* \leq x \tag{16}$$

The equivalence between (13) and (15) can be shown in the same way of Kleene algebras. From (14) to (16), for showing $xa^* \leq x$, it is sufficient to hold that $x + x(a + 1) \leq x$, which can be shown by the assumption. From (16) to (14), we assume $b + x(a + 1) \leq x$. From the assumption, $b \leq x$, $x(a + 1) \leq x$ and $xa^* \leq x$ hold. $ba^* \leq x$ is from $b \leq x$. Remark that if x is either 0 or 1, we have the right-induction law $b + ax \leq x \rightarrow ba^* \leq x$ of Kleene algebras.

We compare to the original Kleene algebras. The proof will be shown later.

Proposition 2. *The right-unfold law $1 + a^*a \leq a^*$ of Kleene algebras is a theorem of monodic tree Kleene algebras but the right-induction law $b + xa \leq x \rightarrow ba^* \leq x$ of Kleene algebras is not a theorem.*   □

The right-unfold law $1 + a^*a \leq a^*$ of Kleene algebras holds according to the axiom $1 + a^*(a + 1) \leq a^*$ with partial right-distributivity (9).

A lazy Kleene algebra by Möller[7] also gives up right-distributivity but does not have right-unfold and right-induction laws.

Lemma 1. *Operations $\cdot$ and * in a monodic tree Kleene algebra are monotone.*

Proof. Assuming $a \leq b$, we show $ac \leq bc$, $ca \leq cb$ and $a^* \leq b^*$. From the fact that $a + b = b$, we have $bc = (a + b)c$. By (8), $ac + bc = bc$ holds and thus $ac \leq bc$. From $a + b = b$, we have $cb = c(a + b)$. By (10), $ca + cb \leq cb$ holds and thus $ca \leq cb$. To obtain $a^* \leq b^*$, we first show $1 + ab^* \leq b^*$ (by (13)). From monotonicity of $+$ and $\cdot$, we have $1 + ab^* \leq 1 + b \cdot b^*$. From (11), $1 + b \cdot b^* \leq b^*$ holds and thus $1 + ab^* \leq b^*$.   □

The theorems below are used in the proof of the completeness theorem.

Lemma 2. *The following are theorems of monodic tree Kleene algebras.*

$$(a+1)^* = a^* \tag{17}$$
$$(a^*)^* = a^* \tag{18}$$
$$1 + aa^* = a^* \tag{19}$$
$$1 + a^*(a+1) = a^* \tag{20}$$
$$(a+b)^* = a^*(ba^*)^* \tag{21}$$
$$(ab)^*a \le a(ba)^* \tag{22}$$

Proof. By (13), to show that $(a+1)^* \le a^*$, it is sufficient to show $1 + (a+1)a^* \le a^*$, which is obtained by distributivity and (11). The inequation $(a^*)^* \le a^*$ can be shown as follows.

$$a^* + aa^* \le a^* \quad \text{(by (11))}$$
$$1 + a^*a^* \le a^* \quad \text{(by (13))}$$
$$(a^*)^* \le a^* \quad \text{(by (13))}$$

By (13), to show $a^* \le 1 + aa^*$, it is sufficient to show $1 + a(1 + aa^*) \le 1 + aa^*$, which is from (11). By (14), to show, $a^* \le 1 + a^*(a+1)$ it is sufficient to show $1 + (1 + a^*(a+1))(a+1) \le 1 + a^*(a+1)$, which is obtained by (12) and monotonicity.

Since the original proof of $(a+b)^* \le a^*(b^*a)^*$ in Kleene algebras does not involve the right-induction law and right-distributivity[5], the above equation also holds in our setting. The inequation $a^*(b^*a)^* \le (a+b)^*$ in (22) can be shown as follows. First, we have $(a^*b)^*a^* \le ((a+b+1)^*(a+b+1))^*(a+b+1)^*$ from monotonicity.

$$
\begin{aligned}
&((a+b+1)^*(a+b+1))^*(a+b+1)^* \\
&\le ((a+b+1)^*(a+b+1+1))^*((a+b+1)^* + 1) \\
&\le ((a+b+1)^*)^*((a+b+1)^* + 1) && \text{(by (12))} \\
&\le ((a+b+1)^*)^* && \text{(by (12))} \\
&\le (a+b+1)^* && \text{(by (18))} \\
&\le (a+b)^* && \text{(by (17))}
\end{aligned}
$$

By (13), to show $(ab)^*a \le a(ba)^*$, it is sufficient to show $a + aba(ba)^* \le a(ba)^*$, which holds by (11).   □

Next, we show the set of regular tree expressions satisfies the axioms of monodic tree Kleene algebras.

Lemma 3. *For two tree languages S and T in $\mathsf{T}_{\Sigma \cup \{\square\}}$, (i) $T^{*\square} \circ_\square S$ is the least fixed point of function $\lambda X.\ S \cup T \circ_\square X$ and (ii) $S \circ_\square T^{*\square}$ is the least fixed point of function $\lambda X.\ S \cup X \circ_\square (T \cup \{\square\})$.*

Proof. In the proof, we write $\circ$ and * for $\circ_\square$ and $^{*\square}$, respectively. (i) $T^* \circ S = S \cup T \circ (T^* \circ S)$ can be shown easily. For tree language α, we assume

$$\alpha = S \cup T \circ \alpha \tag{23}$$

and show that $T^\star \circ S \subseteq \alpha$. For any $n \geq 0$, it is sufficient to show that $T^n \circ S \subseteq \alpha$. For the base case $n = 0$, the lemma holds from (23). Assume the lemma holds for the case $n = i$, and we show the case when $n = i + 1$.

$$T^{i+1} \circ S = (T \cup T \circ T^i) \circ S$$
$$= T \circ S \cup (T \circ T^i) \circ S$$

By inductive hypothesis, we have $T \circ S \subseteq \alpha$. For $(T \circ T^i) \circ S$ by (23), it is sufficient to show $(T \circ T^i) \circ S \subseteq T \circ \alpha$, which can be shown by inductive hypothesis and monotonicity and associativity of $\circ$.

(ii) We can show $S \circ T^\star = S \cup (S \circ T^\star) \circ (T \cup \{\Box\})$ easily. For tree language β, assume

$$\beta = S \cup \beta \circ (T \cup \{\Box\}) \tag{24}$$

and we show $S \circ T^\star \subseteq \beta$. By Proposition 1, we can use another definition of Kleene star. For $n \geq 0$, it is sufficient to show $S \circ T^{\underline{n}} \subseteq \beta$. For the base case $n = 0$, the lemma holds from (24). Assume the lemma holds for $n = i$, we show $n = i + 1$.

$$S \circ T^{\underline{i+1}} = S \circ (T^{\underline{i}} \circ (T \cup \{\Box\}))$$
$$= (S \circ T^{\underline{i}}) \circ (T \cup \{\Box\})$$

The lemma holds from (24), inductive hypothesis and monotonicity of $\circ$. $\square$

We can show that the other axioms from (1) to (10) of monodic tree Kleene algebras are satisfied by tree languages with functions $\cup, \circ_\Box, {}^{\star\Box}$ and constants $\emptyset, \{\Box\}$ via easy observations. Using the lemma above, we have the following theorem.

Theorem 1. *Let Σ be a signature and Γ be a set of substitution constants. For any $\Box \in \Gamma$,*

$$(\text{Reg}(\Sigma, \Gamma), \cup, \circ_\Box, {}^{\star\Box}, \emptyset, \{\Box\})$$

is a monodic tree Kleene algebra. $\square$

Since Lemma 3 does not depend on the regularity of the two languages in the claim, we can obtain the following proposition.

Proposition 3. *Let Σ be a signature and Γ be a set of substitution constants. For any $\Box \in \Gamma$, $(\mathscr{P}(T_{\Sigma \cup \Gamma}), \cup, \circ_\Box, {}^{\star\Box}, \emptyset, \{\Box\})$ is a monodic tree Kleene algebra.* $\square$

Here, we prove Proposition 2 by giving a counterexample. For a tree language L and a substitution constant $\Box$, define $L^{\overline{\star}\Box}$ by $L^{\overline{\star}\Box} = \bigcup_{j \geq 0} L^{\overline{j}, \Box}$ where

$$L^{\overline{0}, \Box} = \{\Box\} \quad \text{and}$$
$$L^{\overline{n+1}, \Box} = L^{\overline{n}, \Box} \circ_\Box L.$$

For tree language $\{f(\Box, \Box)\}$, $\{f(\Box, \Box)\}^{\overline{\star}}$ consists of complete binary trees and we have $\{\Box\} \cup \{f(\Box, \Box)\}^{\overline{\star}} \circ_\Box \{f(\Box, \Box)\} \subseteq \{f(\Box, \Box)\}^{\overline{\star}}$, which corresponds to the assumption of the right-induction law of a Kleene algebra. On the other hand, we can see that $\{f(\square, \square)\}^* \not\subseteq \{f(\square, \square)\}^{\overline{*}}$ since $\{f(\square, \square)\}^*$ contains more terms than complete binary trees.

Next, we introduce matrices on monodic tree Kleene algebras and operations on the matrices. Let K be a monodic tree Kleene algebra and $M(n, K)$ be the class of n by n matrices on K. In the following, we assume E and X are matrices as follows.

$$E = \begin{pmatrix} a & b \\ c & d \end{pmatrix} \qquad X = \begin{pmatrix} x & y \\ z & w \end{pmatrix} \tag{25}$$

The addition and multiplication matrices in $M(n, K)$ is defined in the usual way.

Lemma 4. $M(n, K)$ satisfies axioms (1)–(10).   □

Kleene star is defined essentially in the same way of Kleene algebras.

$$E^* = \begin{pmatrix} a & b \\ c & d \end{pmatrix}^* = \begin{pmatrix} (a + bd^*c)^* & (a + bd^*c)^*bd^* \\ (d + ca^*b)^*ca^* & (d + ca^*b)^* \end{pmatrix}$$

The definition of Kleene star for n by n matrices is inductively given in a similar way of Kleene algebras.

$$\left(\begin{array}{c|c} A & B \\ \hline C & D \end{array} \right)^* = \left(\begin{array}{c|c} (A + BD^*C)^* & (A + BD^*C)^*BD^* \\ \hline (D + CA^*B)^*CA^* & (D + CA^*B)^* \end{array} \right)$$

Lemma 5. For matrices E and X in $M(n, K)$, the matrix E^* satisfies the following monodic tree Kleene algebra axioms (11)–(13) where I is the identity matrix.

$$I + EE^* \leq E^* \tag{26}$$
$$I + E^*(E + I) \leq E^* \tag{27}$$
$$EX \leq X \rightarrow E^*X \leq X \tag{28}$$

Proof. Since in this proof, the axiom (14) is not used, the cases for arbitrary $n \geq 1$ can be shown by induction and thus we give only the base, i.e. $n = 2$. Let E and X be the matrices in (25). The inequation (26) can be written as the following four inequations.

$$1 + a(a + bd^*c)^* + b(d + ca^*b)^*ca^* \leq (a + bd^*c)^*$$
$$a(a + bd^*c)^*bd^* + b(d + ca^*b)^*d^* \leq (a + bd^*c)^*bd^*$$
$$c(a + bd^*c) + d(d + ca^*b)^*ca^* \leq (d + ca^*b)^*ca^*$$
$$1 + c(a + bd^*c)^*bd^* + d(d + ca^*b)^* \leq (d + ca^*b)^*$$

For example, we can show $b(d + ca^*b)^*ca^* \leq (a + bd^*c)^*$ as follows.

$$\begin{aligned} b(d + ca^*b)^*ca^* &\leq bd^*(ca^*bd^*)^*ca^* && \text{(by (21))} \\ &\leq bd^*c(a^*bd^*c)^*a^* && \text{(by (22))} \\ &\leq bd^*ca^*(bd^*ca^*)^* && \text{(by (22))} \\ &\leq (a + bd^*c)a^*(bd^*ca^*)^* \\ &= (a + bd^*c)(a + bd^*c)^* \\ &\leq (a + bd^*c)^* \end{aligned}$$

The rests can be shown in similar ways.

The inequation (27) consists of the following four inequations and each of them can be easily shown.

$$1 + (a + bd^*c)^*(a+1) + (a + bd^*c)^*bd^*c \leq (a + bd^*c)^*$$
$$(a + bd^*c)^*b + (a + bd^*c)^*bd^*(d+1) \leq (a + bd^*c)^*bd^*$$
$$(d + ca^*b)^*ca^*(a+1) + (d + ca^*b)^*c \leq (d + ca^*b)^*c$$
$$1 + (d + ca^*b)ca^*b + (d + ca^*b)^*(d+1) \leq (d + ca^*b)^*$$

For (28), we show that the assumptions $ax + by \leq x$ and $cx + dy \leq y$ imply the following inequations.

$$(a + bd^*c)^*x + (a + bd^*c)^*bd^*y \leq x$$
$$(d + ca^*b)^*ca^*x + (d + ca^*b)^*y \leq y$$

We only show $(a + bd^*c)^*x \leq x$.

$$
\begin{array}{ll}
d^*y \leq y & (dy \leq y \text{ and } (15)) \\
bd^*y \leq x & (by \leq x) \\
bd^*cx \leq x & (cx \leq y) \\
ax + bd^*cx \leq x & (ax \leq x) \\
(a + bd^*c)x \leq x & \\
(a + bd^*c)^*x \leq x &
\end{array}
$$

□

Consequently, the set of matrices on a monodic tree Kleene algebra satisfies all the axioms of a monodic tree Kleene algebra except for (14).

According to Lemmas 4 and 5, we can see that matrices on a monodic tree Kleene algebra has a monodic tree Kleene algebra *like* structure. Some of theorems of monodic tree Kleene algebras also hold in $M(n, K)$.

Lemma 6. *(i) Operations* · *and* * *on* $M(n, K)$ *is monotone. (ii) The equations and inequations (17)–(19) and (21)–(22) hold in* $M(n, K)$. □

A matrix in which any entries are either 0 or 1 is called a 0-1 *matrix*. Althouh the lemma below mentions properties of $M(n, K)$ concerning 0-1 matrices, each statement in the lemma can also be applied to 0-1 vectors.

Lemma 7. *Let* X *be a 0-1 matrix,* P *be a permutation matrix and* A *and* B *be matrices, then the following equations hold where* P^T *is the transpose of* P.

$$X(A + B) = XA + XB \tag{29}$$
$$XA \leq X \rightarrow XA^* \leq X \tag{30}$$
$$B + XA \leq X \rightarrow BA^* \leq X \tag{31}$$
$$AX = XB \rightarrow A^*X = XB^* \tag{32}$$
$$X(AX)^* = (XA)^*X \tag{33}$$
$$(P^T AP)^* = P^T A^* P \tag{34}$$

Proof. Since in this proof, the axiom (14) is only used for the base case and for the inductive step, (31) can be used as an inductive hypothesis. Hence, we give only the base $n = 2$. The first one is obvious. Let E and X be the matrices in (25). To prove (30), we show the following statement.

$$X(A + I) \leq X \rightarrow XA^* \leq X \tag{35}$$

The left-hand side of (35) consists of the following inequations.

$$x(a + 1) + yc \leq x$$
$$xb + y(d + 1) \leq y$$
$$z(a + 1) + wc \leq z$$
$$zb + w(d + 1) \leq w$$

The right-hand side consists the following inequations.

$$x(a + bd^*c)^* + y(d + ca^*b)^*ca^* \leq x$$
$$x(a + bd^*c)^*bd^* + y(d + ca^*b)^* \leq y$$
$$z(a + bd^*c)^* + w(d + ca^*b)^*ca^* \leq z$$
$$z(a + bd^*c)^*bd^* + w(d + ca^*b)^* \leq z$$

For example, $x(a + bd^*c)^* \leq x$ can be shown as follows.

$$yd^* \leq y \quad (y(d + 1) \leq y)$$
$$yd^*c \leq x \quad (yd^* \leq y \text{ and } yc \leq x)$$
$$xbd^*c \leq x \quad (yd^*c \leq x \text{ and } xb \leq y)$$
$$x(a + bd^*c + 1) \leq x \quad (xbd^*c \leq x \text{ and } x(a + 1) \leq x)$$
$$x(a + bd^*c)^* \leq x \quad (x(a + bd^*c + 1) \leq x)$$

The last step is by the axiom (14) for the base case or by the inductive hypothesis of for the induction step. Finally, we can see that $EX \leq X$ implies $E(X+I) \leq X$ and the lemma holds.

The Horn clause (31) is directly obtained from (30).

The simulation law (32) can be shown in the same way of the original proof by Kozen[5], since as we have shown that the Kleene algebra axioms (Lemma 5 and (31)) hold in our setting if X is restricted to a 0-1 matrix.

The shift law for specific case (33), i.e. $X(AX)^* = (XA)^*X$, is obtained from (32) by replacing A with XA and B with AX, respectively.

To prove (34), we show $A^*P = P(P^TAP)^*$. Multiplying P^T from left and the facts that $P^TP = I$ and $PP^T = I$, we obtain $(P^TAP)^* = P^TA^*P$. By Lemma 6, we obtain $(PP^TA)^*P \leq P(P^TAP)^*$. Since P is a 0-1 matrix, by (31) in this lemma, to show $P(P^TAP)^* \leq (PP^TA)^*P$, it is sufficient to show that $P + (PP^TB)^*(PP^TB)P \leq (PP^TB)^*P$. □

4   Subclass of Regular Tree Expressions

In this section, we give subclass of regular tree expressions, called monodic regular tree expressions.

Definition 2. *Let Σ be a signature and $\square$ be a substitution constant. The set* RegExp$(\Sigma, \square)$ *of monodic regular tree expressions is defined as follows.*

1. *The symbol $\mathbf{0}$ is a monodic regular tree expression.*
2. *A term of the form $f(\square, \ldots, \square)$ is a monodic regular tree expression.*
3. *If e_1 and e_2 are monodic regular tree expressions, so are $e_1 + e_2$, $e_1 \cdot e_2$ and e_1^*.*

The set of monodic regular tree expressions is a subclass of regular tree expressions when the multiplication $\cdot$ and the Kleene star $*$ are regarded as $\cdot_\square$ and $*^\square$, respectively. The interpretation of monodic regular tree expressions is given by functions $\cup$, $\circ_\square$ and $*^\square$ on tree languages.

Definition 3. *A tree language L which can be expressed by a monodic regular tree expressions is called* monodic regular.

We denote the set of all monodic regular tree languages by Reg$(\Sigma, \square)$.

Proposition 4. *(i) The set Reg$(\Sigma, \square)$ of monodic regular tree language is closed under functions $\cup$, $\circ_\square$ and $*^\square$. (ii) (Reg$(\Sigma, \square), \cup, \circ_\square, *^\square, \emptyset, \{\square\}$) is a monodic tree Kleene algebra.* $\square$

Example 1. A regular tree expression $f(\square, \square) \cdot_\square (g(\square) + h(\square, \square))$ is monodic but $f(a, c)$ and $(f(\square, \square_1) \cdot_\square a) \cdot_{\square_1} c$ are not. $\square$

Next, we introduce the subclass of tree automata corresponding to monodic regular tree languages. For the definition of behaviors of tree automata, please refer to the books[1,4].

Definition 4 ([1]). *A tree automaton is a tuple $(\Sigma, Q, Q_{final}, \Delta)$ where Σ is a signature, Q is a finite set of states, $Q_{final} \subseteq Q$ is a set of final states and Δ is a set of transition rules. A transition rule has the form either $f(q_1, \ldots, q_n) \to q$ or $q' \to q$ where $f \in \Sigma_n, q_1, \ldots, q_n, q, q' \in Q$.*

Definition 5. *A tree automaton in which the left-hand side of each transition rule has only one kind of states is called* monodic.

In the following, we give a matrix representation of monodic tree automaton $A = (\Sigma, Q, Q_{final}, \Delta)$. Without loss of generality, Q can be written as $\{1, \ldots, n\}$ for some integer $n \geq 1$. Let M_A be a matrix in M$(n + 1, $RegExp$(\Sigma, \square))$ where the (p, q) entry is given by the following formula.

$$\sum \{f(\square, \ldots, \square) \mid f(q, \ldots, q) \to p \in \Delta, \texttt{arity}(f) \geq 1\} \cup$$
$$\{\square \mid q \to p \in \Delta\} \cup \{c \mid c \to p \in \Delta, q = n + 1\}$$

Let v be a $n+1$ vector in which the $n+1$-th row is $\square$ and the others are 0 and u be a $n+1$ vector in which for each $q \in Q_{final}$, the q-th row is $\square$ and the others are 0. The triple (v, M_A, u) is called a *matrix representation* of A where v and u are vectors and M_A is a matrix on the free monodic tree Kleene algebra over Σ and $\square$, i.e. the quotient of monodic regular tree expressions modulo provable equivalence. We call the vector u the *final vector* and v the *initial vector*.

Example 2. Let A_0 be a tree automaton in which Δ consists of the following transition rules

$$f(1,1) \to 1 \quad g(1,1) \to 2 \quad a \to 1 \quad b \to 1 \quad b \to 2$$

and the final state is just 2. Then the matrix representation is as follows.

$$\left(\begin{pmatrix} 0 \\ 0 \\ \square \end{pmatrix}, \begin{pmatrix} f(\square,\square) & 0 & a+b \\ g(\square,\square) & 0 & b \\ 0 & 0 & 0 \end{pmatrix}, \begin{pmatrix} 0 \\ \square \\ 0 \end{pmatrix} \right)$$

Since $Q = \{1,2\}$, the matrix has size $2+1 = 3$. $\square$

Next, we justify the matrix representation of tree automata via regular tree equation systems[1]. Let $X_1, \ldots, X_n$ be variables and $s_{i,j}$ ($1 \le i \le m_j, 1 \le j \le p$) be terms in $T_\Sigma(\{X_1, \ldots, X_n\})$, which is the set of terms with variables $\{X_1, \ldots, X_n\}$. A *regular tree equation system* S is given by the following set of equations.

$$X_1 = s_{1,1} + \cdots + s_{m_1,1}$$
$$\cdots$$
$$X_p = s_{1,p} + \cdots + s_{m_p,p}$$

A *solution of* S is a p-tuple of tree languages $(L_1, \ldots, L_p) \in \mathscr{P}(T_\Sigma)^p$ satisfying the following condition.

$$L_1 = \theta(s_{1,1}) \cup \cdots \cup \theta(s_{m_1,1})$$
$$\cdots$$
$$L_p = \theta(s_{1,p}) \cup \cdots \cup \theta(s_{m_p,p})$$

where θ is a tree substitution $\{(X_1, L_1), \ldots, (X_p, L_p)\}$. For regular tree equation system S, we define $\hat{S} : \mathscr{P}(T_\Sigma)^p \to \mathscr{P}(T_\Sigma)^p$ as $\lambda(L_1, \ldots, L_p).(L'_1, \ldots, L'_p)$ where

$$L'_1 = L_1 \cup \theta(s_{1,1}) \cup \cdots \cup \theta(s_{m_1,1}),$$
$$\cdots$$
$$L'_p = L_p \cup \theta(s_{1,p}) \cup \cdots \cup \theta(s_{m_p,p})$$

and $\theta = \{(X_1, L_1), \ldots, (X_p, L_p)\}$. The order on $\mathscr{P}(T_\Sigma)^p$ is defined component-wise.

Lemma 8 ([1]). *For regular tree equation system* S, *the least fixed-point of* $\hat{S}$ *is the least solution of* S. $\square$

Theorem 2 ([1]). *For any regular tree equations, the least solution is a regular tree language. Conversely, for any regular tree language, there exists a regular tree equations representing the regular tree language.*   □

In the proof of the above theorem, tree automaton $A = (\Sigma, Q, Q_{final}, \Delta)$ with $Q = \{1, \ldots, n\}$ is translated into regular tree equation system S_A consisting of

$$i = \sum \{l \mid l \to i \in \Delta\} \qquad (1 \le i \le n)$$

where states are regarded as variables. For instance, a tree automaton consisting of

$$f(1,2) \to 1 \quad g(2,1) \to 2 \quad a \to 1 \quad b \to 1 \quad b \to 2$$

corresponds to the following regular tree equation system.

$$1 = f(1,2) + a + b$$
$$2 = g(2,1) + b$$

If A is monodic, then we can obtain another definition of $\hat{S}_A$ as

$$\lambda(L_1, \ldots, L_p).\, (L'_1, \ldots, L'_p)$$

where

$$L'_1 \;=\; L_1 \cup \theta_1(s_{1,1}) \cup \cdots \cup \theta_1(s_{p,1}),$$
$$\cdots$$
$$L'_p \;=\; L_p \cup \theta_1(s_{1,p}) \cup \cdots \cup \theta_p(s_{p,p})$$

and $\theta_n = \{(n, L_n)\}$ for $1 \le n \le p$. For a substitution constant $\square$, we have $\theta_n = \{(n, \{\square\})\} \circ \{(\square, L_n)\}$ for $1 \le n \le p$.

Summarizing the observations above, the fixed-point operator $\hat{S}_A$ of the above regular tree equation system S_A can be written as the matrix

$$X = (P + I + C)X = (P + I)X + C$$

where I is the identity matrix of size n, $X = (X_1, \ldots, X_n)^T$, $C = (C_1, \ldots, C_n)^T$, $C_i = \sum\{c \mid c \to i \in \Delta, c \in \Sigma_0\}$ and $P_{i,j} = \sum\{f(\square, \ldots, \square) \mid f(j, \ldots, j) \to i \in \Delta, f \in \Sigma\}$. The matrix $P + I$ corresponds to the fixed-point operator. The least solution is given by the least-fixed point and thus we have $(P + I)^*C = P^*C$. This is the language represented by the tree automaton A. More precisely, the sequence of languages represented by each state of A. Although A can be represented by matrices C and P and final 0-1 vector U, for the discussions below initial vectors also have to be given by 0-1 vectors. Henceforth, we give an initial vector, a final vector and a matrix M_A as follows.

$$\left(\begin{pmatrix} \mathbf{0} \\ \hline 1 \end{pmatrix}, \left(\begin{array}{c|c} P & C \\ \hline \mathbf{0} & 0 \end{array} \right), \begin{pmatrix} U \\ \hline 0 \end{pmatrix} \right)$$

This matrix-represented automaton corresponds to the following regular tree equation system in which entries of C are produced from the new variable x_0.

$$\begin{pmatrix} X \\ x_0 \end{pmatrix} = \begin{pmatrix} 1 \\ \mathbf{0} \end{pmatrix} + \left(\left(\begin{array}{c|c} P & C \\ \hline 0 & 0 \end{array} \right) + I \right) \begin{pmatrix} X \\ x_0 \end{pmatrix}$$

The least-fixed point can be computed as follows.

$$\left(\begin{array}{c|c} P & C \\ \hline 0 & 0 \end{array} \right)^* \begin{pmatrix} \mathbf{0} \\ 1 \end{pmatrix} = \left(\begin{array}{c|c} P^* & P^*C \\ \hline 0 & 1 \end{array} \right) \begin{pmatrix} \mathbf{0} \\ 1 \end{pmatrix} = \begin{pmatrix} P^*C \\ 1 \end{pmatrix}$$

Using the initial vector, we have the following expressions.

$$(U | 0) \begin{pmatrix} P^*C \\ 1 \end{pmatrix}$$

Finally, by the final vector, we can retrieve the language represented by the tree automaton.

In the following, we also deal with languages including substitution constants. This means that we also consider tree automata in which a initial vector may not only be of the form $(1, \mathbf{0})^T$ but also any 0-1 vectors.

Lemma 9. *For a tree automaton A and its matrix representation (v, M_A, u), the language accepted by A is $[u^T M_A^* v]$.*

Proof. (sketch) We regard the tree automaton as a regular tree equation system, then the lemma holds because of Lemmas 3 and 8 and the discussion after Theorem 2. □

Example 3. Let A_0 be the tree automaton considered in Example 2. Then the corresponding expression can be obtained as follows.

$$
\begin{aligned}
M_{A_0}^* &= \left(\begin{array}{cc|c} f(\square,\square) & 0 & a+b \\ g(\square,\square) & 0 & b \\ \hline 0 & 0 & 0 \end{array} \right)^* \\
&= \left(\begin{array}{c|c} (A+BD^*C)^* & (A+BD^*C)^*BD^* \\ (D+CA^*B)^*CA^* & (D+CA^*B)^* \end{array} \right) \\
&= \left(\begin{array}{c|c} A^* & A^*B \\ \hline 0 & 0 \end{array} \right) \\
&= \left(\begin{array}{ccc} f(\square,\square)^* & f(\square,\square)^* f(\square,\square)^* \cdot (a+b) + f(\square,\square)^* \cdot b \\ g(\square,\square) \cdot f(\square,\square)^* & \square & g(\square,\square) \cdot f(\square,\square)^* \cdot (a+b) + b \\ 0 & 0 & 0 \end{array} \right)
\end{aligned}
$$

Finally, we have the following expression.

$$(0 \, \square \, 0) \, M_{A_0}^* \begin{pmatrix} 0 \\ 0 \\ \square \end{pmatrix} = g(\square,\square) \cdot f(\square,\square)^* \cdot (a+b) + b$$

□

The proof of the following theorem follows the original one by Kozen[5] as follows.

1. First, we construct tree automata for given two regular tree expressions.
2. Second, we translate the tree automata for deterministic ones.
3. Then, we minimamize the tree automata.

In the proof, we use the following lemmas.

- Lemma 2 of monodic tree Kleene algebras
- Lemmas 4, 5 and 6 of arbitrary matrices on a monodic tree Kleene algebra
- Lemma 7 of 0-1 matrices on a monodic tree Kleene algebra

Theorem 3. *Let α and β be monodic regular tree expressions such that $[\alpha] = [\beta]$ and $[\alpha] \subseteq T_\Sigma$. Then $\alpha = \beta$ is a theorem of monodic tree Kleene algebras.* $\square$

5   Remarks and Future Work

In this paper, we have not yet considered the independence of the axioms in Definition 1. Since for defining the class of regular tree languages, only one function in the statement of Lemma 3 is needed, the axiom (14) may be redundant. Moreover, the axiom (14) is used in the proof of completeness theorem only in the case that x is a 0-1 matrix. However, the argument of the independence does not affect the soundness and the completeness theorems (Theorems 1 and 3).

After submitting the paper, we learned that almost the same system has been proposed by McIver and Weber, called a *probabilistic Kleene algebra*[8], for analyzing probabilistic distributed systems. In this conference, McIver, Cohen and Morgan show that probabilistic Kleene algebras can be used for protocol verification[9]. A probabilistic Kleene algebra has the same axioms except for it includes the left annihilation law, i.e. $a0 = 0$.

The class of monodic regular tree languages is given by monodic regular tree expressions, i.e. the number of kinds of place-holders is restricted to one. We conjecture that the expressive power of the class coincides with the subclass of regular tree expressions defined below. Let Σ be a signature and a set Γ of substitution constants. The set of *essentially monodic regular tree expressions* is defined as follows.

1. The symbol **0** is an essentially monodic regular tree expression.
2. A term of the form $f(\Box, \ldots, \Box)$ is an essentially monodic regular tree expression for any $\Box \in \Gamma$.
3. If e_1 and e_2 are essentially monodic regular tree expressions, so are $e_1 + e_2$, $e_1 \cdot_\Box e_2$ and $e_1^{*\Box}$ for any $\Box \in \Gamma$.

For dealing with the whole class of regular tree expressions, there may be two directions. The first one is to use modal Kleene algebras[10]. A tree can be encoded with two modalities in a modal Kleene algebra. Another direction is to consider products of two monodic tree Kleene algebras. The whole class of regular tree expressions seems a *many-sorted* monodic tree Kleene algebra.

Acknowledgments

The authors appreciate Georg Struth, who visited us with the grant from the International Information Science Foundation (IISF), for a lot of his valuable comments to this study. We also thank to Yasuo Kawahara and Yoshihiro Mizoguchi for fruitful discussions on this study. This research was supported by Core Research for Evolutional Science and Technology (CREST) Program "New High-performance Information Processing Technology Supporting Information-oriented Society" of Japan Science and Technology Agency (JST).

References

1. Comon, H., Dauchet, M., Gilleron, R., Jacquemard, F., Lugiez, D., Tison, S., Tommasi, M.: Tree automata techniques and applications. Available on: http://www.grappa.univ-lille3.fr/tata/ (1997)
2. Ésik, Z.: Axiomatizing the equational theory of regular tree languages (extended anstract). In Morvan, M., Meinel, C., Krob, D., eds.: STACS. Volume 1373 of Lecture Notes in Computer Science., Springer (1998) 455–465
3. Takai, T., Furusawa, H., Kahl, W.: Reasoning about term rewriting in Kleene categories with converse. In Düntsch, I., Winter, M., eds.: Proceedings of the 3rd Workshop on Applications of Kleene algebera. (2005) 259–266
4. Gécseg, F., Steinby, M.: Tree languages. In: Handbook of formal languages, 3: beyond words. Springer (1997) 1–68
5. Kozen, D.: A completeness theorem for Kleene algebras and the algebra of regular events. In Kahn, G., ed.: Proceedings of the Sixth Annual IEEE Symp. on Logic in Computer Science, LICS 1991, IEEE Computer Society Press (1991) 214–225
6. Kozen, D.: A completeness theorem for Kleene algebras and the algebra of regular events. Information and Computation **110**(2) (1994) 366–390
7. Möller, B.: Lazy Kleene algebra. In Kozen, D., Shankland, C., eds.: MPC. Volume 3125 of Lecture Notes in Computer Science., Springer (2004) 252–273
8. McIver, A., Weber, T.: Towards automated proof support for probabilistic distributed systems. In Sutcliffe, G., Voronkov, A., eds.: LPAR. Volume 3835 of Lecture Notes in Computer Science., Springer (2005) 534–548
9. McIver, A., Cohen, E., Morgan, C.: Using probabilistic Kleene algebra for protocol verification. In: Relations and Kleene Algebra in Computer Science. Volume 4136 of Lecture Notes in Computer Science. (2006)
10. Möller, B., Struth, G.: Modal Kleene algebra and partial correctness. In Rattray, C., Maharaj, S., Shankland, C., eds.: AMAST. Volume 3116 of Lecture Notes in Computer Science., Springer (2004) 379–393

Weak Relational Products

Michael Winter*

Department of Computer Science,
Brock University,
St. Catharines, Ontario, Canada, L2S 3A1
mwinter@brocku.ca

Abstract. The existence of relational products in categories of relations is strongly connected with the representability of that category. In this paper we propose a canonical weakening of the notion of a relational product. Unlike the strong version, any (small) category of relations can be embedded into a suitable category providing all weak relational products. Furthermore, we investigate the categorical properties of the new construction and prove several (weak) versions of propositions well-known for relational products.

1 Introduction

The relational product of two objects A and B in a category of relations is an abstract version of the cartesian product of two sets. It is characterized by an object $A \times B$ together with two projections π and ρ from $A \times B$ to A and B, respectively. A category of relations may provide a relational product for every pair of objects. In this case, it can be shown that the category is representable, i.e. there is an embedding into the category **Rel** of sets and relations. On the other hand, not every reasonable category of relations is representable. This indicates, that one cannot always embed the given structure into a category that provides relational products. This is a major disadvantage of this construction since products are usually needed to model certain concepts by relations such as programming languages and most kinds of logics. Other constructions usually required such as sums and powers, i.e. the counterparts of disjoint unions and powersets, can always be created.

In this paper we propose a canonical weakening of the concept of a relational product, the weak relational product. This will be done within the theory of allegories - a categorical model of relations. We will investigate certain properties of the new construction and compare them to those of relational products. In particular, we are interested in the following list of properties. Notice, that those properties are not necessarily independent.

- *Product in* $\mathrm{MAP}(\mathcal{R})$*:* The given construction may establish a product in the subcategory of mappings (in the sense of category theory). If valid, this

---

* The author gratefully acknowledges support from the Natural Sciences and Engineering Research Council of Canada.

R.A. Schmidt (Ed.): RelMiCS /AKA 2006, LNCS 4136, pp. 417–431, 2006.

property ensures that the corresponding concept is suitable as an abstract version of cartesian products of sets. Therefore, it is essential for any notion of products.

- $\mathcal{R}$ *representable:* A category of relations might be representable, i.e. the morphisms are (up to a suitable mapping) concrete relations between sets. There are non-representable allegories. Depending on the product construction considered the existence of **all** possible products may force the allegory to be representable.
- *Unsharpness property:* The unsharpness property is a violation of an equality in terms of relational products. It was claimed [2] that this property may be important to model certain behavior of concurrent processes.
- *Embedding property:* It might be possible to embed a given allegory into another allegory providing all products of a certain kind. With this property we refer to whether this can be always done.
- *Equational theory:* Since the theory of allegories and several of its extensions are/can be defined as an equational theory it is interesting whether a given concept of products can also be expressed by equations.

In the following table we have summarized the validity of the properties above within the concepts of relational and weak relational products.

Property	Relational Product	Weak Relational Product
Product in $\mathrm{MAP}(\mathcal{R})$	+	+
$\mathcal{R}$ representable	+	-
Unsharpness property	-	+
Embedding property	-	+
Equational theory	+	$-/+^*$

$*$ can be defined by equations in division allegories where all partial identities split

Fig. 1. Properties of relational products

In addition to the properties of Table 1 we are going to prove several (weak) versions of propositions well-known for relational products.

2   Relational Preliminaries

Throughout this paper, we use the following notation. To indicate that a morphism R of a category $\mathcal{R}$ has source A and target B we write $R : A \to B$. The collection of all morphisms $R : A \to B$ is denoted by $\mathcal{R}[A, B]$ and the composition of a morphism $R : A \to B$ followed by a morphism $S : B \to C$ by $R; S$. Last but not least, the identity morphism on A is denoted by $\mathbb{I}_A$.

We recall briefly some fundamentals on allegories [5] and relational constructions within them. For further details we refer to [5,9,10]. Furthermore, we assume that the reader is familiar with the basic notions from category theory such as products and co-products. For unexplained material we refer to [1].

Definition 1. *An allegory $\mathcal{R}$ is a category satisfying the following:*

1. *For all objects A and B the collection $\mathcal{R}[A, B]$ is a meet semi-lattice, whose elements are called relations. Meet and the induced ordering are denoted by $\sqcap$ and $\sqsubseteq$, respectively.*
2. *There is a monotone operation $\smile$ (called converse) such that for all relations $Q : A \to B$ and $R : B \to C$ the following holds: $(Q;R)^{\smile} = R^{\smile};Q^{\smile}$ and $(Q^{\smile})^{\smile} = Q$.*
3. *For all relations $Q : A \to B$, $R, S : B \to C$ we have $Q;(R \sqcap S) \sqsubseteq Q;R \sqcap Q;S$.*
4. *For all relations $Q : A \to B$, $R : B \to C$ and $S : A \to C$ the modular law $Q;R \sqcap S \sqsubseteq Q;(R \sqcap Q^{\smile};S)$ holds.*

An allegory is called a distributive allegory if

5. *the collection $\mathcal{R}[A, B]$ is a distributive lattice with a least element. Join and the least element are denoted by $\sqcup$ and $\bot\!\!\!\bot_{AB}$, respectively.*
6. *For all relations $Q : A \to B$ and objects C we have $Q;\bot\!\!\!\bot_{BC} = \bot\!\!\!\bot_{AC}$.*
7. *For all relations $Q : A \to B$, $R, S : B \to C$ we have $Q;(R \sqcup S) = Q;R \sqcup Q;S$.*

A distributive allegory is called locally complete iff each $\mathcal{R}[A, B]$ is a complete lattice. Finally, a division allegory is a distributive allegory with a binary operation $/$ satisfying the following:

8. *For all relations $R : B \to C$ and $S : A \to C$ there is a relation $S/R : A \to B$ (called the left residual of S and R) such that for all $Q : A \to B$ the following holds: $Q;R \sqsubseteq S \iff Q \sqsubseteq S/R$.*

If $\mathcal{R}[A, B]$ has a greatest element it is denoted by $\top\!\!\!\top_{AB}$.

Notice, that allegories and distributive allegories are defined by equations. The same can be done for division allegories [5].

The left residual can be used to define another residual operation $Q\backslash S := (S^{\smile}/Q^{\smile})^{\smile}$, called the right residual of S and Q. A symmetric version, called the symmetric quotient, of the residuals may be defined as

$$\mathrm{syq}(Q, R) := (Q\backslash R) \sqcap (Q^{\smile}/R^{\smile}).$$

For further properties of relations in allegories we refer to [5,9,10].

An important class of relations is given by mappings.

Definition 2. *Let $Q : A \to B$ be a relation. Then we call*

1. *Q univalent (or functional) iff $Q^{\smile};Q \sqsubseteq \mathbb{I}_B$,*
2. *Q total iff $\mathbb{I}_A \sqsubseteq Q;Q^{\smile}$,*
3. *Q a map (or a mapping) iff Q is univalent and total,*
4. *Q injective iff $Q^{\smile}$ is univalent,*
5. *Q surjective iff $Q^{\smile}$ is total.*

In the next lemma we have summarized some properties of the residuals and the symmetric quotient. Proofs can be found in [5,4,9,10].

Lemma 1. *Let $\mathcal{R}$ be a division allegory and $Q : A \to B$, $R : A \to C$, $S : A \to D$ be relations, and $f : D \to A$ be a mapping. Then we have*

1. $Q; (Q \backslash R) \sqsubseteq R$,
2. $f; \mathrm{syq}(Q, R) = \mathrm{syq}(Q; f^{\smile}, R)$,
3. $\mathrm{syq}(Q, R)^{\smile} = \mathrm{syq}(R, Q)$,
4. $\mathrm{syq}(Q, R); \mathrm{syq}(R, S) \sqsubseteq \mathrm{syq}(Q, S)$,
5. *if $\mathrm{syq}(Q, R)$ is total then equality holds in 4.,*
6. *if $\mathrm{syq}(Q, R)$ is surjective then $Q; \mathrm{syq}(Q, R) = R$.*

In the next lemma we have collected several properties of univalent relations used in this paper. A proof can be found in [9,10].

Lemma 2. *Let $\mathcal{R}$ be an allegory so that $\mathbb{T}_{AA}$ exists, $Q : A \to B$ be univalent, $P : A \to B$, $R, S : B \to C$, $T : C \to B$ and $U : C \to A$. Then we have*

1. $Q; (R \sqcap S) = Q; R \sqcap Q; S$,
2. $(T; Q^{\smile} \sqcap U); Q = T \sqcap U; Q$,
3. $\mathbb{T}_{AA}(Q^{\smile} \sqcap P) \sqcap Q^{\smile} = Q^{\smile} \sqcap P$.

The collection of all mappings of a division allegory $\mathcal{R}$ constitutes a subcategory and is denoted by $\mathrm{MAP}(\mathcal{R})$.

The subcategory of mappings may provide products (in the sense of category theory) for certain objects. As mentioned in the introduction any useful concept of products should establish a categorical product in $\mathrm{MAP}(\mathcal{R})$. Notice, that $\mathcal{R}$ itself may have categorical products. But, contrary to the products in $\mathrm{MAP}(\mathcal{R})$, those products are not suitable to provide an abstract description of pairs. Any allegory is self-dual (i.e. isomorphic to its co-category) by the converse operation $\smile$, which implies that products and co-products coincide. Therefore, they are called bi-products, and they are related to the relational sums defined below, which constitutes the abstract counterpart of a disjoint union.

Definition 3. *Let $\{A_i \mid i \in I\}$ be a set of objects of a locally complete distributive allegory indexed by some set I. An object $\sum_{i \in I} A_i$, together with relations $\iota_j \in \mathcal{R}[A_j, \sum_{i \in I} A_i]$ for all $j \in I$, is called a relational sum of $\{A_i \mid i \in I\}$ iff for all $i, j \in I$ with $i \neq j$ the following holds*

$$\iota_i; \iota_i^{\smile} = \mathbb{I}_{A_i}, \qquad \iota_i; \iota_j^{\smile} = \mathbb{\bot\!\!\!\bot}_{A_i A_j}, \qquad \bigsqcup_{i \in I} (\iota_i^{\smile}; \iota_i) = \mathbb{I}_{\sum_{i \in I} A_i}.$$

$\mathcal{R}$ has (binary) relational sums iff for every (pair) set of objects the relational sum does exist.

The relational sum is a categorical product and co-product, and hence, unique up to isomorphism. In **Rel** the relational sum is given by the disjoint union of sets and the corresponding injection functions.

Definition 4. *Let* $Q : A \to A$ *be a symmetric idempotent relation, i.e.,* $Q^\smile = Q$ *and* $Q;Q = Q$. *An object* B *together with a relation* $R : B \to A$ *is called a splitting of* Q *(or* R *splits* Q*) iff* $R; R^\smile = \mathbb{I}_B$ *and* $R^\smile; R = Q$.

In **Rel** the splitting of Q is given by the set of equivalence classes (note that it is not assumed that Q is reflexive, so the union of the equivalence classes is in general just a subset of A), and R relates each equivalence class to its elements. A splitting is unique up to isomorphism.

The last construction we want to introduce is the abstract counterpart of a power set - the relational power.

Definition 5. *Let* $\mathcal{R}$ *be a division allegory. An object* $\mathcal{P}(A)$, *together with a relation* $\varepsilon : A \to \mathcal{P}(A)$ *is called a relational power of* A *iff*

$$\mathrm{syq}(\varepsilon, \varepsilon) \sqsubseteq \mathbb{I}_{\mathcal{P}(A)} \quad and \quad \mathrm{syq}(R, \varepsilon) \text{ is total}$$

for all relation $R : B \to A$. *If the relational power does exist for any object then* $\mathcal{R}$ *is called a power allegory.*

Notice, that $\mathrm{syq}(\varepsilon, \varepsilon) = \mathbb{I}_{\mathcal{P}(A)}$, and that $\mathrm{syq}(R, \varepsilon)$ is, in fact, a mapping. In **Rel** the relation $e_A := \mathrm{syq}(\mathbb{I}_A, \varepsilon) : A \to \mathcal{P}(A)$ maps each element to the singleton set containing that element. This relation is always (in all allegories) an injective mapping (cf. [10]).

Definition 6. *An allegory is called systemic complete iff it is a power allegory that has relational sums, and in which all symmetric idempotent relations split.*

The univalent part $\mathrm{unp}(R)$ of a relation R was introduced in [9] in the context of (heterogeneous) relation algebras, i.e. division allegories where the order structure is a complete atomic Boolean algebra.

Definition 7. *Let* $\mathcal{R}$ *be a division allegory, and let be* $R : A \to B$ *in* $\mathcal{R}$. *The univalent part of* R *is defined by* $\mathrm{unp}(R) := R \sqcap (R^\smile \backslash \mathbb{I}_B)$.

The following lemma was already proved in [9]. The proof provided there makes use of complements, which are not available in an arbitrary division allegory. Here we provide a complement free proof.

Lemma 3. *Let* $\mathcal{R}$ *be a division allegory and* $R : A \to B$. *Then we have*

1. $\mathrm{unp}(R)$ *is univalent and included in* R,
2. $\mathrm{unp}(\mathrm{unp}(R)) = \mathrm{unp}(R)$,
3. R *is univalent iff* $\mathrm{unp}(R) = R$.

Proof. 1. The second assertion is obvious, and the first is shown by

$$\mathrm{unp}(R)^\smile; \mathrm{unp}(R) = (R \sqcap (R^\smile \backslash \mathbb{I}_B))^\smile; (R \sqcap (R^\smile \backslash \mathbb{I}_B))$$
$$\sqsubseteq R^\smile; (R^\smile \backslash \mathbb{I}_B)$$
$$\sqsubseteq \mathbb{I}_B.$$

2. The inclusion '$\sqsubseteq$' is obvious, and $\text{unp}(R)^\smile; (R^\smile\backslash\mathbb{I}_B) \sqsubseteq R^\smile; (R^\smile\backslash\mathbb{I}_B) \sqsubseteq \mathbb{I}_B$ implies $(R^\smile\backslash\mathbb{I}_B) \sqsubseteq (\text{unp}(R)^\smile\backslash\mathbb{I}_B)$ and hence

$$\text{unp}(R) = \text{unp}(R) \sqcap (R^\smile\backslash\mathbb{I}_B) \sqsubseteq \text{unp}(R) \sqcap (\text{unp}(R)^\smile\backslash\mathbb{I}_B) = \text{unp}(\text{unp}(R)).$$

3. This follows immediately from

$$R^\smile; R \sqsubseteq \mathbb{I}_B \Leftrightarrow R \sqsubseteq (\text{unp}(R)^\smile\backslash\mathbb{I}_B)$$
$$\Leftrightarrow R \sqcap (\text{unp}(R)^\smile\backslash\mathbb{I}_B) = R$$
$$\Leftrightarrow \text{unp}(R) = R. \qquad \square$$

3   Weak Relational Products

A relational product, as defined in the introduction, is also a categorical product in the subcategory of mappings but not necessarily vice versa. The equation $\pi^\smile; \rho = \mathbb{T}_{AB}$ may not be valid. This equation states that the greatest relation is tabular [5]. We will weaken this axiom by requiring that each tabulation is included in this relation.

Definition 8 (Weak relational product). *Let $\mathcal{R}$ be an allegory and A and B objects of $\mathcal{R}$. An object $A \times B$ together with two relations $\pi : A \times B \to A$ and $\rho : A \times B \to B$ is called a weak relational product iff*

(P1) $\pi^\smile; \pi \sqsubseteq \mathbb{I}_A$,

(P2) $\rho^\smile; \rho \sqsubseteq \mathbb{I}_B$,

(P3) $\pi; \pi^\smile \sqcap \rho; \rho^\smile = \mathbb{I}_{A \times B}$,

(P4) $f^\smile; g \sqsubseteq \pi^\smile; \rho$ *for all mappings $f : C \to A$ and $g : C \to B$.*

$\mathcal{R}$ *is called a weak pairing allegory iff a weak relational product for each pair of objects exists.*

Example 1. Consider the concrete allegory with one object $A = \{0, 1\}$ and the four relations $\bot\!\!\!\bot_A := \emptyset, \mathbb{I}_A, \bar{\mathbb{I}}_A := \{(0, 1), (1, 0)\}$ and $\mathbb{T}_A := A \times A$. It is easy to verify that this structure establishes indeed an allegory with exactly two mappings $\mathbb{I}_A$ and $\bar{\mathbb{I}}_A$. It is well-known that the matrices with entries of a (complete) allegory form an allegory. Mappings in our example are matrices with exactly one entry $\mathbb{I}_A$ or $\bar{\mathbb{I}}_A$ in each row and $\bot\!\!\!\bot_A$ otherwise. The pair

$$\pi := \begin{pmatrix} \mathbb{I}_A & \bot\!\!\!\bot_A \\ \mathbb{I}_A & \bot\!\!\!\bot_A \\ \bot\!\!\!\bot_A & \mathbb{I}_A \\ \bot\!\!\!\bot_A & \mathbb{I}_A \end{pmatrix} \qquad \rho := \begin{pmatrix} \mathbb{I}_A & \bot\!\!\!\bot_A \\ \bot\!\!\!\bot_A & \mathbb{I}_A \\ \mathbb{I}_A & \bot\!\!\!\bot_A \\ \bot\!\!\!\bot_A & \mathbb{I}_A \end{pmatrix}$$

establishes a weak relational product. Notice, that if we replace $\mathbb{I}_A$ by the greatest element we obtain the well-known matrix representation of the projections (cf. [9]).

The weak version still establishes a categorical product in $\mathrm{MAP}(\mathcal{R})$, and is, therefore, unique (up to isomorphism).

Theorem 1. *Let $\mathcal{R}$ be an allegory. Then a weak relational product $(A \times B, \pi, \rho)$ is a categorical product of A and B in $\mathrm{MAP}(\mathcal{R})$.*

Proof. Let $(A \times B, \pi, \rho)$ be a weak relational product. By P1, P2 and P3 the relations π and ρ are mappings, and hence in $\mathrm{MAP}(\mathcal{R})$. Let $f : C \to A$ and $g : C \to B$ be mappings. Then we have

$$(f; \pi^\smile \sqcap g; \rho^\smile); \rho = f; \pi^\smile; \rho \sqcap g \qquad\qquad \text{Lemma 2(2)}$$
$$= g$$

where the last equality follows from

$$g \sqsubseteq f; f^\smile; g \qquad\qquad f \text{ is total,}$$
$$\sqsubseteq f; \pi^\smile; \rho \qquad\qquad \text{Axiom P4.}$$

The equality $(f; \pi^\smile \sqcap g; \rho^\smile); \pi = f$ is shown analogously. Furthermore, the following computation shows that $f; \pi^\smile \sqcap g; \rho^\smile$ is a mapping, and hence an element of $\mathrm{MAP}(\mathcal{R})$

$$(f; \pi^\smile \sqcap g; \rho^\smile)^\smile; (f; \pi^\smile \sqcap g; \rho^\smile)$$
$$= (\pi; f^\smile \sqcap \rho; g^\smile); (f; \pi^\smile \sqcap g; \rho^\smile)$$
$$\sqsubseteq \pi; f^\smile; f; \pi^\smile \sqcap \rho; g^\smile; g; \rho^\smile$$
$$\sqsubseteq \pi; \pi^\smile \sqcap \rho; \rho^\smile \qquad\qquad f \text{ and } g \text{ are univalent,}$$
$$= \mathbb{I}_{A \times B} \qquad\qquad \text{Axiom P3}$$

$$(f; \pi^\smile \sqcap g; \rho^\smile); (f; \pi^\smile \sqcap g; \rho^\smile)^\smile$$
$$= (f; \pi^\smile \sqcap g; \rho^\smile); (\pi; f^\smile \sqcap \rho; g^\smile)$$
$$= (f; \pi^\smile \sqcap g; \rho^\smile); \pi; f^\smile \sqcap (f; \pi^\smile \sqcap g; \rho^\smile); \rho; g^\smile \qquad \text{Lemma 2(1)}$$
$$= f; f^\smile \sqcap g; g^\smile \qquad\qquad \text{previous computation}$$
$$\sqsupseteq \mathbb{I}_C \qquad\qquad f \text{ and } g \text{ are total.}$$

Last but not least, let h be a mapping with $h; \pi = f$ and $h; \rho = g$. Then we conclude

$$f; \pi^\smile \sqcap g; \rho^\smile = h; \pi; \pi^\smile \sqcap h; \rho; \rho^\smile$$
$$= h; (\pi; \pi^\smile \sqcap \rho; \rho^\smile) \qquad\qquad \text{Lemma 2(1)}$$
$$= h \qquad\qquad \text{Axiom P3}$$

This completes the proof. $\square$

Notice, that we have also shown that for a weak relational product the product morphism induced by the mappings $f : C \to A$ and $g : C \to B$, i.e. the unique mapping $h : C \to A \times B$ satisfying $h; \pi = f$ and $h; \rho = g$, is actually given by the relation $f; \pi^{\smile} \sqcap g; \rho^{\smile}$.

Unfortunately, we are just able to prove parts of the converse implication.

Lemma 4. *Let $\mathcal{R}$ be an allegory. Then a categorical product $(A \times B, \pi, \rho)$ of A and B in $\mathrm{MAP}(\mathcal{R})$ fulfils the axioms P1, P2, P4 and the inclusion $\sqsupseteq$ of P3.*

Proof. Suppose $(A \times B, \pi, \rho)$ is a categorical product of A and B in $\mathrm{MAP}(\mathcal{R})$. Axioms P1, P2 and the inclusion $\sqsupseteq$ of P3 are trivial since π and ρ are mappings.

Now, let $f : C \to A$ and $g : C \to B$ be mappings. Then there is a unique mapping $h : C \to A \times B$ with $h; \pi = f$ and $h; \rho = g$. We conclude

$$
\begin{aligned}
g &= h; \rho \\
&\sqsubseteq h; (\pi; \pi^{\smile} \sqcap \rho; \rho^{\smile}); \rho \qquad\qquad \text{inclusion } \sqsupseteq \text{ of P3} \\
&\sqsubseteq h; \pi; \pi^{\smile}; \rho \\
&= f; \pi^{\smile}; \rho,
\end{aligned}
$$

which implies $f^{\smile}; g \sqsubseteq f^{\smile}; f; \pi^{\smile}; \rho \sqsubseteq \pi^{\smile}; \rho$ since f is univalent. $\square$

We have not been able to find an example showing that the full converse does not hold. Constructing such an example is a non-trivial problem since the next lemma indicates that the remaining inclusion is not hard to fulfill. In particular, we prove that inclusion in the case the allegory provides a suitable splitting.

Lemma 5. *Let $\mathcal{R}$ be an allegory, and let $(A \times B, \pi, \rho)$ be a categorical product in $\mathrm{MAP}(\mathcal{R})$. Furthermore, assume that there exists an $\in \mathcal{R}$ that splits $\pi; \pi^{\smile} \sqcap \rho; \rho^{\smile}$. Then $(A \times B, \pi, \rho)$ is a weak relational product.*

Proof. By Lemma 4 it remains to show the inclusion $\sqsubseteq$ of P3. Let $R : C \to A \times B$ be the splitting of $\pi; \pi^{\smile} \sqcap \rho; \rho^{\smile}$, and define $\tilde{\pi} := R; \pi$ and $\tilde{\rho} := R; \rho$. We want to show that $(C, \tilde{\pi}, \tilde{\rho})$ is a weak relational product of A and B. Once verified Lemma 1 implies that $(C, \tilde{\pi}, \tilde{\rho})$ is another categorical product of A and B in $\mathrm{MAP}(\mathcal{R})$, and hence isomorphic to $A \times B$. It is easy to verify (cf. [1]) that the isomorphism is given by the two mapping $h : C \to A \times B$ and $k : A \times B \to C$ fulfilling $h; \pi = \tilde{\pi}, h; \rho = \tilde{\rho}, k; \tilde{\pi} = \pi$ and $k; \tilde{\rho} = \rho$. Theorem 1 also shows $k = \pi; \tilde{\pi}^{\smile} \sqcap \rho; \tilde{\rho}^{\smile}$. Furthermore, in [5] it was shown the inverse of an isomorphism in an allegory is its converse so that $h^{\smile} = k$ follows. We conclude

$$
\begin{aligned}
\mathbb{I}_{A \times B} &= k; h & \text{pair of isomorphisms} \\
&= (\pi; \tilde{\pi}^{\smile} \sqcap \rho; \tilde{\rho}^{\smile}); h \\
&= (\pi; \pi^{\smile}; h^{\smile} \sqcap \rho; \rho^{\smile}; h^{\smile}); h \\
&= (\pi; \pi^{\smile} \sqcap \rho; \rho^{\smile}); h^{\smile}; h & \text{Lemma 2(1)} \\
&= (\pi; \pi^{\smile} \sqcap \rho; \rho^{\smile}); k; h \\
&= \pi; \pi^{\smile} \sqcap \rho; \rho^{\smile}
\end{aligned}
$$

In order to show that $(C, \tilde{\pi}, \tilde{\rho})$ is a weak relational product we derive Axiom P1 from

$$\tilde{\pi}^{\smallsmile}; \tilde{\pi} = \pi^{\smallsmile}; R^{\smallsmile}; R; \pi$$
$$= \pi^{\smallsmile}; (\pi; \pi^{\smallsmile} \sqcap \rho; \rho^{\smallsmile}); \pi$$
$$= \mathbb{I}_A \sqcap \pi^{\smallsmile}; \rho; \rho^{\smallsmile}; \pi \qquad \text{Lemma 2(2)}$$
$$\sqsubseteq \mathbb{I}_A,$$

and Axiom P2 analogously. We get the totality of $\tilde{\pi}$ from $\mathbb{I}_C = R; R^{\smallsmile} \sqsubseteq R; \pi; \pi^{\smallsmile}; R^{\smallsmile} = \tilde{\pi}; \tilde{\pi}^{\smallsmile}$ and for $\tilde{\rho}$ analogously. Together with

$$\tilde{\pi}; \tilde{\pi}^{\smallsmile} \sqcap \tilde{\rho}; \tilde{\rho}^{\smallsmile}$$
$$= R; \pi; \pi^{\smallsmile}; R^{\smallsmile} \sqcap R; \rho; \rho^{\smallsmile}; R^{\smallsmile}$$
$$\sqsubseteq R; (\pi; \pi^{\smallsmile}; R^{\smallsmile} \sqcap R^{\smallsmile}; R; \rho; \rho^{\smallsmile}; R^{\smallsmile})$$
$$\sqsubseteq R; (\pi; \pi^{\smallsmile} \sqcap R^{\smallsmile}; R; \rho; \rho^{\smallsmile}; R^{\smallsmile}; R); R^{\smallsmile}$$
$$\sqsubseteq R; (\pi; \pi^{\smallsmile} \sqcap (\pi; \pi^{\smallsmile} \sqcap \rho; \rho^{\smallsmile}); \rho; \rho^{\smallsmile}; (\pi; \pi^{\smallsmile} \sqcap \rho; \rho^{\smallsmile})); R^{\smallsmile}$$
$$= R; (\pi; \pi^{\smallsmile} \sqcap (\pi; \pi^{\smallsmile}; \rho \sqcap \rho); (\rho^{\smallsmile}; \pi; \pi^{\smallsmile} \sqcap \rho^{\smallsmile})); R^{\smallsmile} \qquad \text{Lemma 2(2)}$$
$$\sqsubseteq R; (\pi; \pi^{\smallsmile} \sqcap \rho; \rho^{\smallsmile}); R^{\smallsmile}$$
$$= R; R^{\smallsmile}; R; R^{\smallsmile}$$
$$= \mathbb{I}_C$$

we have shown Axiom P3. Last but not least, the computation

$$\tilde{\pi}^{\smallsmile}; \tilde{\rho} = \pi^{\smallsmile}; R^{\smallsmile}; R; \rho$$
$$= \pi^{\smallsmile}; (\pi; \pi^{\smallsmile} \sqcap \rho; \rho^{\smallsmile}); \rho$$
$$= \pi^{\smallsmile}; \rho \sqcap \pi^{\smallsmile}; \rho \qquad \text{Lemma 2(2)}$$
$$= \pi^{\smallsmile}; \rho$$

implies Axiom P4. $\qquad\qquad\square$

Even though every allegory can be embedded into one providing the splitting required by the last lemma, it is not clear that the given product remains one after embedding.

Now we want to show that for certain allegories the weak relational product can be defined by equations. This seems to be of particular interest because the allegories considered to construct weak relational products (cf. next section) are of that kind.

Lemma 6. *Let $\mathcal{R}$ be a division allegory in which all partial identities split. Then $(A \times B, \pi, \rho)$ is a weak relational product iff the Axioms P1-P3 and*

$$\text{unp}(R)^{\smallsmile}; \text{unp}(S) \sqsubseteq \pi^{\smallsmile}; \rho$$

for all relations $R : C \to A$ and $S : C \to B$ hold.

Proof. The implication $\Rightarrow$ is trivial since $\mathrm{unp}(f) = f$ for all mappings by Lemma 3 (3). For the converse implication assume P1-P4 and let $R : C \to A$ and $S : C \to B$ be arbitrary relations. Now, let $i := \mathbb{I}_C \sqcap \mathrm{unp}(R); \mathrm{unp}(R)^\smallsmile \sqcap \mathrm{unp}(S); \mathrm{unp}(S)^\smallsmile$ and $s : D \to C$ be its splitting. Then $s; \mathrm{unp}(R)$ is univalent since s and R are. Furthermore, this relation is total because $\mathbb{I}_D = s; s^\smallsmile; s; s^\smallsmile = s; i; s^\smallsmile \sqsubseteq s\mathrm{unp}(R); \mathrm{unp}(R)^\smallsmile; s^\smallsmile$. Analogously, we get that $s; \mathrm{unp}(S)$ is a mapping. Notice, that we have $Q = (\mathbb{I}_A \sqcap Q; Q^\smallsmile); Q$ for arbitrary relations $Q : A \to B$ and $i; j = i \sqcap j$ for partial identities $i, j : A \to A$. Proofs of those properties can be found in [5,9,10]. We conclude

$$
\begin{aligned}
&\mathrm{unp}(R)^\smallsmile; \mathrm{unp}(S) \\
&= \mathrm{unp}(R)^\smallsmile; (\mathbb{I}_D \sqcap \mathrm{unp}(R); \mathrm{unp}(R)^\smallsmile); (\mathbb{I}_D \sqcap \mathrm{unp}(S); \mathrm{unp}(S)^\smallsmile); \mathrm{unp}(S) \\
&= \mathrm{unp}(R)^\smallsmile; (\mathbb{I}_D \sqcap \mathrm{unp}(R); \mathrm{unp}(R)^\smallsmile \sqcap \mathbb{I}_D \sqcap \mathrm{unp}(S); \mathrm{unp}(S)^\smallsmile); \mathrm{unp}(S) \\
&= \mathrm{unp}(R)^\smallsmile; i; \mathrm{unp}(S) \\
&= \mathrm{unp}(R)^\smallsmile; s^\smallsmile; s; \mathrm{unp}(S) \\
&= (s; \mathrm{unp}(R))^\smallsmile; s; \mathrm{unp}(S) \\
&\sqsubseteq \pi^\smallsmile; \rho.
\end{aligned}
$$

This completes the proof. $\square$

Example 2. In this example we want to show that, in contrast to relational products, unsharpness may hold for a weak relational product, i.e. there are relations Q, R, S and T with $(Q; \pi^\smallsmile \sqcap R; \rho^\smallsmile); (\pi; S \sqcap \rho; T) \neq Q; S \sqcap R; T$. Our example will show that even under the additional assumption of totality of the relations involved unsharpness is possible. Suppose $\mathcal{R}$ is a weak pairing allegory with a greatest element in $\mathcal{R}[A, B]$ and $\pi^\smallsmile; \rho \neq \mathbb{T}_{AB}$. Then we have

$$
\begin{aligned}
(\mathbb{I}_A; \pi^\smallsmile \sqcap \mathbb{T}_{AB}; \rho^\smallsmile); (\pi; \mathbb{T}_{AB} \sqcap \rho; \mathbb{I}_B) &= (\pi^\smallsmile \sqcap \mathbb{T}_{A(A \times B)}); (\mathbb{T}_{(A \times B)B} \sqcap \rho) \\
&= \pi^\smallsmile; \rho \\
&\neq \mathbb{T}_{AB} \\
&= \mathbb{I}_A; \mathbb{T}_{AB} \sqcap \mathbb{T}_{AB}; \mathbb{I}_B.
\end{aligned}
$$

One important property of relational products is that one can transform any relation into the abstract counterpart of a set of pairs, i.e. by a vector or a left ideal element $\mathbb{T}_{AA}; v = v$. We want to investigate whether this is also possible for weak relational products. Consider the two operations

$$
\tau(R) := \mathbb{T}_{AA}; (\pi^\smallsmile \sqcap R; \rho^\smallsmile) \quad \text{and} \quad \sigma(v) := (\mathbb{T}_{AA}; v \sqcap \pi^\smallsmile); \rho.
$$

τ maps relations to vectors and σ vectors to relations.

Lemma 7. *Let $\mathcal{R}$ be a weak pairing allegory with greatest elements, $R : A \to B$, $v : A \to A \times B$ be a vector, $Q : C \to A$ univalent, and $S : C \to A \times B$. Then we have*

1. $R; \pi^{\smile} \sqcap (Q; \pi^{\smile} \sqcap S); \rho; \rho^{\smile} = Q; \pi^{\smile} \sqcap S$,
2. $\tau(\sigma(v)) = v$,
3. $\sigma(\tau(R)) \sqsubseteq R$ with $'='$ if $R \sqsubseteq \pi^{\smile}; \rho$.

Proof. 1.-2. These properties were shown in [9,10,11] for relational products. The proofs provided there also apply to weak relational products without modifications.

3. Consider the computation

$$\sigma(\tau(R)) = (\mathbb{T}_{AA}; \mathbb{T}_{AA}; (\pi^{\smile} \sqcap R; \rho^{\smile}) \sqcap \pi^{\smile}); \rho$$
$$= (\mathbb{T}_{AA}; (\pi^{\smile} \sqcap R; \rho^{\smile}) \sqcap \pi^{\smile}); \rho$$
$$= (\pi^{\smile} \sqcap R; \rho^{\smile}); \rho \qquad \text{Lemma 2(3)}$$
$$= \pi^{\smile}; \rho \sqcap R \qquad \text{Lemma 2(2)}$$
$$\sqsubseteq R.$$

Obviously, we get $'='$ if $R \sqsubseteq \pi^{\smile}; \rho$. □

4 Creating Weak Relational Products

The main proposition of this section (Corollary 1) states that any (small) allegory can be embedded into a weak pairing allegory. This theorem is based on the fact that the cartesian product of two power sets can be constructed by the power set of the disjoint union of the sets. An abstract version of this proposition is given in the next lemma and summarized by the following diagram.

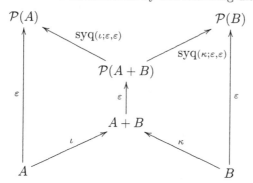

Notice, that the constructed weak relational product is not necessarily a relational product [10].

Lemma 8. *Let $\mathcal{R}$ be an allegory, A and B objects of $\mathcal{R}$ so that the relational sum $(A + B, \iota_{AB}, \kappa_{AB})$ and the relational powers $\mathcal{P}(A)$, $\mathcal{P}(B)$ and $\mathcal{P}(A + B)$ exist. Then $(\mathcal{P}(A + B), \mathrm{syq}(\iota_{AB}; \varepsilon_{A+B}, \varepsilon_A), \mathrm{syq}(\kappa_{AB}; \varepsilon_{A+B}, \varepsilon_B))$ is a weak relational product of $\mathcal{P}(A)$ and $\mathcal{P}(B)$.*

Proof. Axiom P1 follows immediately from

$$\pi^{\smallsmile}; \pi = \mathrm{syq}(\iota; \varepsilon, \varepsilon)^{\smallsmile}; \mathrm{syq}(\iota; \varepsilon, \varepsilon)$$
$$= \mathrm{syq}(\varepsilon, \iota; \varepsilon); \mathrm{syq}(\iota; \varepsilon, \varepsilon) \qquad \text{Lemma 1(3)}$$
$$\sqsubseteq \mathrm{syq}(\varepsilon, \varepsilon) \qquad \text{Lemma 1(4)}$$
$$= \mathbb{I}_{\mathcal{P}(A)}.$$

Axiom P2 is shown analogously. Since $\mathrm{syq}(\iota; \varepsilon, \varepsilon)$ is total by definition we get

$$\pi; \pi^{\smallsmile} = \mathrm{syq}(\iota; \varepsilon, \varepsilon); \mathrm{syq}(\iota; \varepsilon, \varepsilon)^{\smallsmile}$$
$$= \mathrm{syq}(\iota; \varepsilon, \varepsilon); \mathrm{syq}(\varepsilon, \iota; \varepsilon) \qquad \text{Lemma 1(3)}$$
$$= \mathrm{syq}(\iota; \varepsilon, \iota; \varepsilon) \qquad \text{Lemma 1(5)}$$

and $\rho; \rho^{\smallsmile} = \mathrm{syq}(\kappa; \varepsilon, \kappa; \varepsilon)$ analogously. Furthermore, we have

$$\varepsilon; ((\iota; \varepsilon) \backslash (\iota; \varepsilon) \sqcap (\kappa; \varepsilon) \backslash (\kappa; \varepsilon))$$
$$= (\iota^{\smallsmile}; \iota \sqcup \kappa^{\smallsmile}; \kappa); \varepsilon; ((\iota; \varepsilon) \backslash (\iota; \varepsilon) \sqcap (\kappa; \varepsilon) \backslash (\kappa; \varepsilon))$$
$$\sqsubseteq \iota^{\smallsmile}; \iota; \varepsilon; (\iota; \varepsilon) \backslash (\iota; \varepsilon) \sqcup \kappa^{\smallsmile}; \kappa; \varepsilon; (\kappa; \varepsilon) \backslash (\kappa; \varepsilon)$$
$$\sqsubseteq \iota^{\smallsmile}; \iota; \varepsilon \sqcup \kappa^{\smallsmile}; \kappa; \varepsilon \qquad \text{Lemma 1(1)}$$
$$= (\iota^{\smallsmile}; \iota \sqcup \kappa^{\smallsmile}; \kappa); \varepsilon$$
$$= \varepsilon$$

so that $(\iota; \varepsilon) \backslash (\iota; \varepsilon) \sqcap (\kappa; \varepsilon) \backslash (\kappa; \varepsilon) \sqsubseteq \varepsilon \backslash \varepsilon$ follows. Again, the similar inclusion $(\iota; \varepsilon)^{\smallsmile} / (\iota; \varepsilon)^{\smallsmile} \sqcap (\kappa; \varepsilon)^{\smallsmile} / (\kappa; \varepsilon)^{\smallsmile} \sqsubseteq \varepsilon^{\smallsmile} / \varepsilon^{\smallsmile}$ is shown analogously. Together, we conclude

$$\pi; \pi^{\smallsmile} \sqcap \rho; \rho^{\smallsmile} = \mathrm{syq}(\iota; \varepsilon, \iota; \varepsilon) \sqcap \mathrm{syq}(\kappa; \varepsilon, \kappa; \varepsilon)$$
$$= (\iota; \varepsilon) \backslash (\iota; \varepsilon) \sqcap (\iota; \varepsilon)^{\smallsmile} / (\iota; \varepsilon)^{\smallsmile} \sqcap (\kappa; \varepsilon) \backslash (\kappa; \varepsilon) \sqcap (\kappa; \varepsilon)^{\smallsmile} / (\kappa; \varepsilon)^{\smallsmile}$$
$$= \varepsilon \backslash \varepsilon \sqcap \varepsilon^{\smallsmile} / \varepsilon^{\smallsmile}$$
$$= \mathrm{syq}(\varepsilon, \varepsilon)$$
$$= \mathbb{I}_{\mathcal{P})(A+B)}.$$

In order to prove P4 let $f : C \to A$ and $g : C \to B$ be mappings. The relation $\mathrm{syq}(\iota^{\smallsmile}; \varepsilon; f^{\smallsmile} \sqcup \kappa^{\smallsmile}; \varepsilon; g^{\smallsmile}, \varepsilon)$ is a mapping by definition, and we have

$$\mathrm{syq}(\iota^{\smallsmile}; \varepsilon; f^{\smallsmile} \sqcup \kappa^{\smallsmile}; \varepsilon; g^{\smallsmile}, \varepsilon); \pi$$
$$= \mathrm{syq}(\iota^{\smallsmile}; \varepsilon; f^{\smallsmile} \sqcup \kappa^{\smallsmile}; \varepsilon; g^{\smallsmile}, \varepsilon); \mathrm{syq}(\iota; \varepsilon, \varepsilon)$$
$$= \mathrm{syq}(\iota; \varepsilon; \mathrm{syq}(\iota^{\smallsmile}; \varepsilon; f^{\smallsmile} \sqcup \kappa^{\smallsmile}; \varepsilon; g^{\smallsmile}, \varepsilon)^{\smallsmile}, \varepsilon) \qquad \text{Lemma 1(2)}$$
$$= \mathrm{syq}(\iota; \varepsilon; \mathrm{syq}(\varepsilon, \iota^{\smallsmile}; \varepsilon; f^{\smallsmile} \sqcup \kappa^{\smallsmile}; \varepsilon; g^{\smallsmile}), \varepsilon) \qquad \text{Lemma 1(3)}$$
$$= \mathrm{syq}(\iota; (\iota^{\smallsmile}; \varepsilon; f^{\smallsmile} \sqcup \kappa^{\smallsmile}; \varepsilon; g^{\smallsmile}), \varepsilon) \qquad \text{Lemma 1(6)}$$
$$= \mathrm{syq}(\varepsilon; f^{\smallsmile}, \varepsilon)$$
$$= f; \mathrm{syq}(\varepsilon, \varepsilon) \qquad \text{Lemma 1(2)}$$
$$= f.$$

$\mathrm{syq}(\iota^{\smile}; \varepsilon; f^{\smile} \sqcup \kappa^{\smile}; \varepsilon; g^{\smile}, \varepsilon); \rho = g$ is shown analogously. We conclude

$$f^{\smile}; g = f^{\smile}; \mathrm{syq}(\iota^{\smile}; \varepsilon; f^{\smile} \sqcup \kappa^{\smile}; \varepsilon; g^{\smile}, \varepsilon); \rho$$

$$= f^{\smile}; \mathrm{syq}(\iota^{\smile}; \varepsilon; f^{\smile} \sqcup \kappa^{\smile}; \varepsilon; g^{\smile}, \varepsilon); (\pi; \pi^{\smile} \sqcap \rho; \rho^{\smile}); \rho$$

$$\sqsubseteq f^{\smile}; \mathrm{syq}(\iota^{\smile}; \varepsilon; f^{\smile} \sqcup \kappa^{\smile}; \varepsilon; g^{\smile}, \varepsilon); \pi; \pi^{\smile}; \rho$$

$$= f^{\smile}; f; \pi^{\smile}; \rho$$

$$\sqsubseteq \pi^{\smile}; \rho.$$

This completes the proof. □

If the allegory provides splitting of partial identities, the construction described above can be distributed to any pair of objects.

Theorem 2. *Any systemic complete allegory is a weak pairing allegory.*

Proof. Let $\mathcal{R}$ be a systemic complete power allegory, and let A and B be objects of $\mathcal{R}$. By Lemma 8 there is a weak relational product $(\mathcal{P}(A) \times \mathcal{P}(B), \pi, \rho)$ of $\mathcal{P}(A)$ and $\mathcal{P}(B)$. Let be $i := \pi; e_A^{\smile}; e_A; \pi^{\smile} \sqcap \rho; e_B^{\smile}; e_B; \rho^{\smile}$. The relation i is a partial identity, which is shown as follows

$$i = \pi; e_A^{\smile}; e_A; \pi^{\smile} \sqcap \rho; e_B^{\smile}; e_B; \rho^{\smile}$$

$$= \pi; \pi^{\smile} \sqcap \rho; \rho^{\smile} \qquad \text{e_A and e_B are univalent}$$

$$= \mathbb{I}_{\mathcal{P}(A) \times \mathcal{P}(B)}. \qquad \text{Axiom P3}$$

Since $\mathcal{R}$ is systemic complete there is an object C and a relation $s : C \rightarrow \mathcal{P}(A) \times \mathcal{P}(B)$ that splits i. Notice, that s is an injective mapping since i is a partial identity. We want to show that C together with the relations $\tilde{\pi} := s; \pi; e_A^{\smile}$ and $\tilde{\rho} := s; \rho; e_B^{\smile}$ is a weak relational product of A and B.

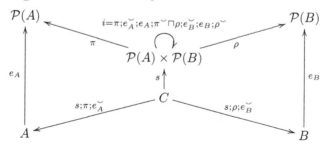

Axiom P1 is shown by

$$\tilde{\pi}^{\smile}; \tilde{\pi} = (s; \pi; e_A^{\smile})^{\smile}; s; \pi; e_A^{\smile}$$

$$= e_A; \pi^{\smile}; s^{\smile}; s; \pi; e_A^{\smile}$$

$$= e_A; \pi^{\smile}; \pi; e_A^{\smile} \qquad \text{s is total and injective}$$

$$\sqsubseteq e_A; e_A^{\smile} \qquad \text{Axiom P1}$$

$$= \mathbb{I}_A, \qquad \text{e_A is total and injective}$$

and Axiom P2 follows analogously. The computation

$$\tilde{\pi}; \tilde{\pi}^{\smile} \sqcap \tilde{\rho}; \tilde{\rho}^{\smile} = s; \pi; e_A^{\smile}; e_A; \pi^{\smile}; s^{\smile} \sqcap s; \rho; e_B^{\smile}; e_B; \rho^{\smile}; s^{\smile}$$

$$= s; (\pi; e_A^{\smile}; e_A; \pi^{\smile} \sqcap \rho; e_B^{\smile}; e_B; \rho^{\smile}); s^{\smile} \qquad \text{Lemma 2(1)}$$

$$= s; i; s^{\smile}$$

$$= s; s^{\smile}; s; s^{\smile}$$

$$= \mathbb{I}_C$$

verifies Axiom P3. In order to prove Axiom P4 we first observe

$$\tilde{\pi}^{\smile}; \tilde{\rho} = e_A; \pi^{\smile}; s^{\smile}; s; \rho; e_B^{\smile}$$

$$= e_A; \pi^{\smile}; i; \rho; e_B^{\smile} \qquad s \text{ splits } i$$

$$= e_A; \pi^{\smile}; (\pi; e_A^{\smile}; e_A; \pi^{\smile} \sqcap \rho; e_B^{\smile}; e_B; \rho^{\smile}); \rho; e_B^{\smile}$$

$$= e_A; (e_A^{\smile}; e_A; \pi^{\smile}; \rho \sqcap \pi^{\smile}; \rho; e_B^{\smile}; e_B); e_B^{\smile} \qquad \text{Lemma 2(2)}$$

$$= e_A; e_A^{\smile}; e_A; \pi^{\smile}; \rho; e_B^{\smile} \sqcap e_A; \pi^{\smile}; \rho; e_B^{\smile}; e_B; e_B^{\smile} \qquad \text{Lemma 2(2)}$$

$$= e_A; \pi^{\smile}; \rho; e_B^{\smile} \sqcap e_A; \pi^{\smile}; \rho; e_B^{\smile} \qquad e_A, e_B \text{ total and injective}$$

$$= e_A; \pi^{\smile}; \rho; e_B^{\smile}.$$

Now, let $f : D \to A$ and $g : D \to B$ be mappings. Then $f; e_A$ and $g; e_B$ are also mappings from C to $\mathcal{P}(A)$ and $\mathcal{P}(B)$, respectively. This implies $e_A^{\smile}; f^{\smile}; g; e_B = (f; e_A)^{\smile}; g; e_B \sqsubseteq \pi^{\smile}; \rho$. We conclude

$$f^{\smile}; g = e_A; e_A^{\smile}; f^{\smile}; g; e_B; e_B^{\smile} \qquad e_A, e_B \text{ total and injective}$$

$$\sqsubseteq e_A; \pi^{\smile}; \rho; e_B^{\smile}$$

$$= \tilde{\pi}^{\smile}; \tilde{\rho},$$

which finally verifies Axiom P4.   □

Since the systemic completion of a small allegory is systemic complete ([5] 2.221 and 2.434) we have shown the main result of this section.

Corollary 1. *Any small allegory may be faithfully represented in a weak pairing allegory.*

This corollary also shows that there are indeed weak pairing allegories in which the weak relational product is not always a relational product. For example, consider the allegory induced by the non-representable McKenzie relation algebra. According to Corollary 1 this allegory can be embedded into a weak pairing allegory. This allegory can not have all relational products since then it would be representable [5], which is a contradiction.

References

1. Asperti A., Longo G.: Categories, Types and Structures. The MIT Press, Cambridge, Massachusetts, London, England (1991)
2. Berghammer R., Haeberer A., Schmidt G., Veloso P.A.S.: Comparing two different approaches to products in abstract relation algebra. Algebraic Methodology and Software Technology, Proc. 3rd Int'l Conf. Algebraic Methodology and Software Technology (AMAST'93), Springer (1994), 167-176.
3. Desharnais, J.: Monomorphic Characterization of n-ary Direct Products. Information Sciences, 119 (3-4) (1999), 275-288
4. Furusawa H., Kahl W.: A Study on Symmetric Quotients. Technical Report 1998-06, University of the Federal Armed Forces Munich (1998)
5. Freyd P., Scedrov A.: Categories, Allegories. North-Holland (1990).
6. Maddux, R.D.: On the Derivation of Identities involving Projection Functions. Logic Colloquium'92, ed. Csirmaz, Gabbay, de Rijke, Center for the Study of Language and Information Publications, Stanford (1995), 145-163.
7. Olivier J.P., Serrato D.: Catégories de Dedekind. Morphismes dans les Catégories de Schröder. C.R. Acad. Sci. Paris 290 (1980), 939-941.
8. Olivier J.P., Serrato D.: Squares and Rectangles in Relational Categories - Three Cases: Semilattice, Distributive lattice and Boolean Non-unitary. Fuzzy sets and systems 72 (1995), 167-178.
9. Schmidt G., Ströhlein T.: Relationen und Graphen. Springer (1989); English version: Relations and Graphs. Discrete Mathematics for Computer Scientists, EATCS Monographs on Theoret. Comput. Sci., Springer (1993).
10. Winter M.: Strukturtheorie heterogener Relationenalgebren mit Anwendung auf Nichtdetermismus in Programmiersprachen. Dissertationsverlag NG Kopierladen GmbH, München (1998)
11. Zierer H.: Relation algebraic domain constructions. TCS 87 (1991), 163-188

Author Index

Lecture Notes in Computer Science

For information about Vols. 1–4055

please contact your bookseller or Springer

Vol. 4106: T.R. Roth-Berghofer, M.H. Göker, H. A. Güvenir (Eds.), Advances in Case-Based Reasoning. XIV, 566 pages. 2006. (Sublibrary LNAI).

Vol. 4104: T. Kunz, S.S. Ravi (Eds.), Ad-Hoc, Mobile, and Wireless Networks. XII, 474 pages. 2006.

Vol. 4099: Q. Yang, G. Webb (Eds.), PRICAI 2006: Trends in Artificial Intelligence. XXVIII, 1263 pages. 2006. (Sublibrary LNAI).

Vol. 4098: F. Pfenning (Ed.), Term Rewriting and Applications. XIII, 415 pages. 2006.

Vol. 4097: X. Zhou, O. Sokolsky, L. Yan, E.-S. Jung, Z. Shao, Y. Mu, D.C. Lee, D. Kim, Y.-S. Jeong, C.-Z. Xu (Eds.), Emerging Directions in Embedded and Ubiquitous Computing. XXVII, 1034 pages. 2006.

Vol. 4096: E. Sha, S.-K. Han, C.-Z. Xu, M.H. Kim, L.T. Yang, B. Xiao (Eds.), Embedded and Ubiquitous Computing. XXIV, 1170 pages. 2006.

Vol. 4095: S. Nolfi, G. Baldassare, R. Calabretta, D. Marocco, D. Parisi, J.C. T. Hallam, O. Miglino, J.-A. Meyer (Eds.), From Animals to Animats 9. XV, 869 pages. 2006. (Sublibrary LNAI).

Vol. 4094: O. H. Ibarra, H.-C. Yen (Eds.), Implementation and Application of Automata. XIII, 291 pages. 2006.

Vol. 4093: X. Li, O.R. Zaïane, Z. Li (Eds.), Advanced Data Mining and Applications. XXI, 1110 pages. 2006. (Sublibrary LNAI).

Vol. 4092: J. Lang, F. Lin, J. Wang (Eds.), Knowledge Science, Engineering and Management. XV, 664 pages. 2006. (Sublibrary LNAI).

Vol. 4091: G.-Z. Yang, T. Jiang, D. Shen, L. Gu, J. Yang (Eds.), Medical Imaging and Augmented Reality. XIII, 399 pages. 2006.

Vol. 4090: S. Spaccapietra, K. Aberer, P. Cudré-Mauroux (Eds.), Journal on Data Semantics VI. XI, 211 pages. 2006.

Vol. 4089: W. Löwe, M. Südholt (Eds.), Software Composition. X, 339 pages. 2006.

Vol. 4088: Z.-Z. Shi, R. Sadananda (Eds.), Agent Computing and Multi-Agent Systems. XVII, 827 pages. 2006. (Sublibrary LNAI).

Vol. 4087: F. Schwenker, S. Marinai (Eds.), Artificial Neural Networks in Pattern Recognition. IX, 299 pages. 2006. (Sublibrary LNAI).

Vol. 4085: J. Misra, T. Nipkow, E. Sekerinski (Eds.), FM 2006: Formal Methods. XV, 620 pages. 2006.

Vol. 4084: M.A. Wimmer, H.J. Scholl, Å. Grönlund, K.V. Andersen (Eds.), Electronic Government. XV, 353 pages. 2006.

Vol. 4083: S. Fischer-Hübner, S. Furnell, C. Lambrinoudakis (Eds.), Trust and Privacy in Digital Business. XIII, 243 pages. 2006.

Vol. 4082: K. Bauknecht, B. Pröll, H. Werthner (Eds.), E-Commerce and Web Technologies. XIII, 243 pages. 2006.

Vol. 4081: A. M. Tjoa, J. Trujillo (Eds.), Data Warehousing and Knowledge Discovery. XVII, 578 pages. 2006.

Vol. 4080: S. Bressan, J. Küng, R. Wagner (Eds.), Database and Expert Systems Applications. XXI, 959 pages. 2006.

Vol. 4079: S. Etalle, M. Truszczyński (Eds.), Logic Programming. XIV, 474 pages. 2006.

Vol. 4077: M.-S. Kim, K. Shimada (Eds.), Geometric Modeling and Processing - GMP 2006. XVI, 696 pages. 2006.

Vol. 4076: F. Hess, S. Pauli, M. Pohst (Eds.), Algorithmic Number Theory. X, 599 pages. 2006.

Vol. 4075: U. Leser, F. Naumann, B. Eckman (Eds.), Data Integration in the Life Sciences. XI, 298 pages. 2006. (Sublibrary LNBI).

Vol. 4074: M. Burmester, A. Yasinsac (Eds.), Secure Mobile Ad-hoc Networks and Sensors. X, 193 pages. 2006.

Vol. 4073: A. Butz, B. Fisher, A. Krüger, P. Olivier (Eds.), Smart Graphics. XI, 263 pages. 2006.

Vol. 4072: M. Harders, G. Székely (Eds.), Biomedical Simulation. XI, 216 pages. 2006.

Vol. 4071: H. Sundaram, M. Naphade, J.R. Smith, Y. Rui (Eds.), Image and Video Retrieval. XII, 547 pages. 2006.

Vol. 4070: C. Priami, X. Hu, Y. Pan, T.Y. Lin (Eds.), Transactions on Computational Systems Biology V. IX, 129 pages. 2006. (Sublibrary LNBI).

Vol. 4069: F.J. Perales, R.B. Fisher (Eds.), Articulated Motion and Deformable Objects. XV, 526 pages. 2006.

Vol. 4068: H. Schärfe, P. Hitzler, P. Øhrstrøm (Eds.), Conceptual Structures: Inspiration and Application. XI, 455 pages. 2006. (Sublibrary LNAI).

Vol. 4067: D. Thomas (Ed.), ECOOP 2006 – Object-Oriented Programming. XIV, 527 pages. 2006.

Vol. 4066: A. Rensink, J. Warmer (Eds.), Model Driven Architecture – Foundations and Applications. XII, 392 pages. 2006.

Vol. 4065: P. Perner (Ed.), Advances in Data Mining. XI, 592 pages. 2006. (Sublibrary LNAI).

Vol. 4064: R. Büschkes, P. Laskov (Eds.), Detection of Intrusions and Malware & Vulnerability Assessment. X, 195 pages. 2006.

Vol. 4063: I. Gorton, G.T. Heineman, I. Crnkovic, H.W. Schmidt, J.A. Stafford, C.A. Szyperski, K. Wallnau (Eds.), Component-Based Software Engineering. XI, 394 pages. 2006.

Vol. 4062: G. Wang, J.F. Peters, A. Skowron, Y. Yao (Eds.), Rough Sets and Knowledge Technology. XX, 810 pages. 2006. (Sublibrary LNAI).

Vol. 4061: K. Miesenberger, J. Klaus, W. Zagler, A.I. Karshmer (Eds.), Computers Helping People with Special Needs. XXIX, 1356 pages. 2006.

Vol. 4060: K. Futatsugi, J.-P. Jouannaud, J. Meseguer (Eds.), Algebra, Meaning, and Computation. XXXVIII, 643 pages. 2006.

Vol. 4059: L. Arge, R. Freivalds (Eds.), Algorithm Theory – SWAT 2006. XII, 436 pages. 2006.

Vol. 4058: L.M. Batten, R. Safavi-Naini (Eds.), Information Security and Privacy. XII, 446 pages. 2006.

Vol. 4057: J.P.W. Pluim, B. Likar, F.A. Gerritsen (Eds.), Biomedical Image Registration. XII, 324 pages. 2006.

Vol. 4056: P. Flocchini, L. Gąsieniec (Eds.), Structural Information and Communication Complexity. X, 357 pages. 2006.